Edited by
Sabu Thomas,
Dominique Durand,
Christophe Chassenieux, and
P. Jyotishkumar

Handbook of
Biopolymer-Based Materials

Related Titles

Lendlein, A., Sisson, A. (Eds.)
Handbook of Biodegradable Polymers
Synthesis, Characterization and Applications

2011
Hardcover
ISBN: 978-3-527-32441-5

Kalia, S., Avérous, L.
Biopolymers
Biomedical and Environmental Applications Series: Wiley-Scrivener

2012
Hardcover
ISBN: 978-0-470-63923-8

Mittal, V. (Ed.)
Renewable Polymers
Synthesis, Processing, and Technology

2012
Hardcover
ISBN: 978-0-470-93877-5

McDermott, A. (Ed.)
Solid State NMR Studies of Biopolymers

2010
Hardcover
ISBN: 978-0-470-72122-3

Loos, K. (Ed.)
Biocatalysis in Polymer Chemistry

2011
Hardcover
ISBN: 978-3-527-32618-1

*Edited by Sabu Thomas, Dominique Durand,
Christophe Chassenieux, and P. Jyotishkumar*

Handbook of Biopolymer-Based Materials

From Blends and Composites to Gels and Complex Networks

Volume 1

WILEY-VCH Verlag GmbH & Co. KGaA

The Editors

Prof. Dr. Sabu Thomas
Mahatma Gandhi University
Centre for Nanosc.a. Nanotech.
Priyadarshini Hills P.O.
Kottayam, Kerala 686-560
India

Prof. Dominique Durand
LUNAM Université du Maine
IMMM UMR CNRS 6283
Dept. Polymères, Colloïdes, Interfaces
1, Avenue Olivier Messiaen
72085 Le Mans cedex 9
France

Prof. Christophe Chassenieux
LUNAM Université du Maine
IMMM UMR CNRS 6283
Dept. Polymères, Colloïdes, Interfaces
1, Avenue Olivier-Messiaen
72085 Le Mans cedex 9
France

Dr. P. Jyotishkumar
INSPIRE Faculty
Department of Polymer Science and
Rubber Technology
Cochin University of Science and
Technology, Kochi-682022

All books published by **Wiley-VCH** are carefully produced. Nevertheless, authors, editors, and publisher do not warrant the information contained in these books, including this book, to be free of errors. Readers are advised to keep in mind that statements, data, illustrations, procedural details or other items may inadvertently be inaccurate.

Library of Congress Card No.: applied for

British Library Cataloguing-in-Publication Data
A catalogue record for this book is available from the British Library.

Bibliographic information published by the Deutsche Nationalbibliothek
The Deutsche Nationalbibliothek lists this publication in the Deutsche Nationalbibliografie; detailed bibliographic data are available on the Internet at <http://dnb.d-nb.de>.

© 2013 Wiley-VCH Verlag GmbH & Co. KGaA, Boschstr. 12, 69469 Weinheim, Germany

All rights reserved (including those of translation into other languages). No part of this book may be reproduced in any form – by photoprinting, microfilm, or any other means – nor transmitted or translated into a machine language without written permission from the publishers. Registered names, trademarks, etc. used in this book, even when not specifically marked as such, are not to be considered unprotected by law.

Print ISBN: 978-3-527-32884-0
ePDF ISBN: 978-3-527-65248-8
ePub ISBN: 978-3-527-65247-1
mobi ISBN: 978-3-527-65246-4
oBook ISBN: 978-3-527-65245-7

Cover Design   Grafik-Design Schulz, Fußgönheim
Typesetting   Thomson Digital, Noida, India
Printing and Binding   Markono Print Media Pte Ltd, Singapore

Contents

Foreword *XIII*
List of Contributors *XV*

Volume 1

1   Biopolymers: State of the Art, New Challenges, and Opportunities *1*
         *Christophe Chassenieux, Dominique Durand, Parameswaranpillai
         Jyotishkumar, and Sabu Thomas*
1.1   Introduction *1*
1.2   Biopolymers: A Niche For Fundamental Research
         in Soft Matter Physics *3*
1.3   Biopolymers: An Endless Source of Applications *4*
1.4   Topics Covered by the Book *5*
1.5   Conclusions *5*
         References *6*

**2   General Overview of Biopolymers: Structure, Properties, and
         Applications** *7*
         Charles Winkworth-Smith and Tim J. Foster
2.1   Introduction *7*
2.2   Plant Cell Wall Polysaccharides *11*
2.2.1   Cellulose *11*
2.2.1.1 Cellulose Extraction *12*
2.2.1.2 Nanocellulose *13*
2.2.1.3 Microfibrillated Cellulose *14*
2.2.1.4 Cellulose Nanowhiskers *14*
2.2.2   Hemicelluloses *15*
2.2.2.1 Galactomannans *15*
2.2.2.2 Konjac Glucomannan *19*
2.2.2.3 Xylan *19*
2.2.2.4 Xyloglucan *20*
2.2.3   Pectins *21*
2.3   Biocomposites *23*
2.3.1   Natural Fiber Composites *23*

2.3.2	Cellulose Composites	25
2.3.3	Cellulose–Polymer Interactions	26
2.3.4	Semi-Solid Composites	27
2.4	Future Outlook	28
	References	29

3	**Biopolymers from Plants**	**37**
	Maria J. Sabater, Tania Ródenas, and Antonio Heredia	
3.1	Introduction	37
3.2	Lipid and Phenolic Biopolymers	38
3.2.1	The Biopolymer Cutin	38
3.2.1.1	Cutin Monomers: Biosynthesis and Physicochemical Characteristics	39
3.2.1.2	Molecular Architecture of Cutin	40
3.2.1.3	Cutin Biosynthesis	41
3.2.2	Lignin	42
3.2.2.1	Monomer Precursors and Chemical Reactivity	42
3.2.2.2	Lignin Biosynthesis	43
3.2.3	Suberin	45
3.2.3.1	Chemical Composition	45
3.2.3.2	Biosynthesis and Fine Structure	47
3.3	Carbohydrate Biopolymers: Polysaccharides	48
3.3.1	Structural Polysaccharides	49
3.3.1.1	Cellulose	49
3.3.1.2	Hemicellulose	54
3.3.1.3	Pectin	57
3.3.2	Storage Polysaccharides	59
3.3.2.1	Starch	59
3.3.2.2	Fructans: Inulin	63
3.3.3	Other: Gums (Guar Gum, Gum Arabic, Gum Karaya, Gum Tragacanth, and Locust Bean Gum)	66
3.4	Isoprene Biopolymers: Natural Rubber	67
3.4.1	*cis*-Polyisoprene	67
3.4.1.1	Occurrence	67
3.4.1.2	Composition, Structure, and Properties	67
3.4.1.3	*cis*-1,4-Polyisoprene Biosynthesis	68
3.4.1.4	Applications	73
3.4.2	*trans*-Polyisoprene	73
3.5	Concluding Remarks	74
	References	75

4	**Bacterial Biopolymers and Genetically Engineered Biopolymers for Gel Systems Application**	**87**
	Deepti Singh and Ashok Kumar	
4.1	Introduction	87
4.1.1	Nucleic Acid Biopolymers: Central Dogma	89

4.2	Microbial Polysaccharides as Biopolymers	90
4.2.1	Synthesis and Applications	90
4.3	Microbial Biopolymers as Drug Delivery Vehicle	92
4.3.1	ε-Poly-L-Lysine (ε-PL) and Its Applications	92
4.3.2	Polyhydroxyalkanoates and Its Applications	92
4.4	Polyanhydrides	93
4.5	Recombinant Protein Polymer Production	94
4.6	Recombinant Genetically Engineered Biopolymer: Elastin	95
4.7	Collagen as an Ideal Biopolymer	97
4.7.1	Microbial Recombinant Collagens: Production in *Pichia Pastoris*	97
4.8	Biopolymers for Gel System	99
4.9	Hydrogels of Biopolymers for Regenerative Medicine	99
4.9.1	Polysaccharide Hydrogels	99
4.9.2	Cellulose-Derived Biopolymers-Based Hydrogels	100
4.9.3	Protein Biopolymers-Based Hydrogels	100
4.10	Supermacroporous Cryogel Matrix from Biopolymers	100
4.10.1	Protein Cryogel	101
4.11	Biopolymers Impact on Environment	102
4.12	Conclusion	103
	References	104

5	**Biopolymers from Animals**	**109**
	Khaleelulla Saheb Shaik and Bernard Moussian	
5.1	Introduction	109
5.2	Chitin and Hyaluronic Acid in the Living World	110
5.3	Milestones in Chitin History	110
5.4	From Trehalose to Chitin	112
5.5	Chitin Synthase	115
5.6	Regulation of Chitin Synthesis in Fungi	117
5.7	Organization of Chitin in the Fungal Cell Wall	118
5.8	Organization of Chitin in the Arthropod Cuticle	119
5.9	Chitin-Organizing Factors	123
5.10	Secretion and Cuticle Formation	126
5.11	Transcriptional Regulation of Cuticle Production	128
5.12	Chitin Synthesis Inhibitors	130
5.13	Noncuticular Chitin in Insects	131
5.14	Chitin as a Structural Element	133
5.15	Application of Chitin	134
5.16	Conclusion	135
	References	135

6	**Polymeric Blends with Biopolymers**	**143**
	Hero Jan Heeres, Frank van Maastrigt, and Francesco Picchioni	
6.1	Introduction	143
6.2	Starch-Based Blends	146

6.2.1	Polymer Selection for Starch Blending	147
6.2.2	Starch Structure	150
6.2.3	Uncompatibilized Blends	152
6.2.4	Compatibilization	155
6.2.5	Composites	157
6.3	Blends with Chitosan (One Amino Group Too Much . . .)	158
6.4	Future Perspectives	161
6.4.1	Biopolymer Plasticization	161
6.4.2	Blend Morphology and Compatibilization	162
6.4.3	Blend Processing: Technological Aspects	163
	References	164

7   Macro-, Micro-, and Nanocomposites Based on Biodegradable Polymers   173

Luc Avérous and Eric Pollet

7.1	Introduction	173
7.2	Biodegradable Polymers	174
7.2.1	Classification	174
7.2.2	Agro-Polymers: The Case of Starch	175
7.2.2.1	Native Starch Structure	175
7.2.2.2	Plasticized Starch	176
7.2.3	Biodegradable Polyesters	177
7.2.3.1	Polyesters Based on Agro-Resources	177
7.2.3.2	Petroleum-Based Polyesters	179
7.3	Biocomposites	181
7.3.1	Generalities	181
7.3.2	The Case of Biocomposites Based on Agro-Polymers	181
7.3.2.1	Cellulose Fiber Reinforcement	181
7.3.2.2	Lignin and Mineral Fillers	182
7.3.3	The Case of Biocomposites Based on Biopolyesters	182
7.3.3.1	Generalities	182
7.3.3.2	The Case of Biocomposites Based on Aromatic Copolyesters	183
7.4	Nanobiocomposites	186
7.4.1	Generalities	186
7.4.2	Nanobiocomposites Based on Agro-Polymers (Starch)	187
7.4.2.1	Whisker-Based Nanobiocomposites	190
7.4.2.2	Starch Nanocrystal-Based Nanobiocomposites	190
7.4.2.3	Nanoclay-Based Nanobiocomposites	190
7.4.3	Nanobiocomposites Based on Biopolyesters	193
7.4.3.1	Poly(lactic acid)-Based Nanobiocomposites	193
7.4.3.2	Polyhydroxyalkanoate-Based Nanobiocomposites	194
7.4.3.3	Polycaprolactone-Based Nanobiocomposites	195
7.4.3.4	Biodegradable Aliphatic Copolyester-Based Nanobiocomposites	197
7.4.3.5	Aromatic Copolyester-Based Nanobiocomposites	199
	References	200

8	**IPNs Derived from Biopolymers** *211*	
	Fernando G. Torres, Omar Paul Troncoso, and Carlos Torres	
8.1	Introduction *211*	
8.2	Types of IPNs *212*	
8.3	IPNs Derived from Biopolymers *214*	
8.3.1	Alginate *215*	
8.3.2	Agarose *215*	
8.3.3	Chitosan *215*	
8.3.4	Starch *216*	
8.3.5	Dextran *217*	
8.3.6	Gum Arabic *217*	
8.3.7	Fibrinogen *217*	
8.3.8	Collagen and Gelatin *217*	
8.3.9	Cellulose and Cellulose Derivatives *218*	
8.3.10	Polyhydroxyalkanoates *218*	
8.3.11	Lactide-Derived Polymers *219*	
8.4	Manufacture of IPNs *220*	
8.4.1	Casting–Evaporation Processing *220*	
8.4.2	Emulsification Cross-Linking Technique *220*	
8.4.3	Miniemulsion/Inverse Miniemulsion Technique *221*	
8.4.4	Freeze Drying Technique *222*	
8.5	Characterization of IPNs *222*	
8.5.1	Morphological and Structural Characterization *222*	
8.5.2	FTIR Spectroscopy *223*	
8.5.3	Mechanical Characterization *224*	
8.5.4	Rheological Characterization *225*	
8.5.5	Swelling Behavior Characterization *225*	
8.5.6	Thermal Characterization *226*	
8.6	Applications of IPNs *226*	
8.6.1	Drug Delivery Applications *226*	
8.6.2	Scaffolds for Tissue Engineering *227*	
8.6.3	Other Biomedical Applications *227*	
8.6.4	Antibacterial Applications *228*	
8.6.5	Sensor, Actuators, and Artificial Muscle Applications *228*	
8.7	Conclusions *229*	
	References *229*	
9	**Associating Biopolymer Systems and Hyaluronate Biomaterials** *235*	
	Deborah Blanchard and Rachel Auzély-Velty	
9.1	Introduction *235*	
9.2	Synthesis and Self-Association of Hydrophobically Modified Derivatives of Chitosan and Hyaluronic Acid in Aqueous Solution *237*	
9.2.1	General Aspects of Association in Amphiphilic Polyelectrolytes *237*	
9.2.2	Synthesis and Behavior in Aqueous Solution of Hydrophobically Modified Water-Soluble Derivatives of Chitosan *239*	

9.2.3	Synthesis and Behavior in Aqueous Solution of Hydrophobically Modified Water-Soluble Derivatives of Hyaluronic Acid *242*
9.3	Design of Novel Biomaterials Based on Chemically Modified Derivatives of Hyaluronic Acid *248*
9.3.1	Nanoassemblies Based on Amphiphilic Hyaluronic Acid *249*
9.3.2	Hydrogels for Cell Biology and Tissue Engineering *254*
9.3.2.1	Strategies for the Cross-Linking of HA to Obtain Scaffolds for Cells *254*
9.3.2.2	Engineering Biological Functionality in Hyaluronic Acid-Based Scaffolds *260*
9.3.2.3	Patterning of Hyaluronic Acid *261*
9.4	Conclusions *271*
	References *271*

10	**Polymer Gels from Biopolymers** *279*
	Esra Alveroglu, Ali Gelir, and Yasar Yilmaz
10.1	Introduction *279*
10.2	Experimental Methods *279*
10.3	Polymerization and Gelation Kinetics *281*
10.3.1	Fluoroprobe–Polymer Interactions *283*
10.3.2	Real-Time Monitoring of Monomer Conversion *286*
10.4	Sol–Gel Transition and Universality Discussion *287*
10.5	Imprinting the Gels *292*
10.6	Heterogeneity of Hydrogels *301*
10.6.1	Effect of Ion Doping on Swelling Properties and Network Structure of Hydrogels *301*
10.6.2	Current Measurements for Searching the Internal Morphology of the Gels *302*
10.7	Ionic p-Type and n-Type Semiconducting Gels *303*
10.7.1	Electrical Properties of Ionic p-Type and n-Type Semiconducting Gels *304*
10.7.2	Polymeric Gel Diodes with Ionic Charge Carriers *305*
10.8	Conclusions *307*
	References *308*

11	**Conformation and Rheology of Microbial Exopolysaccharides** *317*
	Jacques Desbrieres
11.1	Introduction *317*
11.2	Conformation of Polysaccharides *318*
11.3	Secondary Solid-State Structures for Microbial Polysaccharides *318*
11.3.1	No Secondary Solid-State Structure *319*
11.3.2	Single-Chain Conformation *319*
11.3.3	Simple or/and Double Helices *321*
11.3.4	Double Helix *323*
11.3.5	Triple Helix *323*

11.4	Conformation in Solution: Solution Properties and Applications	325
11.4.1	Dextran and Pullulan 325	
11.4.2	Hyaluronan 326	
11.4.3	Xanthan 331	
11.4.4	Succinoglycan 334	
11.5	Gelling Properties in the Presence of Salts 336	
11.5.1	β(1→4)-D-Glucuronan 336	
11.5.2	Polysaccharide 1644 337	
11.5.3	Gellan and Similar Polysaccharides 337	
11.5.4	β(1→3)-D-Glucans 341	
11.5.5	YAS 34 342	
11.6	Conclusions 345	
	References 345	

12 Sulfated Polysaccharides in the Cell Wall of Red Microalgae 351
Shoshana (Malis) Arad and Oshrat Levy-Ontman

12.1	Introduction 351	
12.2	Sulfated Polysaccharides from Red Microalgae – General Overview 352	
12.3	Sulfated Polysaccharides of Red Microalgal Cell Walls: Chemical Aspects 354	
12.4	Proteins in the Cell Wall of Red Microalgae 355	
12.5	Rheology of Red Microalgal Polysaccharide Solutions 356	
12.6	Modifications of the Sulfated Polysaccharides 359	
12.7	Red Microalgal Sulfated Polysaccharide Bioactivities 362	
	References 365	

Volume 2

13 Dielectric Spectroscopy and Thermally Stimulated Current Analysis of Biopolymer Systems 371
Valérie Samouillan, Jany Dandurand, and Colette Lacabanne

14 Solid-State NMR Spectroscopy of Biopolymers 403
Garrick F. Taylor, Phedra Marius, Chris Ford, and Philip T.F. Williamson

15 EPR Spectroscopy of Biopolymers 443
Janez Štrancar and Vanja Kokol

16 X-Ray Photoelectron Spectroscopy: A Tool for Studying Biopolymers 473
Ana Maria Botelho do Rego, Ana Maria Ferraria, Manuel Rei Vilar, and Sami Boufi

17 Light-Scattering Studies of Biopolymer Systems 533
Taco Nicolai and Dominique Durand

18	X-Ray Scattering and Diffraction of Biopolymers 567 Yoshiharu Nishiyama and Marli Miriam de Souza Lima	
19	Large-Scale Structural Characterization of Biopolymer Systems by Small-Angle Neutron Scattering 583 Ferenc Horkay	
20	Microscopy of Biopolymer Systems 611 Changmin Hu and Wenguo Cui	
21	Rheo-optical Characterization of Biopolymer Systems 645 Dagang Liu, Rakesh Kumar, Donglin Tian, Fei Lu, and Mindong Chen	
22	Rheological Behavior of Biopolymer Systems 673 Tao Feng and Ran Ye	
23	Physical Gels of Biopolymers: Structure, Rheological and Gelation Properties 699 Camille Michon	
24	Interfacial Properties of Biopolymers, Emulsions, and Emulsifiers 717 Adamantini Paraskevopoulou and Vassilis Kiosseoglou	
25	Modeling and Simulation of Biopolymer Systems 741 Denis Bouyer	
26	Aging and Biodegradation of Biocomposites 777 Siji K. Mary, Prasanth Kumar Sasidharan Pillai, Deepa Bhanumathy Amma, Laly A. Pothen, and Sabu Thomas	
27	Biopolymers for Health, Food, and Cosmetic Applications 801 Robin Augustine, Rajakumari Rajendran, Uroš Cvelbar, Miran Mozetič, and Anne George	

Index   851

Foreword

Our industrialized world is driven in large part by petroleum. It is an important source of energy and materials, and many of the products that we depend upon for day-to-day living are derived from it. However, the cost of oil and gas has increased dramatically over the past decade and this upward spiral is expected to continue due to increased demand, finite quantities, and unreliable supply chains. Beyond the monetary cost, our petroleum-based economy comes at a significant environmental price that cannot be sustained. As a consequence, research scientists, university professors, university students, technology developers in industry, and government policy makers focus their interest in the future prospects for a world less dependent on fossil fuels, taking steps to reduce greenhouse gas emissions, and efficiently addressing the significant challenges associated with plastic wastes in the global environment. It is now widely recognized that more cost-effective and environmentally benign alternatives to petroleum and the products derived from it will be required in order to realize a future with a sustainable economy and environment. Since you have this book in your hands, chances are that you are one of these actors. The bio-derived polymers discussed in this book and their applications provide part of the solution to these problems.

This book examines the current state of the art, new challenges, opportunities, and applications in the field of biopolymers. It is organized in two volumes morphology, structure, and properties (Chapters 1–12), and characterization and applications (Chapters 13–27). This book summarizes in an edited format and in a comprehensive manner many of the recent technical research accomplishments in the area of biopolymers and their blends, composites, IPNs, and gels from macro- to nanoscale. The handpicked selection of topics and expert contributors make this survey of biopolymers an outstanding resource reference for anyone involved in the field of eco-friendly biomaterials for advanced technologies. It surveys processing–morphology–property relationship of biopolymers, their blends, composites, and gels. The influence of experimental conditions and preparation techniques (processing) on the generation of micro- and nanomorphologies and the dependence of these morphologies on the properties of the biopolymer systems are discussed in detail. The application of various theoretical models for the prediction of the morphologies of these systems is discussed. This book also illustrates the use of biopolymers in health, medicine, food, and cosmetics.

There are already a number of fine texts that comprehensively cover the subject of biopolymers in great detail, but the content of this book is unique. For the first time, a book deals with processing, morphology, dynamics, structure, and properties of various biopolymers and their multiphase systems. It covers an up-to-date record on the major findings and observations in the field of biopolymers.

Grenoble, Institute of Technology *Alain Dufresne*
February 26, 2013

List of Contributors

Esra Alveroglu
Istanbul Technical University
Department of Physics
Maslak
34469 Istanbul
Turkey

Shoshana (Malis) Arad
Ben-Gurion University of the Negev
Department of Biotechnology
Engineering
POB 653 Beer-Sheva 84105
Israel

Robin Augustine
Mahatma Gandhi University
Centre for Nanoscience and
Nanotechnology
Priyadarshini Hills
Kottayam 686560
Kerala
India

Rachel Auzély-Velty
Université Joseph Fourier
Centre de Recherches sur les
Macromolécules Végétales
(CERMAV-CNRS)
601 rue de la Chimie
38041 Grenoble
France

Luc Avérous
Université de Strasbourg
BioTeam/ICPEES-ECPM, UMR 7515
25 rue Becquerel
67087 Strasbourg Cedex 2
France

Deepa Bhanumathy Amma
Bishop Moore College
Department of Chemistry
Mavelikara 690110
Kerala
India

Deborah Blanchard
Université Joseph Fourier
Centre de Recherches sur les
Macromolécules Végétales
(CERMAV-CNRS)
601 rue de la Chimie
38041 Grenoble
France

Ana Maria Botelho do Rego
Technical University of Lisbon
Institute of Nanoscience and
Nanotechnology
Centro de Química-Física Molecular
Av. Rovisco Pais
1049-001 Lisboa
Portugal

List of Contributors

Sami Boufi
University of Sfax
Faculté des Sciences de Sfax
Laboratoire des Sciences des
Matériaux et Environnement
BP 1171-3000 Sfax
Tunisia

Denis Bouyer
Université de Montpellier
Institut Européen des Membranes
2, Place Eugene Bataillon
34967 Montpellier
Cedex 2
France

Christophe Chassenieux
LUNAM Université du Maine
IMMM UMR CNRS 6283
Dept. Polymères, Colloïdes,
Interfaces
1 Avenue Olivier Messiaen
72085 Le Mans Cedex 9
France

Mindong Chen
Nanjing University of Information
Science & Technology
Department of Chemistry
Nanjing 210044
China

Wenguo Cui
The First Affiliated Hospital of
Soochow University
Department of Orthopedics
188 Shizi Street
Suzhou, Jiangsu 215006
China

and

Soochow University
Orthopedic Institute
708 Renmin Road
Suzhou, Jiangsu 215007
China

Uroš Cvelbar
Jožef Stefan Institute
F4 Plasma Laboratory
Jamova 39
1000 Ljubljana
Slovenia

Jany Dandurand
Université Paul Sabatier
CIRIMAT UMR CNRS 5085
Physique des Polymères 3R1B2
118 route de Narbonne
31062 Toulouse cedex 02
France

Jacques Desbrieres
Université de Pau et des Pays de
l'Adour
IPREM
Helioparc Pau Pyrenees
2 Avenue P. Angot
64053 Pau Cedex 09
France

Marli Miriam de Souza Lima
Universidade Estadual de Maringa
Pharmacy Departement LAFITEC
Av. Colombo, 5790 - Zona 07
CEP 87020-900 Maringa
Parana, Brasil

Dominique Durand
LUNAM Université du Maine
IMMM UMR CNRS 6283
Dept. Polymères, Colloïdes,
Interfaces
1 Avenue Olivier Messiaen
72085 Le Mans Cedex 9
France

Tao Feng
Shanghai Institute of Technology
Department of Food Science and
Technology
School of Perfume and Aroma
Technology
120 Caobao Road
Shanghai 200235
China

Ana Maria Ferraria
Technical University of Lisbon
Institute of Nanoscience and
Nanotechnology
Centro de Química-Física Molecular
Av. Rovisco Pais
1049-001 Lisboa
Portugal

Chris Ford
University of Southampton
School of Biological Sciences
Highfield Campus
Southampton SO17 1BJ
UK

Tim J. Foster
University of Nottingham
School of Biosciences
Division of Food Sciences
Sutton Bonington Campus
Loughborough
Leicestershire LE12 5RD
UK

Ali Gelir
Istanbul Technical University
Department of Physics
Maslak
34469 Istanbul
Turkey

Anne George
Medical College Kottayam
Department of Anatomy
Gandhinagar
Kottayam 686008
Kerala
India

and

Center of Excellence for Polymer
Materials and Technologies
Tehnoloski Park 24
1000 Ljubljana
Slovenia

Hero Jan Heeres
University of Groningen
Department of Chemical
Engineering
Nijenborgh 4
9747 Groningen
The Netherlands

Antonio Heredia
Facultad de Ciencias
Departamento de Biología Molecular
y Bioquímica
Campus de Teatinos, s/n
29071 Málaga
Spain

Ferenc Horkay
National Institute of Child Health
and Human Development NICHD
13 South Dr Room 3W16, MSC 5772
Bethesda Md 20892-5772
USA

Changmin Hu
Shanghai Jiao Tong University
School of Biomedical Engineering
and Med-X Research Institute
1954 Hua Shan Road
Shanghai 200030
China

Parameswaranpillai Jyotishkumar
INSPIRE Faculty
Department of Polymer Science and
Rubber Technology
Cochin University of Science and
Technology, Kochi-682022

Vassilis Kiosseoglou
Aristotle University of Thessaloniki
School of Chemistry
Laboratory of Food Chemistry and
Technology
54124 Thessaloniki
Greece

Vanja Kokol
University of Maribor
Faculty of Mechanical Engineering
Institute for Engineering Materials
and Design
Smetanova ul. 17
2000 Maribor
Slovenia

Ashok Kumar
Indian Institute of Technology
Kanpur
Department of Biological Sciences
and Bioengineering
Kanpur 208016
Uttar Pradesh
India

Rakesh Kumar
Birla Institute of Technology, Mesra.
Patna Campus
Department of Applied Chemistry
P.O. - B. V. College, Patna - 800014,
Bihar
India

Colette Lacabanne
Université Paul Sabatier
CIRIMAT UMR CNRS 5085
Physique des Polymères 3R1B2
118 route de Narbonne
31062 Toulouse cedex 02
France

Oshrat Levy-Ontman
Sami Shamoon College of
Engineering
Department of Chemical
Engineering
Beer-Sheva 84100
POB 950
Israel

Dagang Liu
Nanjing University of Information
Science & Technology
Department of Chemistry
Nanjing 210044
Ningliu Rd 219
China

Fei Lu
Nanjing University of Information
Science & Technology
Department of Chemistry
Nanjing 210044
China

Phedra Marius
University of Southampton
School of Biological Sciences
Highfield Campus
Southampton SO17 1BJ
UK

Siji K. Mary
Bishop Moore College
Department of Chemistry
Mavelikara 690110
Kerala
India

Camille Michon
AgroParisTech
UMR1145 Ingénierie Procédés
Aliments
AgroParistech/Inra/Cnam
1 avenue des Olympiades
91300 MASSY
France

Bernard Moussian
University of Tübingen
Interfaculty Institute for Cell Biology
Department of Animal Genetics
Auf der Morgenstelle 28
72076 Tübingen
Germany

Miran Mozetič
Jožef Stefan Institute
F4 Plasma Laboratory
Jamova 39
1000 Ljubljana
Slovenia

and

Center of Excellence for Polymer
Materials and Technologies
Tehnoloski Park 24
1000 Ljubljana
Slovenia

Taco Nicolai
LUNAM Université du Maine
IMMM UMR CNRS 6283
Dept. Polymères, Colloïdes,
Interfaces
1 Avenue Olivier Messiaen
72085 Le Mans Cedex 9
France

Yoshiharu Nishiyama
Centre de Recherches sur les
Macromolécules Végétales
(CERMAV-CNRS),
BP53, 38041 Grenoble cedex 9,
France

Adamantini Paraskevopoulou
Aristotle University of Thessaloniki
School of Chemistry
Laboratory of Food Chemistry and
Technology
54124 Thessaloniki
Greece

Francesco Picchioni
University of Groningen
Department of Chemical
Engineering
Nijenborgh 4
9747 Groningen
The Netherlands

Eric Pollet
Université de Strasbourg
LIPHT-ECPM, EAc (CNRS) 4379
25 rue Becquerel
67087 Strasbourg Cedex 2
France

Laly A. Pothen
Bishop Moore College
Department of Chemistry
Mavelikara 690110
Kerala
India

Rajakumari Rajendran
Mahatma Gandhi University
Centre for Nanoscience and
Nanotechnology
Priyadarshini Hills
Kottayam 686560
Kerala
India

Manuel Rei Vilar
Université Paris Diderot
ITODYS UMR CNRS 7086
Bat. Lavoisier
15 rue J. A. De Baïf
75025 Paris cedex 13
France

Tania Ródenas
Instituto de Tecnología Química
UPV-CSIC
Universidad Politécnica de Valencia-
Consejo Superior de Investigaciones
Científicas
Av. Los Naranjos, s/n
46022, Valencia (Spain)

Maria J. Sabater
Instituto de Tecnología Química
UPV-CSIC
Universidad Politécnica de Valencia-
Consejo Superior de Investigaciones
Científicas
Av. Los Naranjos, s/n
46022, Valencia (Spain)

Khaleelulla Saheb Shaik
University of Tübingen
Interfaculty Institute for Cell Biology
Department of Animal Genetics
Auf der Morgenstelle 28
72076 Tübingen
Germany

Valérie Samouillan
CIRIMAT UMR CNRS 5085
Physique des Polymères 3R1B2
118 route de Narbonne
31062 Toulouse cedex 02
France

Prasanth Kumar Sasidharan Pillai
Bishop Moore College
Department of Chemistry
Mavelikara 690110
Kerala
India

Deepti Singh
Indian Institute of Technology
Kanpur
Department of Biological Sciences
and Bioengineering
Kanpur 208016
Uttar Pradesh
India

Janez Štrancar
Jožef Štefan Institute
Department of Solid State Physics
EPR Center
Laboratory of Biophysics
Jamova 39
1000 Ljubljana
Slovenia

Garrick F. Taylor
University of Southampton
School of Biological Sciences
Highfield Campus
Southampton SO17 1BJ
UK

Sabu Thomas
Mahatma Gandhi University
Centre for Nanoscience and
Nanotechnology
Priyadarshini Hills
Kottayam 686560
Kerala
India

Donglin Tian
Nanjing University of Information
Science & Technology
Department of Chemistry
Nanjing 210044
China

Carlos Torres
Catholic University of Peru
Department of Mechanical
Engineering
Av. Universitaria 1801, San Miguel
Lima 32
Peru

Fernando G. Torres
Catholic University of Peru
Department of Mechanical
Engineering
Av. Universitaria 1801, San Miguel
Lima 32
Peru

Omar Paul Troncoso
Catholic University of Peru
Department of Mechanical
Engineering
Av. Universitaria 1801, San Miguel
Lima 32
Peru

Frank van Mastrigt
University of Groningen
Department of Chemical
Engineering
Nijenborgh 4
9747 Groningen
The Netherlands

Philip T.F. Williamson
University of Southampton
Faculty of Natural & Environmental
Sciences
Highfield Campus
Southampton SO17 1BJ
UK

Charles Winkworth-Smith
University of Nottingham
School of Biosciences
Division of Food Sciences
Sutton Bonington Campus
Loughborough, Leicestershire LE12 5RD
UK

Ran Ye
Biosystems Engineering & Soil
Science
University of Tennessee
2506 E.J. Chapman Drive
Knoxville
TN 37996-4531
USA

Yasar Yilmaz
Istanbul Technical University
Department of Physics
Maslak
34469 Istanbul
Turkey

1
Biopolymers: State of the Art, New Challenges, and Opportunities

Christophe Chassenieux, Dominique Durand, Parameswaranpillai Jyotishkumar, and Sabu Thomas

1.1
Introduction

The term "biopolymers" usually describes polymers produced in a natural way by living species. Their molecular backbones are composed of repeating units of saccharides, nucleic acids, or amino acids and sometimes various additional chemical side chains contributing also to their functionalities.

If the largest part of biopolymers is extracted from biomass, such as polysaccharides from cellulose and proteins from collagen or milk, biopolymers can also be produced from biomonomers using conventional chemical processes as polylactic acid, or directly in microorganisms or genetically modified organisms, as polyhydroxyalkanoates. The genetic manipulation of microorganisms brings a tremendous potentiality for the biotechnological production of biopolymers with tailored properties quite suitable for high-value medical applications such as tissue engineering and drug delivery.

Throughout history, biopolymers have been mainly used by mankind as food, or for making clothing and furniture. Since the industrial time, fossil fuels such as oil are the greatest source in the development and manufacture of almost every commercial product, such as the plastic, which is currently used at a very large scale. But these fuels are not unlimited resources, and environmental concerns over all aspects of using fossil fuels for production and energy must be taken into account. We must act in a sustainable manner, which means that the resources must be consumed at a rate such that they can be restored by natural cycles of our planet. Therefore, today, the renewable nature of biopolymers leads them to a renaissance and a new interest. In the last 20 years, this interest in sustainable products has driven the development of new biopolymers from renewable feedstocks. Biopolymers have to compete with polymers derived from fossil fuel not only because of their functional properties but also in terms of cost. In this respect, biopolymers are competitive when the price of oil is high and the price of feedstocks, such as starch from corn, is low.

In addition, the biodegradability of biopolymers gives them a specific advantage for the environmental concerns, for example, single-use packaging in food, automotive, or electronics industries [1].

The bionanocomposites deserve a special attention because they form a fascinating interdisciplinary area that brings together biology, materials science, and nanotechnology. Generally, polymer nanocomposites are the result of the combination of polymers and inorganic/organic fillers at the nanometer scale. The extraordinary versatility of these new materials comes from the large choice of biopolymers and fillers available, such as clays, cellulose whiskers, and metal nanoparticles. These new materials have been elaborated thanks to the development of new powerful techniques such as electrospinning [2]. In these materials, the interaction between fillers at the nanometer scale acts as a bridge in the polymer matrix that leads to the enhancement of the mechanical properties of nanocomposites with respect to conventional microcomposites [3]. But, bionanocomposites also add a new dimension because they are biocompatible and/or biodegradable materials. So, they are gradually absorbed and/or eliminated by the body. Their degradation is mainly due to hydrolysis or mediated by metabolic processes. Therefore, nanocomposites present a great interest for biomedical technologies such as tissue engineering, medical implants, dental applications, and controlled drug delivery. Nevertheless, the spread out of these valuable bionanocomposites in our everyday life can only be achieved provided they are easily accessible to consumers in terms of volume. Cellulose whiskers may soon be a challenge for nanoclays that are being used as traditional nanofillers for many applications [4], since they are now produced on an industrial scale.

Food products are also usually made of nanostructured materials based on biopolymers, and the elaboration of nanoparticles based on proteins and/or polysaccharides has recently revolutionized the world of biocompatible and degradable natural biological materials [5]. Therefore, the toolbox that micro- and nanotechnologies offer provides new opportunities for product and process innovations in the food industry. The control of the process and functionality at the nanoscale leads to more sustainable food production. This approach allows the development of nutrient delivery systems with healthy and/or less caloric value nutrients, sensors, and diagnostic devices that can monitor and ensure the safety of food products throughout the food chain. At least, various enhanced packaging concepts extend the shelf life of fresh products or indicate quality deterioration of the packaged product. However, for consumers, the general feeling toward foods that are associated with these new technologies and more particularly nanotechnologies is not totally positive. Thus, it is imperative to develop a good communication of the applications of nanotechnologies that allows the consumers to make an informed decision whether or not they would like to have the benefits of certain applications of nanotechnologies, or whether they do not accept certain risks.

All these driving forces act as stimuli to develop new materials based on biopolymers, and there are many opportunity areas such as industrial, medical, food, consumer products, and pharmaceutical applications for which biopolymers act as stabilizers, thickeners, binders, dispersants, lubricants, adhesives, drug-delivery agents, and so on.

1.2
Biopolymers: A Niche For Fundamental Research in Soft Matter Physics

If, for over half a century, the study of biopolymers has been nearly the reserved field of biochemists and molecular biologists, during the last decade the soft matter physics community has seized this research field. Its purpose is not only for pure intellectual curiosity but also for modeling and understanding various mechanisms involved in the soft matter field, and for the consequences that a better understanding of these biopolymers might lead to. In fact, the fundamental physics underlying the biopolymer behavior and the techniques applied for their study are often similar. With the development of new powerful X-ray sources, new microscopies (cryo-TEM, ultra speed confocal scanning laser microscopy), and the advent of single-molecule techniques [6], polymer physicists are now strongly active in this field and there exists a strong collaboration between biologists and physicists.

For example, biologists can create specific mutations to design molecules for specific studies of the role played by specific groups located at precise points along the chain, for understanding by example the influence of a particular residue on the folding–unfolding process of biopolymer and its influence on the mechanical properties, which can be measured by pulling with an AFM tip [7]. Also, the understanding of the biological molecular machines allows designing synthetic molecules to perform analogous tasks [8].

More generally, most biological macromolecular assemblies are predominantly made from mixtures of stiff biopolymers, and our cells, muscles, and connective tissue owe their remarkable mechanical properties to these complex biopolymer networks. The understanding of their incessant assembly, disassembly, restructuring, active and passive mechanical deformation needs a lot of theoretical modeling efforts because if flexible polymer behavior is well depicted in literature, the stiffness of these biopolymers and the resulting anisotropic networks that lead to smart mechanical and dynamical properties are far from being understood.

The proteins, which are the major biomacromolecules in our body and play a fundamental role in making our body work, give us another example. As it is well known, proteins have a strong tendency to self-assemble after denaturation and this quasiuniversal mechanism, valuable for all the proteins, can lead to three generic structures, particulate gels around the isoelectric point, isolated amyloid fibrils, and spherulites far away from the isoelectric point [9,10]. Therefore, proteins have appeared as good model systems for understanding and modeling various self-assembling mechanisms, and more particularly the competition between processes of aggregation, gelation, and phase separation that play a major role in the self-assembly of most complex systems. So, a good control of the macroscopic phase separation during the protein self-assembly by kinetically trapping the structure at a particular stage of the process allows to create a large variety of new arrested structures. The linear and nonlinear rheological behaviors of these matrices and the transport properties of various probes inside are still actively investigated for understanding the relationships between these properties and the structures of the matrices. Such fundamental researches also have societal and medical interests because the aggregation of misfolded protein molecules into so-called amyloid fibrils is directly implicated in many diseases such as Alzheimer's disease. More pleasant, the food industry also utilizes the self-assembling ability of proteins such as the beta-lactoglobulin, a major protein component of milk, to texture foods such as yogurt by forming gels.

The mixtures of biopolymer are well known to display very rich phase behaviors. The understanding of the underlying physics of these phase behavior and of the rheology–morphology relationships of the resulting phases also constitutes a challenge of interest and importance. Such mixtures can also be used as efficient stabilizers for gas bubbles generated using high-throughput devices [11]. Since biomedical applications are targeted, the stability of the bubbles and their monodispersity are key points to be addressed, which can be achieved efficiently by combining biopolymer/protein mixtures and microfluidics.

Functional biopolymer nanoparticles or microparticles formed by heat treatment of globular protein–ionic polysaccharide electrostatic complexes under appropriate conditions are another example of the high potential of biopolymer mixtures [12]. Such biopolymer particles can be used as encapsulation and delivery systems, fat mimetics, lightening agents, or texture modifiers.

1.3
Biopolymers: An Endless Source of Applications

The trend observed for physics also holds for chemistry since biopolymers have become new building blocks from the point of view of macromolecular chemistry in the last decade. This fact owes much to the emergence of new polymerization techniques such as controlled radical polymerization (CRP) and to "Click" chemistry. Processing of these chemical alternatives allows a very good control of the macromolecular architecture, the molar mass distribution, and the functionality of the macromolecules. Block copolymers involving a biopolymer block are nowadays of an easier access, rendering possible the study of their self-assembly in bulk and/or in a selective solvent of one of the blocks. The specificities of the biopolymer block in terms of bioactivity, biocompatibility, and biodegradability allow targeting application fields such as pharmaceutical science for which the self-assemblies (polymersomes, micellar aggregates, microgels, etc.) are used as drug delivery systems with potential targeting properties [13]. In bulk, it should also be said that block copolymers based on a biopolymer block are pretty promising. Actually, the strong segregation force exhibited by most biopolymer blocks with respect to synthetic ones results in an efficient microphase separation despite the small polymerization degrees of each of the blocks. Well-organized thin films with periodicity as small as 5 nm have been obtained for the first time for which applications to soft electronics are expected [14].

1.4
Topics Covered by the Book

This book reviews the recent accomplishments obtained in terms of preparation, characterization, and applications of biopolymers. In the first chapters, general overviews and descriptions of the main concepts and issues regarding the most important biopolymer families are introduced. The synthesis of various biopolymers through the processing of genetic engineering tools or by nature itself is described in Chapters 4, 5, and 11. In each case, the impact of different synthetic conditions on the characteristics of the macromolecules is described in close correlation with their potential from the applicative point of view.

In the same way, Chapters 6–10 focus on the use of biopolymers as material. Each chapter focuses on one type of material such as polymer blends, macro- to nanocomposites, interpenetrated networks (IPN), or hydrogels. In most cases, the elaboration of biomaterials can be rationalized through a modelization approach as depicted in Chapter 24. The resulting practical cases of biopolymer used in our everyday life are described in Chapter 26. Their main advantages in terms of biodegradation, recycling, and life cycle are discussed in Chapter 25.

Biopolymers display interesting properties, not only in bulk but also in solution and at interfaces, as shown in Chapter 23.

The remaining chapters mainly focus on an experimental technique that can be used for gathering information on biopolymers from the point of view of their structure, dynamics, at different length and timescales in bulk or at interfaces. In each case, the background for understanding the technique is given, practical cases are described, and the limits of the technique are discussed. In Chapter 12, the ability of dielectric techniques to access the molecular mobility and chain dynamics of biopolymers and biological systems is addressed. These studies can be nicely complemented at other timescale by running NMR or EPR measurements as shown in Chapters 13 and 14.

Chapters 16–19 describe the use of scattering techniques and of microscopy for investigating the structure of biopolymers at several but complementary length scales. Structural but also mechanical properties of the biopolymers may also be derived from rheological measurements as described in Chapters 20–22. When an accurate knowledge of the chemical composition of the extreme layer of biopolymer surfaces is needed, XPS analysis can be used as detailed in Chapter 15.

1.5
Conclusions

Academic researchers involved in the biopolymer areas often work across various disciplines, physics, soft matter, chemistry, biochemistry, and biology and have developed skills that enable them to transfer their knowledge from one field to another. Thus, these skills should enable them to face some great challenges including the understanding of the physics of life, the nanoscale design of functional smart materials, the directed assembly of extended structures with targeted properties, and the emergence of physics far from equilibrium. The gap between nature and scientist know-how regarding "tailor-made" biopolymers is still wide, and a biomimetic approach of biopolymer synthesis may need tremendous development of specific genetic engineering tools. Furthermore, one should really have concerns regarding the life cycle of materials involving biopolymers, which are not always their single component, in order to avoid what we are currently facing with their actual homologues based on fossil fuels.

References

1 Johansson, C., Bras, J., Mondragon, I., Nechita, P., Plackett, D., Simon, P. et al. (2012) Renewable fibers and bio-based materials for packaging applications – a review of recent developments. *Bioresources*, **7** (2), 2506–2552.

2 Schiffman, J.D. and Schauer, C.L. (2008) A review: electrospinning of biopolymer nanofibers and their applications. *Polym. Rev.*, **48** (2), 317–352.

3 Faruk, O., Bledzki, A.K., Fink, H.P., and Sain, M. (2012) Biocomposites reinforced with natural fibers: 2000–2010. *Prog. Polym. Sci.*, **37** (11), 1552–1596.

4 Paul, D.R. and Robeson, L.M. (2008) Polymer nanotechnology: nanocomposites. *Polymer*, **49**, 3187–3204.

5 Sundar, S., Kundu, J., and Kundu, S.C. (2010) Biopolymeric nanoparticles. *Sci. Technol. Adv. Mater.*, **11** (1), 014104.

6 Deniz, A.A., Mukhopadhyay, S., and Lemke, E.A. (2008) Single-molecule biophysics: at the interface of biology, physics and chemistry. *J. R. Soc. Interface*, **5** (18), 15–45.

7 Alessandrini, A. and Facci, P. (2005) AFM: a versatile tool in biophysics. *Meas. Sci. Technol.*, **16** (6), R65–R92.

8 Kay, E.R., Leigh, D.A., and Zerbetto, F. (2007) Synthetic molecular motors and mechanical machines. *Angew. Chem. Int. Ed.*, **46** (1–2), 72–191.

9 Durand, D., Gimel, J.C., and Nicolai, T. (2002) Aggregation, gelation and phase separation of heat denatured globular proteins. *Physica A – Stat. Mech. Appl.*, **304** (1–2), 253–265.

10 Foegeding, E.A. and Davis, J.P. (2011) Food protein functionality: a comprehensive approach. *Food Hydrocoll.*, **25** (8), 1853–1864.

11 Park, J.I., Tumarkin, E., and Kumacheva, E. (2010) Small, stable, and monodispersed bubbles encapsulated with biopolymers. *Macromol. Rapid Commun.*, **31**, 222–227.

12 Jones, O.G. and McClements, D.J. (2010) Functional biopolymer particles: design, fabrication, and applications. *Compr. Rev. Food Sci. Food Saf.*, **9** (4), 374–397.

13 Schatz, C. and Lecommandoux, S. (2010) Polysaccharide-containing block copolymers: synthesis, properties and applications of an emerging family of glycoconjugates. *Macromol. Rapid Commun.*, **31**, 1664–1684.

14 Cushen, J., Otsuka, I., Bates, C., Halila, S., Fort, S., Rochas, C. et al. (2012) Oligosaccharide/silicon-containing block copolymers with 5 nm features for lithographic applications. *ACS Nano*, **6**, 3424–3433.

2
General Overview of Biopolymers: Structure, Properties, and Applications
Charles Winkworth-Smith and Tim J. Foster

2.1
Introduction

Biopolymer research is at the intersection of two new major branches of science, that of nanotechnology and environmental sciences. As the awareness of global warming increases and the cost of oil continues to rise, renewable biomaterials are gaining importance. This is being undertaken from a strong baseline as there has been significant research over the past four decades investigating the properties of these materials and looking for areas of exploitation in their use in structuring products. There are a number of industrial uses of biopolymers, ranging from coatings and adhesives, and the increased use in blends in the area of bioplastics, to cosmetics, personal care, pharmaceutical, and food products. The functionality in these products depends on a number of factors: the solvent type/quality and amount; the role of the biopolymer, that is, surface activity in emulsions and foams or the bulk structuring through imparting viscosity or gelation; the process employed to create the structure; and the interaction with other formulation materials. Underpinning the precise functionality is knowledge of the biopolymer structure. The biopolymers in most use are polysaccharides and proteins, and therefore their constituent sugars and amino acids, their sequence, and contribution to hydrodynamic three-dimensional structure in the solvent determine the overall polymeric functionality. Both classes of biopolymers are reviewed in detail in good "handbooks" designed to provide the reader with enough theoretical background of the individual types of polysaccharides or proteins to be able to understand their functionality and how they can be used [1,2]. In this chapter, we will provide an overview of these materials, some very recent and exciting developments, a focus on a particularly underutilized yet highly abundant polysaccharide, and an outlook for the future.

In order to begin, we can look at the two classes of biopolymers separately but also look at areas of commonality where developments in analytical techniques and extraction/refining of the materials might provide sources of innovation to the area.

Polysaccharides are attained from botanical sources such as seeds, grains, fruits, tubers, plant cell walls, or as exudates from tissue wounding (galactomannans, xyloglucans, starches, β-glucans, pectins, glucomannans, celluloses, gum arabic, gum tragacanth, karaya gum), from seaweeds (carrageenans, alginates, agarose, furcellaran), from animal sources (chitosan, hyaluronan), and from bacterial fermentations (xanthan, gellan, dextran, curdlan) and fungi (scleroglucan/schizophyllan). The plural use of these names is purposeful, as these subclasses of polysaccharides vary in molecular structure as a function of the specific source, plant maturity, the part of the plant used, and the method of extraction. To exemplify this, we can take a look at the galactomannans, which have a backbone structure of β-D-1,4-linked mannose with side chains of α-D-galactose, linked to C-6 of the mannan backbone. Depending on seed source, fenugreek, guar, tara, and locust bean gums have mannose:galactose ratios of 1:1, 2:1, 3:1, and 4:1, respectively, which affects their functionality; for example, locust bean gum gels through a freeze–thaw process, whereas guar gum does not. There are also inter- and intramolecular distributions of the galactose side chains, providing varying degrees of blockiness [3], which can be fractionated as a function of temperature of dissolution [4], which also determines the extent of interaction with other polysaccharides [5,6], requiring the galactose-free regions for the interactions to be stable [7]. Further details of galactomannans are provided in Section 2.2.2.1. Such complexity is not restricted to the galactomannans and is common among all of these polysaccharides provided by nature, and has had recent focus in studies on pectin [8,9], carrageenan [10], and cellulose derivatives [11].

Proteins used in industrial applications are also derived from a variety of sources, with animal (milk, meat, egg, and gelatin), plant (soy, pea, wheat (gluten), and potato), and fungal (hydrophobins) proteins being major areas of focus and innovation. Advances also exist in the development of algal proteins [12], and developments in biotechnology allow the use of genetic tools for controlled protein expression to permit scale-up opportunities in bacterial and fungal fermentations [13]. Proteins also share a trait with polysaccharides in that there is a wide variety of structural complexity in such nature-derived materials. Milk proteins, for example, may be classified as either casein or whey proteins, but within this subclass there are numerous subfractions (casein: phosphoproteins αS1-, αS2-, β-, and κ-casein; whey: β-lactoglobulin, α-lactalbumin, serum albumin, immunoglobulin, caseinomacropeptide) that have their own particular functionality, derived from their primary to tertiary/quaternary structures. This field of expertise is a great example of innovation in the area of biopolymers, where whey protein, for example, was a waste by-product from the cheese industry, and often used as animal feed or fertilizer. A recently published book entitled "Dairy-Derived Ingredients: Food and Nutraceutical Uses" [14] outlines the important developments in the industry, and scientific understanding has led to innovative use of whey protein as a versatile structuring agent. The most exciting developments in the past two decades have been in the study of cold-set β-lactoglobulin where the protein is preheated and then gelation induced by changing the ionic environment [15–19], where properties also depend on protein concentration [20,21]. The technology has also been shown to be transferable to other globular proteins, for example, soy [22], with excellent reviews of the area covering the state of the art [23–25], and can be used to structure emulsion products [26–28] and to control texture in mixed biopolymer systems [29]. New structures and rheologies are also being developed using β-lactoglobulin gels [30–32].

Polysaccharides and proteins are often used together in product formulations, and there are two common modes of interaction: either associative or segregative phase separation. Typically, the associative separation is through electrostatic interactions of oppositely charged biopolymers, creating simple or complex coacervates [33], which have the application potential for micro- and nanoencapsulation, multilayer structures, formation and stabilization of emulsions/foams (surface activity), formation of new gels, recovery of proteins from industrial by-products, clarifying agents, particulate gels (e.g., fat replacement), and packaging and edible films. The greater potential, however, is in the area of segregative phase separation, where the biopolymers demix and phase concentrate, and the microstructure is determined by formulation control of relative phase volumes and through processing effects. The phase-separated mixtures perform as water-in-water (w/w) emulsions [34] and therefore shear forces can be used to shape the dispersed phase. In addition, if one of the biopolymers is surface active, inclusions can be incorporated to create additional hierarchical complexity (e.g., oil-in-water-in-water emulsions [35–37]). While this discipline is maturing, there are still interesting developments, including phase separation of similar polysaccharides (κ- and ι-carrageenan [38]), phase concentration to anisotropic liquid crystalline phase [39], the positioning of particles at the interface of w/w emulsions [40], and the control of phase separation inside water droplets in water-in-oil emulsions [41].

Persin *et al.* [42] have highlighted challenges and opportunities in polysaccharide research and technology, which include the identification of new polysaccharide sources, possibly through developments in understanding the biosynthesis of natural hierarchical structures in the cell wall or storage organs, and through improvements in characterization methods to develop further an understanding of polysaccharide interactions with other (bio)polymers. Disassembly of what nature provides, partial/full fractionation of the constituent polymers, control of crystalline materials, and modifications linked to reassembly utilizing new process technologies will provide a continuation of the innovation in turning nature's materials into product opportunities. Indeed, some of these new insights are being attained at the interface of product technologies, for example, food and textiles, where blends of biopolymers are being used to modify fiber texture [43] and the use of different solvents to modify food polymers [44].

Therefore, a deeper understanding of the starting material, provided by nature, will enable future developments to be achieved, and in the next section we take a close look at the hierarchical structure of the plant cell wall, as there is a huge increase in cellulose-based research. Cellulose is the main building block of the plant cell wall. Cellulose fibers have been used for millennia to make rope, reinforced mud bricks, and textiles. The introduction of oil-based plastics made many of these uses obsolete. However, cellulose has found a renewed interest in "green" composite materials. Cellulose is considered to be an almost inexhaustible source of raw material with a total annual biomass production of about 1.5×10^{12} t [45].

Composites are a combination of a strong reinforcing, load-carrying material embedded in a matrix that acts as a binder and maintains the position and orientation of the reinforcement material. While both components retain their own chemical and physical properties, together they produce a material with unique qualities they would be unable to produce on their own. There are many natural composites including wood, bone, and teeth [46]. There are many historical examples of composites such as reinforced mud walls, concrete, and combinations of wood, bone, and animal glue. The first major industrial composites were glass fiber-reinforced resins, invented in the 1930s. This provided a strong, lightweight alternative to wood and metal to build boats and aircraft. By the 1970s, as better plastic resins and stronger reinforcing fibers including aramid fiber (better known as Kevlar, developed by DuPont) and carbon fiber were developed, new composites were invented. Recently, the push toward more environment-friendly materials and processes has resulted in a large increase in the research of bio-based composites.

One of the main focuses of bio-based composites is natural fibers. Natural fibers are essentially composed of the plant cell walls, which are themselves natural composites. Cells of higher plants are able to withstand an internal osmotic pressure of between 0.1 and 3.0 MPa (1 MPa = 145 psi) [47]. The pressure rigidifies the cells by creating tension in the walls. Despite the large variety in cell wall structure among all types of plants, a high cellulose content is common to all. Typically, 35–50% of plant dry weight is cellulose. Cellulose is normally embedded in a matrix primarily composed of hemicelluloses and lignin, which compose 20–35 and 5–30%, respectively, of plant dry weight [48]. In a few cases, such as cotton bolls, cellulose is present in a nearly pure state. Hemicelluloses are comprised of several different polysaccharides such as xylans, xyloglucans, glucomannans, galactomannans, and galactoglucomannans [49,50]. All higher plants contain the same general polysaccharides, but in different proportions. Lignins are amorphous polymers consisting mainly of aromatic units such as guaiacyl, syringyl, and phenylpropane [51]. Plant cell walls are comprised of the middle lamella and the primary and secondary cell walls. The primary cell wall is formed while the cell is growing. It defines not only the rate of growth of the plant cells but also its size and shape. The primary cell wall is composed of 90% polysaccharide and 10% proteins (glycoproteins) [52]. The secondary cell wall is formed after the cell is fully grown and is much thicker, more rigid, and stronger than the primary cell wall. The middle lamella is the outermost layer, comprised mainly of pectins that glue the adjacent plant cells together. The degree of polymerization (DP) [49] and the degree of crystallinity of cellulose in the secondary cell wall are also higher.

In addition to solid composites such as films and foams, there is also a growing interest in using plant cell wall material to create soft solid structures, particularly in the food, cosmetic, and pharmaceutical industries driven by their low cost, sustainability, and inferred naturalness [53]. There are also many health benefits. A high intake of dietary fiber, traditionally defined as the portions of plant foods that are resistant to digestion by human digestive enzymes (i.e., polysaccharides and lignin), appears to significantly lower the risk of developing coronary heart disease, stroke, hypertension, diabetes, obesity, and certain gastrointestinal diseases [54]. Recommended intakes for healthy adults are between 20 and 35 g/day [55]. However, many people consume very low levels; for instance, the majority of people in the United States consume less than half the recommended levels of dietary fiber daily [56]. It would therefore be useful to increase the use of cellulose and other cell wall polysaccharides in food systems.

2.2
Plant Cell Wall Polysaccharides

2.2.1
Cellulose

Cellulose was first discovered by the French chemist Anselme Payen in 1838. He described a resistant fibrous solid that remains behind after treatment of various plant tissues with acids and ammonia. He determined the molecular formula to be $C_6H_{10}O_5$, an isomer of starch. Cellulose has since been established to be a high molecular weight homopolymer of β-1,4-linked anhydro-D-glucose units. Due to the bond angles of the acetyl oxygen bridges, every second anhydroglucose ring is rotated 180° in the plane. This means that the repeating unit of cellulose is a dimer of glucose, cellobiose (Figure 2.1). The structure is stabilized by an intramolecular hydrogen bond network. This network makes cellulose a relatively stable polymer, which does not readily dissolve in aqueous solvents and has no melting point [57]. Each chain possesses directional asymmetry with one end having a reducing functional group and the other end nonreducing [58].

The size of the cellulose molecule is often described by its degree of polymerization. The DP strongly depends on the source of cellulose [59]. The degree of polymerization also depends on processing and extraction method.

Naturally, cellulose does not occur as an isolated molecule but is found in the form of fibrils. Cellulose is synthesized in the cell as individual molecules that undergo self-assembly at the site of biosynthesis [60]. Approximately 36 molecules are then assembled into larger units known as elementary fibrils (protofibrils) [58]. These are

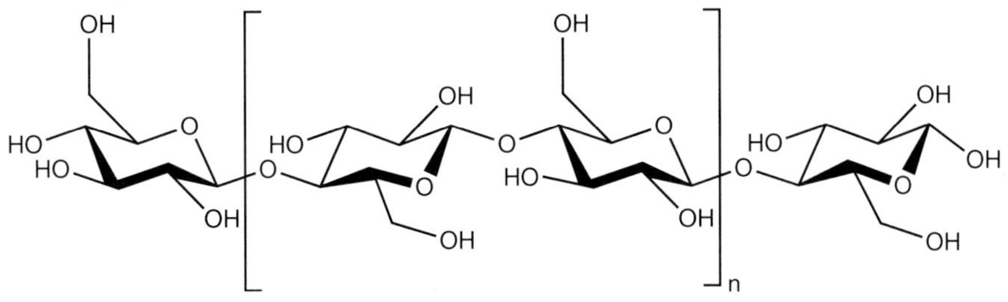

Figure 2.1 Chemical structure of cellulose.

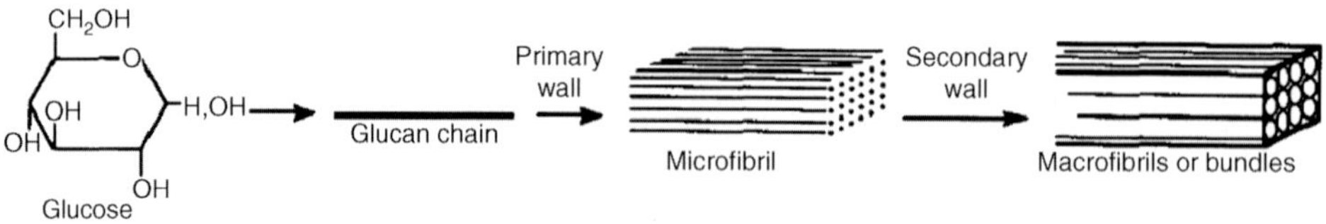

Figure 2.2 Steps in the assembly of native cellulose [62].

then packed into larger units known as microfibrils, which are then assembled into large macroscopic cellulose fibres (Figure 2.2). Microfibrils have cross dimensions ranging from 2 to 20 nm depending on the source of cellulose. The aggregation occurs primarily via van der Waals forces. An important feature of cellulose is its crystalline nature. This means the cellulose chains have a structured order. The component molecules of each individual microfibril are packed tightly, which prevents penetration by enzymes and even small molecules such as water. The supramolecular structure of the cellulose fiber is crystalline but there are some regions that are amorphous as well as some irregularities such as kinks and twists of the microfibrils and voids such as surface micropores. The effect of structural heterogeneity is that the fibers are at least partially hydrated by water and there is some penetration by larger molecules such as enzymes [61].

In the ordered crystalline regions, the cellulose chains are tightly packed together and held in place by strong hydrogen bonding (in-plane) and van der Waals interactions (between planes). The molecular orientation in the cellulose chains and therefore their packing arrangement can vary widely depending on the source, method of extraction, or treatment of the cellulose, resulting in a variety of allomorphs. The allomorph most commonly found in nature is cellulose I. Native cellulose is composed of two suballomorphs Iα and Iβ. Cellulose Iα has a one-chain triclinic structure and is predominant in primitive organisms. Cellulose Iβ has monoclinic unit cells and is predominant in higher plants [59]. Native cellulose has parallel chain alignment. This refers to the chain direction as regards to the reducing and nonreducing ends of the polymer. Regenerated or mercerized cellulose II has antiparallel chains [63] in a two-chain unit cell. Cellulose III can be generated by adding liquid ammonia to cellulose I or cellulose II producing cellulose III$_1$ and III$_2$, respectively. Heat treatment of cellulose III$_1$ and III$_2$ leads to cellulose IV$_1$ and IV$_2$ [64]. These can then be reverted back to their original cellulose. While cellulose I is the most commonly found form in nature, it is the least thermodynamically stable. Cellulose III is the most stable form [59].

2.2.1.1 Cellulose Extraction

Cellulose extraction is generally difficult as it does not melt and is not soluble in either water or common organic solvents. This is due to the hydrogen bond network and its partially crystalline structure. Currently, the most important industrial-scale extraction process of cellulose is the viscose process. The viscose process is more than 100 years old. Cellulose, often from wood pulp or cotton linters, is treated with sodium hydroxide and then carbon disulfide. The resulting product, dissolved in sodium hydroxide, is cellulose xanthogenate forming a thick solution called viscose. The viscose is then forced through very small openings into an acid that coagulates the regenerated cellulose. Depending on the size of the openings, fibers (rayon) or films (cellophane) can be produced. The viscose process has some environmental concerns due to the hazardous by-products such as CS_2, H_2S, and heavy metals. The cuprammonium process is another route for producing regenerated cellulose but also has environmental problems [65] and is no longer widely used.

An alternative to the viscose method is the Lyocell process. Lyocell was first manufactured in 1987 and has been produced commercially since 1991. It is a much more environment-friendly process:

- Preparation of a homogeneous concentrated solution (dope) of the starting cellulose (dissolving pulp) in an N-methylmorpholine N-oxide (NMMO)–water mixture.
- Extrusion of the highly viscous spinning dope at elevated temperatures through an air gap into a precipitation bath (dry jet–wet spinning process).
- Coagulation of the cellulose fiber in the precipitation bath.
- Washing, drying, and post treatment of the cellulose fiber.
- Recovery of the NMMO from the precipitation and washing baths.

The method is adapted from Ref. [65]. A major benefit of the Lyocell process is that it is capable of dissolving cellulose without derivatization, complexation, or special activation [66].

Research on ionic liquids has seen enormous growth recently due to the increased interest in green chemistry. Ionic liquids are often referred to as "green" solvents [67]. They are organic salts that exist as liquids at relatively low temperatures (<100 °C) [68]. Ionic liquids are also known as ionic fluids, molten salts, fused salts, or neoteric solvents. Some studies have shown that ionic liquids such as 1-butyl-3-methylimidazolium chloride (BMIMCl or [C4mim]Cl) and 1-allyl-3-methyl-imidazolium chloride (AMIMCl) can dissolve cellulose, whether it is refined or natural, without causing derivatization [69]. Microwave heating can significantly accelerate its dissolution [70]. Solutions containing up to 25% cellulose can be obtained by using [C4mim]Cl as the solvent at 70 °C (heated by microwave pulses), although compositions between 5 and 10% are more readily prepared [71]. Cellulose can then be precipitated from the ionic liquid by adding water, ethanol, or acetone.

Recently, Tatarova *et al.* [72] have shown that solutions of LiCl, urea, and water exert a swelling effect on regenerated cellulosics due to the propensity for formation of Li–cellulose coordination complexes. Urea acts as a cosolvent for the LiCl. This may prove a useful alternative to alkali swelling treatments.

2.2.1.2 Nanocellulose

Within the past 15 years, there has been a growing interest in nanocellulose. There are two major classes of nanocellulose, micro (or nano) fibrillated cellulose and cellulose nanowhiskers (nanocrystals), although there are many differing terminologies that can lead to confusion. Both have at least one dimension in the nanoscale (1–100 nm). The two classes of nanocellulose are distinguished by their method of preparation. Cellulose nanowhiskers are prepared using strong acid while nanofibrillated cellulose (NFC) is mainly prepared by mechanical homogenization.

2.2.1.3 Microfibrillated Cellulose

Microfibrillated cellulose (MFC) (also termed nanofibrillated cellulose) can be extracted from cellulose fibrils using a variety of mechanical processes including high-pressure homogenization, grinding/refining, cryocrushing, high-intensity ultrasonic treatments, and microfluidization [73]. These processes generate high shear that cleaves the cellulose fibrils along its longitudinal axis resulting in a greatly increased surface area. The long, flexible fibrils have lateral dimensions between 10 and 100 nm and length generally in the micrometer scale and consist of both the crystalline and amorphous domains [74]. The long fibrils result in a weblike structure [75]. The elastic modulus of single microfibrils from tunicate has recently been measured using atomic force microscopy with values between 145 and 150 GPa [76].

MFC was first extracted from wood in 1983 by Turbak *et al.* [77] and Herrick *et al.* [78]. The high-energy requirements, however, meant that there was no large-scale production of MFC. The energy required to produce MFC in the early 1980s was as high as 30 000 kWh/t. Using different pretreatment methods, this has been reduced by as much as 98% to 500 kWh/t [79]. Pretreatments that have been developed include acid hydrolysis, enzymatic hydrolysis, prebeating/grinding, TEMPO oxidation, and carboxymethylation [80–82].

Due to cellulose's hydrophilic nature, one of the major problems encountered with MFC is hornification (agglomeration upon drying). This results in the MFC having to be used either in a never dried state or chemically modified. The use of never dried suspensions is ultimately undesirable due to high shipping costs, large storage facilities, and its propensity for bacterial spoilage [73]. Different routes that have been investigated to produce a redispersible powdered MFC include carboxymethylation, grafting of acetyl moieties, or silylation (which also improves compatibilization with hydrophobic matrices) [74]. All these methods attempt to limit hydrogen bonding between fibrils by introducing steric hindrance or electrostatic groups. The original MFC structure may also be preserved using a variety of cryo techniques such as high-pressure freezing (vitrification), freeze fracturing, or freeze etching [85].

2.2.1.4 Cellulose Nanowhiskers

Cellulose nanowhiskers have many appealing intrinsic properties such as their nanoscale dimensions, high surface area, unique morphology, low density, and mechanical strength (the axial Young's modulus is theoretically stronger than that of steel and similar to that of Kevlar [86]). Cellulose whiskers also have the concomitant benefit of being easily chemically modified. Cellulose whiskers have many potential applications such as tablet binders, texturizing agents, fat replacers, additives in paper (e.g., security paper), and in nanocomposites.

Cellulose nanowhiskers are isolated from cellulose microfibrils by the treatment with an acid. The microfibrils consist of crystalline regions surrounded by disordered amorphous regions. The acid hydrolyzes the noncrystalline regions faster than the crystalline areas that have a higher resistance to hydrolysis [87]. Needle-shaped particles remain as a residue. The resulting suspension is then diluted with water and washed with successive centrifugations. The remaining suspension is then dialyzed against distilled water to remove any free acid molecules from the dispersion. Additional steps such as filtration, differential centrifugation, and ultracentrifugation can also be used [88,89]. Generally, the longer the hydrolysis, the shorter the nanowhiskers and the narrower the size polydispersity [90].

The type of acid used will impart different properties to the whiskers. The two most commonly used acids are sulfuric acid and hydrochloric acid. Phosphoric acid [91] and hydrobromic acid have also been used. The concentration of sulfuric acid is typically about 65% (wt) used between room temperature and 70 °C. The hydrolysis time can be as little as 30 min up to leaving it overnight, depending on the temperature used [58]. Hydrolysis using hydrochloric acid generally uses an acid concentration of 2.5–4 N at reflux temperature and for a variable time depending on the source of the cellulose. Sulfuric acid produces negatively charged whiskers that are more stable in aqueous solution than whiskers from hydrochloric acid that have no charge [92].

Depending on the acid used, the whiskers are able to self-assemble into nematic liquid crystalline structures at high concentrations [93]. The chiral nematic structure can even be preserved after total water removal, making it possible to form iridescent films of the nanowhiskers, increasing the range of potential applications [94].

Cellulose whiskers can be made from a variety of sources such as valonia, cotton, wood pulp, sugar beet pulp, tunicate, and bacterial cellulose. Table 2.1 shows the range of sizes of whiskers from different sources. The aspect ratio is the length-to-width (L/w) ratio.

2.2.2
Hemicelluloses

Hemicelluloses are characterized by a β-1,4-linked backbone with an equatorial configuration at C-1 and C-4 [95]. Hemicelluloses have a tendency to hydrogen bond with cellulose chains due to their cellulose-like conformation. They can usually be extracted with alkaline treatment. Hemicelluloses cover a broad range of structurally and physiochemically different polysaccharides. Polysaccharides often grouped together as hemicelluloses include xylans, mannans, and xyloglucans. Hemicelluloses are more abundant in the secondary cell wall than primary cell walls [96].

2.2.2.1 Galactomannans
Galactomannans are a group of polysaccharides that have a β-1,4-linked mannan backbone with different levels of galactose substitution. Galactomannans are structurally important components of the cell wall as well as an important source

Table 2.1 Examples of the length (L) and width (w) of cellulose nanowhiskers from various sources obtained by different techniques [58].

Source	L (nm)	w (nm)	Technique
Bacteria	100–1000	10–50	TEM
	100–1000	5–10 × 30–50	TEM
Cotton	100–150	5–10	TEM
	70–170	~7	TEM
	200–300	8	TEM
	255	15	DDL
	150–210	5–11	AFM
Cotton linter	100–200	10–20	SEM-FEG
	25–320	6–70	TEM
	300–500	15–30	AFM
MCC	35–265	3–48	TEM
	250–270	23	TEM
	~500	10	AFM
Ramie	150–250	6–8	TEM
	50–150	5–10	TEM
Sisal	100–500	3–5	TEM
	150–280	3.5–6.5	TEM
Tunicate		8.8 × 18.2	SANS
	1160	16	DDL
	500–1000	10	TEM
	1000–3000	15–30	TEM
	100–1000	15	TEM
	1073	28	TEM
Valonia	>1000	10–20	TEM
Softwood	100–200	3–4	TEM
	100–150	4–5	AFM
Hardwood	140–150	4–5	AFM

of storage polysaccharides. They are typically obtained from the endosperms of the seeds of leguminous plants. Some typical galactomannans include

fenugreek gum, M:G = 1 : 1;
guar gum, M:G = 2 : 1;
tara gum, M:G = 3 : 1;
locust bean gum, M:G = 4 : 1.

Galactomannan solubility is due to the presence of the galactose side units that prevent the mannan backbone from forming aggregates [97]. The mannose chains must have at least 12% galactose substitution to be water soluble. Ivory nut mannan contains about 95% mannose with few galactose side units and is therefore insoluble in water but can be solubilized in 1% sodium hydroxide.

Galactomannans are nonionic and thus are unaffected by ionic strength or pH but degrade at extreme pH.

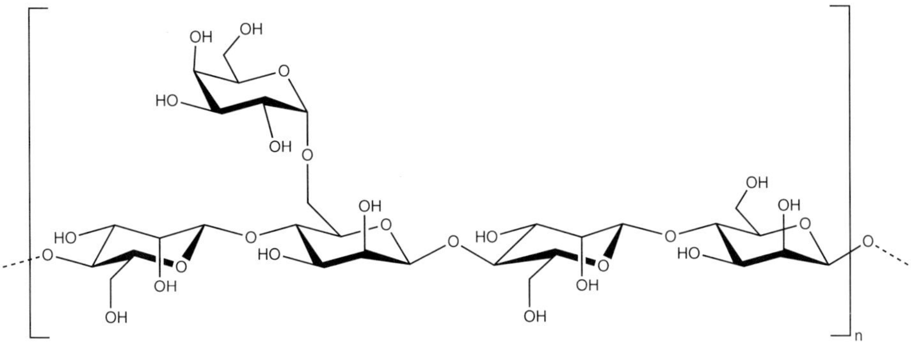

Figure 2.3 Chemical structure of locust bean gum.

Locust Bean Gum Locust bean gum (LBG), also known as carob bean gum, is obtained from the seeds of the carob tree (*Ceratonia siliqua*), found mostly in Mediterranean regions. It has a β-1,4-linked mannose backbone. Approximately every fourth mannose unit is substituted with a 1,6-linked α-galactose residue (Figure 2.3). The galactose side chains are unevenly distributed. This results in there being smooth regions of the mannose backbone, which are able to self-associate with other LBG molecules. LBG therefore needs heating to 60–90 °C to break the network of self-associations. This can make the determination of molecular weight problematic. A reduction in water activity or solution temperature can increase the amount of aggregations, which results in the formation of a 3D network. This is particularly useful in ice cream production. The weak gel structure can help impart excellent meltdown resistance in ice cream with a smooth texture without giving a slimy mouthfeel. Ice creams stabilized with LBG contain significantly smaller ice crystals than ice creams produced without stabilizers. LBG is also used in cream cheese to bind water and produce a spreadable texture without imparting sliminess.

LBG and κ-carrageenan gels are used as alternatives to gelatin gels. They have a higher melting point than gelatin, which can be advantageous in hot countries. LBG also acts synergistically with xanthan gum. Xanthan interacts with the unsubstituted regions of the mannan backbone. At low concentrations, there is a synergistic increase in viscosity. At higher concentrations, soft, elastic gels are formed [98].

Guar Guar gum is derived from guar beans (*Cyamopsis tetragonolobus*). India is the major producer, accounting for 80% of the world's total production. Guar has a β-1,4-linked mannose backbone with a galactose side chain approximately every two mannose units (Figure 2.4). Guar is able to fully hydrate in cold water. It does not form gels but does show good stability to freeze–thaw cycles and retards ice crystal growth.

Guar produces high-viscosity solutions making it useful as a thickener. Only small amounts are needed, making it a very cost-effective alternative to cornstarch. Guar is also used in baked goods to increase dough yield, improve texture, and prevent

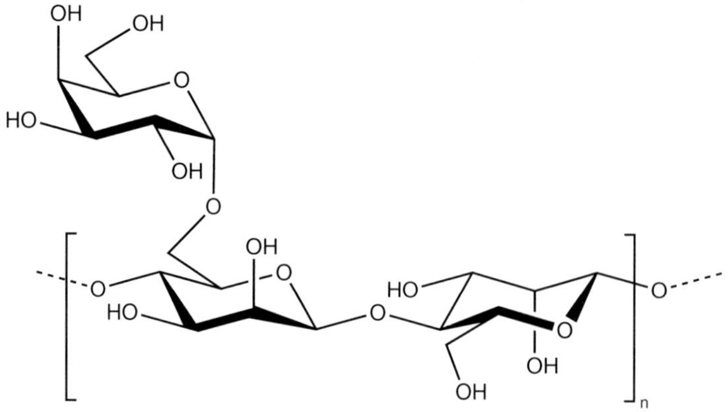

Figure 2.4 Chemical structure of guar.

syneresis. In addition to the food industry, guar is used in textiles, pharmaceuticals, and cosmetics. Recently, guar has found a new use as an additive to the fluids used in fracking gas (hydraulic fracturing, a technique used to force liquids at high pressure into shale rock to release natural gas).

Fenugreek India is the largest producer of fenugreek gum (*Trigonella foenum-graecum*), also known as "methi" in Hindi. The galactomannan is found in the endosperm of the fenugreek seeds. Fenugreek gum has only been used industrially since 1990. It has a β-1,4-linked mannose backbone that is fully substituted with galactose residues (Figure 2.5). Due to the high galactose content, fenugreek is the most water-soluble galactomannan.

Fenugreek is currently used mainly as a health additive to lower blood sugar and reduce cholesterol levels [99,100].

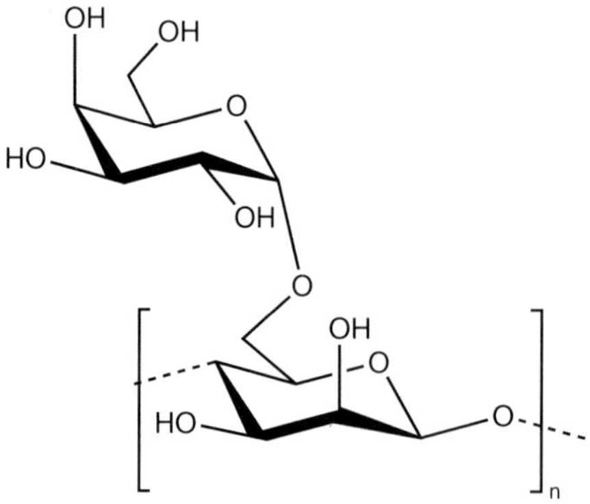

Figure 2.5 Chemical structure of fenugreek.

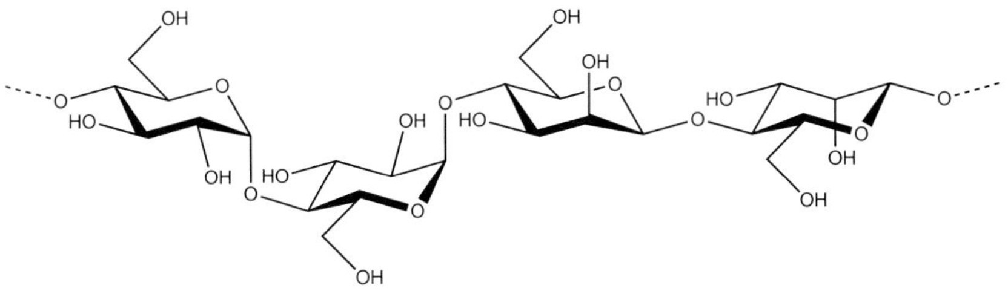

Figure 2.6 Chemical structure of konjac glucomannan.

2.2.2.2 Konjac Glucomannan

Konjac glucomannan is the major storage polysaccharide from the tuber *Amorphophallus konjac*, which grows in eastern tropical regions. It has a β-1,4-linked mannose and glucose backbone in a ratio of 1.6 : 1 (Figure 2.6) with an acetyl group about every 19 glucose residues [101]. Konjac readily dissolves in water producing a highly viscous pseudoplastic liquid. The polymer remains in solution due to the presence of the acetyl groups. If konjac is treated with alkali, the acetyl groups will be removed, resulting in konjac becoming water insoluble. Konjac is a nonionic polymer and thus is acid stable and tolerant to high levels of salt. Konjac produces synergistic gels with xanthan and κ-carrageenan. Some applications for konjac include healthy/slim foods, noodles, fruit jellies, thickeners, and edible films.

2.2.2.3 Xylan

Xylans have a β-1,4-linked xylose backbone. Xylans may be substituted with α-1,2-linked glucuronosyl and 4-O-methyl glucuronosyl residues and are often referred to as glucuronoxylans (Figure 2.7). They are the dominant hemicellulose in the secondary walls of dicots. Xylans may also contain arabinose residues attached

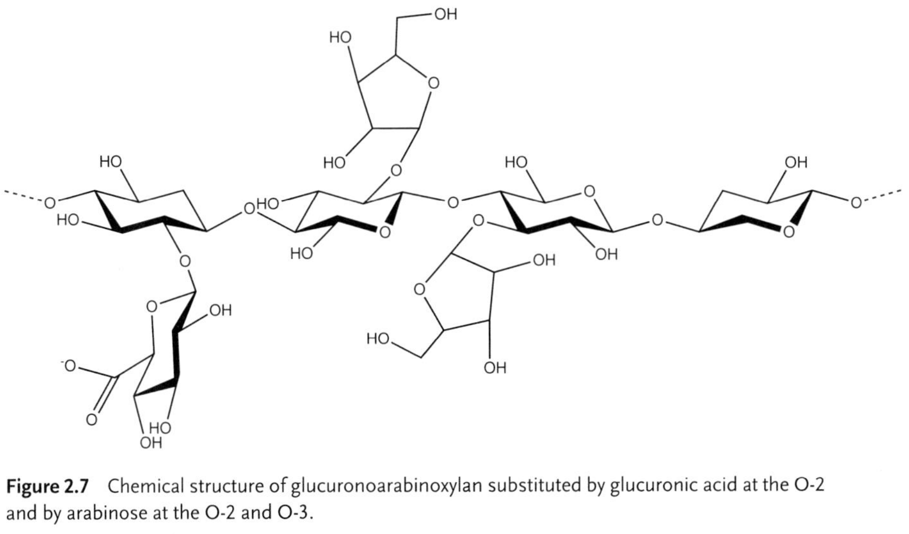

Figure 2.7 Chemical structure of glucuronoarabinoxylan substituted by glucuronic acid at the O-2 and by arabinose at the O-2 and O-3.

to the xylose backbone. Arabinoxylans are the primary noncellulosic polysaccharides in the primary cell walls of monocots (e.g., grasses). Xylan does not have a repeated structure [95]. Xylans are predominantly found in softwoods (10–15%), hardwoods (10–35%), and annual plants such as oat spelts (35–40%) [102].

2.2.2.4 Xyloglucan

Xyloglucans are a family of hemicelluloses with a cellulose-like β-1,4-linked glucan backbone highly substituted with α-D-linked xylopyranosyl residues attached at O-6 (Figure 2.8). Some xylose branches are further substituted at O-2 by combinations of galactopyranose, fucopyranose, arabinofuranose, and O-acetyl residues. Xyloglucans are made of repetitive units, generally described using a one-letter code denoting the different side chains [95]:

G – unbranched glucose residue;
X – α-D-Xyl-(1–6)-Glc;
L – xylose residue substituted with β-Gal;
S – xylose residue substituted with α-L-arabinofuranosyl (α-l-Araf);
F – a Gal residue substituted at O-2 with α-L-Fuc.

The most common core repeating units are XXGG and XXXG. The most widely studied xyloglucan is that from tamarind seed having a repeating unit of XXXG.

X-ray fiber diffraction indicates that the main chain has a flat ribbon-like conformation, similar to that of crystalline cellulose chains [103]. The side chains fold tightly onto the main chain surface. In aqueous solution, the main chain has a twisted conformation, so the side chains are unable to fold as tightly onto the main chain [104].

In the plant cell wall, xyloglucan coats the cellulose microfibrils, limits their aggregation, and connects them via cross-links. These cross-links can directly

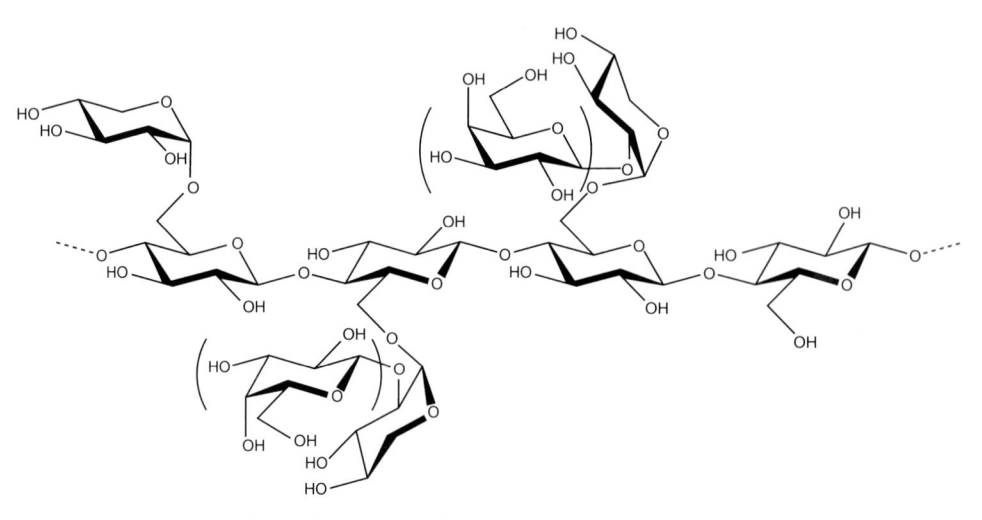

Figure 2.8 Chemical structure of xyloglucan with an XLLG configuration.

influence the mechanical properties of the cell wall [105]. *Xyloglucan endotransglycosylase* (XET) has been implicated in both wall-loosening and wall-strengthening roles [106].

Xyloglucan (extracted from tamarind kernel powder) is currently used as a sizing agent in textiles, especially in Asia. It improves yarn strength during weaving and imparts smoothness and stiffness to fabrics [107]. It is also used as a replacement for starch and galactomannans in papermaking but is not widely implemented [108]. The xyloglucan helps paper formation and increases strength as well as reduces fiber flocculation [109–111].

2.2.3
Pectins

Along with hemicelluloses, pectins help form part of the cell wall matrix. Pectins are present in the primary cell wall and the middle lamella where they bind cells together. Pectins are a group of polysaccharides that are rich in galacturonic acid (GalA). There are three major structural domains of pectin, homogalacturonan (HGA), rhamnogalacturonan-I (RG-I), and rhamnogalacturonan-II (RG-II) (Figure 2.9). These can be linked to form a pectin network throughout the primary cell wall matrix and the middle lamella.

HGA is a linear homopolymer of α-1,4-linked D-galacturonic acid and is thought to contain about 100–200 GalA residues [112]. The GalA residues can be methylesterified and O-acetylated. Sugar beet root and potato tubers have a particularly large amount of acylated GalA [113]. The GalA residues may be substituted with xylose to form xylogalacturonan (XGA) [114]. HGA tends to be insoluble, and thus is hard to extract [52].

RG-I consists of as many as 100 repeats of the disaccharide 1,2-α-L-rhamnose–1,4-α-D-galacturonic acid [115]. RG-I is abundant and heterogeneous and generally thought to be glycosidically bonded to HGA domains. Many of the rhamnose residues have side chains. The side chains can range from 1 glycosyl residue up to 50 or more. The predominant side chains contain linear and branched α-L-Ara$f$ and/or β-D-galactopyranosyl (Gal$p$) residues, although their relative proportions and chain lengths may differ depending on the plant source [116,117]. The highly branched nature of RG-I has led to it being known as the hairy region of pectin, in contrast to HGA domains that are known as the smooth region [114].

RG-II is not structurally related to RG-I. It has an HGA backbone, nine or more residues long. It has side chains composed of 11 different sugars [114]. RG-II is the only boron-containing polysaccharide that can be isolated from a biological source [118]. Boron is an essential microelement for plant growth. Boron deficiency can lead to disorganized cell expansion and cell walls with abnormal morphology [119]. O'Neill *et al.* propose that dRG-II-B (a borate ester cross-linked dimer of RG-II) is an essential component of the cross-linked pectin matrix. The covalent cross-links of the primary cell wall may result in the formation of a macromolecular complex composed of RG-II, RG-I, and HGA, linked by glycosidic bonds. This matrix may participate in regulating the rate of cell growth and

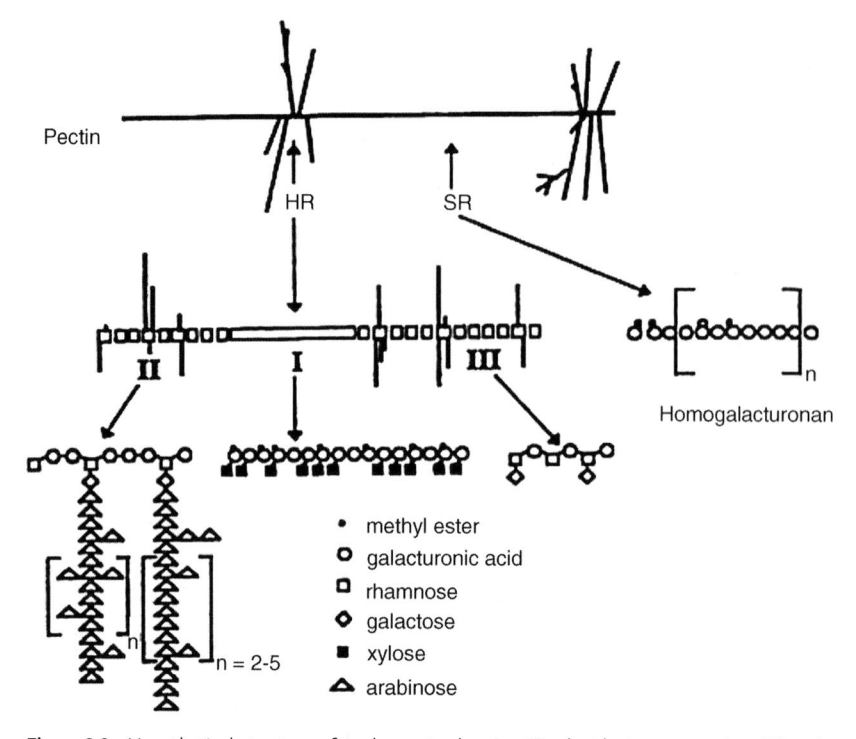

Figure 2.9 Hypothetical structure of apple pectin showing (I) xylogalacturonan region, (II) region with arabinan side chains, and (III) rhamnogalacturonan region making up the hairy region (source: *Handbook of Hydrocolloids*).

may create wall domains that control the rate at which enzymes, polysaccharides, biologically active oligosaccharides, and even lower molecular weight compounds pass through the wall [120].

Many of the galacturonic acid groups along the main chain are esterified with methoxy groups. The percentage of galacturonic acids that are esterified is referred to as the degree of esterification (DE). The pattern of the distribution of methoxy groups is also important and is often described as blockiness [121,122]. Commercially, pectins are generally categorized according to their methoxy content:

- high-methoxy (HM) pectin – DE higher than 50;
- low-methoxy (LM) pectin – DE lower than 50.

HM pectin forms gels under conditions of low pH (below 3.5) and high solids content (above 55%). HM pectin is typically used for jam making [123]. The number of methoxy groups determines the speed at which the gels will set. The rate of gelation is decreased as more methoxy groups are removed. As the pH is reduced, gel strength and setting temperature will decrease [98]. If the pH is reduced to a point where the setting temperature and preparation temperature are similar, the pectin tends to pregel. This results in a nonhomogeneous weaker gel. It will also be more susceptible to syneresis. This can be solved by preparing pectin mixtures at higher pH and then acidifying when ready.

LM pectin forms gels in the presence of divalent cations, most commonly calcium. The reactivity to calcium increases with decreasing DE. LM pectins are often used to produce low-sugar jams. They are also often used in milk products where they utilize the calcium present to enhance viscosity and stabilize emulsions [124,125]. If HM pectins are reacted with ammonia, normally in an aqueous alcohol slurry at ambient temperature, a LM amidated pectin can be produced. Amidated pectin requires less calcium to gel than conventional LM pectins.

Pectin is present in nearly all terrestrial plants but is most abundant in vegetables and fruits. The major source is the rind of citrus fruit. Apple pomace and sugar beet pulp are also widely used as sources of pectin. This is often waste material from another industry, for example, apple pomace from cider production. Pectin is generally extracted from plant material using hot aqueous mineral acid. The pectin is then isolated from the solution. One of the main challenges for producers is to retain a high molecular weight as pectin is very susceptible to degradation either by enzymes already present in the plant material or by heat during drying and subsequent processing [98].

Jams have been made for centuries by adding fruits containing high levels of pectin. Pectin has been industrially produced since the early twentieth century. Pectin is now also used in other food and personal care products as a thickener, stabilizer, and gelling agent. Pectin also forms excellent films

2.3
Biocomposites

2.3.1
Natural Fiber Composites

Natural fibers are cellulose microfibrils embedded in an amorphous matrix of lignin and hemicelluloses. Common fibers include jute, flax, and sisal. They have many benefits as they are biodegradable, renewable, strong, and abundantly available, and have low density and low cost. They are also much less abrasive than many fibers currently used in composites.

There are major difficulties in using natural fibers as reinforcing fillers. Natural fibers start degrading at about 240 °C and thus cannot be processed at high temperature. The fibers also absorb moisture and may have a moisture content of between 5 and 10% [126]. This can lead to poor processability. The fibers can also be biodegraded by microorganisms and are susceptible to UV light [127]. Natural fibers may also not function well as a reinforcing component due to poor adhesion at the fiber–matrix interface. They also aggregate in a hydrophobic polymer matrix. This can be counteracted by appropriate pretreatments with suitable additives such as stearic acid, mineral oil, or maleated ethylene (generally at a concentration of about 1%), which reduces fiber–fiber interaction [127]. To improve adhesion between the fibers and matrix polymers, compatibilizers or coupling agents may also be used, such as silane, zirconate, or titanate. Sodium alginate and sodium hydroxide have also been used with banana and coir fibers to increase adhesive bonding and thus improve their ultimate tensile strength [128]. Chemical grafting is another route to improving the physical properties of fibers by attaching suitable additives such as vinyl monomers to the surface of the fiber, which improves the bonding between fiber and matrix [129].

Many different parameters influence the strength of composites, including the volume fraction of fibers, fiber aspect ratio, fiber–matrix adhesion, stress transfer at the interface, and orientation [127].

Natural fiber composites have been produced with a variety of matrix materials. Many that are in use now, for instance, in the automotive industry, rely on oil-based matrix resins such as polypropylene, polyethylene, and polyvinyl chloride (PVC) [130–133]. In the pursuit of completely renewable and biodegradable composites, other matrix materials are currently being researched. Those of interest include polylactic acid, cellulose esters, polyhydroxybutyrates, starch, and lignin-based plastics [134]. Few are commercially available and often have high cost, poor processability, and, as with most polymers from natural sources, low moisture stability. Polylactic acid (PLA), which is derived from starch by fermentation, is perhaps one of the most promising renewable polymers [135]. Bodros et al. [136] have shown that PLA/flax composites have a specific tensile strength and Young's modulus close to that of fiber glass polyester composites and have greater tensile strength and modulus than those of similar polypropylene/flax composites.

One answer to achieving improved compatibility between fiber and matrix is to utilize cellulose as the matrix material itself. All-cellulose composites were initially developed by Nishino et al. [137], where the cellulose pulp was dissolved in lithium chloride/N,N-dimethylacetamide (LiCl/DMAc). The cellulose solution then formed the surrounding matrix for aligned ramie fibers. The solution was then coagulated with methanol and dried. The composite had longitudinal strength as high as 480 MPa as well as extremely good thermal stability. One of the major difficulties in producing all cellulose composites is the high viscosity of concentrated cellulose solutions. An alternative to using a cellulose solution as the matrix material is to only partially dissolve fibers such as flax or ramie [138].

An alternative to using natural fibers is to use purified primary cell wall fragments. This would have the advantage that cheap waste sources of cellulosic material such as sugar beet pulp or vegetable waste could be used for composite manufacture. Hepworth and Bruce [139] used swede root as the source of cell wall material, which was ground into a fine paste. Lipid membranes were destroyed with 1% detergent and some of the pectins were removed with 0.5 M HCl to break up the cells. The cell wall fragments were then bound together in a matrix of polyvinyl alcohol (PVA). The composite had a tensile stiffness of 5.4 GPa and a strength of 70 MPa, which compares favorably to epoxy and phenolic composites reinforced with randomly oriented vegetable fibers. They did, however, comment that if the cellulose microfibrils were completely extracted and aligned, then a higher strength could probably be achieved. This was later done using chemical extraction of the noncellulosic components of the cell wall with 2% sodium hydroxide [139,140]. The purified cellulose suspension was then passed through a homogenizer to separate the microfibrils. They used four different matrix materials: PVA, acrylic polymer, epoxy, and hemicellulose. The hemicellulose chosen was locust bean gum. Composites made with LBG showed higher tensile strength and stiffness than pure LBG. The LBG composites had a lower stiffness compared to PVA or epoxy but had higher strength and stiffness compared to flax/epoxy composites.

2.3.2
Cellulose Composites

Despite the many attractive properties of natural fibers, there are still major drawbacks. The noncellulosic components of natural fibers often impart some of the most undesirable aspects. It is often useful therefore to create composites with pure cellulose. Lignin is first removed with alkali (usually NaOH) and then treated with acid to remove any hemicellulose and pectin [73]. The cellulose is also often bleached.

Both cellulose nanowhiskers and MFC are prepared in water. To create nanocomposite films, the suspensions are mixed with a polymer that has been previously dissolved in water and then the liquid is evaporated. Due to the inherently high sensitivity of these polymers to humidity, storage is difficult [75]. Water also induces a strong plasticizing effect and greatly affects the properties of the film [141].

To investigate the effect of moisture on the dynamical mechanical properties of cellulose composites, Dammstrom *et al.* [142] produced composite films with bacterial cellulose and glucuronoxylan from aspen wood chips. Humidity scans using dynamic mechanical analysis (DMA) showed no softening for the pure bacterial cellulose sample but there was a pronounced softening at 85% RH for pure glucuronoxylan. The composite of the two polymers also showed a decrease in modulus but at a slightly lower RH as compared to pure glucuronoxylan. The glucuronoxylan acts as a plasticizer but also changes the spatial organization of the cellulose fibers although there are no clear interactions between the cellulose and hemicellulose.

The use of cellulose nanowhiskers may improve the water transition properties of xylan films. Saxena and Ragauskas [143] have shown that xylan films reinforced with 10% sulfonatednanowhiskers had a 74% reduction in specific water transmission properties compared to control xylan films. There was also a 362% improvement in xylan films reinforced with 10% softwood kraft pulps. The high crystallinity of the nanowhiskers and the dense composite structure formed due to a rigid hydrogen-bonded network lead to a film that has reduced moisture transmission properties [144]. Nanowhisker-reinforced xylan films also have excellent tensile strength [145]. The use of plasticizers such as xylitol or sorbitol improves the mechanical properties of the films by conferring flexibility and workability [146].

Nanowhisker-reinforced hemicellulose films have also been produced using konjac glucomannan [147]. The addition of nanowhiskers to konjac, plasticized by glycerol, induced the formation of fiber-like structures with lengths of several millimeters, although the differences in film structure did not appear to be related to the thermal properties of the films.

Recently, Azeredo et al. [148] have used nanowhiskers to reinforce mango puree to produce edible films. The nanowhiskers increased the tensile strength and Young's modulus in respect to pure puree films as well as improved the water barrier properties of the films [148].

2.3.3
Cellulose–Polymer Interactions

Both hemicelluloses and pectins are likely to bind to cellulose in the plant cell wall but there is as yet no complete consensus as to what the binding mechanisms are. Xyloglucan is known to bind to cellulose but there is some debate as to whether it depends on molecular weight or side chain composition. The presence of fucose was suggested by Levy et al. [103] to create a flat conformation that enabled xyloglucan to bind to cellulose. This was further confirmed by binding experiments with fucosylated (pea) and nonfucosylated (tamarind seed) xyloglucans [103,149]. However, the role of fucose in facilitating xyloglucan–cellulose interactions was later cast into doubt by Vanzin et al. [150] by using the *Arabidopsis* mutant *mur2* that eliminates xyloglucan fucosylation in all major plant organs. Despite the lack of fucose, the *mur2* plants showed a normal growth and wall strength. Lima et al. [151] have since shown with *in vitro* experiments that the binding capacity of xyloglucan is improved when the molecular weight of the polymer is decreased by enzymatic hydrolysis and that the branching with fucose seems not to be a key factor in binding. Fucose though may still have some role in strengthening the cell wall structure.

Whitney et al. [152] first developed the method of modeling plant cell wall interactions using the Gram-negative bacterium *Acetobacter aceti* ssp. *xylinum* that synthesizes pure, highly crystalline cellulose I as an extracellular polysaccharide. The cellulose forms a thick pellicle that floats on the surface of the medium. Viewed using deep-etch freeze-fracture TEM, the cellulose ribbons appear to form a randomly oriented network. Different polysaccharides can be added to the medium. In the presence of tamarind xyloglucan, the bacterial cellulose also forms a network similar in structure to that of plant cell walls [153]. Whitney et al. [154] found that galactose-depleted xyloglucan is likely to self-associate, leading to phase separation into xyloglucan-rich and cellulose-rich phases with limited direct binding between the two polymer types. This led them to conclude that galactose content has a greater effect on composites than other xyloglucan variants with fucose substitution only acting as a secondary modulator. These results are consistent with the work done on *Arabidopsis* mutants.

The *A. xylinum* system has also been used for other cell wall polysaccharides such as mannans. In the presence of konjac glucomannan, a structure is formed that shows considerable heterogeneity [155]. Micrographs show regions of apparent cross-linking of cellulose ribbons by glucomannan and areas of glucomannan network within a cellulose network. The cellulose acts as a template where the mannan residues along the polymer chains adopt a cellulosic conformation. Konjac, as well as low-galactose galactomannans, dramatically reduces the cellulose crystallinity. Mannans with a low galactose content are also more effective at disrupting cellulose organization. This suggests that the galactose side chains present a significant barrier to incorporation within cellulose fibrils. Due to the much more polydisperse fine structure of galactomannans, as opposed to xyloglucan, the long-range alignment of fibrils is not as pronounced. In secondary cell walls, both gluco- and galactomannans are significant components. The mannan-based polymers are able to help the coalescence/densification of cellulose and thus provide a stronger composite compared to xyloglucan in the primary cell wall where flexibility is the major requirement.

Many of the models of cell wall organization suggest that the cellulose–hemicellulose network is independent of the one formed by pectins [156]. However, Zykwinska *et al.* [157] have demonstrated that pectin is able to bind *in vitro* to cellulose microfibrils. The neutral pectin side chains are likely to enable noncovalent cellulose binding rather than the pectin backbone domains [158]. Commercial citrus pectins that have a very low number of neutral side chains, due to their harsh acidic extraction, do not bind to cellulose [159].

2.3.4
Semi-Solid Composites

There are many semi-solid composites particularly in food systems where a gel network is embedded with a filler. The filler enhances the rheological properties and textural characteristics of the system. Common fillers in the food industry include meat fibers, starch granules, or emulsified oil droplets [160]. Inorganic fillers such as glass fiber or hydroxyapatite clay may be used in non-food applications. The size, shape, strength, and phase volume of the filler all influence its reinforcing effect. When suitably incorporated, the filler should improve the mechanical properties of the system due to load transfer from the matrix to the filler particle. One of the most common fillers in the food industry is microcrystalline cellulose (MCC). MCC has been shown to improve gelatin gel strength [161,162]. By applying torque to the setting gels, Koh and Kasapis [162] found they could highly orient the MCC fibers resulting in a higher storage modulus of 1.5% gelatin gels due to an enhancement in network strength.

MFC has outstanding rheological properties. MFC suspensions possess a classical pseudoplastic (shear thinning) behavior that is common to many polymer solutions [81] and have a solid-like viscoelastic behavior. The rheological properties are not affected by either temperature or pH although an increase in ionic strength does increase both the G' and G'' moduli of MFC suspensions [163]. MFC may also be able to increase viscosity or gel strength of LM pectin. Agoda-Tandjawa *et al.* [164] added LM pectin to a suspension of MFC, which had a synergistic effect on the rheological properties of the composites. Addition of calcium to the mixtures induced gelation, which was enhanced by the presence of sodium, resulting in the formation of a stronger cellulose/LM pectin gel.

Since the early 1990s, MCC has been used as a fat replacer. The MCC is first subjected to severe mechanical attrition such as high-pressure homogenization, which breaks it down into colloidal crystalline aggregates. It is then codried with a hydrocolloid such as carboxymethyl cellulose, which aids later redispersion and forms a network that evenly distributes the particles [165]. When added to aqueous medium (generally above 5% solids content) the insoluble submicron-sized crystals disperse to create a stable thixotropic gel that has a creamy mouthfeel, opacity, and body [166]. MCC also enhances the gelling properties of galactomannans [167].

Ang and Miller [168,169] first reported the effect of powdered cellulose on the viscosity of polymer solutions. Powdered cellulose was able to increase the viscosity of guar gum, carboxymethyl cellulose, and 0.1–0.3% (w/v) xanthan solutions [169]. Recently, Day et al. [170] added xanthan to rehydrated carrot cell wall particles (CWPs) [168,170]. The addition of xanthan to CWPs at concentrations lower than 1% (w/w) influenced the rheological behavior of the CWP dispersions due to the increase in the viscoelastic properties of the continuous phase, but this was not the case for concentrations higher than 3% (w/w) as the viscoelastic behavior of the mixtures was dominated by the CWP particle network. There have been several other studies using cell wall dispersions from a variety of horticultural sources such as tomato, apple, and carrot that have all demonstrated the importance of interactions between the solid particles and the deformability of the fully packed particles [171–176].

2.4
Future Outlook

Nature provides an array of biopolymers, with differences in molecular structures and subsequent properties, which can be controlled through the choice of formulation and process. Indeed, the introduction of new process options can provide new functionalities from conventional biopolymers [177]. An increased focus on the bio-based economy will drive innovations and a lending of technologies to produce new cosmetic, pharmaceutical, textile, and food structures. Polymer modifications and new insights into polymer interactions will allow high-performance, functional fibers and surface-active particles to be formed [178], which may be incorporated into films with defined surface, permeability, sorption, and barrier properties. From a food industry perspective, new functional fibers may enable extension of meat products, to help satisfy the growing meat consumption globally, and may provide new structures for improved digestion control to tackle the growing problem of obesity worldwide, through structures that are functional and interact with the body at different parts of the GI tract, with possibilities of controlled encapsulation and release of micro/macronutrients. The future for biopolymer research and technological innovations is bright.

References

1. Phillips, G.O. and Williams, P.A. (2009) *Handbook of Hydrocolloids*, Woodhead Publishing.
2. Phillips, G.O. and Williams, P.A. (2011) *Handbook of Food Proteins*, Woodhead Publishing.
3. McCleary, B.V., Clark, A.H., Dea, I.C.M., and Rees, D.A. (1985) The fine structure of carob and guar galactomannans. *Carbohydr. Res.*, **139**, 237–260.
4. Richardson, P.H. and Norton, I.T. (1998) Gelation behaviour of concentrated locust bean gum solutions. *Macromolecules*, **31**, 1575–1583.
5. Mannion, R.O., Melia, C.D., Launay, B., Cuvelier, G., Hill, S.E., Harding, S.E., and Mitchell, J.R. (1992) Xanthan/locust bean gum interactions at room temperature. *Carbohydr. Polym.*, **19**, 91–97.
6. Lundin, L. and Hermansson, A.M. (1995) Influence of locust bean gum on the rheological behaviour and microstructure of K-κ-carrageenan. *Carbohydr. Polym.*, **28**, 91–99.
7. Dea, I.C.M., Clark, A.H., and Mccleary, B.V. (1986) Effect of the molecular fine structure of galactomannans on their interaction properties – the role of unsubstituted sides. *Food Hydrocolloids*, **1**, 129–140.
8. Williams, M.A.K., Buffet, G.M.C., Foster, T.J., and Norton, I.T. (2001) Simulation of endo-PG digest patterns and implications for the determination of pectin fine structure. *Carbohydr. Res.*, **334**, 243–250.
9. Williams, M.A.K., Buffet, G.M.C., and Foster, T.J. (2002) Analysis of partially methyl-esterified galacturonic acid oligomers by capillary electrophoresis. *Anal. Biochem.*, **301**, 117–122.
10. Helbert, W. (2012) Analysis and modification of carrageenan structure using enzymes, in *Gums and Stabilisers for the Food Industry* (eds P.A. Williams and G.O. Phillips), RSC Publishing.
11. Sullo, A., Wang, Y., Koschella, A., Heinze, T., and Foster, T.J. (2013) Self-association of novel mixed 3-mono-O-alkyl cellulose: effect of the hydrophobic moieties ratio. *Carbohydr. Polym.*, in press.
12. Chronakis, I.S. and Madsen, M. (2011) Algal proteins, in *Handbook of Food Proteins* (eds G.O. Phillips and P.A. Williams), Woodhead Publishing.
13. Yeh, C.M., Wang, J.P., and Su, F.S. (2007) Extracellular production of a novel ice structuring protein by *Bacillus subtilis* – a case of recombinant food peptide additive production. *Food Biotechnol.*, **21**, 119–128.
14. Corredig, M. (2009) *Dairy-Derived Ingredients: Food and Nutraceutical Uses*, Woodhead Publishing.
15. Barbut, S. and Foegeding, E.A. (1993) Ca^{2+}-induced gelation of pre-heated whey protein isolate. *J. Food Sci.*, **58**, 867–871.
16. Hongsprabhas, P. and Barbut, S. (1997) Protein and salt effects on Ca^{2+}-induced cold gelation of whey protein isolate. *J. Food Sci.*, **62**, 382–385.
17. Hongsprabhas, P., Barbut, S., and Marangoni, A.G. (1999) The structure of cold-set whey protein isolate gels prepared with Ca^{++}. *Food Sci. Technol. – Leb.*, **32**, 196–202.
18. Veerman, C., Baptist, H., Sagis, L.M.C., and van der Linden, E. (2003) A new multistep Ca^{2+}-induced cold gelation process for beta-lactoglobulin. *J. Agric. Food Chem.*, **51**, 3880–3885.
19. Ako, K., Nicolai, T., and Durand, D. (2010) Salt-induced gelation of globular protein aggregates: structure and kinetics. *Biomacromolecules*, **11**, 864–871.
20. Alting, A.C., Hamer, R.J., Dekruif, C.G., and Visschers, R.W. (2003) Cold-set globular protein gels: interactions, structure and rheology as a function of protein concentration. *J. Agric. Food Chem.*, **51**, 3150–3156.
21. Mudgal, P., Daubert, C.R., and Foegeding, E.A. (2009) Cold-set thickening mechanism of β-lactoglobulin at low pH: concentration effects. *Food Hydrocolloids*, **23**, 1762–1770.
22. Maltais, A., Remondetto, G.E., Gonzalez, R., and Subirade, M. (2005) Formation of soy protein isolate cold-set gels: protein and salt effects. *J. Food Sci.*, **70**, C67–C73.
23. Alting, A.C., de Jongh, H.H.J., Visschers, R.W., and Simons, J. (2002) Physical and chemical interactions in cold gelation of food proteins. *J. Agric. Food Chem.*, **50**, 4682–4689.
24. Foegeding, E.A. and Davis, J.P. (2011) Food protein functionality: a comprehensive approach. *Food Hydrocolloids*, **25**, 1853–1864.

25 Nicolai, T., Britten, M., and Schmitt, C. (2011) β-Lactoglobulin and WPI aggregates: formation, structure and applications. *Food Hydrocolloids*, **25**, 1945–1962.
26 Farrer, D., Finlayson, R. M., Foster, T. J., Pelan, E. G., Russel, A. R., and Thomas, A. (2003) Instant emulsion. WO03/053149.
27 Line, V.L.S., Remondetto, G.E., and Subirade, M. (2005) Cold gelation of beta-lactoglobulin oil-in-water emulsions. *Food Hydrocolloids*, **19**, 269–278.
28 Rosa, P., Sala, G., van Vliet, T., and van de velde, F. (2006) Cold gelation of whey protein emulsions. *J. Texture Stud.*, **37**, 516–537.
29 De Jong, S., Klok, H.J., and van der Velde, F. (2009) The mechanism behind microstructure formation in mixed whey protein–polysaccharide cold-set gels. *Food Hydrocolloids*, **23**, 755–764.
30 van Riemsdijk, L.E., Snoeren, J.P.M., van der Goot, A.J., Boom, R.M., and Hamer, R.J. (2011) New insights on the formation of colloidal whey protein particles. *Food Hydrocolloids*, **25**, 333–339.
31 Purwanti, N., Smiddy, M., van der Goot, A.J., de Vries, R., Alting, A., and Boom, R. (2011) Modulation of rheological properties by heat-induced aggregation of whey protein solution. *Food Hydrocolloids*, **25**, 1482–1489.
32 Kroes-Nijboer, A., Venema, P., and van der Linden, E. (2012) Fibrillar structures in food. *Food Funct.*, **3**, 221–227.
33 Turgeon, S.L., Schmitt, C., and Sanchez, C. (2007) Protein–polysaccharide complexes and coacervates. *Curr. Opin. Colloid Interface Sci.*, **12**, 166–178.
34 Foster, T.J., Underdown, J., Brown, C.R.T., Ferdinando, D.P., and Norton, I.T. (1997) Emulsion behaviour of non-gelled biopolymer mixtures, in *Food Colloids: Proteins, Lipids and Polysaccharides* (eds E. Dickinson and B. Bergenstahl), RSC Publishing.
35 Aronson, M. P., Brown, C. R. T., Chatfield, R. J., and Willis, E. (1999) Detergent compostions.WO099/42548.
36 Brown, C. R. T., Carew, P. S., Eklund, J. C., and Fairley, P. (2002) Shear gel compositions.WO02/36086.
37 Foster, T.J. (2007) Structure design in the food industry, in *Product Design and Engineering: Raw Materials, Additives and Applications*, vol. **2** (eds U. Brockel, W. Meier, and G. Wagner), Wiley-VCH Verlag GmbH, Weinheim.
38 Lundin, L.O., Odic, K., and Foster, T.J. (1999) Phase separation in mixed carrageenan systems, in *Supramolecular and Colloidal Structures in Biomaterials and Biosubstrates* (ed. V. Prakash), CFTRI, Mysore, India.
39 Boyd, M.J., Hampson, F.C., Jolliffe, I.G., Dettmar, P.W., Mitchell, J.R., and Melia, C.D. (2009) Strand-like phase separation in mixtures of xanthan gum with anionic polyelectrolytes. *Food Hydrocolloids*, **23**, 2458–2467.
40 Firoozmand, H., Murray, B.S., and Dickinson, E. (2009) Interfacial structuring in a phase-separating mixed biopolymer solution containing colloidal particles. *Langmuir*, **25**, 1300–1305.
41 Fransson, S., Loren, N., Altskar, A., and Hermansson, A.M. (2009) Effect of confinement and kinetics on the morphology of phase separating gelatin–maltodextrin droplets. *Biomacromolecules*, **10**, 1446–1453.
42 Persin, Z., Stana-Kleinschek, K., Foster, T.J., van dam, J.E.G., Boeriu, C.G., and Navard, P. (2011) Challenges and opportunities in polysaccharides research and technology: the EPNOE views for the next decade in the areas of materials, food and health care. *Carbohydr. Polym.*, **84**, 22–32.
43 Wendler, F., Persin, Z., Stana-Kleinschek, K., Reischl, M., Ribitsch, V., Bohn, A., Fink, H.P., and Meister, F. (2011) Morphology of polysaccharide blend fibers shaped from NaOH, N-methylmorpholine-N-oxide and 1-ethyl-3-methylimidazolium acetate. *Cellulose*, **18**, 1165–1178.
44 Koganti, N., Mitchell, J.R., Ibbett, R.N., and Foster, T.J. (2011) Solvent effects on starch dissolution and gelatinization. *Biomacromolecules*, **12**, 2888–2893.
45 Klemm, D., Heublein, B., Fink, H.P., and Bohn, A. (2005) Cellulose: fascinating biopolymer and sustainable raw material. *Angew. Chem., Int. Ed.*, **44**, 3358–3393.
46 Fratzl, P. and Weinkamer, R. (2007) Nature's hierarchical materials. *Prog. Mater. Sci.*, **52**, 1263–1334.
47 Somerville, C., Bauer, S., Brininstool, G., Facette, M., Hamann, T., Milne, J.,

Osborne, E., Paredez, A., Persson, S., Raab, T., Vorwerk, S., and Youngs, H. (2004) Toward a systems approach to understanding plant cell walls. *Science*, **306**, 2206–2211.

48 Lynd, L.R., Weimer, P.J., van Zyl, W.H., and Pretorius, I.S. (2002) Microbial cellulose utilization: fundamentals and biotechnology. *Microbiol. Mol. Biol. Rev.*, **66**, 506–577.

49 Spencer, F.S. and Maclachlan, G.A. (1972) Changes in molecular weight of cellulose in pea epicotyl during growth. *Plant Physiol.*, **49**, 58–63.

50 Herth, W. and Meyer, Y. (1977) Ultrastructural and chemical analysis of wall fibrils synthesized by tobacco mesophyll protoplasts. *Biol. Cellulaire*, **30**, 33–40.

51 Moran, J.I., Alvarez, V.A., Cyras, V.P., and Vazquez, A. (2008) Extraction of cellulose and preparation of nanocellulose from sisal fibers. *Cellulose*, **15**, 149–159.

52 McNeil, M., Darvill, A.G., Fry, S.C., and Albersheim, P. (1984) Structure and function of the primary cell walls of plants. *Annu. Rev. Biochem.*, **53**, 625–663.

53 Foster, T.J. (2011) Natural structuring with cell wall materials. *Food Hydrocolloids*, **25**, 1828–1832.

54 Anderson, J.W., Baird, P., Davis, R.H., Jr., Ferreri, S., Knudtson, M., Koraym, A., Waters, V., and Williams, C.L. (2009) Health benefits of dietary fiber. *Nutr. Rev.*, **67**, 188–205.

55 Marlett, J.A., Mcburney, M.I., and Slavin, J.L. (2002) Position of the American Dietetic Association: health implications of dietary fiber. *J. Am. Diet. Assoc.*, **102**, 993–1000.

56 Park, Y., Hunter, D.J., Spiegelman, D., Bergkvist, L., Berrino, F., van den Brandt, P.A., Buring, J.E., Colditz, G.A., Freudenheim, J.L., Fuchs, C.S., Giovannucci, E., Goldbohm, R.A., Graham, S., Harnack, L., Hartman, A.M., Jacobs, D. R., Kato, I., Krogh, V., Leitzmann, M.F., Mccullough, M.L., Miller, A.B., Pietinen, P., Rohan, T.E., Schatzkin, A., Willett, W.C., Wolk, A., Zeleniuch-Jacquotte, A., Zhang, S.M.M., and Smith-Warner, S.A. (2005) Dietary fiber intake and risk of colorectal cancer – a pooled analysis of prospective cohort studies. *J. Am. Med. Assoc.*, **294**, 2849–2857.

57 Kroonbatenburg, L.M.J., Kroon, J., and Northolt, M.G. (1986) Chain modulus and intramolecular hydrogen bonding in native and regenerated cellulose fibres. *Polym. Commun.*, **27**, 290–292.

58 Habibi, Y., Lucia, L.A., and Rojas, O.J. (2010) Cellulose nanocrystals: chemistry, self-assembly, and applications. *Chem. Rev.*, **110**, 3479–3500.

59 Collinson, S.R. and Thielemans, W. (2010) The catalytic oxidation of biomass to new materials focusing on starch, cellulose and lignin. *Coord. Chem. Rev.*, **254**, 1854–1870.

60 Brown, R.M. and Saxena, I.M. (2000) Cellulose biosynthesis: a model for understanding the assembly of biopolymers. *Plant Physiol. Biochem.*, **38**, 57–67.

61 Stone, J.E., Treiber, E., and Abrahams, B. (1969) Accessibility of regenerated cellulose to solute molecules of a molecular weight of 180 to 2×10^6. *Tappi*, **52**, 108–110.

62 Delmer, D. P. and Amor, Y. (1995) Cellulose biosynthesis. *Plant Cell*, **7**, 987–1000.

63 Langan, P., Nishiyama, Y., and Chanzy, H. (1999) A revised structure and hydrogen-bonding system in cellulose II from a neutron fiber diffraction analysis. *J. Am. Chem. Soc.*, **121**, 9940–9946.

64 Zugenmaier, P. (2001) Conformation and packing of various crystalline cellulose fibers. *Prog. Polym. Sci.*, **26**, 1341–1417.

65 Fink, H.P., Weigel, P., Purz, H.J., and Ganster, J. (2001) Structure formation of regenerated cellulose materials from NMMO-solutions. *Prog. Polym. Sci.*, **26**, 1473–1524.

66 Franks, N. E. (1980) Regenerated cellulose containing cross linked sodium lignate or sodium lignosulfonate. US4215212.

67 el Seoud, O.A., Koschella, A., Fidale, L.C., Dorn, S., and Heinze, T. (2007) Applications of ionic liquids in carbohydrate chemistry: a window of opportunities. *Biomacromolecules*, **8**, 2629–2647.

68 Seddon, K.R. (1997) Ionic liquids for clean technology. *J. Chem. Technol. Biotechnol.*, **68**, 351–356.

69 Zhu, S.D., Wu, Y.X., Chen, Q.M., Yu, Z. N., Wang, C.W., Jin, S.W., Ding, Y.G., and Wu, G. (2006) Dissolution of cellulose

with ionic liquids and its application: a mini-review. *Green Chem.*, **8**, 325–327.
70 Varma, R.S. and Namboodiri, V.V. (2001) An expeditious solvent-free route to ionic liquids using microwaves. *Chem. Commun.*, 643–644.
71 Swatloski, R.P., Spear, S.K., Holbrey, J.D., and Rogers, R.D. (2002) Dissolution of cellulose with ionic liquids. *J. Am. Chem. Soc.*, **124**, 4974–4975.
72 Tatarova, I., Manian, A.P., Siroka, B., and Bechtold, T. (2010) Nonalkali swelling solutions for regenerated cellulose. *Cellulose*, **17**, 913–922.
73 Moon, R.J., Martini, A., Nairn, J., Simonsen, J., and Youngblood, J. (2011) Cellulose nanomaterials review: structure, properties and nanocomposites. *Chem. Soc. Rev.*, **40**, 3941–3994.
74 Andresen, M., Johansson, L.-S., Tanem, B.S., and Stenius, P. (2006) Properties and characterization of hydrophobized microfibrillated cellulose. *Cellulose*, **13**, 665–677.
75 Siqueira, G., Bras, J., and Dufresne, A. (2009) Cellulose whiskers versus microfibrils: influence of the nature of the nanoparticle and its surface functionalization on the thermal and mechanical properties of nanocomposites. *Biomacromolecules*, **10**, 425–432.
76 Iwamoto, S., Kai, W., Isogai, A., and Iwata, T. (2009) Elastic modulus of single cellulose microfibrils from tunicate measured by atomic force microscopy. *Biomacromolecules*, **10**, 2571–2576.
77 Turback, A. F., Snyder, F. W., and Sandberg, K. R. (1983) Microfibrillated cellulose, a new cellulose product: properties, uses and commercial potential. *J. Appl. Polym. Sci. Appl. Polym. Symp.* **37**, 815–827.
78 Herrick, F. W., Casebier, R. L., Hamilton, J. K., and Sandberg, K. R. (1983) Microfibrillated Cellulose: Morphology and accessibility. *J. Appl. Polym. Sci. Appl. Polym. Symp.* **37**, 797–813.
79 Aulin, C., Gallstedt, M., Larsson, T., Salmen, L., and Lindtrom, T. (2011) Novel nanocellulose-based barriers. *Abstract Papers of the American Chemical Society*, **241**, record number 184.
80 Klemm, D., Kramer, F., Moritz, S., Lindstrom, T., Ankerfors, M., Gray, D., and Dorris, A. (2011) Nanocelluloses: a new family of nature-based materials. *Angew. Chem., Int. Ed.*, **50**, 5438–5466.
81 Paakko, M., Ankerfors, M., Kosonen, H., Nykanen, A., Ahola, S., Osterberg, M., Ruokolainen, J., Laine, J., Larsson, P.T., Ikkala, O., and Lindstrom, T. (2007) Enzymatic hydrolysis combined with mechanical shearing and high-pressure homogenization for nanoscale cellulose fibrils and strong gels. *Biomacromolecules*, **8**, 1934–1941.
82 Spence, K.L., Venditti, R.A., Rojas, O.J., Habibi, Y., and Pawlak, J.J. (2011) A comparative study of energy consumption and physical properties of microfibrillated cellulose produced by different processing methods. *Cellulose*, **18**, 1097–1111.
83 Eyholzer, C., Bordeanu, N., Lopez-Suevos, F., Rentsch, D., Zimmermann, T., and Oksman, K. (2010) Preparation and characterization of water-redispersible nanofibrillated cellulose in powder form. *Cellulose*, **17**, 19–30.
84 Eyholzer, C., Lopez-Suevos, F., Tingaut, P., Zimmermann, T., and Oksman, K. (2010) Reinforcing effect of carboxymethylated nanofibrillated cellulose powder on hydroxypropyl cellulose. *Cellulose*, **17**, 793–802.
85 Zimmermann, T., Tingaut, P., Eyholzer, C., and Richter, K. (2010) Applications of nanofibrillated cellulose in polymer composites. *Proceedings of the International Conference on Nanotechnology for the Forest Products Industry.* Espoo, Finland, 27–29 Sept.
86 Tashiro, K. and Kobayashi, M. (1991) Theoretical evaluation of three-dimensional elastic constants of native and regenerated celluloses – role of hydrogen bonds. *Polymer*, **32**, 1516–1530.
87 Favier, V., Chanzy, H., and Cavaille, J.Y. (1995) Polymer nanocomposites reinforced by cellulose whiskers. *Macromolecules*, **28**, 6365–6367.
88 Elazzouzi-Hafraoui, S., Nishiyama, Y., Putaux, J.L., Heux, L., Dubreuil, F., and Rochas, C. (2008) The shape and size distribution of crystalline nanoparticles prepared by acid hydrolysis of native cellulose. *Biomacromolecules*, **9**, 57–65.
89 Bai, W., Holbery, J., and Li, K.C. (2009) A technique for production of

nanocrystalline cellulose with a narrow size distribution. *Cellulose*, **16**, 455–465.

90 Beck-Candanedo, S., Roman, M., and Gray, D.G. (2005) Effect of reaction conditions on the properties and behavior of wood cellulose nanocrystal suspensions. *Biomacromolecules*, **6**, 1048–1054.

91 Lin, N., Huang, J., and Dufresne, A. (2012) Preparation, properties and applications of polysaccharide nanocrystals in advanced functional nanomaterials: a review. *Nanoscale*, **4**, 11, 3274–3294.

92 Araki, J., Wada, M., Kuga, S., and Okano, T. (1998) Flow properties of microcrystalline cellulose suspension prepared by acid treatment of native cellulose. *Colloid Surf. A*, **142**, 75–82.

93 Revol, J.F., Bradford, H., Giasson, J., Marchessault, R.H., and Gray, D.G. (1992) Helicoidal self-ordering of cellulose microfibrils in aqueous suspension. *Int. J. Biol. Macromol.*, **14**, 170–172.

94 Revol J. F., Godbout, L., and Gray, D. G. (1998) Solid self-assembled films of cellulose with chiral nematic order and optically variable properties. *Journal of Pulp and Paper Science*, **24**, 146–149.

95 Scheller, H.V. and Ulvskov, P. (2010) Hemicelluloses. *Annu. Rev. Plant Biol.*, **61**, 263–289.

96 Caffall, K.H. and Mohnen, D. (2009) The structure, function, and biosynthesis of plant cell wall pectic polysaccharides. *Carbohydr. Res.*, **344**, 1879–1900.

97 Garti, N., Madar, Z., Aserin, A., and Sternheim, B. (1997) Fenugreek galactomannans as food emulsifiers. *Food Sci. Technol. – Leb.*, **30**, 305–311.

98 Phillips, G.O. and Williams, P.A. (eds) (2000) *Handbook of Hydrocolloids*, Woodhead Publishing.

99 Sharma, R.D. (1986) Effect of fenugreek seeds and leaves on blood glucose and serum insulin responses in human subjects. *Nutr. Res.*, **6**, 1353–1364.

100 Sharma, R.D., Raghuram, T.C., and Rao, N. S. (1990) Effects of fenugreek seeds on blood glucose and serum lipids in type I diabetes. *Eur. J. Clin. Nutr.*, **44**, 301–306.

101 Williams, M.A.K., Foster, T.J., Martin, D. R., Norton, I.T., Yoshimura, M., and Nishinari, K. (2000) A molecular description of the gelation mechanism of konjac mannan. *Biomacromolecules*, **1**, 440–450.

102 Hettrich, K., Fischer, S., Schroder, N., Engelhardt, J., Drechsler, U., and Loth, F. (2006) Derivatization and characterization of xylan from oat spelts. *Macromol. Symp.*, **232**, 37–48.

103 Levy, S., York, W.S., Stuikeprill, R., Meyer, B., and Staehelin, L.A. (1991) Simulations of the static and dynamic molecular conformations of xyloglucan – the role of the fucosylated side chain in surface specific side chain folding. *Plant J.*, **1**, 195–215.

104 Umemura, M. and Yuguchi, Y. (2005) Conformational folding of xyloglucan side chains in aqueous solution from molecular dynamics simulation. *Carbohydr. Res.*, **340**, 2520–2532.

105 Ebringerova, A. (2006) Structural diversity and application potential of hemicelluloses. *Macromol. Symp.*, **232**, 1–12.

106 Eklof, J.M. and Brumer, H. (2010) The XTH gene family: an update on enzyme structure, function, and phylogeny in xyloglucan remodeling. *Plant Physiol.*, **153**, 456–466.

107 Zhou, Q., Rutland, M.W., Teeri, T.T., and Brumer, H. (2007) Xyloglucan in cellulose modification. *Cellulose*, **14**, 625–641.

108 Shankaracharya, N.B. (1998) Tamarind – chemistry, technology and uses: a critical appraisal. *J. Food Sci. Technol. – Mysore*, **35**, 193–208.

109 Christiernin, M., Henriksson, G., Lindstrom, M.E., Brumer, H., Teeri, T.T., Lindstrom, T., and Laine, J. (2003) The effects of xyloglucan on the properties of paper made from bleached kraft pulp. *Nord. Pulp Pap. Res. J.*, **18**, 182–187.

110 Lima, D.U., Oliveira, R.C., and Buckeridge, M.O. (2003) Seed storage hemicelluloses as wet-end additives in papermaking. *Carbohydr. Polym.*, **52**, 367–373.

111 Yan, H.W., Lindstrom, T., and Christiernin, M. (2006) Some ways to decrease fibre suspension flocculation and improve sheet formation. *Nord. Pulp Pap. Res. J.*, **21**, 36–43.

112 Zhan, D.F., Janssen, P., and Mort, A.J. (1998) Scarcity or complete lack of single rhamnose residues interspersed within the homogalacturonan regions of citrus pectin. *Carbohydr. Res.*, **308**, 373–380.

113 Pauly, M. and Scheller, H.V. (2000) O-Acetylation of plant cell wall polysaccharides: identification and partial characterization of a rhamnogalacturonan

O-acetyl-transferase from potato suspension-cultured cells. *Planta*, **210**, 659–667.

114 Willats, W.G.T., Mccartney, L., Mackie, W., and Knox, J.P. (2001) Pectin: cell biology and prospects for functional analysis. *Plant Mol. Biol.*, **47**, 9–27.

115 Albersheim, P., Darvill, A.G., O'Neill, M.A., Schols, H.A., and Voragen, A.G.J. (1996) An hypothesis: the same six polysaccharides are components of the primary cell walls of all higher plants. *Pectins Pectinases*, **14**, 47–55.

116 Lerouge, P., O'Neill, M.A., Darvill, A.G., and Albersheim, P. (1993) Structural characterisation of endo-gylcanase-generated oligoglycosyl side chains of rhamnogalacturonan-I. *Carbohydr. Res.*, **243**, 359–371.

117 Ralet, M.C., Lerouge, P., and Quemener, B. (2009) Mass spectrometry for pectin structure analysis. *Carbohydr. Res.*, **344**, 1798–1807.

118 Kobayashi, M., Matoh, T., and Azuma, J. (1996) Two chains of rhamnogalacturonan II are cross-linked by borate–diol ester bonds in higher plant cell walls. *Plant Physiol.*, **110**, 1017–1020.

119 O'Neill, M.A., Warrenfeltz, D., Kates, K., Pellerin, P., Doco, T., Darvill, A.G., and Albersheim, P. (1996) Rhamnogalacturonan-II, a pectic polysaccharide in the walls of growing plant cell, forms a dimer that is covalently cross-linked by a borate ester – *in vitro* conditions for the formation and hydrolysis of the dimer. *J. Biol. Chem.*, **271**, 22923–22930.

120 Baronepel, O., Gharyal, P.K., and Schindler, M. (1988) Pectins as mediators of wall porosity in soybean cells. *Planta*, **175**, 389–395.

121 Daas, P.J.H., Meyer-Hansen, K., Schols, H.A., de Ruiter, G.A., and Voragen, A.G.J. (1999) Investigation of the non-esterified galacturonic acid distribution in pectin with endopolygalacturonase. *Carbohydr. Res.*, **318**, 135–145.

122 Winning, H., Viereck, N., Norgaard, L., Larsen, J., and Engelsen, S.B. (2007) Quantification of the degree of blockiness in pectins using ^{1}H NMR spectroscopy and chemometrics. *Food Hydrocolloids*, **21**, 256–266.

123 Pilgrim, G.W., Walter, R.H., and Oakenfull, D.G. (1991) Jam, jellies and preserves, in *The Chemistry and Technology of Pectins* (ed. R.H. Walter), Academic Press, San Diego, CA.

124 Glahn, P.E. (1982) Hydrocolloid stabilization of protein suspensions at low pH. *Prog. Food Nutr. Sci.*, **6**, 171–177.

125 Benzion, O. and Nussinovitch, A. (1997) Physical properties of hydrocolloid wet glues. *Food Hydrocolloids*, **11**, 429–442.

126 Riccieri, J.E., de Carvalho, L.H., and Vazquez, A. (1999) Interfacial properties and initial step of the water sorption in unidirectional unsaturated polyester/vegetable fiber composites. *Polym. Compos.*, **20**, 29–37.

127 Saheb, D.N. and Jog, J.P. (1999) Natural fiber polymer composites: a review. *Adv. Polym. Technol.*, **18**, 351–363.

128 Mani, P. and Satyanarayana, K.G. (1990) Effects of the surface treatments of lignocellulosic fibers on their debonding stress. *J. Adhes. Sci. Technol.*, **4**, 17–24.

129 Ellis, W.D. and O'dell, J.L. (1999) Wood–polymer composites made with acrylic monomers, isocyanate, and maleic anhydride. *J. Appl. Polym. Sci.*, **73**, 2493–2505.

130 Keener, T.J., Stuart, R.K., and Brown, T.K. (2004) Maleated coupling agents for natural fibre composites. *Compos. Part A: Appl. Sci. Manuf.*, **35**, 357–362.

131 Abdelmouleh, M., Boufi, S., Belgacem, M.N., and Dufresne, A. (2007) Short natural-fibre reinforced polyethylene and natural rubber composites: effect of silane coupling agents and fibres loading. *Compos. Sci. Technol.*, **67**, 1627–1639.

132 Zheng, Y.-T., Cao, D.-R., Wang, D.-S., and Chen, J.-J. (2007) Study on the interface modification of bagasse fibre and the mechanical properties of its composite with PVC. *Compos. Part A: Appl. Sci. Manuf.*, **38**, 20–25.

133 Mohanty, A.K., Misra, M., and Drzal, L.T. (2002) Sustainable bio-composites from renewable resources: opportunities and challenges in the green materials world. *J. Polym. Environ.*, **10**, 19–26.

134 Oksman, K., Skrifvars, M., and Selin, J.F. (2003) Natural fibres as reinforcement in polylactic acid (PLA) composites. *Compos. Sci. Technol.*, **63**, 1317–1324.

135 Graupner, N., Herrmann, A.S., and Muessig, J. (2009) Natural and man-made cellulose fibre-reinforced poly(lactic acid)

(PLA) composites: an overview about mechanical characteristics and application areas. *Compos. Part A: Appl. Sci. Manuf.*, **40**, 810–821.
136 Bodros, E., Pillin, I., Montrelay, N., and Baley, C. (2007) Could biopolymers reinforced by randomly scattered flax fibre be used in structural applications? *Composites Science and Technology*, **67**, 462–470.
137 Nishino, T., Matsuda, I., and Hirao, K. (2004) All-cellulose composite. *Macromolecules*, **37**, 7683–7687.
138 Gindl, W. and Keckes, J. (2005) All-cellulose nanocomposite. *Polymer*, **46**, 10221–10225.
139 Hepworth, D.G. and Bruce, D.M. (2000) The mechanical properties of a composite manufactured from non-fibrous vegetable tissue and PVA. *Compos. Part A: Appl. Sci. Manuf.*, **31**, 283–285.
140 Bruce, D.M., Hobson, R.N., Farrent, J.W., and Hepworth, D.G. (2005) High-performance composites from low-cost plant primary cell walls. *Compos. Part A: Appl. Sci. Manuf.*, **36**, 1486–1493.
141 de Rodriguez, N.L.G., Thielemans, W., and Dufresne, A. (2006) Sisal cellulose whiskers reinforced polyvinyl acetate nanocomposites. *Cellulose*, **13**, 261–270.
142 Dammstrom, S., Salmen, L., and Gatenholm, P. (2005) The effect of moisture on the dynamical mechanical properties of bacterial cellulose/glucuronoxylannanocomposites. *Polymer*, **46**, 10364–10371.
143 Saxena, A. and Ragauskas, A.J. (2009) Water transmission barrier properties of biodegradable films based on cellulosic whiskers and xylan. *Carbohydr. Polym.*, **78**, 357–360.
144 Saxena, A., Elder, T.J., and Ragauskas, A.J. (2011) Moisture barrier properties of xylan composite films. *Carbohydr. Polym.*, **84**, 1371–1377.
145 Saxena, A., Elder, T.J., Pan, S., and Ragauskas, A.J. (2009) Novel nanocellulosic xylan composite film. *Compos. Part B: Eng.*, **40**, 727–730.
146 Peng, X.-W., Ren, J.-L., Zhong, L.-X., and Sun, R.-C. (2011) Nanocomposite films based on xylan-rich hemicelluloses and cellulose nanofibers with enhanced mechanical properties. *Biomacromolecules*, **12**, 3321–3329.
147 Mikkonen, K.S., Mathew, A.P., Pirkkalainen, K., Serimaa, R., Xu, C., Willfor, S., Oksman, K., and Tenkanen, M. (2010) Glucomannan composite films with cellulose nanowhiskers. *Cellulose*, **17**, 69–81.
148 Azeredo, H.M.C., Mattoso, L.H.C., Wood, D., Williams, T.G., Avena-Bustillos, R.J., and Mchugh, T.H. (2009) Nanocomposite edible films from mango puree reinforced with cellulose nanofibers. *J. Food Sci.*, **74**, N31–N35.
149 Hayashi, T., Ogawa, K., and Mitsuishi, Y. (1994) Characterization of the adsorption of xyloglucan to cellulose. *Plant Cell Physiol.*, **35**, 1199–1205.
150 Vanzin, G. F., Madson, M., Carpita, N. C., Raikhel, N. V., Keegstra, K., and Reiter, W. D. (2002) The mur2 mutant of *Arabidopsis thaliana* lacks fucosylatedxyloglucan because of a lesion in the fucosyltransferase AtFUT1. Proceedings of the National Academy of Sciences of the United States of America, **99**, 3340–3345.
151 Lima, D.U., Loh, W., and Buckeridge, M.S. (2004) Xyloglucan–cellulose interaction depends on the sidechains and molecular weight of xyloglucan. *Plant Physiol. Biochem.*, **42**, 389–394.
152 Whitney, S.E.C., Brigham, J.E., Darke, A.H., Reid, J.S.G., and Gidley, M.J. (1995) *In vitro* assembly of cellulose/xyloglucan networks – ultrastructural and molecular aspects. *Plant J.*, **8**, 491–504.
153 Whitney, S.E.C., Gidley, M.J., and Mcqueen-Mason, S.J. (2000) Probing expansin action using cellulose/hemicellulose composites. *Plant J.*, **22**, 327–334.
154 Whitney, S. E. C., Wilson, E., Webster, J., Bacic, A., Reid, J. S. G., and Gidley, M. J. (2006) Effects of structural variation in xyloglucan polymers on the interactions with bacterial cellulose. *American Journal of Botany*, **93**, 1402–1414.
155 Whitney, S.E.C., Brigham, J.E., Darke, A.H., Reid, J.S.G., and Gidley, M.J. (1998) Structural aspects of the interaction of mannan-based polysaccharides with bacterial cellulose. *Carbohydr. Res.*, **307**, 299–309.
156 Cosgrove, D.J. (2000) Expansive growth of plant cell walls. *Plant Physiol. Biochem.*, **38**, 109–124.

157 Zykwinska, A.W., Ralet, M.C.J., Garnier, C.D., and Thibault, J.F.J. (2005) Evidence for *in vitro* binding of pectin side chains to cellulose. *Plant Physiol.*, **139**, 397–407.

158 Zykwinska, A., Gaillard, C., Buleon, A., Pontoire, B., Garnier, C., Thibault, J.-F., and Ralet, M.-C. (2007) Assessment of *in vitro* binding of isolated pectic domains to cellulose by adsorption isotherms, electron microscopy, and X-ray diffraction methods. *Biomacromolecules*, **8**, 223–232.

159 Chanliaud, E. and Gidley, M.J. (1999) In vitro synthesis and properties of pectin/*Acetobacter xylinus* cellulose composites. *Plant J.*, **20**, 25–35.

160 Mavrakis, C. and Kiosseoglou, V. (2008) The structural characteristics and mechanical properties of biopolymer/mastic gum microsized particles composites. *Food Hydrocolloids*, **22**, 854–861.

161 Kasapis, S. (1999) The elastic moduli of the microcrystalline cellulose–gelatin blends. *Food Hydrocolloids*, **13**, 543–546.

162 Koh, L.W. and Kasapis, S. (2011) Orientation of short microcrystalline cellulose fibers in a gelatin matrix. *Food Hydrocolloids*, **25**, 1402–1405.

163 Agoda-Tandjawa, G., Durand, S., Berot, S., Blassel, C., Gaillard, C., Garnier, C., and Doublier, J.L. (2010) Rheological characterization of microfibrillated cellulose suspensions after freezing. *Carbohydr. Polym.*, **80**, 677–686.

164 Agoda-Tandjawa, G., Durand, S., Gaillard, C., Garnier, C., and Doublier, J. L. (2012) Rheological behaviour and microstructure of microfibrillated cellulose suspensions/low-methoxyl pectin mixed systems. Effect of calcium ions. *Carbohydrated Polymers*, **87**, 1045–1057.

165 Lucca, P.A. and Tepper, B.J. (1994) Fat replacers and the functionality of fat in foods. *Trends Food Sci. Technol.*, **5**, 12–19.

166 Samir, M., Alloin, F., and Dufresne, A. (2005) Review of recent research into cellulosic whiskers, their properties and their application in nanocomposite field. *Biomacromolecules*, **6**, 612–626.

167 Newman, R.H. and Hemmingson, J.A. (1998) Interactions between locust bean gum and cellulose characterized by ^{13}C NMR spectroscopy. *Carbohydr. Polym.*, **36**, 167–172.

168 Ang, J.F. (1991) Water retention capacity and viscosity effect of powdered cellulose. *J. Food Sci.*, **56**, 1682–1684.

169 Ang, J.F. and Miller, W.B. (1991) Multiple functions of powdered cellulose as a food ingredient. *Cereal Food World*, **36**, 558–564.

170 Day, L., Xu, M., Oiseth, S.K., Lundin, L., and Hemar, Y. (2010) Dynamic rheological properties of plant cell-wall particle dispersions. *Colloid Surf. B*, **81**, 461–467.

171 Kabbert, R., Goworek, S., and Kunzek, H. (1997) Preparation of single-cell material from enzymatically disintegrated apple tissue – changes in structure and functional properties. *Z. Lebensm. Unters Forsch. A*, **205**, 380–387.

172 Kotcharian, A., Kunzek, H., and Dongowski, G. (2004) The influence of variety on the enzymatic degradation of carrots and on functional and physiological properties of the cell wall materials. *Food Chem.*, **87**, 231–245.

173 Kunzek, H., Kabbert, R., and Gloyna, D. (1999) Aspects of material science in food processing: changes in plant cell walls of fruits and vegetables. *Z. Lebensm. Unters Forsch. A*, **208**, 233–250.

174 Kunzek, H., Muller, S., Vetter, S., and Godeck, R. (2002) The significance of physicochemical properties of plant cell wall materials for the development of innovative food products. *Eur. Food Res. Technol.*, **214**, 361–376.

175 Pickardt, C., Dongowski, G., and Kunzek, H. (2004) The influence of mechanical and enzymatic disintegration of carrots on the structure and properties of cell wall materials. *Eur. Food Res. Technol.*, **219**, 229–239.

176 Vetter, S. and Kunzek, H. (2002) Material properties of processed fruit and vegetables. II. Water hydration properties of cell wall materials from apples. *Eur. Food Res. Technol.*, **214**, 43–51.

177 Sereno, N.M., Hill, S.E., and Mitchell, J.R. (2007) Impact of the extrusion process on xanthan gum behaviour. *Carbohydr. Res.*, **342**, 1333–1342.

178 Murray, B.S., Durga, K., Yusoff, A., and Stoyanov, S.D. (2011) Stabilization of foams and emulsions by mixtures of surface active food-grade particles and proteins. *Food Hydrocolloids*, **25**, 627–638.

3
Biopolymers from Plants
Maria J. Sabater, Tania Ródenas, and Antonio Heredia

3.1
Introduction

Nowadays, the extraordinary growth of economic activities worldwide has led to the use of biologically derived polymers (biopolymers) in different economic sectors for achieving real sustainable growth [1]. Biopolymers are renewable polymers produced by living organisms whose main feature is their biodegradability. That means that these macromolecules will decompose simply by soil microorganisms and natural weathering processes giving recyclable soil nutrients without leaving any harmful residues behind [2–6].

Biopolymers play an important role in both the natural world and modern industrial economies. In fact, some natural polymers produced by microorganisms, plants, or animals manipulate essential biological information (proteins and nucleic acids), while other polymers (polysaccharides) provide fuel for cell activities and serve as structural elements in living systems [1–7]. Besides their importance in biology (e.g., lipids, carbohydrate biopolymers, etc.), their use has also been established over time for numerous applications in the packaging, food, paper, textile industries, medicine, agriculture, and so on. (e.g., chitin, starch, cellulose, natural rubber, etc.), so their implementation in important economic sectors has contributed to create a true sustainable industry around them [1–8]. Biopolymers also have the potential to cut carbon emissions and reduce CO_2 quantities in the atmosphere, because the CO_2 released when they degrade can be reabsorbed by crops grown to replace them.

In this chapter, we will put aside polymers produced by animals (e.g., chitin, etc.) and microorganisms (e.g., polyhydroxyalkanoates, etc.), focusing exclusively on biopolymers obtained and generated by plants [9,10]. Semisynthetic and/or synthetic biopolymers such as polyglycolic acid (biodegradable polymer prepared from glycolic acid) or polylactic acid (polyester obtained mainly through bacterial fermentation followed by several chemical transformations) will also not be discussed in this chapter.

Biopolymers from plants are thus naturally occurring macromolecules or renewable materials, often composite, obtained from biomass that can be grown year on year indefinitely (from agricultural nonfood crops), hence contrasting with the traditional polymers (mainly derived from petrochemicals), which will eventually run out [6].

This chapter has been organized into six main sections, three of them specifically devoted to provide key fundamental aspects of these macromolecules. Biopolymers, mainly structural biopolymers, have been compiled and treated on the basis of their chemical composition and their function in nature, for example, lipid (cutin), phenolic molecules associated with lipids (lignin and suberin), carbohydrates (structural and storage carbohydrates and gums), and isoprene biopolymers. Further subdivisions cover aspects related to composition, structure, and natural occurrence. In some cases, a more detailed treatise of the biosynthetic aspects of these materials, especially for cutin, lignin, cellulose, starch, and so on, has been addressed and emphasized. When possible, interesting and actual applications for these macromolecules have also been provided.

3.2
Lipid and Phenolic Biopolymers

There are abundant literature references about these biopolymers, for which we have tried to include the most relevant ones in this section. However, in global terms, there are no studies on these biopolymers as a whole, but at individual level, due precisely to their different chemical composition. Note that the classification into lipid and phenolic biopolymers is based on the predominance of two components: (1) a lipidic component represented by aliphatic chains and (2) an aromatic component represented by phenolic molecules. In accordance to this, (a) cutin is mainly constituted by a lipidic component, (b) lignin shows a prevalence of a polyphenolic component, whereas (c) suberin shows a mixture of both components. These three major biopolymers are listed and discussed below.

3.2.1
The Biopolymer Cutin

Epidermal cells of aerial parts of higher plants such as leaves, fruits, and nonwoody stems are covered by a continuous extracellular membrane of soluble and polymerized lipids called cuticle or cuticular membrane. The structure and composition of the cuticle vary largely among plants, organs, and growth stages [11], but it is basically composed of a cutin matrix with waxes embedded in (intracuticular waxes) and deposited on the surface of the matrix (epicuticular waxes) (Figure 3.1). Cuticular waxes are a complex mixture of n-alkanes, long-chain alcohols and fatty acids, and terpenoids. Thus, cuticle can be considered as the interface between plant and environment.

Plant cuticular material occurs in large amounts in both natural and agricultural plant communities: between 180 and 1500 kg/ha. Considering that the weight of an isolated cuticle ranges from 2000 $\mu g/cm^2$ (fruit cuticles) to 450–800 $\mu g/cm^2$ (leaf cuticles), 40–80% of it corresponds to cutin that can be considered as the major lipid plant polymer.

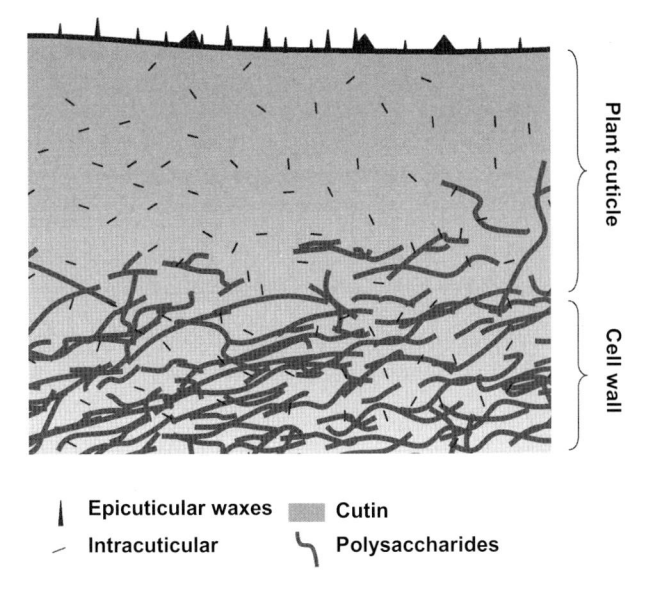

Figure 3.1 Scheme of a transverse section of the plant cuticle. Cutin is the matrix where the other cuticular components are deposited: epicuticular wax layer on the surface, intracuticular wax molecules randomly distributed, and in the inner part polysaccharide fibrils from the epidermal cell wall.

Plant cutin is the main support of cuticle and, together with cuticular waxes, is involved in waterproofing leaves and fruits of higher plants, regulating gas exchange between plant and environment, and minimizing the deleterious impact of pathogens. Despite the complex nature of this biopolymer, significant progress in chemical composition, molecular architecture, and biosynthesis has been made in the past. As mentioned earlier, cutin is the major constituent (40–80% dry weight) of the plant cuticle and, from a chemical point of view, is defined as a polymeric network of epoxidized polyhydroxylated C_{16} and C_{18} fatty acids cross-linked by ester bonds [11,12]. Cutin can be depolymerized by cleavage of the ester bonds by alkaline hydrolysis, transesterification, and other methods [12]. These chemical methods yield monomers as well as their derivatives, depending on the reagent employed. Figure 3.2 shows the major cutin monomers present in plants. 9- or 10,16-dihydroxyhexadecanoic acid and 16-hydroxyhexadecanoic acid are the major components of C_{16} cutins. Only in some cases, 16-hydroxy-10-oxo-C_{16} acid and 16-oxo-9 or 10-hydroxy C_{16} acid are present. Major components of the C_{18} family are 18-hydroxy-9,10-epoxyoctadecanoic acid and 9,10,18-trihydroxyoctadecanoic acid together with their monounsaturated homologues.

3.2.1.1 Cutin Monomers: Biosynthesis and Physicochemical Characteristics

The biosynthetic pathway of cutin monomers, located in epidermal cells, was elucidated with great effort in the past [13,14]. Of both monomer families, the biosynthesis of the C_{18} family has been more thoroughly studied.

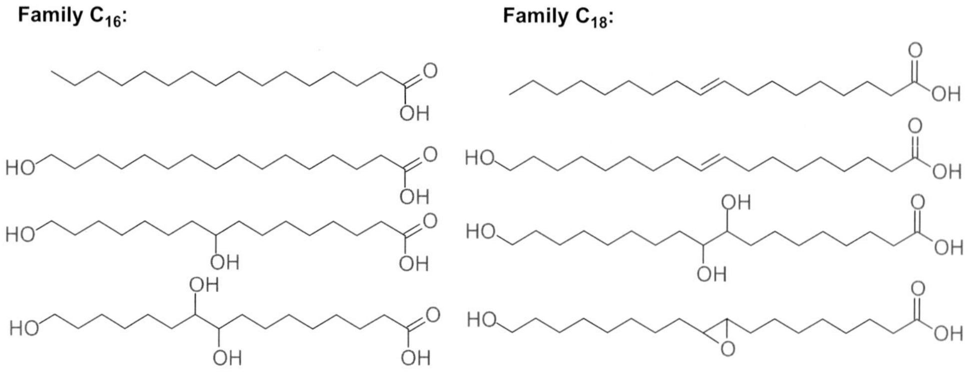

Family C$_{16}$:

Family C$_{18}$:

Figure 3.2 Chemical structures of the main monomers of plant cutin.

Pioneering work on this subject showed that oxygenated octadecanoic fatty acids (Figure 3.2) are derived from oleic acid (and also from linoleic acid) by ω-hydroxylation and epoxidation of the double bond followed by hydrolysis of the epoxide [13,15]. It has been suggested that the epoxidation step is catalyzed by a cytochrome P450-dependent enzyme, while the hydrolysis occurs by the action of specific esterases or hydrolases located in the epidermal cells [14]. Molecular and biochemical studies of the enzymes involved in the above-mentioned pathway have revealed that, for example, in *Vicia sativa* microsomes, oleic acid was converted into 9,10-epoxyoctadecanoic acid, which could then be hydrolyzed to the diol 9,10-dihydroxyoctadecanoic acid by an epoxide hydrolase [16,17]. Enantioselectivity of this epoxidation reaction was also studied: the epoxides (9R,10S) and (9S,10R) were produced in a 9:1 ratio [18]. An enzyme system also located in the microsomal fraction of epidermal cells was shown to generate 9,10,18-trihydroxyoctadecanoic acid and 18-hydroxy-9,10-epoxyhexadecanoic acid by ω-hydroxylation of both the epoxide and the diol [17]. Further studies have demonstrated the involvement of a cytochrome P450-dependent ω-hydroxylase in these reactions. The expression in yeast of this cytochrome P450 CYP94A1 gene from *V. sativa* catalyzed the ω-hydroxylation of both saturated and unsaturated fatty acids [19]. For a comprehensive review on the current picture of cutin monomer biosynthesis, see Ref. 20.

On the other hand and more recently, the physicochemical properties of aqueous dispersions of these hydroxy and polyhydroxy fatty acids have already been documented [21,22]. Bipolar amphiphilic fatty acids (the so-called bolaamphiphilic fatty acids), such as 16-hydroxyhexadecanoic acid, are not well dispersed in solution, whereas endogenous cutin monomer such as 9(10),16-dihydroxyhexadecanoic acid shows lumps of aggregated membranes [22]. These characteristics can be attributed to most of the cutin monomers shown in Figure 3.2.

3.2.1.2 Molecular Architecture of Cutin

Cutin biopolymer is refractory to chemical degradation; its depolymerization can only be achieved under severe chemical conditions in basic media or, in nature, by the action of bacterial and fungal esterases. The major cutin monomers shown in Figure 3.2 were identified by chemical degradation of the biopolymer and their covalent linkages elucidated based on their chemical reactivity [13]. An important number of analytical studies have demonstrated that, while only half of the mid-chain hydroxyl groups in the polymer are involved in side chain ester cross-linking, most of the primary hydroxyl groups are involved in ester bonds and, consequently, there are a very low number of unesterified carboxyl functional groups.

In spite of the significant amount of information on cutin molecular analysis, some aspects of its molecular architecture are quite elusive. Soluble products obtained after depolymerization do not always represent the real situation in the intact polymer; therefore, the recent use of nondestructive techniques has provided useful complementary information. Our current knowledge on plant cutin structure and physicochemical properties indicates that this biopolymer can be considered a cross-linked, amorphous, and hydrophobic polyester, mainly inert from a chemical point of view. Cutin is also described as a glassy polymer with high heat capacity and low water and gas permeabilities and, from a biomechanical perspective, it shows viscoelastic behavior at physiological temperature. These characteristics provide cutin the properties of a biological barrier polymer. For a more detailed description of the biophysical characteristics of cutin, the interested reader is referred to Refs [23,24].

It is noticeable that in the cuticle of some species, once all the wax and cutin components have been removed, there is some remaining residual material. This depolymerization-resistant residue is thought to represent cutin monomers held together by non-ester bonds. The residue, significantly large in weight in some species, is named cutan [25]. It has been suggested that cutan may be ubiquitous in plants since it has been reported to be a constituent in fossilized plant cuticles, terrestrial sediments, and coals covering a large part of the geological past [26]. Although the cuticles of some species appear to completely lack cutans, in a number of species the two biopolymers, cutin and cutan, may occur in any ratio differing in their relative abundance at different stages of cuticle development [26]. The structure and monomeric composition of cutan isolated from leaf cuticles of *Agave americana* and *Clivia miniata* have been elucidated in the past few years [25,27]. Using Fourier transform infrared and ^{13}C nuclear magnetic resonance spectroscopic analyses, X-ray diffraction, and exhaustive ozonolysis, it was concluded that the unsaponifiable polymeric core material consists of an amorphous three-dimensional network of polymethylenic chains linked by ether bonds with the presence of double bonds and free carboxyl groups.

3.2.1.3 Cutin Biosynthesis

In spite of the importance of this biopolyester, the exact elucidation of its biosynthesis pathway is still unclear. Significant efforts in the past decade looking for the gene that codifies the protein involved in the polymerization process do not provide clear information. Until now, this protein, if exists, is missing. For a detailed information of the research done in this field, see the exhaustive review in Ref. [20].

On the other hand, learning about the chemical properties of cutin monomers could be key to determine the unresolved mechanisms of plant cutin genesis. Cutin monomers have bifunctional chemical groups with potential to bind, a property that, according to polymer science, indicates that they are able to generate a nonlinear, amorphous, and cross-linked polymer. In the resulting polymer, relatively polar groups such as ester bonds would not be statistically significant when compared to abundant aliphatic methylene chains ($-CH_2-$). The rotational freedom of methylene chains may allow a multitude of different conformations to the polymer, although the significant amount of in-chain hydroxyl groups will constrain chain linearity and weak chain–chain interactions. Moreover, the types of functional groups and their location in cutin monomers, especially primary and secondary hydroxyl groups, confer the monomers self-assembly properties. Cutin monomers, under the right orientation and at a given molecular density, generate structures based on short-range interactions such as hydrogen bonding and other weak interactions between monomers followed by chemical polymerization. The self-assembly and self-polymerization properties of cutin monomers under specific chemical conditions have been recently described [28–31], opening a new and promising field of study to complement current knowledge on cutin architecture and synthesis. In this sense, involvement of self-assembled polyhydroxy fatty acid particles (i.e., *cutinsomes*) in tomato fruit cutin formation at early stages of development excluding the participation of any protein was recently confirmed with antibodies raised against these supramolecular particles [32].

3.2.2
Lignin

The cell wall of higher plants has a hierarchical complex structure. It is built up of two parts: the primary cell wall and the secondary cell wall, which is further divided into three different layers. The primary cell wall is the first to be laid down when the cell is formed and where, in epidermal plant cells, the biopolymer cutin described above is developed. It is composed of randomly oriented polysaccharide microfibrils, which allow for expansion of the cell during plant growth and development. The secondary cell wall, developed when the plant growth stops, occupies the greatest volume of the cell wall and influences the properties of the cell and the macroscopic properties of the plant [33]. On this secondary cell wall, the biopolymer lignin is formed in deep interaction with the polysaccharides contributing to many physiological functions such as structural support, cementing polysaccharides, and the resistance to pathogen degradation [34]. These functions have their origin in the hydrophobicity and macromolecular stiffness of the lignin structure.

3.2.2.1 Monomer Precursors and Chemical Reactivity
Lignin is one of the most abundant natural polymers on Earth. It can be considered a phenolic biopolymer, formed by the enzymatic polymerization of three phenolic alcohols, known as monolignols, *p*-coumaryl, coniferyl, and sinapyl alcohols (Figure 3.3), derived from the deamination pathway of the phenylalanine and

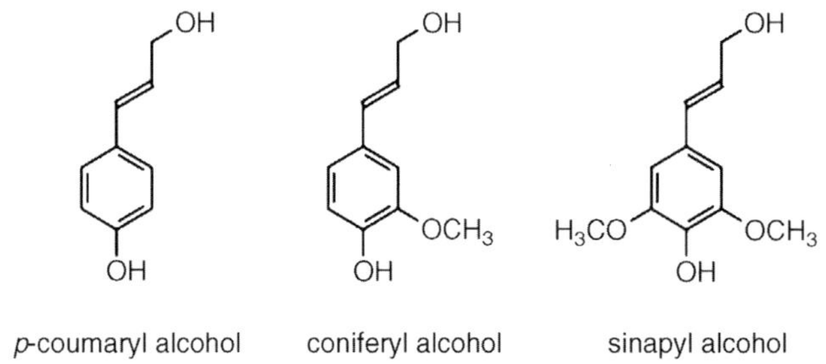

Figure 3.3 Phenolic alcohol precursors of plant lignin.

tyrosine amino acids. However, the native structure of lignin remains unclear because no method has been established to isolate lignin from the plant cell wall in its native state. This is due to the difficulty in isolation originated by the high cross-linking of lignin with other lignin molecules or polysaccharides.

Both lignin isolated from softwood [35] and synthetic lignins obtained by free radical polymerization have revealed that many kinds of intermolecular linkages exist in lignin [36]. The lignification process encompasses the biosynthesis of monolignols, their transport to the cell wall, and further polymerization into the final supramolecule. Bond formation is thought to result from oxidative (radical-mediated) coupling between a monolignol and the growing polymer. The oxidative coupling between monolignols can result in the formation of several different interunit linkages (Figure 3.4).

In native lignins, 8-O-4-linkages are the most abundant, whereas in lignins formed *in vitro* by mixing coniferyl alcohol, hydrogen peroxide, and peroxidase, high percentages of 8–8 and 8–5 linkages are found [37–39]. The most abundant linkage is the β-O-4 ether bond, which accounts for 40–60%. Other major linkages include β-5-phenylcoumaran (10%), 5–5 biphenyl (5–10%), 5-O-diphenyl ether (5%), and β-pinoresinol ($\leq$5%).

After the formation of monolignols, the next step is the oxidative coupling of lignin monomers into lignin polymers. Laccase and horseradish peroxide are the two main enzymes catalyzing the polymerization. While the formation of precursors is intracellular, the polymerization process takes place outside the cell. Among all the steps of lignin biosynthesis, the polymerization is the least defined as will be discussed below. Each monolignol is oxidized into phenoxyl radicals with different resonance structures [40]. The oxidation, thus, can lead to different types of linkages and further final structures of lignin. Lignification, in this sense, happens through free radical coupling reactions first described by Freudenberg [40]. It is assumed that the distribution of linkage types is determined by the relative reaction rates between free radicals and variations in the microenvironment.

3.2.2.2 Lignin Biosynthesis

As mentioned earlier, the lignifications process follows the sequence: biosynthesis and transport of monolignols to the cell wall and further polymerization into the final molecule. At present, there are two models for coupling radicals to produce a

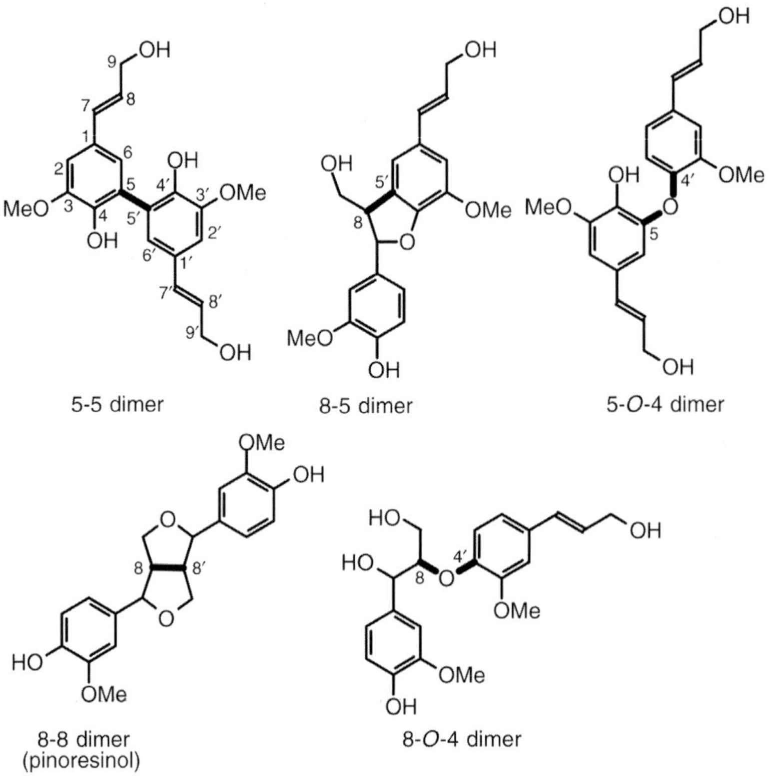

Figure 3.4 Major interunit linkages found in native plant lignins.

functional lignin molecule. The first, the so-called random coupling model, was published during the early studies on the structure of lignin. It is focused on the hypothesis that lignin formation proceeds through coupling of the individual monolignols to the growing lignin polymer in a random mode [41–43]. These polymerized dehydrogenated lignin-like polymers (DHPs) are usually synthesized using peroxidase/hydrogen peroxide from *in vitro* experiments. Under this molecular scenario, both the amount and type of individual phenolics available at the lignification site and the normal chemical coupling properties regulate lignin formation.

The more recent second model, the so-called dirigent protein model, indicates that lignification is under a strict regulation of specialized proteins that control the formation of individual bonds [44,45]. Thus, dirigent proteins direct the coupling of two monolignol radicals, producing a dimer with a single regio- and stereoconfiguration. These dimers, known as lignans, are commonly found in many plants. The key argument for this new model is the belief that nature would not leave the formation of such an important molecule as lignin "to the only way to explain the high chance" [45]. It is argued that the only way to explain and justify the high proportion of 8-O-4 linkages in lignin would be through regulation by specific dirigent proteins.

3.2.3
Suberin

Suberin is a plant biopolymer essentially deposited in specific cell wall locations in external and internal tissues during plant development, such as the periderm of bark and tubers, but its process of formation, the suberization, also takes place in plants when specific organs of plants need to form an effective diffusion barrier, as occurs after wounding [20,46–49].

The suberin ultrastructure presents alternating electron-dense opaque and translucent lamellae after visualization by transmission electron microscopy (TEM) [50].

The lighter bands seem to consist of suberin aliphatics and associated waxes, while the darker lamellae are composed of the polyaromatic domains [51,52]. On the other hand, microscopic studies showed that the secondary cell wall accounts for most of the thickness and contains aliphatic suberin and some of the polyaromatic compounds.

The periderm is a three-layered structure formed during the secondary growth of plant cells to replace the epidermis in stems, branches, and roots. Suberin plays important physiological roles in protection against environmental aggressions and pathogen attack and in the control of temperature and water loss [53,54]. It is known that the heavily suberized epidermal cells minimize mechanical freezing injury to some fruits by maintaining a low constant moisture level in tree trunks through the winter [55]. The thickness and water permeability of suberized tissues also have important biophysical and biomechanical properties [56].

3.2.3.1 Chemical Composition

From a chemical and structural point of view, the term suberin is ambiguous. Suberin has a two-domain structure with an ester-bound polyaliphatic domain and a nonesterified, lignin-like, polyaromatic domain [57–59]. Some authors apply the same concept only for the ester-bound domain, mainly formed by aliphatic compounds with the presence of some aromatics [60].

The aliphatic domain of suberin is composed of a complex network of esterified substituted long-chain α,ω-dicarboxylic acids, ω-hydroxy acids, fatty acids, and alcohols. In addition, glycerol and hydroxycinnamic acid derivatives constitute the so-called polyaromatic domain [50,61,62].

Suberin aliphatic acids have chain lengths of 16, 18, and 22 carbons; the C_{16} and C_{22} are mostly saturated, while the C_{18} compounds have mid-chain unsaturation, and hydroxyl and epoxy functional groups [63]. Monomers up to C_{32} have also been found [64,65]. Dicarboxylic acids and ω-hydroxy fatty acids are the major components. Chemically speaking, these monomers are similar to cutin monomers with the additional characteristic that longer chain lengths and greater amounts of dicarboxylic acids are present in the suberin biopolymer. Figure 3.5 shows some structural examples of these aliphatic monomers.

Some suberins have very low proportions of mid-chain oxidized monomers [66], while in some suberins the major components such as epoxy acids and diacids constitute over 40% of the monomers, as has been reported in cork suberin

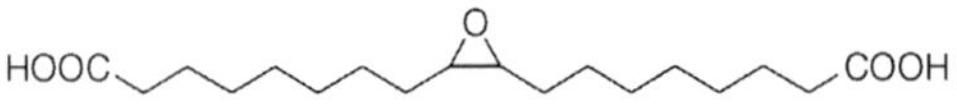

3 Biopolymers from Plants

9,10-epoxy-1,18-dioic-C_{18} acid

1,18-dioic-$C_{18:1}$ acid

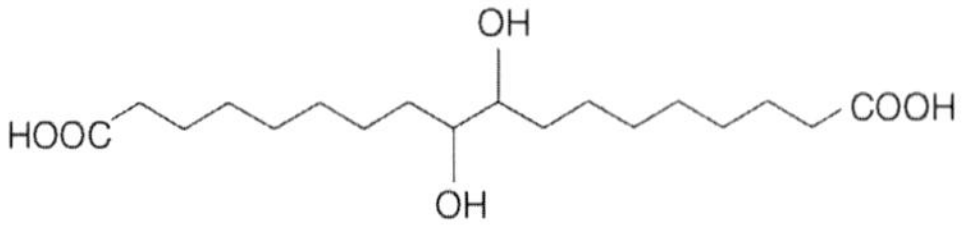

9,10-dihydroxy-1,18-dioic-C_{18} acid

Figure 3.5 Some suberin aliphatic monomers.

aliphatics [65] and birch outer bark [67]. Of the unsaturated compounds, present as the major monomers of potato suberin, octadec-9-en-1,18-dioic and 18-hydroxyoctadecen-9-oic acids, both *cis*- and *trans*-isomers have been detected [60].

Some unusual compounds have also been reported as components of some suberins such as *Ribes* species stem suberin that has been found to contain up to 10% of α,ω-diols [65]. Also, 2-hydroxy fatty acids have been found as minor components of the endodermal cell wall suberin of the primary roots of maize [68] and Norway spruce (*Picea abies*) [69].

Glycerol is one interesting and ubiquitous monomer of suberin and can account for up to 26% of the total monomer mixture, but is often left unanalyzed due to analytical methodologies. In potato suberin, the content of glycerol is about 22% of the released monomers [60].

On the other hand, aromatic hydroxycinnamates found after depolymerization of suberin polyester are commonly ferulic acid and coumaric acid. Moreover, caffeic acid, vanillin, benzoic acid, and hydroxycinnamic amines such as tyramine have also been found [70]. Finally, it is interesting to indicate that pentacyclic triterpenoids have been detected among the suberin depolymerizate samples of some species [71].

After the removal of soluble extractives and ester-bound suberin polyaliphatics, the polyaromatic residue with significant residual cell wall polysaccharides remains from suberized tissues. In cork, this residue is about 40% of the extractive-free suberized cell walls and about 70% of potato natural periderm [70]. Solid-state NMR analysis of isolated potato suberin-enriched material showed that roughly 50% of the content was composed of polysaccharides while the rest was comprised of a 2 : 1 ratio of aromatic and aliphatic structures [72].

It is important to indicate that the structural studies of the polyaromatic domain of suberin have given contradictory results depending on the methodology used. The suberin aromatic domain has been proposed to have a lignin-like structure [73]. A good number of studies revealed that suberin polyaromatics have a low methoxyl content [70]. In this sense, the degradation methods applied to the aromatic domain have been adopted from the lignin literature. They include the use of nitrobenzene oxidation to release mostly vanillin and *p*-hydroxybenzaldehyde. Thioacidolysis is another method used for the characterization of the aromatic domain of suberin in potato [70]. In this case, the typical β-O-4-guaiacyl and syringyl units were detected.

3.2.3.2 Biosynthesis and Fine Structure

Despite the physiological importance of suberin, its biosynthesis and deposition, as in the case of the biopolyester cutin discussed earlier, remains unclear. As suberin deposition occurs in specific tissues, and is induced by wounding and stress caused by environmental factors, it is the final result of a complex regulation process. Nevertheless, early parts of the biosynthetic pathway are common for both polymers, cutin and suberin. Fatty acid elongation and ω-oxidation represent two characteristic processes of suberin biosynthesis. The different variations in suberin monomer composition between species require different pathways, leading to mid-chain oxidation or unsaturation [74]. The biosynthesis of suberin aromatic compounds is even less understood, but apparently derives from phenylalanine, which is the amino acid precursor of the phenylpropanoid pathway biosynthesis [50]. Wound-induced suberization of potato disks has been found to reach a constant level in 7–11 days. This model has also been used to investigate suberin biosynthesis [75,76]. Alkyl ferulates have been found to start accumulating in healing potato periderm 3–5 days after wounding [77]. Other authors, using NMR approach [78], reported accumulation of both hydroxycinnamic acids and monolignols at day 1 after wounding, which stabilized after reaching a maximum in approximately 5 days. Suberin-associated waxes are deposited with a similar time course to suberin aliphatic compounds [79].

It is important to note that, in addition to aliphatic polyester and polyaromatic structures, suberized tissues contain large amounts of nonpolymeric material, extracted using organic solvents, called suberin waxes [70]. These compounds play a critical role in waterproofing the periderm and suberized tissues in combination with the suberin macromolecular structure.

Concerning the structure, the overall macromolecular assembly of suberin is not yet totally understood [50,61,79]. *In planta*, suberin is an insoluble biopolymer that must be depolymerized by ester-cleaving reactions and other strong degradative methods. From a good number of studies, some structural pieces have been obtained after partial depolymerization reactions producing oligomers [22,71] and solid-state NMR studies of intact polymer samples have revealed important information about the structure and linkages present in potato suberin, the plant tissue used as model for these purposes [78]. Taken together, these studies yielded useful information to draw a putative model of this complex and composite biopolymer. Different authors disagree in secondary details and in this chapter the model proposed by Pereira and coworkers will be briefly described [62,79]. In this model (Figure 3.6), a monolayer of α,ω-fatty acid monomers linked by glycerol on both sides presents the framework of the structure. Thus, the glycerol aliphatic structure is surrounded by polyaromatic structures, the phenol ferulic acid being the linkage between the domains. The translucent aliphatic lamellae and opaque lamellae of aromatics, observed at high resolution by transmission electron microscopy, appear connected by linear chains made of ω-hydroxy acids, which cross through the

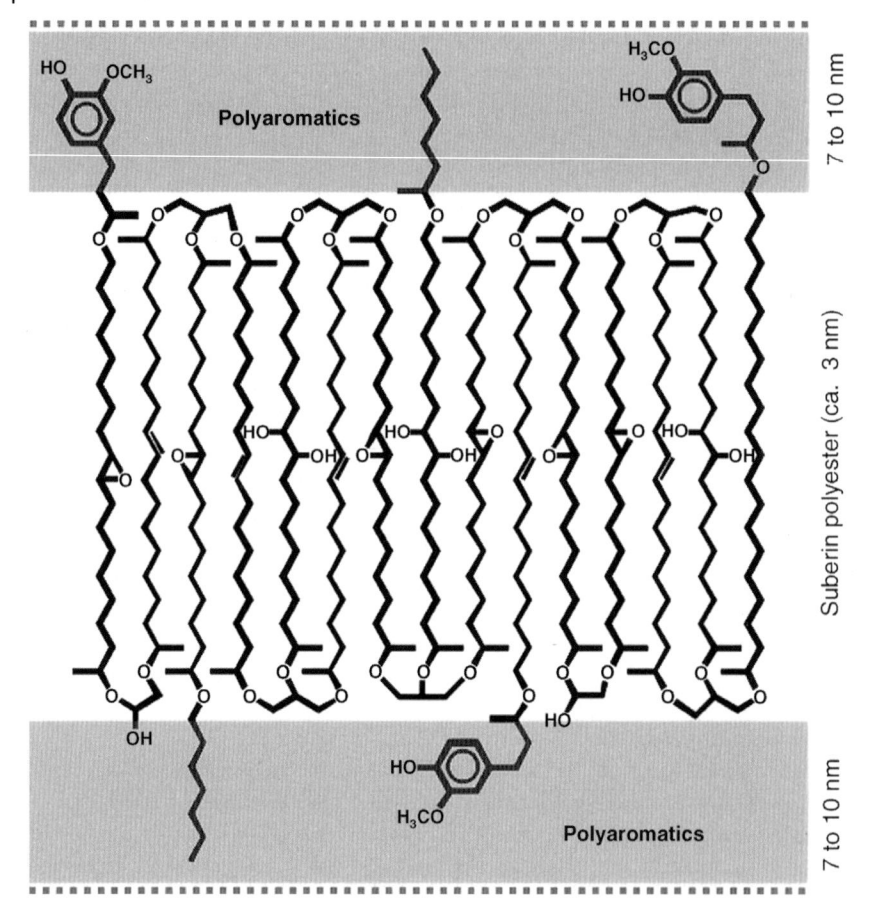

Figure 3.6 Proposed poly(acylglycerol) polyester model for suberin [52].

polyaromatic layers connecting the repeating lamellae of the overall suberin structure. The model also takes into account that intra- and intermolecular bonds could exist at the mid-chain positions of oxygen-containing substituted groups [61].

3.3
Carbohydrate Biopolymers: Polysaccharides

Polysaccharides are biopolymers made up of chains of monosaccharides (sugars) that are linked together by glycosidic bonds formed by condensation reactions. The linkage of monosaccharides into chains creates chains of greatly varying length, ranging from chains of just two monosaccharides (a disaccharide) to the polysaccharides, which consist of many thousands of sugars.

As a group, the polysaccharides play diverse and important roles within the biology of life processes, although many of them have found important commercial applications. According to their function in nature, they can be divided into two broad groups: structural and storage polysaccharides. Finally, other polysaccharides with industrial interest are the gums, which are generally used in foods, pharmaceuticals, and material science as gels, emulsifiers, adhesives, lubricants, and so on and will also be treated in a separate section.

3.3.1
Structural Polysaccharides

3.3.1.1 Cellulose

Composition and Occurrence The cell wall is composed of polysaccharides and proteins. In addition, some cells have walls impregnated with lignin. In all cases, the polysaccharides constitute the major part of the wall. The wall polysaccharides are often classified into cellulose, hemicelluloses, and pectin. These three types are present in almost all cell walls in varying proportions [80]. In fact, in nature, cellulose is generally found in close association with hemicelluloses, pectin, water, wax, proteins, lignin, and mineral substances [81].

Cellulose is a polysaccharide consisting of a linear chain of several hundred to over 10 000 $\beta(1 \rightarrow 4)$-linked D-glucose units organized in more or less crystalline microfibrils [80,82]. It is the most abundant naturally occurring linear $\beta(1-4)$-glucan on Earth and the principal structural component of the primary cell walls of higher plants (about 40–50%), and some green algae and fungi (oomycetes, although it is rarely found in fungi) [83].

In the animal kingdom, it is produced by a solitary group of primitive vertebrates, the tunicates. Among the prokaryotes, cellulose is produced by some bacteria but has not been found among the archaeans [83]. It is also the main component of a variety of natural fibers such as cotton (the most pure form of cellulose), bast fibers, and leaf fibers. Wood pulp, cotton fibers, and cotton linters are the most important sources of cellulose for commercial processes [84].

It is also produced by certain types of bacteria such as *Acetobacter*, *Pseudomonas*, *Rhizobium*, *Agrobacterium*, and *Sarcina* species and can be synthesized as a continuous film by cultivating the bacteria in a glucose medium [85].

Structure and Properties Cellulose is a highly crystalline polymer completely insoluble in water and most organic solvents, chiral, and biodegradable [86]. Cellulose is derived from D-glucose units, which condense through $\beta(1-4)$-glycosidic bonds leading to a straight chain polymer (Figure 3.7).

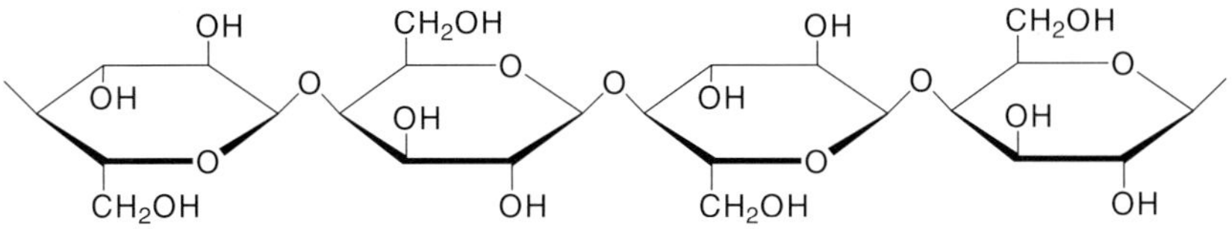

Figure 3.7 Structure of cellulose.

Unlike starch no coiling or branching occurs, so the molecule adopts an extended and rather stiff rod-like conformation, aided by the equatorial conformation of the glucose residues. The multiple hydroxyl groups on the glucose from one chain form hydrogen bonds with oxygen molecules on the same or on a neighboring chain, holding the chains firmly together side-by-side and affording *microfibrils* with high tensile strength. This strength is important in cell walls, where the *microfibrils* are meshed into a carbohydrate matrix, conferring rigidity to plant cells. Many properties of cellulose depend on its chain length or degree of polymerization.

Cellulose is soluble in strong mineral acids, sodium hydroxide solution, and metal complex solutions of copper(II)–ammonia (Schweizer's reagent) and copper(II) diamine [86d]. Under strong acidic conditions, it is hydrolyzed completely to D-glucose whereas very mild hydrolysis produces hydrocellulose with shorter chains, lower viscosity, and tensile strength. Due to high extents of intramolecular and intermolecular hydrogen bonding, cellulose is not thermoplastic but decomposes before melting [86b]. Cellulose forms esters (e.g., cellulose nitrate, cellulose phosphate, cellulose acetate, cellulose propionate, cellulose butyrate, etc.) and ethers (e.g., carboxymethyl cellulose, methyl cellulose, hydroxylpropyl cellulose, etc.) [86a–c].

Plant-derived cellulose is usually contaminated with hemicellulose, lignin, pectin, and other substances, while microbial cellulose is quite pure, has a much higher water content, and consists of long chains.

Cellulose Biosynthesis Cellulose biosynthesis is one of the most important biochemical processes in plant biology. The enzymatic steps leading to cellulose synthesis in embryophytes are not well understood, albeit it is assumed that cellulose synthesis requires both chain initiation and propagation reactions, and the two processes are separate [83].

There is a general agreement that cellulose synthesis in higher plants takes place at the plasma membrane by membrane-bound complexes or rosette terminal complexes (RTCs). Rossettes are thought to be functional in the plasma membrane only, albeit this does not mean, however, that cellulose biosynthesis does not begin in intracellular compartments [83d]. In fact, this is rarely considered as a possibility, essentially because strictly linear chains of β(1–4)-glucan tend to form microfibrillar structures, and such structures have never been observed in the intracellular compartments of higher plants. Nonetheless, the possibility that cellulose biosynthesis begins in the Golgi apparatus in higher plants cannot be ruled out [84,85]. Indeed, noncrystalline chains of β(1–4)-glucan may be synthesized in the Golgi by catalytic subunits of the cellulose synthase that have not yet been assembled in the form of rosettes.

These rosettes or RTCs are hexameric protein structures with a sixfold symmetry, approximately 25 nm in diameter, that contain the cellulose synthase enzymes (CesA) that synthesize the individual cellulose chains [83c,d]. As previously stated, the products of (1–4)-β-D-synthase preparation are microfibrillar. It is important to keep in mind that the process of assembly of individual β(1–4)-glucan chains as microfibrils is poorly understood. It may be occurring spontaneously through the formation of interchain hydrogen bonds when a number of chains sufficient to form an elementary microfibril such as those observed in primary walls (about 3 nm diameter) are in close proximity. Such autoassembly may not be possible if the synthesized chains are produced by spatially separated individual catalytic subunits distributed randomly (e.g., along the Golgi membrane). Data obtained in *Arabidopsis* support the hypothesis that the rosette structures are required for the formation of microfibrils [83e]. In another hypothesis, cellulose biosynthesis could be initiated in the Golgi apparatus in the form of oligosaccharides that are too short to form microfibrillar strings. Another possible explanation for the absence of formation of cellulosic microfibrils in the Golgi is that the β(1–4)-glucan chains may interact in the organelle with other compounds, for instance, other carbohydrates or an aglycone that would prevent microfibril formation. In conclusion, cellulose biosynthesis is considered as occurring exclusively at the plasma membrane because of the lack of experimental evidence for the formation of the polymer in intracellular compartments.

Each RTC floats in the cell's plasma membrane and contains at least three different cellulose synthases, encoded by CesA genes, in an unknown stoichiometry. The cellulose synthases belong to a group of *glycosyltransferases* (GTs) with repetitive action patterns. They are classified in glycosyltransferase family 2 among the current 77 families of these enzymes [83,86a–c].

Cellulose synthase catalytic subunits polymerize β(1–4)-glucan chains from the activated sugar donor UDP-glucose. This substrate is synthesized by the cytosoluble enzyme UDP-glucose pyrophosphorylase from UTP and glucose-1-phosphate. Sucrose synthase, which can synthesize UDP-glucose from sucrose and UDP, has been proposed to be directly involved in cellulose biosynthesis by channeling UDP-glucose to the cellulose synthase catalytic subunit(s) (Figure 3.8) [87,88].

The existence of a membrane-bound form of sucrose synthase potentially involved in this process has been reported, and this may make cellulose polymerization more efficient [87,88]. In addition, several reports describe the localization of sucrose synthase in sites where cellulose synthesis is high, either close to the plasma membrane or in cell walls [89,90]. However, UDP-glucose is not only the direct substrate of glycosyltransferases such as cellulose synthase, but also a key precursor for the different nucleotide sugars and the corresponding noncellulosic cell wall carbohydrates [91].

It is admitted that UDP-glucose is used directly by the cellulose synthase catalytic subunits to elongate cellulose in a repetitive manner, leading to chains that consist of 800–10 000 glucosyl units depending on the origin of cellulose [92]. But before this polymerization takes place, cellulose biosynthesis must be initiated and this may require a primer (Figure 3.8). This primer has been suggested to be sitosterol-β-glucoside [93]. Nonetheless, the role of sitosterol-β-glucoside as a primer for cellulose biosynthesis remains to be firmly demonstrated *in vivo*.

How are Cellulose Chains Polymerized?

1) The glycosyltransferase responsible for cellulose polymerization uses UDP-glucose as monosaccharide donor. In this substrate, the glucosyl residue linked

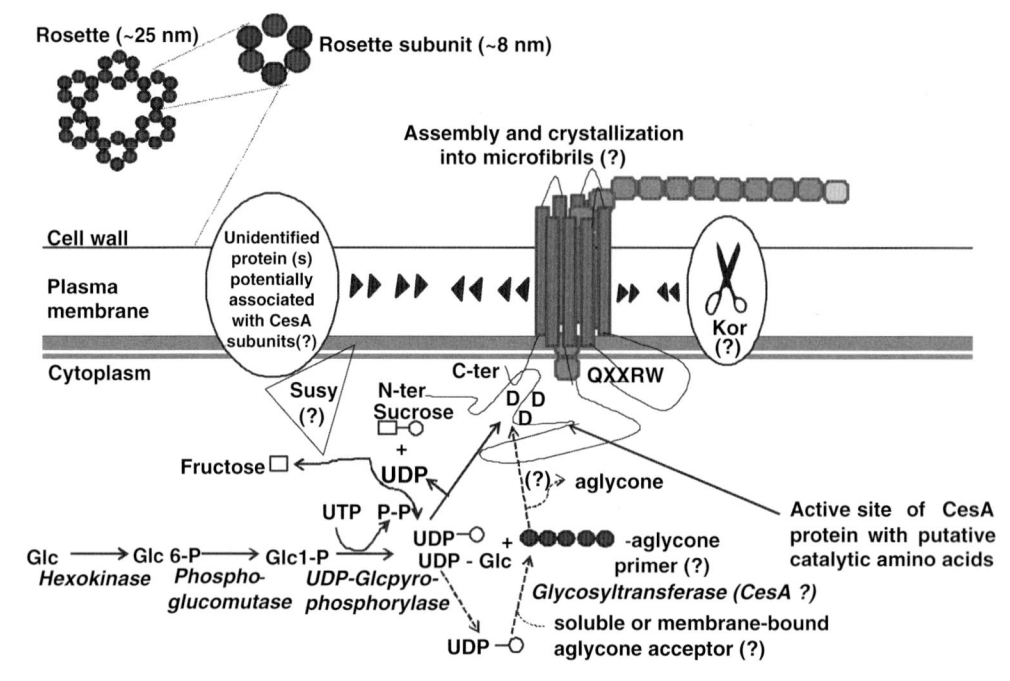

Figure 3.8 Hypothetical model of a cellulose synthase (CesA) protein at the plasma membrane of an embryophyte cell.

   to the pyrophosphate moiety of the molecule is in the α-configuration while cellulose contains exclusively β-linkages. For this reason, the enzyme and its corresponding mechanism is described as being "inverting," with the presumed involvement of a single nucleophilic substitution at the anomeric carbon of the glucosyl unit of the donor [94].
2) The stereochemistry of the β(1–4)-glycosidic bonds that link the glucosyl units in cellulose induces the formation of a regular twofold screw axis along the glucan chain. Thus, monomers are rotated by 180° with respect to their neighbors [94]. For this reason, cellobiose (disaccharide derived from the condensation of two glucose molecules linked in a β(1 → 4) bond) rather than glucose is sometimes considered as the repeating unit of cellulose.
3) Results based on the specific labeling of the reducing ends of growing cellulose chains strongly suggested a propagation of the polysaccharide from its non-reducing end (see Figure 3.8), which is consistent with the mode of elongation of other carbohydrate polymers such as chitin [95].
4) A model describing the polymerization by the addition of the monomers directly to the carbohydrate chains has been proposed, although a hypothetical model in which the sugars are first transferred to Ser or Thr residues of the enzyme has also been proposed [96]. Despite this, there is no experimental evidence for either of these possibilities.

On which Side of the Plasma Membrane Does Catalysis Occur? [88,97,98] All cellulose synthases in embryophytes are integral plasma membrane proteins and consistent with this they all have multiple transmembrane helices.

The predicted topology of cellulose synthase catalytic subunits suggests the existence of eight transmembrane helices that anchor the proteins in the plasma membrane [97]. This even number of helices means that the N-terminal and C-terminal ends of the catalytic subunits face the same side of the plasma membrane. The protein also contains a large domain predicted to be soluble and that carries the observed D,DD,QXXRW motif involved in the catalytic event and common to all putative cellulose synthase catalytic subunits [88,98]. This domain is typically expected to face the cytoplasmic side of the plasma membrane due to the localization of UDP-glucose pyrophosphorylase, which is primarily involved in the biosynthesis of the UDP-glucose substrate of cellulose synthase (Figure 3.8).

There is no known transporter of UDP-glucose in the plasma membrane that would allow the transfer of the substrate across the membrane. Despite these observations, the possibility that the catalysis of cellulose polymerization occurs on the extracellular side of the plasma membrane cannot be completely ruled out. In this case, the N-terminal and C-terminal ends as well as the catalytic domain of the enzyme would be exposed in the cell wall, which is theoretically possible [98].

Thus, catalysis at the cytoplasmic side of the membrane implies that the elongating cellulose chains are translocated across the membrane to reach the cell wall (Figure 3.8). In this respect, we can find several hypothetical models. Among them there is one that predicts that the eight transmembrane helices of each individual cellulose synthase catalytic subunit assemble in the membrane to delimit a pore through which the elongating glucan chains are transported. In the models involving a pore-type structure, the polymerization reaction itself would provide the energy required for the movement and extrusion of the cellulose chains. In this respect, there is no strong experimental evidence supporting any of the proposed models, and the mechanism of translocation of the cellulose chains across the plasma membrane has yet to be elucidated.

Applications Cellulose and its derivatives are widely used in textiles, food, membranes, films, and paper and wood products. Cellulose is also used in medicines (e.g., wound dressing, suspension agents, composites, making rope, mattress, netting, upholstery, muslin, thermal and acoustic insulation, etc.) [86a–c]. Cellulose synthesized from bacteria is highly crystalline with finer diameter fibrils forming fibrous mats with greater surface area than plant-derived cellulose. Bacterial cellulose is used in the production of speaker diaphragms in headphones [86c]. Cellulose nitrate is used in gum cotton and celluloid [86a]. Cellulose esters such as acetates are used as plastic materials, fibers in textiles, and coatings [86a,b]. Cellulose xanthates are employed in the viscose rayon process [86a]. Unlike cellulose, cellulose ethers, for example, carboxymethyl cellulose, methyl cellulose, hydroxypropyl cellulose, and so on, are soluble in water and form films for various applications. These derivatives are also used as plastic materials. Methyl cellulose is used as a food thickening agent and as an ingredient in adhesives, inks, and textile finishing formulations [86b].

3.3.1.2 Hemicellulose

Composition and Occurrence Hemicelluloses are a group of polysaccharides found in association with cellulose and lignin in plant cell walls [99–101]. Their role is to provide a linkage between lignin and cellulose [100]. Hemicelluloses are characterized by having a backbone of β(1–4)-linked sugars with an equatorial linkage configuration [100b,c]. They have much lower molecular weight than cellulose with very low degree of polymerization (generally below 200) [99,100].

Their primary structure depends on the plant source of the polymer. Next to cellulose, hemicelluloses are the most abundant natural polymer in the biosphere and represent about 20–30% of wood mass. They are predominantly located in the primary and secondary walls of vegetable cells [100,102].

Structure and Properties Hemicelluloses exist in amorphous form. Unlike cellulose, they are generally very soluble in water. They are hydrolyzed to saccharides (e.g., glucose, mannose, arabinose, galactose, and xylose), acetic acid, and uronic acid. In fact, hemicelluloses can be divided into two categories: (a) *cellulosans* and (b) *polyuronides* [100,102].

a) *Cellulosans* include the hemicelluloses made of sugar monomers, hexosans (e.g., mannan, galactan, and glucan) and pentosans (e.g., xylan and arabinan).
b) *Polyuronides* are hemicelluloses containing large amounts of hexuronic acids and some methoxy, acetyl, and free carboxylic groups.

Nonetheless, the predominant constituents of hemicelluloses are 1,4-β-D-xylans, 1,4-β-D-mannans, araban (composed of arabinose units), and 1,3- and 1,4-β-D-galactans (complex mixture of monosaccharides) [100,102].

Among all the hemicelluloses, xyloglucan (a cellulosan) is the most abundant hemicellulose in primary walls of spermatophytes, except for grasses. It is composed of a linear glucan backbone with regular side chain additions of xylose and galactose and sometimes fucose and arabinose. Its basic structure has many different variations. Indeed, the branching pattern of xyloglucan is of both functional and taxonomic significance [100c].

In summary, hemicelluloses are a heterogeneous group of polysaccharides and the term, which was coined at a time when the structures were not well understood and biosynthesis was completely unknown, is therefore archaic. In fact, various researchers have suggested that it should not be used. Alternative terms such as cross-linking glycans have been proposed [103], but that has other problems since it is not obvious that cross-linking is a major and common feature of the hemicelluloses.

Nevertheless, most workers in the field still use the term hemicelluloses as a convenient denominator for a group of wall polysaccharides that are characterized by being neither cellulose nor pectin and by having β(1 → 4)-linked backbones of glucose, mannose, or xylose. All these glycans have the same equatorial configuration at C_1 and C_4 and hence the backbones have significant structural similarity (Figure 3.9) [100b].

It has been suggested that hemicelluloses should be used to describe only polysaccharides with this configuration.

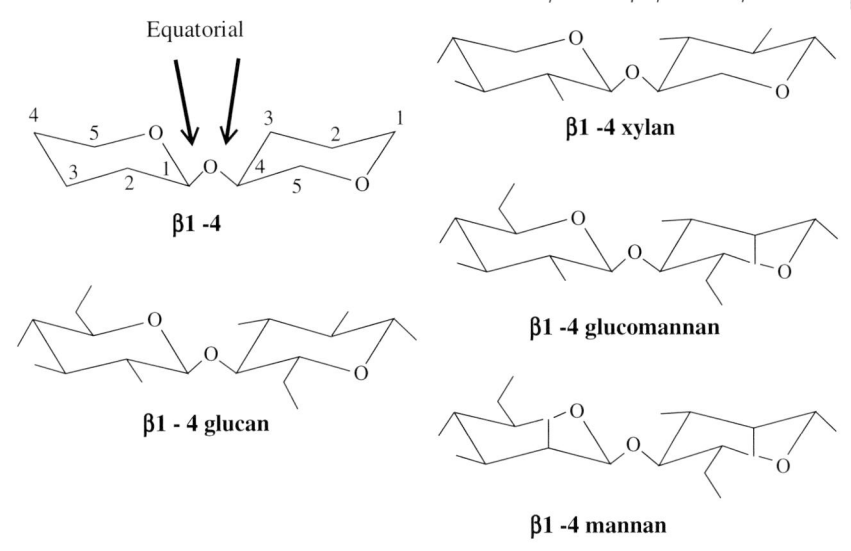

Figure 3.9 Hemicellulose repeating disaccharides. Hemicelluloses are characterized by a β(1–4)-linked backbone at C_1 and C_4.

Hemicellulose Biosynthesis It is known that the pathways of starch, cellulose, mixed-linkage glucan, and hemicellulose biosyntheses have a common metabolic origin, namely, they utilize nucleotide sugars as the source of carbohydrate for glucan polymerization. These pathways are known to involve the interconversion of sugars and sugar phosphates, resulting in the formation of nucleotide sugars that are the essential precursors of the glucans [101]. Thus, nucleotide diphosphate sugars (mostly UDP derivatives) are used as the sugar donors for polysaccharide synthases.

Although the synthesis of matrix polysaccharides occurs in the endoplasmic reticulum and Golgi bodies and vesicles, the synthesis of such nucleotide diphosphate sugars occurs in the cytosol [101]. Since membranes are generally impermeable to sugar nucleotides, there must be some transport mechanism by which they are able to cross. Interestingly, sucrose synthase, which is believed to play a role in the production of UDP-glucose for glucan synthesis, has been shown to be associated with the plasma membrane [87], and thus may channel carbon directly from sucrose to cellulose synthases in the plasma membrane. Unlike protein synthesis, which takes place along the residues of mRNA, there is no similar template for polysaccharide synthesis.

Thus, there are many intriguing questions about the regulation of polysaccharide synthesis and the coordination of synthesis of heteropolymers that involve the action of more than one synthase [104]. In this respect, most of the available evidence suggests that sugar residues are added directly from the sugar nucleotide to a growing chain rather than via a lipid intermediate. However, in the case of maize root mucilage there is clear evidence that the polysaccharide is made as a glycoprotein and here the protein is acting as a primer, with the involvement of glycolipid intermediates [105]. However, the mucilage is not a cell wall polysaccharide and would not necessarily be synthesized in the same way [105c].

So it appears that a number glycosyltransferases is involved in the polysaccharide synthesis of hemicelluloses, provided the involvement of proteinaceus primers or lipid intermediates is traditionally considered unlikely. The number of glycosyltransferases required to make a complex hemicellulose remains unknown, and although a number of glycosyltransferases have been characterized we are still relatively ignorant compared with our knowledge of the biosynthesis of other plant products.

Nonetheless, some glycosyltransferases that have been characterized in determined species are [101] as follows:

1) *xyloglucan synthases* involved in the synthesis of xyloglucans:
    a) glucosyltransferases (e.g., in cell membranes isolated from etiolated pea tissue, the incorporation of glucose from UDP-D-[^{14}C]-glucose units into polysaccharides has been demonstrated [106a,b]);
    b) xylosyltransferases (this enzyme uses UDP-xylose as a substrate to add xylosyl side chain residues to the C_6 position of a glucosyl residue [106c]);
    c) galactosyltransferases (e.g., a galactan synthase responsible for the biosynthesis of a pectic β-1,4-galactan has also been characterized from the mung bean; this enzyme had an apparent pH optimum of 6.5 in the presence of Mg^{2+} [106d]);
    d) fucosyltransferases (e.g., the fucosyltransferase enzyme isolated from the microsomal fraction of primary roots from 6-day-old radish seedlings transfers L-fucose from GDP-L-Fuc to O-2 of an α-L-arabinofuranosyl residue [106e]);
    e) arabinosyltransferases (most arabinosyltransferases that have been characterized are involved in the pectic arabinan synthesis (1–2, 1–3, and 1–5 linkages) or O-glycosylation of hydroxyproline-rich proteins and arabinogalactan proteins [106f]);
2) *xylan synthases*, which are also involved in secondary wall xylan synthesis;
3) *glucuronoarabinoxylan synthases*, which are involved in the glucuronoarabinoxylan synthesis;
4) *β-1,3-1,4-glucan synthases* involved in the mixed-linkage glucan synthesis; and
5) *glucomannan synthases*.

Hemicelluloses are synthesized by glycosyltransferases located in the Golgi membranes. Many glycosyltransferases needed for biosynthesis of xyloglucans and mannans are known.

In contrast, the biosynthesis of xylans and β(1–3,1–4)-glucans remains very elusive, and recent studies have led to more questions than answers [100c,101,106,107].

Applications In contrast to other plant-derived polysaccharides such as cellulose, starch, pectins, and gums, hemicelluloses do not have a comparable commercial value. However, they are a subject of importance since they can affect the extraction of cellulose and also make significant contributions to wood and fiber quality [101].

Nonetheless, hemicelluloses and their derivatives are important in industry due to their useful chemical properties and biological activity. Isolated hemicelluloses are used as food additives, thickeners, emulsifiers, gelling agents, adhesives, and adsorbents [100,102].

Saccharides obtained from hemicellulose may be converted to useful primary chemical substances, such as acetone, butanol, ethanol, furfural, xylitol, 2-methylfuran, furan, lysine, furfuryl alcohol, glutamic acid, hydroxymethylfurfural, levulinic acid, polyols, and so on [100,102].

Hydrogenation of xylose produces xylitol that is used as a sweetener [100,102]. Furfural produced by dehydration of xylose is a very useful precursor for the synthesis of 2-methylfuran, tetrahydrofuran, adiponitrile (hence nylon), furfuryl alcohol, tetrahydrofurfuryl alcohol, glutamic acid, polyurethane foams, resins and plastics, nitrogenated and halogenated derivatives used in drugs, and so on [100,102].

3.3.1.3 Pectin

Composition, Structure, and Occurrence Pectins (together with hemicelluloses) constitute the matrix in which cellulose microfibrils are embedded. Thus, the interactions between the different polysaccharides ensure the strong yet dynamic and flexible properties of the cell wall [108].

The term "pectin" covers a group of acidic heteropolysaccharides with distinct structural domains. These polysaccharides appear to be linked together to form a pectic network throughout a plant. Briefly, they are mainly composed of three major components such as homogalacturonan (HG), rhamnogalacturonan I (RG-I) and rhamnogalacturonan II (RG-II), and xylogalacturonan (XGA) [108,109].

For example, HG is a homopolymer of α-1,4-linked D-galacturonic acid (d-GalUA) that is partially methyl esterified at O-6 and partially acetylated at O-3 and O-4 of GalUA [108c], whereas XGA consists of a HG backbone that it is substituted with single β-1,3-Xyl residues or such residues substituted with a few additional xylose residues (Figure 3.10).

RG-I consists of the skeletal disaccharide repeating unit [-2-L-Rhaα1–4-D-GalUAa1-], with branching oligosaccharides attached to O-3 and/or O-4 of L-rhamnose (l-Rha) residues in the backbone [108c].

RG-II has the most complicated structure of linear α-1,4-linked D-GalUAs with four characteristic side chains containing rare sugar residues such as L-aceric acid (l-AceA), D-apiose (d-Api), 3-deoxy-D-lyxo-2-heptulosaric acid (Dha), and so on [98c].

In general, the content of pectin varies depending on environment, tissue, and species [108a,110]. For example, in *Arabidopsis* (*Arabidopsis thaliana*) leaves, the cell wall contains approximately 50% (w/w) pectin. Similarly, the ratio of HG, XGA, RG-I, and RG-II is also variable, but typically HG is the most abundant polysaccharide, constituting about 65% of the pectin, while RG-I constitutes 20–35% [108a,111]. XGA and RG-II are minor components, each constituting less than 10% [108a,111,112].

Especially, HG and RG-II are well known to be involved in the strengthening of the wall. The mechanical properties of HG and RG-II have been reviewed and described thoroughly in other publications [108a,113]. HG plays additional roles besides pure

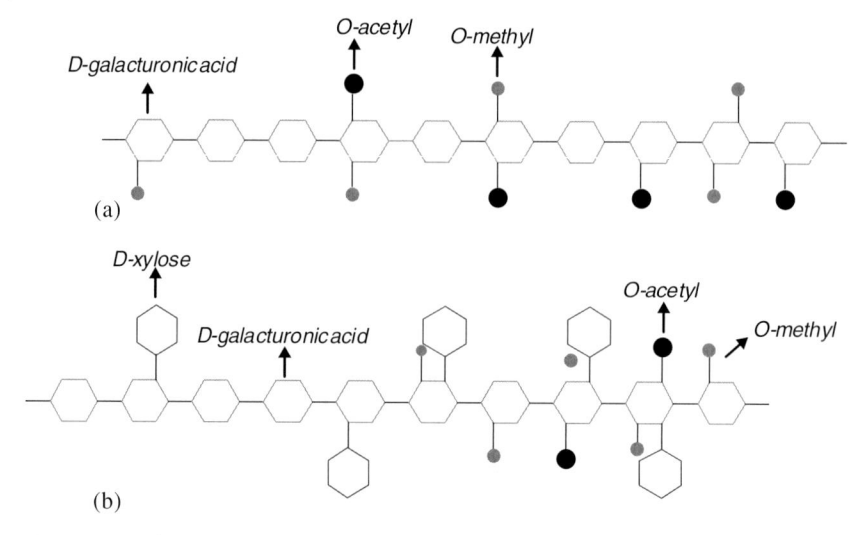

Figure 3.10 Schematic structure of pectins: (a) homogalacturonan and (b) xylogalacturonan.

mechanical support. For example, it is known that plant pathogens cause degradation of pectin, and oligogalacturonides (such as α-1,4-linked oligomers of GalUA) are well established to be part of a signaling cascade that senses wall degradation upon pathogen attack [108a,109,114].

Anyway, the main role of pectin is to provide physical strength to the plant and to provide a barrier against the outside environment. Besides this, pectins are thought to have a role in cell adhesion, cell porosity, and signaling pathogen invasion in cells. Moreover, they are also considered to be associated with cell division and cell expansion, which is corroborated by the fact that the level of pectin biosynthesis is high during cell differentiation and cell growth ([108c] and references cited therein). Nonetheless, the structure and function of pectins are, however, still largely unknown. One of the reasons for this is that the molecular mechanism of pectin biosynthesis is not well understood [109].

Pectins localize mainly at the primary cell wall matrix and middle lamella [108a]. Two major sources of pectins are apple pomace and orange peel.

Pectin Biosynthesis Matrix polysaccharides are assumed to be synthesized in Golgi vesicles, even though it cannot be excluded that some initial steps take place in the endoplasmic reticulum or that some assembly steps take place in the wall. Nevertheless, the evidence for Golgi localization is strong. Indeed, the few pectic biosynthetic enzymes that have been identified have all been shown to be located in the Golgi apparatus [108a].

Given the complexity of pectin structures, it is obvious that a large number of enzymes are required to synthesize these polysaccharides. With reasonable assumption regarding the substrate specificity of the biosynthetic enzymes, Mohnen et al. have predicted that 67 different glycosyltransferases, methyltransferases, and acetyltransferases are required [111]. Among them, only three of these enzymes have been unambiguously identified.

Most of the predicted enzymes are GTs that transfer one glycosyl residue from a nucleotide sugar donor to the nonreducing end of an oligo- or polysaccharide acceptor. The majority of the GTs are likely to be type II membrane proteins with an N-terminal transmembrane anchor and a catalytic domain in the Golgi lumen. However, multimembrane spanning GTs are also present in the Golgi vesicles (e.g., mannan synthase) and it is possible that some are involved in pectin biosynthesis [115].

Transglycosylases that transfer whole oligosaccharide chains from one polysaccharide to an acceptor polysaccharide are well known in starch and xyloglucan biosynthesis. While no transglycosylase activity acting on pectin has been reported, it is conceivable that such enzymes exist [116]. In addition to GTs, methyltransferases, acetyltransferases, and possibly transglycosylases, the biosynthesis of pectin requires feruloyltransferases, albeit here the donor substrate for feruloylation is not known [116].

Pectin biosynthesis has been studied for many years and HG biosynthesis *in vitro* was demonstrated already in 1965 [117]. Biosynthesis of arabinan and galactan was also demonstrated long ago [118].

In summary, the study of pectin biosynthesis is challenging and progress in this area has been slow.

Applications Pectin is soluble in water and forms viscous solutions. The importance of pectins in the food industry is due to its ability to form gels in the presence of Ca^{2+} ions or a solute at low pH. A number of factors that include pH, presence of other solutes, molecular size, degree of methoxylation, number or arrangement of side chains, and charge density on the molecule influence the gelation of pectin [118d] (See http://www.ippa.info/applications_for_pectin.htm.).

In the food industry, pectins are used in jams, jellies, frozen foods, and more recently in low-calorie foods as a fat and/or sugar replacement. In the pharmaceutical industry, it is used to reduce blood cholesterol levels and gastrointestinal disorders. Pectins are also used in edible films, paper substitute, foams, and plasticizers [118d] (See http://www.ippa.info/applications_for_pectin.htm.).

3.3.2
Storage Polysaccharides

3.3.2.1 **Starch**
Composition and Occurrence Starch is the principal carbohydrate reserve of plants where it can be stored in the form of small granules. It occurs in many plant parts, but it is most abundant in storage tissues, such as seeds and the various underground storage organs [119–122]. This storage starch is prepared in amyloplasts from imported sugars and is used to supply carbon for growth and development of heterotrophic tissues and to meet demands seasonally and during times of stress.

There is also a transient starch that provides reserves that cycle daily and is made from newly assimilated carbon in choloroplasts of cells that carry out photosynthetic carbon reduction. Transient starch does not appear to be essential for plant growth and development. Both storage and transient starch usually are a mix of amylose and amylopectin. The ratio of amylose and amylopectin in starch varies as a function of the source, age, and so on. The latter is the predominant component in storage starch [119–122].

Structure and Properties Amylose molecules are long unbranched chains of about 840–22 000 α-D-glucopyranose units linked by α(1–4)-glycosidic bonds [120]. Amylopectin, which makes up 70–80% of most granules, is a highly branched α-1,4,α-1,6-linked polymer (branched α(1–4)-glucan with α(1–6)-glucan branches every 19–25 units) (Figure 3.11) [120–124].

Clustering of the branch points in the amorphous lamellae enables the polymer chains to line up in a parallel fashion [125]. This chain organization forms the basis of the semicrystalline structure of the starch granule [126]. In contrast to amylopectin, amylose is an essentially linear polysaccharide with less than 1% of branch points (Figure 3.11).

The structure of starch is highly complex. Native starch granules contain alternating hard (crystalline) and soft (amorphous) material that is together composed of the long glucose chain molecules amylose and amylopectin. The ratio of these molecules varies according to botanical source and influences some important starch granule characteristics. All starch granules share some basic features, which are useful for identification and classification (size, shape, surface characteristics, presence of the hilum, reaction with iodine, etc.) [122].

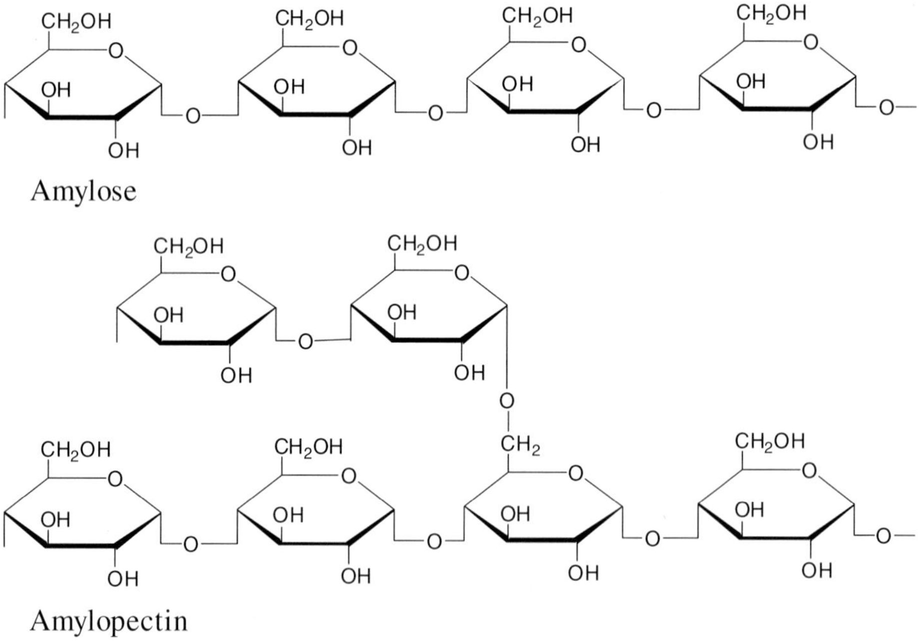

Figure 3.11 Structures of amylose and amylopectin.

Starch is not soluble in water at ambient temperature. However, in hot water its granules gelatinize to form an opalescent dispersion [123]. It is degraded by a number of enzymes such as amylase and amyloglucosidase.

The role of storage starch in plants can be regarded as physiologically equivalent to that of glycogen in animals [124]. For example, 50–80% of the dry weight of cereal, pea, and bean seeds, cassava roots, and potato tubers consists of starch. In these organs, starch synthesis and degradation occur during distinct development periods, which may be separated by months or even years. In starch-storing seeds, starch synthesis occurs during most of the period of growth and maturation, whereas starch degradation occurs during the onset of germination, providing carbon for the initial growth of the seeding [124]. It occurs as water-insoluble granules in the roots, seeds, tubers, and stems of various plants, including corn, maize, wheat, rice, millet, barley, potatoes, arrowroot, and cassava [120–123].

Starch Biosynthesis Starch is an important reserve carbohydrate found in almost all plant species and is deposited as granules in the chloroplast of green leaves (transitory starch) and in amyloplasts of tubers, roots, and seeds (storage starch) [127].

Each starch granule has a highly organized structure defined by the succession of crystalline and amorphous lamellae [128]. In plant storage organs, starch biosynthesis takes place within the amyloplast and is the result of different reactions such as synthesis (polymerization of glucosyl residues), rearrangement, and degradation, in which various starch synthases, transferases and disproportionating enzymes, and hydrolytic enzymes, respectively, play key roles (Figure 3.12).

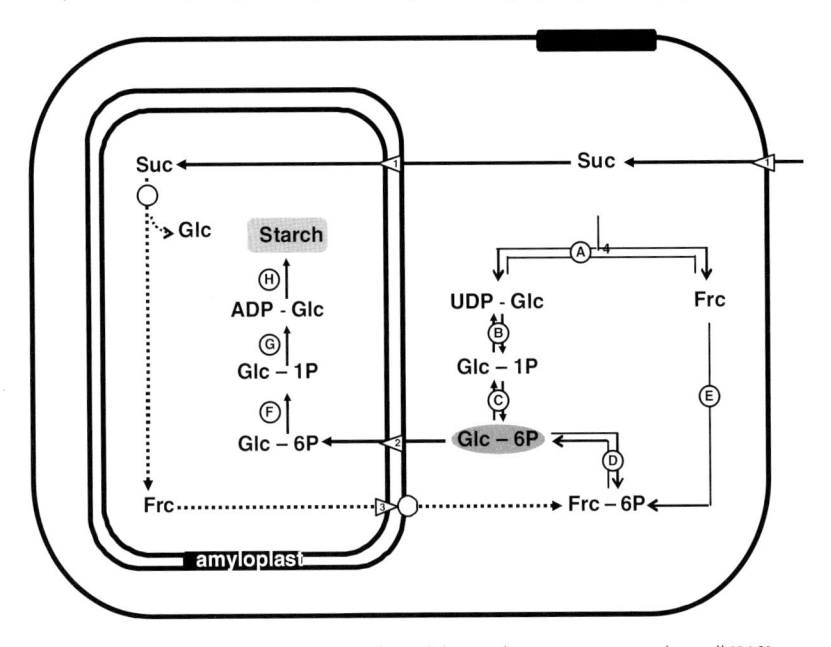

Figure 3.12 Schematic representation of starch biosynthesis in a potato tuber cell [128].

Sucrose is the starting point of starch biosynthesis, which is converted into hexose phosphates in the cytoplasm. In potato, glucose-6-phosphate (Glc6P) is transported into the amyloplast [129]. There, it is first converted to glucose-1-phosphate (Glc1P) by phosphoglucomutase (PGM) and subsequently to ADP-glucose (ADP-Glc) by ADP-glucose pyrophosphorylase (AGPase) (Figure 3.12) [130]. ADP-Glc serves as a substrate for the different starch synthase isoforms, some of which are mainly present in the soluble phase or stroma (SS isoforms), and others associated with the granule.

The exclusively granule-bound starch synthase (GBSSI) catalyzes the formation of amylose [131]. Furthermore, it was shown that GBSSI can also contribute to amylopectin synthesis in potato, pea, and the unicellular alga *Chlamydomonas reinhardtii* [132].

All starch synthases elongate glucan chains by transferring the glucose moiety of ADP-Glc to the nonreducing end of α-1,4-linked glucans. The branching enzymes (BEs) cleave α-1,4 linkages and form α-1,6 linkages. Schwall et al. in 2000 showed that antisensing these enzymes in potato led to less branching of the starch [133].

Additional enzymes such as disproportionating enzymes (D-enzymes) cleave and rejoin α-1,4 linkages in starch polymers and the debranching enzymes (DBEs) hydrolyze the α-1,6 linkage at branch points. In this case, the model shows that amylopectin is synthesized first and amylose is formed later. Two possible mechanisms for amylose biosynthesis have been proposed:

i) the amylopectin-primed pathway, and
ii) the maltooligosaccharide (MOS)-primed pathway [132c,134].

First evidence for the amylopectin-primed pathway was provided by Van de Wal et al. in 1998 [132c], who showed that amylose could be synthesized *in vitro* by cleavage of amylopectin molecules in mutant *Chlamydomonas* starches. It is postulated that amylopectin is cleaved by the action of hydrolytic enzymes such as α-amylase. An alternative explanation is that GBBSI has a dual activity, initially working as a polymerase and at a particular time point as a hydrolase. Also in higher plants, the amylopectin-primed pathway seems to take place. However, in this case it is more difficult to demonstrate amylopectin priming, due to the much lower GBSSI activity in the granules.

For the second pathway, Denyer et al. showed that amylose could be synthesized *in vitro* in pea granules by the elongation of small soluble glucans or MOS, which can diffuse into the granules [134c]. It is estimated that the MOS concentration is sufficiently high in starch-producing plant organs, thereby providing sufficient acceptor substrate for amylose synthesis [135]. The MOS pool may be replenished continuously by the action of DBE and other hydrolytic enzymes. Van de Wal in 2000 has suggested that the two pathways can occur side-by-side. Depending on the conditions in the plant, amylose biosynthesis may switch between the two mechanisms.

In summary, storage starch synthesis occurs exclusively within plastids. The number and shape of granules per plastid vary considerably between organs and between species [119]. Starch granules grow by the addition of new material onto the surface. The enzymes of amylopectin synthesis produce nascent amylopectin molecules at the surface, probably primarily by elongating and then branching chains projecting from the matrix. It seems likely that as amylopectin is elaborated at the granule surface it spontaneously crystallizes to form new matrix material. In contrast, amylose is probably synthesized inside the matrix formed by amylopectin crystallization by an enzyme that is tightly bound within the matrix [119].

Applications Starch is not generally suitable for plastic and fiber products due to its highly branched structure and permeability [121]. However, starch finds extensive applications in foods, as thickeners, inert diluent for drugs, adhesives, and sizing or glazing agent in the paper and textile industries [120–123].

Starch nitrate is used as explosive, whereas starch acetate forms films and is used as a paper and textile sizing agent [123]. Starch ethers, hydroxylethyl ether in particular, are used as adhesives [123]. Starch allyl ethers are used as air-drying coatings [123]. There is a considerable interest in graft copolymers of starch due to their use as biodegradable packing materials and agricultural mulches. For example, starch cross-linked with epichlorohydrin is used in the treatment of rice granules to make them more resistant to breakdown in canned soups.

Amylose is used for making edible films, whereas amylopectin is used for textile sizing and finishing and as a food thickener.

3.3.2.2 Fructans: Inulin

Composition, Occurrence, and Structure Fructans are all oligosaccharides that contain at least two adjacent fructose units. One glucose molecule can be present but it is not necessary [136,137].

In contrast to starch, fructans are soluble *in vivo*. However, *in vitro* it is impossible to dissolve the same amounts as those occurring *in vivo* [138]. This discrepancy is not well understood but it cannot be ruled out that fructans reach a crystalline or glassy state within the plant. The fructose moieties in fructans have the furanose conformation, permitting more flexibility compared to the more common pyranose configuration (e.g., starch) [138,139].

Essentially five structurally different types of fructans can be distinguished in higher plants: inulin, levan, mixed levan, inulin neoseries, and levan neoseries [138].

Inulin-type fructan is a fructose polymer that has mostly or exclusively the $\beta(2-1)$-fructosyl–fructose linkage, whereas levan-type fructan consists mostly or exclusively of $\beta(2-6)$-fructosyl–fructose linkage. Thus, both these fructan types are essentially linear molecules but a low degree of branching can occur through $\beta(2-6)$ linkages (in the case of inulin) or $\beta(2-1)$ linkages (in levan). In case the terminal glucose molecule is absent (Fn-type fructans), the terms inulo-*n*-oses ($\beta(2-1)$ linkages) and levan-*n*-oses ($\beta(2-6)$ linkages) are used; these are reducing molecules in contrast to the regular type fructans (G–Fn) [138].

Inulin is the simplest and most investigated fructan. This type of fructan occurs mainly in dicot species, including chicory (*Cichorium intybus*), the roots of which are used for commercial inulin extraction. Therefore, inulin is found in many plants as a storage carbohydrate being typically located in roots or rhizomes. The

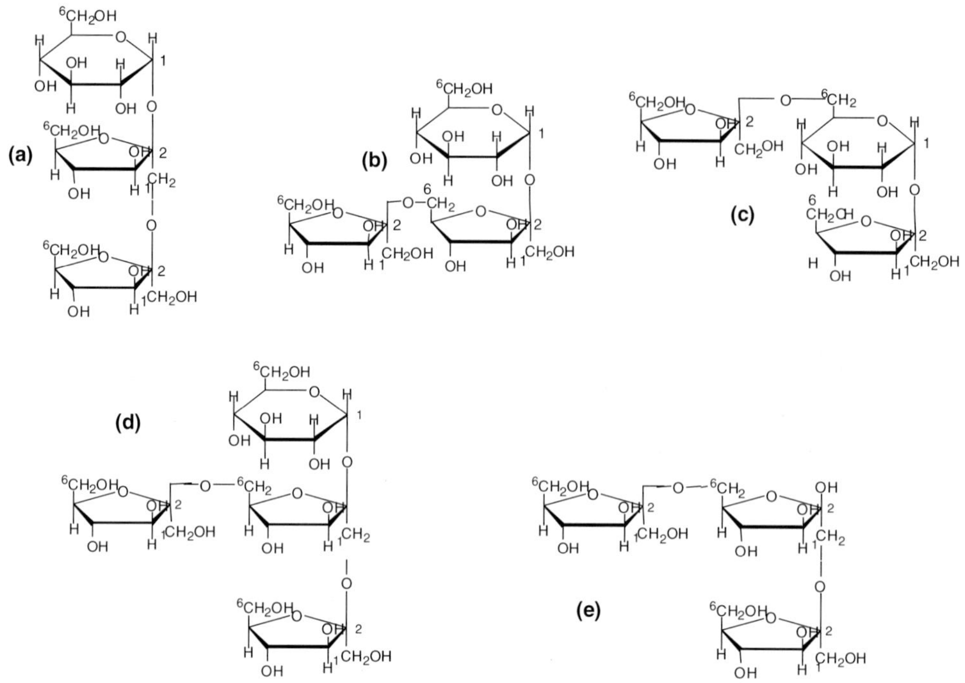

Figure 3.13 Chemical structures of some basic fructan oligosaccharides: (a) 1-kestose; (b) 6-kestose, (c) neokestose; (d) bifurcose; and (e) mixed-type F3 fructan.

smallest representatives of this series (inulin-type fructan) are 1-kestose (G–Fn type, Figure 3.13) and inulobiose (Fn type, Figure 3.13).

Linear levan-type fructan is predominant only in a few grass species such as *Dactylis glomerata* [140]. The smallest representatives of this series are 6-kestose (Figure 3.13; G–Fn type) and levanbiose (Fn type). Most grass species, however, store mixed-type fructans, of which bifurcose is a representative.

In the recent years, a new Fn-type fructan has been found in chicory roots, in which the fructose chain emerges from a levanbiose molecule, proceeding via β(2–1) linkages only [141]. The trisaccharide contains one β(2–1) linkage and one β(2–6) linkage and thus is identical to bifurcose lacking a glucose molecule (Figure 3.13).

Fructans in which fructose chains occur on both ends of a glucose molecule (Fm–G–Fn) are neokestose-based fructans (Figure 3.13). Both inulin neoseries [142] and levan neoseries [143] can be detected, depending on the type of linkage between adjacent fructose moieties.

It is important to notice that plant species capable of fructan biosynthesis did not lose their capacity for starch metabolism [138]. In this regard, fructans cannot be regarded as an absolute alternative for starch. Sometimes similar amounts of fructans and starch occur within one tissue, for example, within the bulbs of *Lachenalia minima* [144].

Although starch cannot be detected within the reserve organs of typical fructan-containing dicot species such as *C. intybus*, *Helianthus tuberosus*, and so on [145], starch is present in leaves of *H. tuberosus* and in flowers of *Campanula rapunculoides*, and starch biosynthesis is initiated during nodule formation on chicory leaves [146].

Despite major advances being made in the elucidation of the metabolism of fructans, their precise physiological function remains a subject of debate. The most documented role is that of a long-term reserve carbohydrate; however, two other functions are often quoted: first as a cryoprotectant and second as an osmotic regulator [136].

Inulin Biosynthesis Without any doubt, inulin is the most well-known and studied fructan type. For these reasons, we concentrate on inulin metabolism in this chapter.

In recent years, significant progress has been made in the elucidation of inulin metabolism in Asteraceae [136,147]. The process began when Edelman and Jefford suggested that in Jerusalem artichoke, the synthesis proceeded with the concerted action of two fructosyltransferases: (a) 1-SST (sucrose:sucrose 1-fructosyltransferase) that catalyzed the formation of 1-kestose and (b) 1-FFT (fructan:fructan 1-fructosyltransferase) that utilizes the formed 1-kestose as substrate for further chain elongation into inulin-type fructans [136,147] (Equations 3.1 and 3.2):

$$G - F + G - F \rightarrow G - F - F + G \tag{3.1}$$

$$G - F - (F)_n + G - F - (F)_m \rightarrow G - F - (F)_{n+1} + G - F - (F)_{m-1} \tag{3.2}$$

In this reaction, sucrose can act only as an acceptor, and not as a donor. The reaction is reversible, and sucrose and high degree polymerization inulins are recognized as being the best acceptors. This hypothesis was heavily criticized as it was formulated without *in vitro* evidence for the synthesis of inulin beyond a DP > 3, with no published evidence for the existence of the key enzyme SST, without use of completely purified enzymes, and without using physiologically acceptable substrate conditions [148].

However, this criticism inspired others to demonstrate the *de novo* synthesis of inulin with DP > 3, using pure enzymes under physiological conditions, and this resulted in products that were identical to those found in the plants from which the enzymes or genes originated. Both enzymes (1-SST and 1-FFT) and genes were isolated from Jerusalem artichoke [149], chicory [150], and globe artichoke [151].

Indeed, at the enzyme level, it was shown that the first enzyme (1-SST), by using a long incubation time (96 h), could also produce oligosaccharides with a DP > 3. The concerted action of these enzymes was proved *in vitro* using *Helianthus*, chicory, and *Cynara* enzymes [149,152,153]. In addition to inulin from the 1-kestose series, inulin can also be formed from the neokestose series – the type of fructan found in onions. In this case, a third enzyme is needed, 6G-FFT (fructan:fructan 6G-fructosyltransferase), which catalyzes the transfer of a fructose residue of 1-kestose to C_6 of the glucose moiety of sucrose, forming neokestose (F–G–F) [149,152,153].

A model for the biosynthesis of GFn- and Fn-type inulins by 1-SST and 1-FFT in the vacuole suggests that sucrose is imported and/or synthesized in the cytosol. Inulin biosynthesis occurs in the vacuole by 1-SST and 1-FFT. 1-SST produces G and

1-kestose (G–F–F) from G–F. Besides 1-kestose, 1-FFT can also use higher degree of polymerization GFn-type inulins as donor and acceptor. Inulo-*n*-oses are biosynthesized by 1-FFT using F as an acceptor in the first step.

Applications Inulin forms a white, crystalline powder that is as sweet as sucrose. The inulin molecule is a small, inert polysaccharide that readily passes through the digestive system and remains neutral to cellular activity. Because it is not absorbed by the body, it is used to sweeten foods consumed by diabetic patients. In fact, it has also been part of daily human diet for several centuries.

Its flavor ranges from bland to subtly sweet (approximately 10% sweetness of sugar/sucrose) so that it can be used to replace sugar, fat, and flour (See http://www.functionalingredientsmag.comn.). This is particularly advantageous because inulin contains a quarter to a third of the food energy of sugar or other carbohydrates and a ninth to a sixth of the food energy of fat.

Inulin increases calcium absorption and possibly magnesium absorption, while promoting the growth of intestinal bacteria [154–156]. It is considered a form of soluble fiber and is sometimes categorized as a prebiotic.

Due to the body's limited ability to process polysaccharides, inulin has minimal increasing impact on blood sugar, and – unlike fructose – is not insulinemic and does not raise triglycerides, making it suitable for diabetics and potentially helpful in managing blood sugar-related illnesses [156].

It is present in many regularly consumed vegetables, fruits, and cereals, including leek, onion, garlic, wheat, chicory, artichoke, and banana. Industrially, inulin is obtained from chicory roots, and is used as a functional food ingredient that offers a unique combination of interesting nutritional properties and important technological benefits.

3.3.3
Other: Gums (Guar Gum, Gum Arabic, Gum Karaya, Gum Tragacanth, and Locust Bean Gum)

The plant gums are generally polysaccharide or polysaccharide derivatives that hydrate in hot or cold water to form viscous solutions or dispersants [157]. Gums are generally used in foods, pharmaceuticals, and material science as gels, emulsifiers, adhesives, flocculants, binders, films, and lubricants. Guar gum is a seed extract containing mannose with galactose branches every second unit. It is used in food, papermaking, and petroleum production. Gum arabic is a plant exudate and consists of a highly branched polymer of galactose, rhamnose, arabinose, and glucuronic acid. Gum arabic is used in food and pharmaceutical industries and in printing inks. Gum karaya is a plant exudate and consists of galactose, rhamnose, and partially acetylated glucuronic acid repeat units. Gum karaya is used in food and pharmaceutical industries and in paper and printing. Gum tragacanth is used in foods, cosmetics, printing, and pharmaceuticals. Locust bean gum is a seed extract that consists of mannose along the main chain with galactose branches every fourth unit. This gum is used in foods, papermaking, textile sizing, and cosmetics [157].

3.4
Isoprene Biopolymers: Natural Rubber

Polyisoprenoids represent one typical class of biopolymers produced by living organisms and exhibit unique structures and features. Natural rubber is one example of polyisoprenoids and one of the most familiar and easily available biopolymers. Because of its valuable characteristics, natural rubber is the most abundantly used biopolymer, together with cellulose, lignin, and starch. Microbiologists, biochemists, and molecular geneticists have extensively studied its structure, biosynthesis, and degradation since the end of the nineteenth century and efforts are being made to improve its production.

3.4.1
cis-Polyisoprene

3.4.1.1 Occurrence
Natural rubber is a strategically important raw material produced by plants, used in thousands of products, including hundreds of medical devices. Natural rubber was originally derived from latex, a milky colloid produced by some plants [158]. Latex containing appreciable quantities of rubber occurs in certain plant species belonging to the Moraceae, Euphorbiaceae, Apocynaceae, and Compositae families. Although more than 2000 species are known to produce natural rubber, currently there are two important commercial sources, *Hevea brasiliensis* (the Brazilian rubber tree) and *Parthenium argentatum* Gray (guayule) [158].

The rubber latex from many species is harvested by tapping the rubber tree, that is, making an incision in the trunk and collecting the sap freely oozing out the ducts. The raw polymer is recovered from the latex by coagulation and drying, yielding high molecular weight (1 million g/mol).

3.4.1.2 Composition, Structure, and Properties
Despite extensive research, the exact nature of natural rubber is still unknown. Early X-ray diffraction studies showed that the purified form of natural rubber is *cis*-polyisoprene, which can also be produced synthetically [159].

Chemically, *cis*-polyisoprene is a linear long-chain polymer of repeating isoprene (2-methyl-1,3-diene) units with a specific gravity of 0.93 at 20 °C [158] (Figure 3.14).

Later, Tanaka, by using ^{1}H NMR and ^{13}C NMR spectroscopy, showed that the second and third units of *Hevea* rubber are *trans*, followed by the repetitive *cis* enchainment [160]. The terminal groups are believed to be CH_2OH (or a fatty acid ester) (Figure 3.14). The presence of cyclized polyisoprene sequences in natural rubber was also detected. Other chemical groups, termed "abnormal," were also identified (aldehydes, epoxides, and amines); however, their origin remains unknown [160].

Natural rubber is soft, sticky, and thermoplastic (a polymer that turns to a liquid when heated and freezes to a very glassy state when cooled sufficiently). It has low tensile strength and low elasticity, but these properties can be dramatically altered by cross-linking the polymer chains. This process, carried out with sulfur, is known as

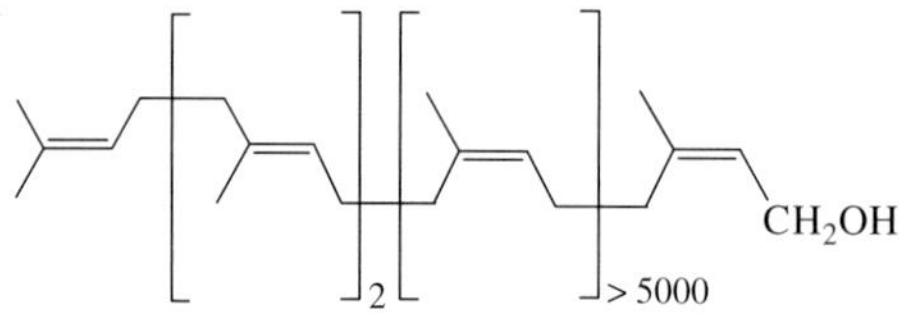

Figure 3.14 Structure of natural rubber: *cis*- and *trans*-1,4-polyisoprene.

"vulcanization" [161]. The key chemical modification of this process is that sulfide bridges are created between adjacent chains (Figure 3.15).

This cross-linking makes rubber nonsticky and improves its tensile strength. The material is no longer thermoplastic. Vulcanized natural rubber exhibits an excellent combination of properties, including elasticity, resilience, abrasion resistance, efficient heat dispersion, and impact resistance. These attractive physical properties of vulcanized rubber have revolutionized its applications [158d,[161].

Natural rubber exhibits other interesting physical and chemical properties. For example, rubber strain crystallizes and owing to the presence of a double bond in each repeat unit, it is sensitive to ozone cracking [162]. Besides this, there are two main solvents for rubber: turpentine and naphtha (petroleum). Despite this, rubber does not dissolve easily, so the material must be finely divided by shredding prior to its immersion. An ammonia solution can be used to prevent the coagulation of raw latex while it is being transported from its collection site (See http://www.npi.gov.au/substances/ammonia/index.html.).

3.4.1.3 *cis*-1,4-Polyisoprene Biosynthesis

Although efforts have been made by generations of polymer scientists to produce high *cis*-1,4-polyisoprene in laboratory and on industrial scale, at the present no synthetic *cis*-1,4-polyisoprene is able to mimic the performance of natural rubber [163a,b]. In fact, synthetic high *cis*-polyisoprene has inferior properties (e.g., tensile strength, tack, etc.) [163c]. This is mainly due to the 100% *cis* microstructure, together with high molecular weights and broad molecular weight distributions of the *cis*-polyisoprene, and the presence of proteins and the so-called abnormal groups [160a,163d,e]. This renders natural rubber an essential renewable resource material.

The *in vitro* biosynthesis of natural rubber has been demonstrated, albeit at the milligram scale. The elucidation of natural rubber biosynthesis in terms of polymer chemical principles may lead to a synthetic strategy for 100% *cis*-polyisoprene [164].

The biosynthesis of natural rubber *in vivo* is catalyzed by a rubber transferase enzyme (EC 2.5.1.20, *cis*-prenyltransferase) bound to rubber particles in the latex serum [158b,165–167]. The monomer is isopentenyl pyrophosphate (IPP).

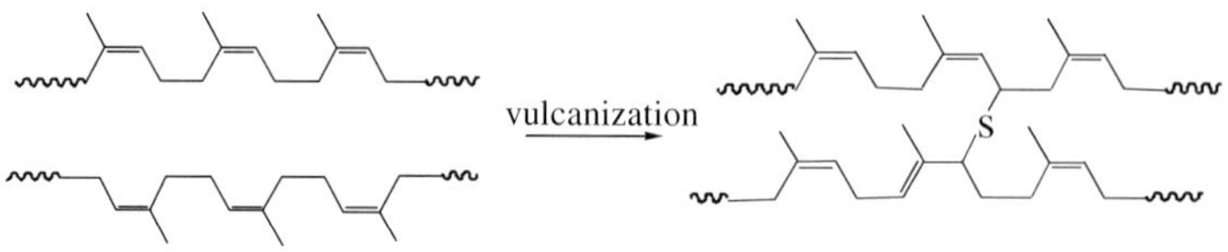

Figure 3.15 Schematic representation of the cross-link involved in the vulcanization process (in reality multiple cross-links are formed in the three dimensions).

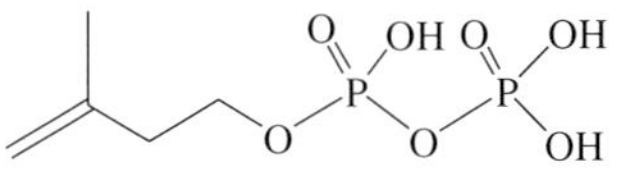

Figure 3.16 Structure of isopentenyl phosphate IPP at pH 7.

Figure 3.16 shows the structure of IPP, an adduct of pyrophosphoric acid ($H_4P_2O_7$), hereafter abbreviated as HPP, and isoprene (IP).

It is generally assumed that sugars are utilized as the main source of carbon to rubber formation (a pathway that was thought to involve at least 17 steps from simple sugar). In this context, the conversion from acetate to IPP has been found in *Hevea* latex serum and guayule extracts. In these cases, a supply of ATP is required for these reactions, and glycolysis is generally assumed to account for this [168].

Polymerization takes place within the boundary phospholipid biomembrane monolayer, at the active sites of the rubber transferase enzyme, and between the nonpolar rubber particles and the aqueous medium [169,170]. The rubber transferase enzyme contains glycosylated hydrophilic regions that mediate the access of incoming hydrophilic building blocks and hydrophobic regions that mediate their placement into the membrane.

The monomer, IPP, which is associated *in vivo* with monovalent cations (K^+, Na^+, or NH_4^+), is isomerized to 1,1-dimethylallyl pyrophosphate (DMAPP) by IPP isomerase. This allylic diphosphate initiates the chain growth. On the other hand, specific *trans*-prenyltransferases catalyze the reaction between DMAPP and 1–3 IPP units to form oligomeric allylic pyrophosphates (APPs), all of which may function as initiators [171]. Figure 3.17 shows the structure of the allylic oligoisoprene pyrophosphates.

IPP and APPs are termed "substrate" and "cosubstrate," respectively, in the biochemical literature, whereas these entities would be termed as "monomer" and "initiator" by polymer chemical terminology. Other cosubstrates/initiators include oligomers of DMAPP such as geranyl pyrophosphate (GPP), farnesyl pyrophosphate (FPP), and geranylgeranyl pyrophosphate (GGPP). Enzymatic activity requires the presence of divalent cations such as Mg^{2+} or Mn^{2+}, called activity cofactors (see Figure 3.17) [165,166b,172].

The substrate, cosubstrates, and cofactors are hydrophilic, while the rubber product is hydrophobic. The enzyme is amphiphilic and as previously stated is located at the interface between the rubber particles and the aqueous phase of the latex. Figure 3.18 shows a flow diagram of natural rubber biosynthesis from isopentenyl diphosphate as reproduced from the biochemical literature [171a].

As can be deduced from the figure, the rubber transferase *cis*-prenyltransferase binds an allylic diphosphate to initiate a new rubber molecule and then elongates the polymer by *cis*-1,4-polymerization of isopentenyl units derived from IPP. Allylic PPs larger than DMAPP such as geranyl diphosphate or farnesyl diphosphate can be used as initiators by rubber transferase, as well as geranylgeranyl diphosphate. All of them are synthesized by soluble *trans*-prenyltransferases.

In this process, one molecule of pyrophosphoric acid (HPP or its salts) is generated in each step; that is, the process is a combination of chain growth and polycondensation, as suggested by Yokozawa et al. [173]. Moreover, all reactions shown in the figure require a divalent cation cofactor such as magnesium.

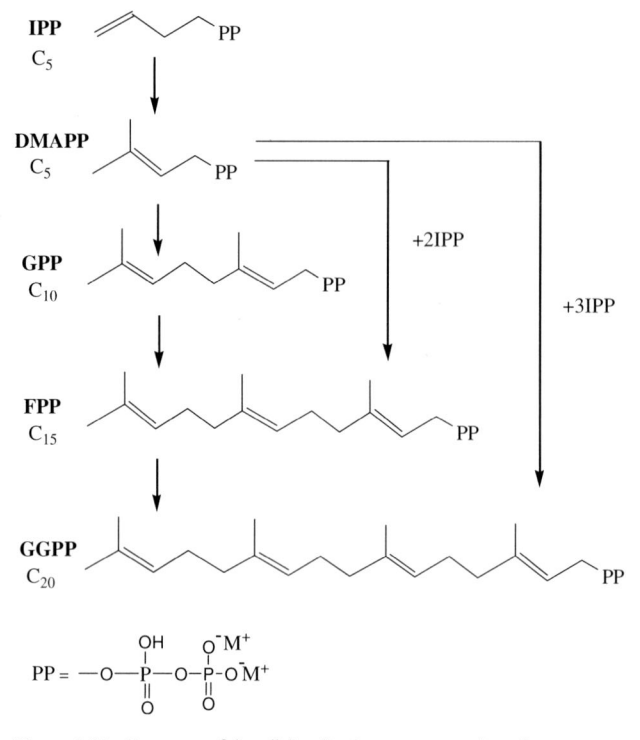

Figure 3.17 Structure of the allylic oligoisoprene pyrophosphate initiators.

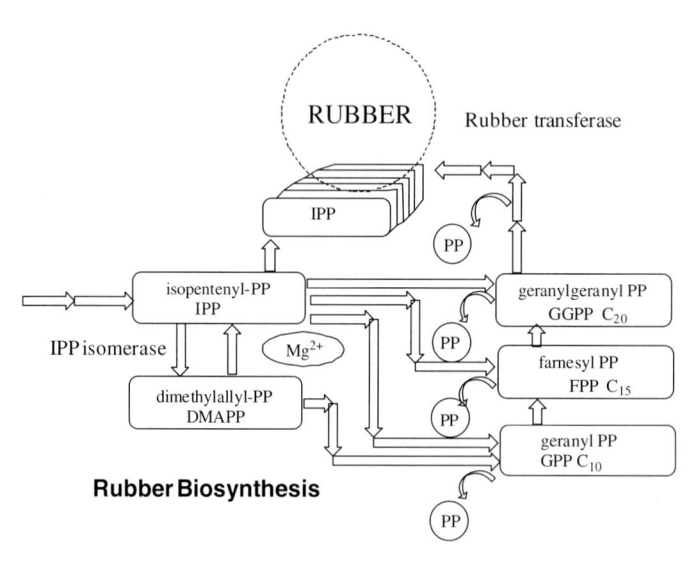

Figure 3.18 Flow diagram of natural rubber biosynthesis as represented. (Reproduced from the biochemical literature [164a.])

It is interesting to note that molecular weights and molecular weight distributions are species dependent; however, little is known about the control of these parameters *in vivo*. In this regard, and according to the polymer chemical literature, broad/multimodal molecular weight distributions are due to branching and/or cross-linking by acid-catalyzed cyclization, or by abnormal functional groups (aldehydes, epoxides, and amines) [160a,174].

The Natural Rubber Biosynthesis: A Natural Living Carbocationic Polymerization (NLCP) Mechanism? While the basic steps of natural rubber biosynthesis are known from biochemical point of view, the understanding of the process in terms of synthetic polymer chemical principles is still incompletely understood. The structures of the intermediates involved in this process are consistent with a carbocationic polymerization mechanism and accordingly it has been proposed that the elementary steps of natural rubber biosynthesis can be described in terms of initiation, propagation, and termination, that is, a natural living carbocationic polymerization (Figure 3.19) [175–178].

According to the polymer chemical terminology, the different allylic pyrophosphates are initiators, the IPP is the monomer, and the rubber transferase in association with the divalent cation cofactors is the coinitiator.

Initiation The polymerization starts by an *initiation*, which involves two events: (a) assisted ionization of the carbon–oxygen bond of the initiator (e.g., GPP, etc.) yielding an allylic cation (resonance stabilized primary/tertiary cation) plus pyrophosphate counteranion and (b) coordination of the enzyme plus cofactor and pyrophosphate "protecting group" to give a carbocation (Figure 3.19).

The ionization at the chain end is favored by resonance stabilization of the allylic carbocation and the increasing entropy of the system (see Figure 3.19) [175–178]. This mechanism is unlikely with the IPP monomer because the cleavage of the carbon–oxygen bond would lead to an energetically unfavorable primary cation. Thus, the vinylidene group of IPP adds to the allylic carbocation yielding a tertiary carbocation that via proton elimination regenerates the trisubstituted allylic pyrophosphate. This mechanism applies to the formation of *trans*-1,1-dimethylallylic initiators, catalyzed by *trans*-prenyltransferase, as well as to the incorporation of the first *cis*-unit, that is, initiation, catalyzed by *cis*-prenyltransferase.

In natural rubber biosynthesis, the ionization is mediated by specific transferases assisted by inorganic cofactors.

Propagation Propagation is a repetitive cation formation/IPP addition/proton loss sequence as occurs in initiation [179]. So, the allylic terminus is ionized (or activated) by Y, and then it reacts with the incoming monomer to give an intermediate tertiary carbocation, which via HX loss regenerates exclusively the allylic chain end (Figure 3.19). This sequence of events sustains propagation by the same mechanism that prevails in initiation. Thus, NLCP can be viewed as a controlled/living carbocationic polymerization process that keeps a dynamic equilibrium between different chemical species (dormant and active species) combined with a controlled

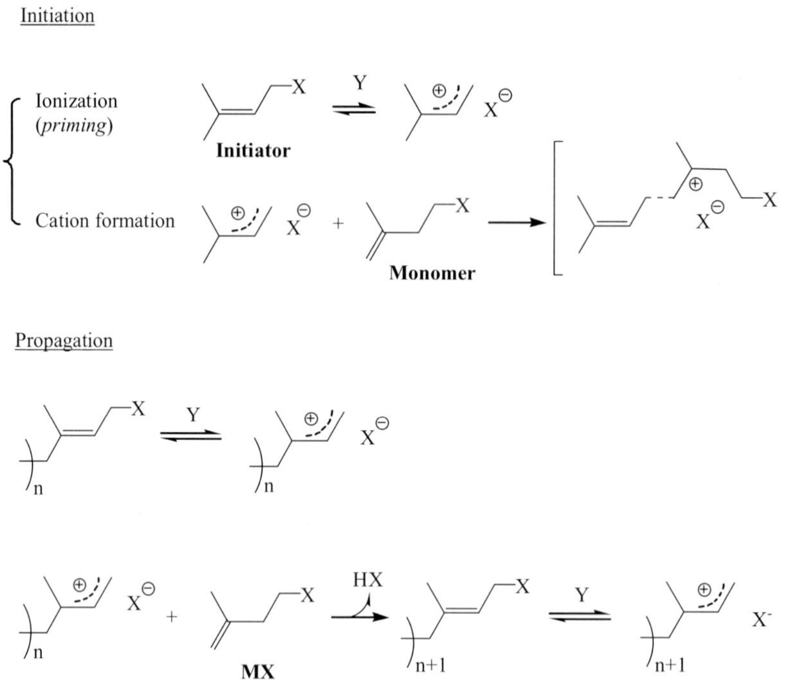

Initiation

Ionization (priming) / *Cation formation*

Figure 3.19 Proposed natural living carbocationic polymerization mechanism [158b, 179a].

loss of HX. Importantly, the HX loss after each monomer incorporation must regenerate exclusively the dormant allylic chain end (Figure 3.19) [180].

Termination Termination in isoprenoid biosynthesis was proposed to be due to the position of specific large motifs (see Figure 3.19) that stopped the chain growth at a specific length [180]. This is an attractive suggestion for the formation of short isoprenoids, but not for *Hevea* rubber that has ~5000 repeat units. With increasing chain length, the natural rubber molecule becomes increasingly hydrophobic and extends beyond the enzyme pocket; thus, chain length regulation by this mechanism is inconceivable.

According to the biochemical literature, chain termination occurs when the enzyme puts aside the rubber molecule. If this detachment (termination) occurs at a specific chain length regulated by steric factors, uniform natural rubber would arise [180]. However, natural rubber exhibits broad molecular weight distribution. The proposed chain activation involving the enzyme and cofactors implies that the enzyme is released after each propagation step. Evidently, the polymerization rate is governed by the rate of IPP addition relative to that of activation of the dormant rubber molecule. The random activation and migration/docking of the rubber molecule to the activating site of the enzyme lead to broad molecular weight distributions.

The natural rubber molecule contains OH end groups, which most likely arise by hydrolytic cleavage of terminal polymer–pyrophosphate linkages. The chain end may also react with fatty acids, which leads to ester end groups. It is unknown whether hydrolytic termination occurs *in vivo* or during processing. At physiological pH, the pyrophosphate end group is a quite stable dianion [181]. It has been argued that termination by hydrolysis does not occur *in vivo*, because this would prevent the formation of high molecular weight rubber. Specific molecular weight regulation depending on the species is plausible due to physical factors, such as the size of the latex particle and/or stabilization by specific proteins in the plant.

3.4.1.4 Applications

The use of rubber is widespread, ranging from household to industrial products, entering the production stream at the intermediate stage or as final products. Tires and tubes are the largest consumers of rubber. The remaining 44% is taken up by the general rubber goods (GRG) sector, which includes all products, except tires and tubes [178].

Other significant uses of rubber are door and window profiles, hoses, belts, matting, flooring, and dampeners (antivibration mounts) for the automotive industry in what is known as the "under the bonnet" products (See http://unctad.org/infocomm/anglais/rubber/uses.htm.). Gloves (medical, household, and industrial) and toy balloons are also large consumers of rubber, although the type of rubber used is that of the concentrated latex. Significant tonnage of rubber is used as adhesives in many manufacturing industries and products, although the two most noticeable are the paper and the carpet industries. Rubber is also commonly used to make rubber bands and pencil erasers. Many aircraft tires and inner tubes are still made of natural rubber due to the high cost of certification for aircraft use of synthetic replacements [182].

In addition, rubber produced as a fiber, sometimes called *elastic*, has significant value for use in the textile industry because of its excellent elongation and recovery properties [182]. For these purposes, manufactured rubber fiber is made as either extruded round fibers or rectangular fibers that are cut into strips from the extruded film. Because of its low dye acceptance, feel, and appearance, the rubber fiber is either covered by yarn of another fiber or directly woven with other yarns into the fabric. In the early 1900s, for example, rubber yarns were used in foundation garments [182].

While rubber is still used in textile manufacturing, its low tenacity limits its use in lightweight garments because latex lacks resistance to oxidizing agents and is damaged by aging, sunlight, oil, and perspiration. Seeking a way to address these shortcomings, the textile industry has turned to neoprene (polymer form of chloroprene), a type of synthetic rubber, as well as another more commonly used elastomer fiber, spandex (also known as elastane), because of their superiority to rubber in both strength and durability [183].

3.4.2
trans-Polyisoprene

Some species (e.g., *Palaquium gutta* trees) produce an inelastic natural latex called gutta-percha, which is composed of *trans*-1,4-polyisoprene, a structural isomer of lower molecular weight, which has similar, but not identical properties to *cis*-polyisoprene [184]. The similar more resinous balata latex is produced from *Mimusops balata*. The term gutta usually refers to the *trans* hydrocarbon from either source.

Finally, chicle, a mixture of low molecular weight *cis*- and *trans*-polyisoprenes in the approximate relative proportion of 1 : 2, together with acetone soluble resins, is obtained from *Achras sapota*, a tropical American fruit tree principally from Yucatán, Guatemala, and other regions of Central America. Chicle is obtained as pinkish to reddish brown pieces and as previously stated it contains both rubber and gutta-percha.

3.5
Concluding Remarks

Biopolymers from plants are biomass-based raw materials that are abundant in nature and important for life, with essential or beneficial functions such as storage of carbon, nutrients, and energy. In addition, many biopolymers are structural components of cells, tissues, and whole organisms. Besides this, they have interesting properties for diverse applications. In fact, the possibility to produce them from renewable resources makes biopolymers interesting candidates to industry. Indeed, due to the growing environmental consciousness, they are becoming a real alternative to petroleum-based polymers.

It is foreseeable that in the not too distant future biopolymers will replace the traditional plastics in an increasing number of applications. Moreover, it is likely that plants will be engineered in order to produce biopolymers in economically viable quantities, or even by modifying enzymes plants will produce novel materials with improved properties.

Basic and applied research has already revealed much information on the enzymatic systems involved in biosynthesis and degradation as well as on the properties of biopolymers, herein our interest in collecting these data series in this chapter. Thus, this chapter provides a thorough overview of the structure, occurrence, properties, biosynthesis, and technical applications of the most representative biopolymers generated by plants: lignin, suberin, cutin, cellulose, hemicellulose, pectin, starch, inulin, gums, and natural rubber.

We hope that this chapter will be helpful to many biologists, chemists, scientists, and other experts in a wide variety of different disciplines, in academia and in industry.

Acknowledgments

Financial support by Consolider-Ingenio 2010 (Project MULTICAT), Spanish MICINN (Project MAT2011-28009) and Generalitat Valenciana (Project PROMETEO/2008/130) are gratefully acknowledged.

References

1 Plackett, D. (ed.) (2011) *Biopolymers: New Materials for Sustainable Films and Coatings*, John Wiley & Sons, Ltd, Chichester, UK.

2 Steinbüchel, A. (ed.) (2003) *Biopolymers: General Aspects and Special Applications*, vol. **10**, Wiley-VCH Verlag GmbH, Weinheim.

3 Kasapis, S., Norton, I.T., and Ubbink, J.B. (eds) (2009) *Modern Biopolymer Science: Bridging the Divide Between Fundamental Treatise and Industrial Application*, Academic Press, London, UK.

4 Sharma, S.K., Ackmez Mudhoo, J.H.C., and Kraus, G.A. (eds) (2011) *A Handbook of Applied Biopolymer Technology: Synthesis, Degradation and Applications*, RSC Green Chemistry, Royal Society of Chemistry.

5 Steinbüchel, A. and Marchessault, R.H. (eds) (2005) *Biopolymers for Medical and Pharmaceutical Applications: Humic Substances, Polyisoprenoids, Polyesters, and Polysaccharides*, 1st edn, Wiley-VCH Verlag GmbH, Weinheim.

6 Kaplan, D.L. (ed.) (1998) *Biopolymers from Renewable Resources (Macromolecular Systems – Materials Approach)*, Springer, Berlin.

7 Gross, R.A. and Scholz, C. (eds) (2001) *Biopolymers from Polysaccharides and Agroproteins*, Oxford University Press, USA.

8 Steinbüchel, A. and Doi, Y. (eds) (2005) *Biotechnology of Biopolymers: From Synthesis to Patents*, Wiley-VCH Verlag GmbH, Weinheim.

9 Mohanty, A.K., Manjusri Misra, M., and Drzal, L.T. (eds) (2005) *Natural Fibers, Biopolymers, and Biocomposites*, CRC Press, Boca Raton, FL.

10 Vandamme, E.J., De Baets, S., and Steinbüchel, A. (eds) (2002) *Polysaccharides II: Polysaccharides from Eukaryotes, Biopolymers*, vol. **6**, 1 edn, Wiley-VCH Verlag GmbH, Weinheim.

11 Riederer, M. and Müller, C. (eds) (2006) *Biology of the Plant Cuticle*, Blackwell, Oxford.

12 Walton, T.J. (1990) Waxes, cutin and suberin. *Methods Plant Biochem.*, **4**, 105–158.

13 Kolattukudy, P.E. (1996) Biosynthetic pathways of cutin and waxes, and their sensitivity to environmental stresses, in *Plant Cuticles: An Integrated Functional Approach* (ed. G. Kerstein), BIOS Scientific Publishers, Oxford, pp. 83–108.

14 Kolattukudy, P.E. (2001) Polyesters in higher plants. *Adv. Biochem. Eng. Biotechnol.*, **71**, 1–49.

15 Nip, M., Tegelaar, E.W., de Leeuw, J., Schenck, P.A., and Holloway, P.J. (1986) A new non-saponifiable highly aliphatic and resistant biopolymer in plant cuticles. Evidence from pyrolysis and ^{13}C-NMR analysis of present-day and fossil plants. *Naturwissenschaften*, **73**, 579–585.

16 Kolattukudy, P., Rogers, L., and Larson, J.D. (1981) Enzymatic reduction of fatty acids and α-hydroxy fatty acids. *Methods Enzymol.*, **71**, 263–275.

17 Pinot, F., Salaun, J.P., Bosch, H., Lesot, A., Mioskowski, C., and Durst, F. (1992) Omega-hydroxylation of Z9-octadecenoic, Z9,10-epoxystearic and 9,10-dihydroxystearic acids by microsomal cytochrome P450 systems from *Vicia sativa*. *Biochem. Biophys. Res. Commun.*, **184**, 183–193.

18 Pinot, F., Bosch, H., Salaün, J.P., Durst, F., Mioskowski, C., and Hammock, B.D. (1997) Epoxide hydrolase activities in the microsomes and the soluble fraction from *Vicia sativa* seedlings. *Plant Physiol. Biochem.*, **35**, 103–110.

19 Pinot, F., Benveniste, I., Salaün, J.P., Loreau, O., Noel, J.P., Schreiber, L., and Durst, F. (1999) Production *in vitro* by the cytochrome P450 CYP94A1 of major C18 cutin monomers and potential messengers in plant–pathogen interactions: enantioselectivity studies. *Biochem. J.*, **342**, 27–32.

20 Pollard, M., Beisson, F., Li, Y.H., and Ohlrogge, J.B. (2008) Building lipid barriers: biosynthesis of cutin and suberin. *Trends Plant Sci.*, **13**, 236–246.

21 Douliez, J.P., Barrault, J., Jerome, F., Heredia, A., Navailles, L., and Nallet, F. (2005) Glycerol derivatives of cutin and suberin monomers: synthesis and self-assembly. *Biomacromolecules*, **6**, 30–34.

22. Heredia-Guerrero, J.A., Benítez, J.J., and Heredia, A. (2008) Self-assembled polyhydroxy fatty acids vesicles: a mechanism for plant cutin synthesis. *Bioessays*, **30**, 273–277.
23. Heredia, A. (2003) Biophysical and biochemical characteristics of cutin, a plant barrier biopolymer. *Biochem. Biophys. Acta*, **1620**, 1–7.
24. Domínguez, E., Heredia-Guerrero, J.A., and Heredia, A. (2011) The biophysical design of plant cuticles: an overview. *New Phytol.* doi: 10.1111/j.1469-8137.2010.03553.x.
25. Villena, J.F., Domínguez, E., Stewart, D., and Heredia, A. (1999) Characterization and biosynthesis of non-degradable polymers in plant cuticles. *Planta*, **208**, 181–187.
26. Tegelaar, E.W., Kerp, H., Visscher, H., Schenck, P.A., and de Leeuw, J.W. (1991) Bias of the paleobotanical record as a consequence of variation in the chemical composition of higher vascular plant cuticles. *Paleobiology*, **17**, 133–144.
27. Villena, J.F., Casado, C.G., Luque, P., and Heredia, A. (1996) Structural characteristics of a resistant biopolymer isolated from plant cuticles. *J. Exp. Bot.*, **47**, 56.
28. Benítez, J.J., Heredia-Guerrero, J.A., and Heredia, A. (2007) Self-assembly of carboxylic acids and hydroxyl derivatives on mica. A qualitative AFM study. *J. Phys. Chem. C*, **111**, 9465–9470.
29. Benítez, J.J., Heredia-Guerrero, J.A., Serrano, F.M., and Heredia, A. (2008) The role of hydroxyl groups in the self-assembly of long chain alkylhydroxyl carboxylic acids on mica. *J. Phys. Chem. C*, **112**, 16968–16972.
30. Heredia, A., Heredia-Guerrero, J.A., Domínguez, E., and Benítez, J.J. (2009) Cutin synthesis: a slippery paradigm. *Biointerphases*, **4**, 1–3.
31. Heredia-Guerrero, J.A., San-Miguel, M.A., Sansom, M.S.P., Heredia, A., and Benítez, J.J. (2009) Chemical reactions in 2D: self-assembly and self-esterification of 9(10),16-dihydroxypalmic acid on mica surface. *Langmuir*, **25**, 6869–6874.
32. Domínguez, E., Heredia-Guerrero, J.A., Benítez, J.J., and Heredia, A. (2010) Self-assembly of supramolecular lipid nanoparticles in the formation of plant biopolyester cutin. *Mol. Biosyst.*, **6**, 948–950.
33. Hill, C. (2006) *Wood Modification: Chemical, Thermal and Other Processes*, John Wiley & Sons, Ltd, Chichester, UK.
34. Mohanty, A.K.M., Drzal, L.T., Selke, S.E., Harte, B.R., and Hinrichsen, G. (2005) Natural fibers, biopolymers, and biocomposites: an introduction, in *Natural Fibers, Biopolymers, and Biocomposites* (eds A.K.M. Mohanty and L.T. Drzal), Taylor & Francis Group, pp. 1–36.
35. Li, K. (1996) Synthesis of lignin–carbohydrate model compounds and neolignans. Department of Wood Science and Forestry Products, Virginia Polytechnic Institute and State University, Blacksburg, VA.
36. Sarkanen, K.V. and Ludwig, C.H. (eds) (1971) *Lignins: Occurrence, Formation, Structure and Reactions*, Wiley-Interscience, New York.
37. Nimz, H.H. and Ludemann, H.D. (1976) Carbon 13 nuclear magnetic resonance spectra of lignins: lignin and DHP acetates. *Holzforschung*, **30**, 33–40.
38. Terashima, N., Atalla, R.H., Ralph, S.A., Landucci, L.L., Lapierre, C., and Monties, B. (1995) New preparations of lignin polymer models under conditions that approximate cell well lignification. I. Synthesis of novel lignin polymer models and their structural characterization by ^{13}C NMR. *Holzforschung*, **49**, 521–527.
39. Chen, C.L. (1998) Characterization of milled wood lignins and dehydrogenative polymerisates from monolignols by carbon-13 NMR spectroscopy, in *Lignin and Lignan Biosynthesis* (eds N.G. Lewis and S. Sarkanen), American Chemical Society, Washington, DC, pp. 255–275.
40. Freudenberg, K. (1964) The formation of lignin in the tissue and *in vitro*, in *The Formation of Wood in Forest Trees* (ed. H.M. Zimmermann), Springer, New York, pp. 203–218.
41. Harkin, J.M. (1967) Lignin: a natural polymeric product of phenol oxidation, in *Oxidative Coupling of Phenols* (eds W.I.

Taylor and A.R. Battersby), Marcel Dekker, New York, pp. 243–321.
42. Freudenberg, K. and Neish, A.C. (eds) (1968) *Constitution and Biosynthesis of Lignin*, Springer, Berlin.
43. Adler, E. (1977) Lignin chemistry: past, present and future. *Wood Sci. Technol.*, **11**, 169–218.
44. Lewis, N.G. and Davin, L.B. (1998) The biochemical control of monolignol coupling and structure during lignan and lignin biosynthesis, in *Lignin and Lignan Biosynthesis* (eds N.G. Lewis and S. Sarkanen), American Chemical Society, Washington, DC, pp. 334–361.
45. Davin, L.B. and Lewis, N.G. (2000) Dirigent proteins and dirigent sites explain the mystery of specificity of radical precursor coupling in lignan and lignin biosynthesis. *Plant Physiol.*, **123**, 453–461.
46. Kolattukudy, P.E. (2001) Suberin from plants, in *Biopolymers* (eds A. Steinbüchel and Y. Doi), Wiley-VCH Verlag GmbH, Weinheim, pp. 41–73.
47. Sabba, R.P. and Lulai, E.C. (2002) Histological analysis of the maturation of native and wound periderm in potato (*Solanum tuberosum* L.) tuber. *Ann. Bot.*, **90**, 1–10.
48. Bernards, M.A. and Razem, F.A. (2001) The poly(phenolic) domain of potato suberin: a non-lignin cell wall biopolymer. *Phytochemistry*, **57**, 1115–1122.
49. Lulai, E.C. (2007) Skin-set, wound healing and related defects, in *Potato Biology and Biotechnology: Advances and Perspectives* (ed. D. Vreugdenhil), Elsevier, pp. 471–500.
50. Graça, J. and Santos, S. (2007) Suberin: a biopolyester of plants' skin. *Macromol. Biosci.*, **7**, 128–135.
51. Bernards, M.A., Lopez, M.L., Zajicek, J., and Lewis, N.G. (1995) Hydroxycinnamic acid-derived polymers constitute the polyaromatic domain of suberin. *J. Biol. Chem.*, **270**, 7382–7386.
52. Soliday, C.L., Kolattukudy, P.E., and Davis, R.W. (1979) Chemical and ultrastructural evidence that waxes associated with the suberin polymer constitute the major diffusion barrier to water vapor in potato tuber (*Solanum tuberosum* L.). *Planta*, **146**, 607–614.
53. Kolattukudy, P.E. and Dean, B.B. (1974) Structure, gas chromatographic measurement, and function of suberin synthesized by potato tuber tissue slices. *Plant Physiol.*, **54**, 116–121.
54. Lulai, E.C. and Corsini, D.L. (1998) Differential deposition of suberin phenolic and aliphatic domains and their roles in resistance to infection during potato tuber (*Solanum tuberosum* L.) wound-healing. *Physiol. Mol. Plant Pathol.*, **53**, 209–222.
55. Paroschy, J.H., Meiering, A.G., Peterson, R.L., Hostetter, G., and Neff, A. (1980) Mechanical winter injury in grapevine trunks. *Am. J. Enol. Vitic.*, **31**, 227–232.
56. Ryser, U. and Holloway, P.J. (1985) Ultrastructure and chemistry of soluble and polymeric lipids in cell walls from seed coats and fibres of *Gossypium* species. *Planta*, **163**, 151–163.
57. Bernards, M.A. and Lewis, N.G. (1998) The macromolecular aromatic domain in suberized tissue: a changing paradigm. *Phytochemistry*, **47**, 915–933.
58. Schreiber, L., Franke, R., and Hartmann, K. (2005) Wax and suberin development of native and wound periderm of potato (*Solanum tuberosum* L.) and its relation to periderm transpiration. *Planta*, **220**, 520–530.
59. Kolattukudy, P.E. (2001) Polyesters in higher plants, in *Advances in Biochemical Engineering/Biotechnology* (eds W. Babel and A. Steinbüchel), Springer, Berlin, pp. 1–49.
60. Graça, J. and Pereira, H. (2000) Suberin structure in potato periderm: glycerol, long-chain monomers, and glyceryl and feruloyl dimers. *J. Agric. Food Chem.*, **48**, 5476–5483.
61. Bernards, M.A. (2002) Demystifying suberin. *Can. J. Bot.*, **80**, 227–240.
62. Pereira, H. (2007) *Cork: Biology, Production and Uses*, Elsevier Science, Burlington, MA.
63. Graça, J. and Santos, S. (2006) Glycerol-derived ester oligomers from cork suberin. *Chem. Phys. Lipids*, **144**, 96–107.
64. Kolattukudy, P.E., Kronman, K., and Poulose, A.J. (1975) Determination of structure and composition of suberin from the roots of carrot, parsnip,

rutabaga, turnip, red beet, and sweet potato by combined gas–liquid chromatography and mass spectrometry. *Plant Physiol.*, **55**, 567–573.

65 Holloway, P.J. (1983) Some variations in the composition of suberin from the cork layers of higher plants. *Phytochemistry*, **22**, 495–502.

66 Franke, R., Briesen, I., Wojciechowski, T., Faust, A., Yephremov, A., Nawrath, C., and Schreiber, L. (2005) Apoplastic polyesters in *Arabidopsis* surface tissues – a typical suberin and a particular cutin. *Phytochemistry*, **66**, 2643–2658.

67 Ekman, R. (1983) The suberin monomers and triterpenoids from the outer bark of *Betula verrucosa* Ehrh. *Holzforschung*, **37**, 205–211.

68 Zeier, J., Ruel, K., Ryser, U., and Schreiber, L. (1999) Chemical analysis and immunolocalisation of lignin and suberin in endodermal and hypodermal/rhizodermal cell walls of developing maize (*Zea mays* L.) primary roots. *Planta*, **209**, 1–12.

69 Graça, J. and Pereira, H. (2010) Methanolysis of bark suberins: analysis of glycerol and acid monomers. *Phytochem. Anal.*, **11**, 45–51.

70 Graça, J. (2010) Hydroxycinnamates in suberin formation. *Phytochem. Rev.*, **9**, 85–91.

71 Lopes, M.H., Gil, A.M., Silvestre, A.J., and Neto, C.P. (2000) Composition of suberin extracted upon gradual alkaline methanolysis of *Quercus suber* L. cork. *J. Agric. Food Chem.*, **48**, 383–391.

72 Garbow, J.R., Ferrantello, L.M., and Stark, R.E. (1989) ^{13}C nuclear magnetic resonance study of suberized potato cell wall. *Plant Physiol.*, **90**, 783–787.

73 Lapierre, C., Pollet, B., and Négrel, J. (1996) The phenolic domain of potato suberin: structural comparison with lignins. *Phytochemistry*, **42**, 949–953.

74 Franke, R. and Schreiber, L. (2007) Suberin – a biopolyester forming apoplastic plant interfaces. *Curr. Opin. Plant Biol.*, **10**, 252–259.

75 Bernards, M.A. and Lewis, N.G. (1992) Alkyl ferulates in wound healing potato tubers. *Phytochemistry*, **31**, 3409–3412.

76 Kolattukudy, P.E. (1981) Structure, biosynthesis and biodegradation of cutin and suberin. *Annu. Rev. Plant Physiol.*, **32**, 539–567.

77 Enstone, D.E., Peterson, C.A., and Ma, F. (2003) Root endodermis and exodermis: structure, function and responses to environment. *J. Plant Growth Regul.*, **21**, 335–351.

78 Yan, B. and Stark, R.E. (2000) Biosynthesis, molecular structure, and domain architecture of potato suberin: a ^{13}C NMR study using isotopically labeled precursors. *J. Agric. Food Chem.*, **48**, 3298–3304.

79 Ranathunge, K., Schreiber, L., and Franke, R. (2011) Suberin research in the genomics era. New interest for an old polymer. *Plant Sci.*, **180**, 399–413.

80 (a) Crawford, R.L. (1981) *Lignin Biodegradation and Transformation*, John Wiley & Sons, Inc., New York. (b) Harholt, J., Suttangkakul, A., and Scheller, H.V. (2010) Biosynthesis of pectin. *Plant Physiol.*, **153**, 384–395.

81 Hassid, W.Z. (1944) Transformation of the carbohydrates. *Annu. Rev. Biochem.*, **13**, 59–92.

82 Nishiyama, Y., Langan, P., and Chanzy, H. (2002) Crystal structure and hydrogen-bonding system in cellulose Iβ from synchrotron X-ray and neutron fiber diffraction. *J. Am. Chem. Soc.*, **124** (31), 9074–9082.

83 (a) *Cellulose*. In Encyclopaedia Britannica Online (accessed October 2012). (b) Faure, D. (2002) The family-3 glycoside hydrolases: from housekeeping functions to host–microbe interactions. *Appl. Environ. Microbiol.*, **68** (4), 1485–1490. (c) Guerriero, G., Fugelstad, J., and Bulone, V. (2010) Chitin synthases from *Saprolegnia* are involved in tip growth and represent a potential target for anti-oomycete drugs. *J. Int. Plant Biol.*, **52** (2), 161–175. (d) Arioli, T., Peng, L., Betzner, A.S., Burn, J., Wittke, W., Herth, W., Camilleri, C., Höfte, H., Plazinski, J., Birch, R., Cork, A., Glover, J., Redmond, J., and Williamson, R.E. (1998) Molecular analysis of cellulose biosynthesis in *Arabidopsis*. *Science*, **279**, 717–720. (e) Coutinho, P.M., Deleury, E., Davies, G.J., and Henrissat, B. (2003) An

evolving hierarchical family classification for glycosyltransferases. *J. Mol. Biol.*, **328** (2), 307317.

84 Carpita, N.C. and McCann, M.C. (2010) The maize mixed-linkage (1–3),(1–4)-β-D-glucan polysaccharide is synthesized at the Golgi membrane. *Plant Physiol.*, **153**, 1362.

85 (a) Gibeaut, D.M. and Carpita, N.C. (1993) Synthesis of (1–3),(1–4)-β-D-glucan in the Golgi apparatus of maize coleoptiles. *Proc. Natl. Acad. Sci. USA*, **90**, 1362. (b) Singh, S.K. and Gross, R.A. (2001) Chapter 1, in *Biopolymers from Polysaccharides and Agroproteins, ACS Symposium Series*, American Chemical Society, Washington, DC, pp. 2–40.

86 (a) Gilbert, R.D. and Kadla, J.F. (1998) Polysaccharides – cellulose, in *Biopolymers from Renewable Sources* (ed. D.L. Kaplan), Springer, New York, pp. 47–90. (b) Stevens, M.P. (1999) *Polymer Chemistry*, Oxford University Press, New York, p. 484. (c) Hon, D.N.-S. (1996) in *Polysaccharides in Medicinal Applications* (ed. S. Dumitriu), Marcel Dekker, New York, p. 87. (d) Kauffman, G.B. (1993) Rayon: the first semi-synthetic fiber product. *J. Chem. Educ.*, **70** (11), 887.

87 Amor, Y., Haigler, C.H., Johnson, S., Wainscott, M., and Delmer, D.P. (1995) A membrane-associated form of sucrose synthase and its potential role in synthesis of cellulose and callose in plants. *Proc. Natl. Acad. Sci. USA*, **92** (20), 9353–9357.

88 Guerriero, G., Fugelstad, J., and Bulone, V. (2010) What do really know about cellulose biosynthesis in higher plants? *J. Int. Plant Biol.*, **52** (2), 161.

89 (a) Salnikov, V.V., Grimson, M.J., Delmer, D.P., and Haigler, C.H. (2001) Sucrose synthase localizes to cellulose synthesis sites in tracheary elements. *Phytochemistry*, **57** (6), 823–833. (b) Albrecht, G., and Mustroph, A. (2003) Localization of sucrose synthase in wheat roots: increased *in situ* activity of sucrose synthase correlates with cell wall thickening by cellulose deposition under hypoxia. *Planta*, **217**, 252–260. (c) Persia, D., Cai, G., Del Casino, D., Faleri, C., Willemse, M.T.M., and Cresti, M. (2008) Sucrose synthase is associated with the cell wall of tobacco pollen tubes. *Plant Physiol.*, **147** (4), 1603–1618.

90 Geisler-Lee, J., Geisler, M., Coutinho, P.M., Segerman, B., Nishikubo, N., Takahashi, J., Aspeborg, H., Derbi, S., Master, E., Andersson-Gunneras, S., Sundberg, B., Karpinski, S., Teeri, T.T., Kleczkowski, L.A., Henrissat, B., and Mellerowicz, E.J. (2006) Poplar carbohydrate-active enzymes. Gene identification and expression analyses. *Plant Physiol.*, **140** (3), 946–962.

91 Seifert, G.J. (2004) Nucleotide sugar interconversions and cell wall biosynthesis: how to bring the inside to the outside. *Curr. Opin. Plant Biol.*, **7** (3), 277–284.

92 Klemm, D., Heublein, B., Fink, H.-P., and Bohn, A. (2005) Cellulose: fascinating biopolymer and sustainable raw material. *Angew. Chem., Int. Ed.*, **44** (22), 3358–3393.

93 Peng, L.C., Kawagoe, Y., Hogan, P., and Delmer, D. (2002) Sitosterol-β-glucoside as primer for cellulose synthesis in plants. *Science*, **295** (4), 147–150.

94 Lairson, L.L., Henrissat, B., Davies, G.J., and Withers, S.G. (2008) Glycosyltransferases: structures, functions, and mechanisms. *Annu. Rev. Biochem.*, **77**, 521–555.

95 Imai, T., Watanabe, T., Yui, T., and Sugiyama, J. (2003) The directionality of chitin biosynthesis: a revisit. *Biochem. J.*, **374**, 755–760.

96 Albersheim, P., Darvill, A., Roberts, K., Staehelin, L.A., and Varner, J.E. (1997) Do the structures of cell wall polysaccharides define their mode of synthesis? *Plant Physiol.*, **113**, 1–3.

97 Delmer, D.P. (1999) Cellulose biosynthesis: exciting times for a difficult field of study. *Annu. Rev. Plant Physiol. Plant Mol. Biol.*, **50**, 245–276.

98 Campbell, J.A., Davies, G.J., Bulone, V., and Henrissat, B. (1997) A classification of nucleotide-diphospho-sugar glycosyltransferases based on amino acid sequence similarities. *Biochem. J.*, **326**, 929–939.

99 Collins, P.M. and Ferrier, R.J. (1995) *Monosaccharides. Their Chemistry and*

Their Roles in Natural Products, John Wiley & Sons, Inc., New York, p. 594.

100 (a) Popa, V.I. (1996) in *Polysaccharides in Medicinal Applications* (ed. S. Dumitriu), Marcel Dekker, New York, p. 107. (b) Harholt, J., Suttangkakul, A., and Scheller, H.V. (2010) Biosynthesis of pectin. *Plant Physiol.*, **153**, 384–395. (c) Scheller, H.V., and Ulskov, P. (2010) Hemicelluloses. *Annu. Rev. Plant. Biol.*, **61**, 263–289.

101 Gregory, A. and Bolwell, G.P. (1999) Hemicellulose biosynthesis, in *Comprehensive Natural Products Chemistry*, vol. 3 (ed. E. Pinto), Academic Press, pp. 599–615.

102 Singh, S.K. and Gross, R. (2001) *Biopolymers from Polysaccharides and Agroproteins*, American Chemical Society, Washington, DC.

103 (a) Vanzin, G.F., Madson, M., Carpita, N.C., Raikhel, N.V., and Keegstra, K. (2002) The *mur2* mutant of *Arabidopsis thaliana* lacks fucosylated xyloglucan because of a lesion in fucosyltransferase AtFUT1. *Proc. Natl. Acad. Sci. USA*, **99** (5), 3340–3345. (b) Vergara, C.E. and Carpita, N.C. (2001) Beta-D-glycan synthases and the CesA gene family: lessons to be learned from the mixed-linkage (1→3),(1→4)beta-D-glucan synthase. *Plant Mol. Biol.*, **47** (12), 145–160.

104 Smith, C.J. (1993) in *Plant Biochemistry and Molecular Biology* (eds P.J. Lea and R.C. Leegoodf), John Wiley & Sons, Ltd, Chichester, UK, pp. 73–111.

105 (a) Green, J.R. and Northcote, D.H. (1978) The structure and function of glycoproteins synthesized during slime-polysaccharide production by membranes of the root-cap cells of maize (*Zea mays*). *Biochem. J.*, **170**, 599–608. (b) Green, J.R. and Northcote, D.H. (1979) Polyprenyl phosphate sugars synthesized during slime-polysaccharide production by membranes of the root-cap cells of maize (*Zea mays*). *Biochem. J.*, **178**, 661–671. (c) Waldron, K.W. and Brett, C.T. (1985) in *Biochemistry of Plant Cell Walls* (eds C.T. Brett and J.R. Hillman), Cambridge University Press, London, p. 79.

106 (a) White, A.R., Xin, Y., and Pezeshk, V. (1993) Xyloglucan glucosyltransferase in Golgi membranes from Pisum sativum (pea). *Biochem. J.*, **294**, 231. (b) White, A.R., Xin, Y., and Pezeshk, V. (1993) Separation of membranes from semiprotoplasts of suspension-cultured sycamore maple (*Acer pseudoplatanus*) cells. *Physiol. Plant.*, **87**, 31. (c) Campbell, R.E., Brett, C.T., and Hillman, J.R. (1988) A xylosyltransferase involved in the synthesis of a protein-associated xyloglucan in suspension-cultured dwarf-French-bean (Phaseolus vulgaris) cells and its interaction with a glucosyltransferase. *Biochem. J.*, **253**, 795. (d) Brickell, L.S. and Grant Reid, J.S. (1995) in *Pectins and Pectinases: Progress in Biotechnology*, vol. 14 (eds J Visser and A.G.J. Voragan), Elsevier, Amsterdam, p. 127. (e) Misawa, H., Tsumuraya, Y., Hashimoto, K., and Hashimoto, Y. (1996) α-L-Fucosyltransferases from Radish Primary Roots. *Plant Physiol.*, **110**, 665. (f) Bolwell, G.P., Dalessandro, G., and Northcote, D.H. (1985) Loss of polygalacturonic acid synthase activity during differentiation in sycamore. *Phytochemistry*, **24**, 699. (g) Gibeaut, D.M. and Carpita, N.C. (1994) Biosynthesis of plant cell wall polysaccharides. *FASEB J.*, **8** (12), 904–915.

107 Bolwell, G.P. (1993) Dynamic aspects of the plant extracellular matrix. *Int. Rev. Cytol.*, **146**, 261–324.

108 (a) Harholt, J., Suttangkakul, A., and Scheller, H.V. (2010) Biosynthesis of pectin. *Plant Physiol.*, **153**, 384–395. (b) Willats, W.G.T., McCatney, L., Mackie, W., and Knox, J.P. (2001) Pectin: cell biology and prospects for functional analysis. *Plant Mol. Biol.*, **47**, 9–27. (c) Hemantaranjan, A. (ed.) (2004) *Advances in Plant Physiology*, vol. 7, Dev Publishers & Distributors, (New Delhi, DEL, India).

109 Ridley, B.L., O'Neill, M.A., and Mohnen, D. (2001) Pectins: structure, biosynthesis, and oligogalacturonide-related signaling. *Phytochemistry*, **57** (6), 929–967.

110 Zablackis, E., Huang, J., Muller, B., Darvill, A.G., and Albersheim, P. (1995) Characterization of the cell-wall polysaccharides of *Arabidopsis thaliana* leaves. *Plant Physiol.*, **107**, 1129–1138.

111 (a) Mohnen, D. (2007) Pectin structure and biosynthesis. *Curr. Opin. Plant Biol.*, **11** (3), 266–277. (b) Mohnen, D., Bar-Peled, L., and Somerville, C. (2008) Cell wall synthesis, in *Biomass Recalcitrance: Deconstruction the Plant Cell Wall for Bioenergy* (ed. M. Himmel), Blackwell Publishing Ltd., pp. 94–187. (c) Bacic, A. (2006) Breaking an impasse in pectin biosynthesis. *Proc. Natl. Acad. Sci. USA*, **103** (15), 5639–5645.

112 Zandleven, J., Beldman, G., Bosveld, M., Schols, H.A., and Voragen, A.G.J. (2007) Xylogalacturonan exists in cell walls from various tissues of *Arabidopsis thaliana*. *Phytochemistry*, **68** (8), 1219–1226.

113 (a) Ryden, P., Sugimoto-Shirasu, K., Smith, A.C., Findlay, K., Reiter, W.D., and McCann, M.C. (2003) Tensile properties of *Arabidopsis* cell walls depend on both a xyloglucan cross-linked microfibrillar network and rhamnogalacturonan II–borate complexes. *Plant Physiol.*, **132**, 1033–1040. (b) Caffall, K.H. and Mohnen, D. (2009) The structure, function, and biosynthesis of plant cell wall pectic polysaccharides. *Carbohydr. Res.*, **344** (14), 1879–1900.

114 (a) D'Ovidio, R., Mattei, B., Roberti, S., and Bellimcampi, D. (2004) Polygalacturonases, polygalacturonase-inhibiting proteins and pectic oligomers in plant–pathogen interactions. *Biochim. Biophys. Acta*, **1696** (2), 237–244. (b) Kohorn, B.D., Johansen, S., Shishido, A., Todorova, T., Martinez, R., Defeo, E., and Obregon, P. (2009) Pectin activation of MAP kinase and gene expression is WAK2 dependent. *Plant J.*, **60** (6), 974–986.

115 Dhugga, K.S., Barreiro, R., Whitten, B., Stecca, K., Hazebroeck, J., Randhawa, G.S., Dolan, M., Kinney, A.J., Tomes, D., Nichols, S., and Anderson, P. (2004) Guar seed β-mannan synthase is a member of the cellulose synthase super gene family. *Science*, **303** (5656), 363–366.

116 Scheller, H.V., Jensen, J.K., Sørensen, S.O., Harholt, J., and Geshi, N. (2007) Biosynthesis of pectin. *Physiol. Plant.*, **129**, 283–295.

117 Villemez, C.L., Lin, T.Y., and Hassid, W.Z. (1965) Biosynthesis of the polygalacturonic acid chain of pectin by a particulate enzyme preparation from *Phaseolus aureus* seedlings. *Proc. Natl. Acad. Sci. USA*, **54** (6), 1626–1632.

118 (a) McNab, J.M., Villemez, C.L., and Albersheim, P. (1968) Biosynthesis of galactan by a particulate enzyme preparation from *Phaseolus aureus* seedlings. *Biochem. J.*, **106**, 355–360. (b) Odzuck, W. and Kauss, H. (1972) Biosynthesis of pure araban and xylan. *Phytochemistry*, **11** (8), 2489–2494. (c) Bolwell, G.P. and Northcote, D.H. (1981) Control of hemicellulose and pectin synthesis during differentiation of vascular tissue in bean (*Phaseolus vulgaris*) callus and in bean hypocotyl. *Planta*, **152** (3), 225–233. (d) Thakur, B.R., Sing, R.K., and Handa, A.K. (1997) Chemistry and uses of pectin – a review. *Crit. Rev. Food Sci. Nutr.*, **37** (1), 47.

119 Roberts, K. (2007) *Handbook of Plant Science*, John Wiley & Sons, Ltd, Chichester, UK.

120 Collins, P. and Ferrier, R. (1995) *Monosaccharides*, John Wiley & Sons, Inc., New York, p. 463.

121 Stevens, M.P. (1999) *Polymer Chemistry*, Oxford University Press, New York, 489 pp.

122 (a) Kaplan, D.L. (1998) in *Biopolymers from Renewable Resources* (ed. D.L. Kaplan), Springer, New York, p. 1. (b) Torrence, R. and Barton, H. (2006) Biology of starch, in *Ancient Starch Research*, Left Coast Press Inc., Walnut Creek, CA.

123 Shogren, R.L. (1998) in *Biopolymers from Renewable Sources* (ed. D.L. Kaplan), Springer, New York, p. 30.

124 Thompson, D.B. (2000) On the non-random nature of amylopectin branching. *Carbohydr. Polym.*, **43** (3), 223–239.

125 (a) Smith, A.M., Denyer, K., and Martin, C. (1997) The synthesis of the starch granule. *Annu. Rev. Plant Physiol. Plant Mol. Biol.*, **48**, 67–87. (b) Ball, S.G., Van de Wal, M.H.B.J., and Visser, R.G.F. (1998) Progress in understanding the biosynthesis of amylose. *Trends Plant. Sci.*, **3** (12), 462–467.

126 Myers, A.M., Morell, M.K., James, M.G., and Ball, S.G. (2000) Recent progress toward understanding biosynthesis of the amylopectin crystal. *Plant Physiol.*, **122** (4), 989–998.

127 Kossmann, J. and Lloyd, J. (2000) Understanding and influencing starch biochemistry. *Crit. Rev. Plant Sci.*, **19** (3), 171–226.

128 Kok-Jacon, G.A., Ji, Q., Vincken, J.-P., and Visser, R.G.F. (2003) Towards a more versatile α-glucan biosynthesis in plants. *J. Plant Physiol.*, **160** (7), 765–777.

129 Kammerer, B., Fischer, K., Hilpert, B., Schubert, S., Gutensoh, M., Weber, A., and Flügge, U. (1998) Molecular characterization of a carbon transporter in plastids from heterotrophic tissues: the glucose 6-phosphate/phosphate antiporter. *Plant Cell*, **10**, 105–118.

130 (a) Tauberger, E., Fernie, A.R., Emmermann, M., Renz, A., Kossmann, J., Willmitzer, L., and Trethewey, R.N. (2000) Antisense inhibition of plastidial phosphoglucomutase provides compelling evidence that potato tuber amyloplasts import carbon from the cytosol in the form of glucose-6-phosphate. *Plant J.*, **23** (1), 43–53. (b) Müller-Röber, B., Sonnewald, U., and Willmitzer, L. (1992) Inhibition of ADP-glucose pyrophosphorylase in transgenic potatoes leads to sugar storing tubers and influences tuber formation and the expression of tuber storage protein genes. *EMBO J.*, **11** (4), 1229–1238. (c) Stark, D.M., Timmerman, K.P., Barry, G.F., Preiss, J., and Kishore, G.M. (1992) Regulation of the amount of starch in plant tissues by ADP glucose pyrophosphorylase. *Science*, **258**, 287–292.

131 Kuipers, A.G.J., Jacobsen, E., and Visser, R.G.F. (1994) Formation and deposition of amylose in the potato tuber starch granule are affected by the reduction of granule-bound starch synthase gene expression. *Plant Cell*, **6**, 43–52.

132 (a) Baba, T., Yoshii, M., and Kainuma, K. (1987) Acceptor molecule of granular-bound starch synthase from sweet-potato roots. *Starch*, **39** (2), 52–56. (b) Denyer, K., Clarke, B., Hylton, C., Tatge, H., and Smith, A.M. (1996) The elongation of amylose and amylopectin chains in isolated starch granules. *Plant J.*, **10** (6), 1135–1143. (c) Van de Wal, M.H.B.J., D'Hulst, C., Vincken, J.-P., Buléon, A., Visser, R.G.F., and Ball, S. (1998) Amylose is synthesized *in vitro* by extension of and cleavage from amylopectin. *J. Biol. Chem.*, **273** (35), 22232–22240.

133 Schwall, G.P., Safford, R., Westcott, R.J., Jeffcoat, R., Tayal, A., Shi, Y.-C., Gidley, M.J., and Jobling, S.A. (2000) Production of very-high-amylose potato starch by inhibition of SBE A and B. *Nat. Biotechnol.*, **18**, 551–554.

134 (a) Mouille, G., Maddelein, M.-L., Libessart, N., Talaga, P., Decq, A., Delrue, B., and Ball, S. (1996) Preamylopectin processing: a mandatory step for starch biosynthesis in plants. *Plant Cell*, **8** (8), 1353–1366. (b) Ball, S., Van de Wal, M.H.B.J., and Visser, R.G.F. (1998) Progress in understanding the biosynthesis of amylose. *Trends Plant Sci.*, **3** (12), 462–467. (c) Denyer, K., Waite, D., Motawia, S., Moller, B.L., and Smith, A.M. (1999) Granule-bound starch synthase I in isolated starch granules elongates malto-oligosaccharides processively. *Biochem. J.*, **340**, 183–191.

135 Denyer, K., Johnson, P., Zeeman, S., and Smith, A.M. (2001) The control of amylose synthesis. *J. Plant Physiol.*, **158**, 479–487.

136 (a) Frank, A. and De Leenheer, L. (2005) Polysaccharides and polyamides, in *The Food Industry. Properties, Production and Patents* (eds A. Steinbüchel and S.K. Rhee), Wiley-VCH Verlag GmbH, Weinheim. (b) Norio, S., Noureddine, B., and Shuichi, O. (eds) (2007) *Recent Advances in Fructooligosaccharides Research*, Research Signpost, Kerala, India.

137 Lewis, D.H. (1993) Nomenclature and diagrammatic representation of oligomeric fructans: a paper for discussion. *New Phytol.*, **124** (4), 583–594.

138 Van Laere, A. and Van den Ende, W. (2002) Inulin metabolism in dicots: chicory as a model system. *Plant Cell Environ.*, **25** (6), 803–813.

139 Pollock, C.J. and Chatterton, N.J. (1988) in *The Biochemistry of Plants*, vol. 14 (ed. J. Preiss), Academic Press, San Diego, CA, pp. 109–139.

140 Bonnet, G.D., Sims, I.M., Simpson, R.J., and Cairns, A.J. (1997) Structural diversity of fructan in relation to the taxonomy of the Poaceae. *New Phytol.*, **136** (1), 11–17.

141 Timmermans, J.W., Slaghek, T., Iizuka, M., De Roover, J., Van Laere, A., and Van den Ende, W.J. (2001) Isolation and structural analysis of new fructans produced by chicory. *Carbohydr. Chem.*, **20**, 375–395.

142 Shiomi, N. (1989) Properties of fructosyltransferases involved in the synthesis of fructan in liliaceous plants. *J. Plant Physiol.*, **134**, 151–155.

143 Chatterton, N.J. and Harrison, P.A. (1993) Fructans in crested wheatgrass leaves. *J. Plant Physiol.*, **160** (8), 843–849.

144 Orthen, B. (2001) Sprouting of the fructan- and starch-storing geophyte *Lachenalia minima*: effects on carbohydrate and water content within the bulbs. *Physiol. Plant.*, **113** (3), 308–314.

145 Pollock, C.J. (1986) Fructans and the metabolism of sucrose in vascular plants. *New Phytol.*, **104** (1), 1–24.

146 Pieron, S., Boxus, P., and Dekegel, D. (1998) Histological study of nodule morphogenesis from *Cichorium intybus* L. leaves cultivated *in vitro*. *In Vitro Cell. Dev. Biol.*, **34** (2), 87–93.

147 Edelman, J. and Jefford, T.G. (1968) The mechanism of fructosan metabolism in higher plants as exemplified in "*Helianthus tuberosus*". *New Phytol.*, **67** (3), 517–531.

148 Cairns, A.J. (1993) Evidence for the *de novo* synthesis of fructan by enzymes from higher plants: a reappraisal of the SST/FFT model. *New Phytol.*, **123** (1), 15–24.

149 Koops, A. and Jonker, H. (1996) Purification and characterization of the enzymes of fructan biosynthesis in tubers of *Helianthus tuberosus* Colombia. II. Purification of sucrose:sucrose 1-fructosyltransferase and reconstitution of fructan synthesis *in vitro* with purified sucrose:sucrose 1-fructosyltransferase and fructan:fructan 1-fructosyltransferase. *Plant Physiol.*, **110**, 1167–1175.

150 Van den Ende, W. and Van Laere, A. (1996) Carbohydrate reserves *in* plants: synthesis and regulation. *Planta*, **200**, 335–342.

151 Hellwege, E.M., Raap, M., Gritscher, D., Wilmitzer, L., and Heyer, H.G. (1998) Differences in chain length distribution of inulin from *Cynara scolymus* and *Helianthus tuberosus* are reflected in a transient plant expression system using the respective 1-FFT cDNAs. *FEBS Lett.*, **427** (1), 25–28.

152 Van den Ende, W., Van Wonterghem, D., Dewil, E., Verhaert, P., De Loof, A., and Van Laere, A. (1996) Carbohydrate Reserves in *Plants*: Synthesis and Regulation. *Physiol. Plant.*, **98**, 455.

153 Heyer, A.G. (1999) Production of modified carbohydrates in transgenic plants. *Bundesgesundheitsblatt*, **43** (2), 94–98.

154 Abrams, S., Griffin, I., Hawthorne, K., Liang, L., Gunn, S., Darlington, G., and Ellis, K. (2005) A combination of prebiotic short- and long-chain inulin-type fructans enhances calcium absorption and bone mineralization in young adolescents. *Am. J. Clin. Nutr.*, **82** (2), 471–476.

155 Coudray, C., Demigné, C., and Rayssiguier, Y. (2003) Effects of dietary fibers on magnesium absorption in animals and humans. *J. Nutr.*, **133** (1), 1–4.

156 (a) Niness, K.R. (1999) Inulin and oligofructose: what are they? *J. Nutr.*, **129** (7), 1402–1406. (b) Kaur, N. and Gupta, A. K. (2002) Applications of inulin and oligofructose in health and nutrition. *J. Biosci.*, **27** (7), 703.

157 Singh, S.K. and Gross, R.A. (2001) Overview: introduction to polysaccharides, agroproteins, and poly(amino acids), in *Biopolymers from Polysaccharides and Agroproteins, ACS Symposium Series* (eds R. Gross *et al.*), American Chemical Society, Washington, DC.

158 (a) Mathew, N.M. (2001) Natural rubber, in *Rubber Technologist's Handbook*, vol. **1** (eds S.K. De and J.R. White), Rapra Technology, 2001, p. 560. (b) Puskas, J.E., Gautriaud, E., Deffieux, A., and Kennedy, J.P. (2006) Natural rubber biosynthesis – a living carbocationic polymerization? *Prog. Polym. Sci.*, **31** (6), 533–548. (c) Punetha, A., Muthukumaran, J., Hemron, A.J., Arumugam, N., Jayakanthan, M., and Sundar, D. (2010) Towards understanding the regulation of rubber biosynthesis: insights into the initiator and elongator enzymes. *J. Bioinf. Seq. Anal.*, **2** (1), 1–10.

(d) Eirich, F.R. (ed.) (1978) *Science and Technology of Rubber*, Academic Press, New York.

159 Nyburg, S.C. (1954) A statistical structure for crystalline rubber. *Acta Crystallogr.*, **7** (5), 385–392.

160 (a) Tanaka, Y. (1989) Structure and biosynthesis mechanism of natural polyisoprene. *Prog. Polym. Sci.*, **14**, 339–371. (b) Tanaka, B., Kawahara, S., Aik-Hwee, E., Shiba, K., and Ohya, N. (1995) Initiation of biosynthesis in *cis*-polyisoprenes. *Phytochemistry*, **39** (4), 779–784.

161 Brydson, J.A. (1978) *Rubber Chemistry*, Applied Science Publishers Ltd, London.

162 (a) Carraher, C.E., Jr. (2000) *Seymour/Carraher's Polymer Chemistry*, 5th edn, revised and expanded, Marcel Dekker, New York. (b) Kauffman, G.B. and Seymour, R.B. (1990) Elastomers. I. Natural rubber. *J. Chem. Educ.*, **67** (5), 422. (c) Lake, G.J. and Mente, P.G. (1993) Ozone cracking and protection of elastomers at high and low temperatures. *J. Nat. Rubber Res.*, **7**, 1.

163 (a) Morton, M. (ed.) (1973) *Rubber Technology*, Van Nostrand Reinhold, New York. (b) Puskas, J.E. (2000) Diene based elastomers, in *Handbook of Elastomers, Plastics Engineering*, vol. 61 (eds A.K. Bhomick and H.L. Stephens), Marcel Dekker, New York, p. 817. (c) Schoenberg, E., Marsh, H.A., Walters, S.J., and Saltman, W.M. (1979) Polyisoprene. *Rubber Chem. Technol.*, **52**, 526. (d) Eng, A.H. and Ong, E.L. (2000) Hevea natural rubber, in *Handbook of Elastomers, Plastics Engineering*, vol. 61 (eds A.K. Bhomick and H.L. Stephens), Marcel Dekker, New York, p. 29. (e) McIntyre, D. and Stephens, H.L. (eds) (2001) *Handbook of Elastomers, Plastics Engineering*, vol. 61, Marcel Dekker, New York, p. 1.

164 (a) Castillon, J. and Cornish, K. (1999) Regulation of initiation and polymer molecular weight of *cis*-1,4-polyisoprene synthesized *in vitro* by particles isolated from *Parthenium argentatum* (Gray). *Phytochemistry*, **51**, 43. (b) Cornish, K., Castillon, J., and Scott, D.J. (2000) Rubber molecular weight regulation, *in vitro*, in plant species that produce high and low molecular weights *in vivo*. *Biomacromolecules*, **1**, 632. (c) Archer, B.L. and Audley, B.G. (1987) New aspects of rubber biosynthesis. *Bot. J. Linn. Soc.*, **94**, 181.

165 Tanaka, Y., Aik-Hwee, E., Ohya, N., Nishiyama, N., Tangpakdee, J., and Kawahara, S. (1996) Initiation of rubber biosynthesis in *Hevea brasiliensis*: characterization of initiating species by structural analysis. *Phytochemistry*, **41** (6), 1501–1505.

166 (a) Cornish, K. (1993) The separate roles of plant *cis* and *trans* prenyl transferase in *cis*-1,4-polyisoprene biosynthesis. *Eur. J. Biochem.*, **218**, 267–271. (b) Cornish, K. and Backhaus, R.A. (1990) Rubber transferase activity in rubber particles of guayule. *Phytochemistry*, **29** (12), 3809–3813. (c) Madhavan, S., Greenblatt, G.A., Foster, M.A., and Benedict, C.R. (1989) Isopentenyl pyrophosphate incorporation into polyisoprene in extracts from *Guayule*. *Plant Physiol. Biochem.*, **89**, 506.

167 (a) Archer, B.L., Audley, B.G., Cockbain, E.G., and McSweeney, G.P. (1963) The biosynthesis of rubber. *J. Biochem.*, **89**, 565–574. (b) Cornish, K. and Siler, D.J. (1996) Characterization of *cis*-prenyl transferase activity localized in a buoyant fraction of rubber particles from *Ficus elastica* latex. *Plant Physiol. Biochem.*, **34**, 377.

168 For early steps of isoprenoid biosynthesis in eukaryotes, see the pioneering work of Bloch, Lynen, and Cornforth in (a) Spurgeon, S.L. and Porter, J.W. (1981) in *Biosynthesis of Isoprenoid Compounds*, vol. 1 (eds J.W., Porter and S.L. Spurgeon), John Wiley & Sons, Inc., New York, pp. 1–46. (b) Qureshi, N. and Porter, J.W. (1981) in *Biosynthesis of Isoprenoid Compounds*, vol. 1 (eds J.W. Porter and S.L. Spurgeon), John Wiley & Sons, Inc., New York, pp. 47–93. (c) Bloch, K. (1992) Sterol molecule: structure, biosynthesis, and function. *Steroids*, **57**, 378.

169 Cornish, K., Wood, D.F., and Windle, J.J. (1999) Rubber particles from four different species, examined by transmission electron microscopy and electron-paramagnetic-resonance spin

labeling, are found to consist of a homogeneous rubber core enclosed by a contiguous, monolayer biomembrane. *Planta*, **210** (1), 85–96.

170 (a) Cornish, K. (2001) Biochemistry of natural rubber, a vital raw material, emphasizing biosynthetic rate, molecular weight and compartmentalization, in evolutionarily divergent plant species. *Nat. Prod. Rep.*, **18**, 182–189. (b) Cornish, K. (2001) Similarities and differences in rubber biochemistry among plant species. *Phytochemistry*, **57** (7), 1123–1134.

171 (a) Castillon, J. and Cornish, K. (1999) Regulation of initiation and polymer molecular weight of *cis*-1,4-polyisoprene synthesized *in vitro* by particles isolated from *Parthenium argentatum* (Gray). *Phytochemistry*, **51** (1), 43–51. (b) Cornish, K. and Castillon, J. (2000) Rubber molecular weight regulation, *in vitro*, in plant species that produce high and low molecular weights *in vivo*. *Biomacromolecules*, **1** (4), 632–641. (c) Archer, B.L., Audley, B.G., and Bot, J.L. (1987) New aspects of rubber biosynthesis. *Bot. J. Linn. Soc.*, **94**, 181–196. (d) Cornish, K. (1993) The separate roles of plant *cis* and *trans* prenyl transferase in *cis*-1,4-polyisoprene biosynthesis. *Eur. J. Biochem.*, **218**, 267–271.

172 (a) Madhavan, S., Greenblatt, G.A., Foster, M.A., and Benedict, C.R. (1989) Stimulation of isopentenyl pyrophosphate incorporation into polyisoprene in extracts from guayule plants (*Parthenium argentatum* Gray) by low temperature and 2-(3,4-dichlorophenoxy) triethylamine. *Plant Physiol.*, **89** (2), 506–511. (b) Scott, D.J., Da Costa, B.M.T., Espy, S.C., Keasling, J.D., and Cornish, K. (2003) Activation and inhibition of rubber transferases by metal cofactors and pyrophosphate substrates. *Phytochemistry*, **64**, 123.

173 (a) Yoon, K.R., Lee, K.B., Chi, Y.S., Yun, W.S., Joo, S.W., and Choi, I.S. (2003) Surface-initiated, enzymatic polymerization of biodegradable polyesters. *Adv. Mater*, **15**, 2063. (b) Yokozawa, T. and Yokohama, A. (2005) Chain-growth polycondensation for well-defined condensation polymers and polymer architecture. *Chem. Rec.*, **5** (1), 47–57.

174 (a) Eng, A.H. and Ong, E.L. (2000) Hevea natural rubber, in *Handbook of Elastomers, Plastics Engineering*, vol. 61 (eds A.K. Bhowmick and H.L. Stephens), Marcel Dekker, New York, p. 29. (b) McIntyre, D., Stephens, H.L., Schloman, W.W., and Bhowmick, A.K. (2001) Guayule rubber, in *Handbook of Elastomers, Plastics Engineering*, vol. 61 (eds A.K. Bhowmick and H.L. Stephens), Marcel Dekker, New York, p. 1.

175 Archer, B.L. and Audley, B.G. (1967) Biosynthesis of rubber. *Adv. Enzymol. Relat. Subj. Biochem.*, **29**, 221.

176 McMullen, A.I. (1963) The polynucleotide double helix as a versatile assembly template during polymer biogenesis. *J. Theor. Biol.*, **5** (1), 127–141.

177 (a) Kaneko, Y. and Kadokawa, J.I. (2005) Vine-twining polymerization: a new preparation method for well-defined supramolecules composed of amylose and synthetic polymers. *Chem. Rec.*, **5** (1), 36–46. (b) Mejias, L., Schollmeyer, D., Sepulveda-Boza, S., and Ritter, H. (2003) Cyclodextrins in polymer synthesis: enzymatic polymerization of a 2,6-dimethyl-beta-cyclodextrin/2,4-dihydroxyphenyl-4′-hydroxybenzylketone host–guest complex catalyzed by horseradish peroxidase (HRP). *Macromol. Biosci.*, **3**, 395. (c) Kadokawa, J.I., Kaneko, Y., Nagase, S.I., Takahashi, T., and Tagaya, H. (2002) Vine-twining polymerization: amylose twines around polyethers to form amylose-polyether inclusion complexes. *Chemistry*, **8** (15), 3321–3326. (d) Sahoo, S.K., Nagarajan, R., Chakraborty, S., Samuelson, L.A., Kumar, J., and Cholli, A.L. (2002) Variation in the structure of conducting polyaniline with and without the presence of template during enzymatic polymerization: a solid-state NMR study. *J. Macromol. Sci.: Pure Appl. Chem.*, **A39**, 1223. (e) Kadokawa, J.I., Kaneko, Y., Nakaya, A., and Tagaya, H. (2001) Copolymerization of propene and 1,2,4-trivinylcyclohexane by a $MgCl_2$-supported $TiCl_4$ catalyst. *Macromolecules*, **34**, 6536–6538.

178 Scott, D.J., Da Costa, B.M.T., Espy, S.C., Keasling, J.D., and Cornish, K. (2003) Activation and inhibition of rubber transferases by metal cofactors and pyrophosphate substrates. *Phytochemistry*, **64** (1), 123–134.

179 (a) Puskas, J.E. and Kaszas, G. (2003) Carbocationic polymerization, in *Encyclopedia of Polymer Science and Technology*, vol. 5 (ed. J.I. Kroschwitz), Wiley–Interscience, New York, p. 382. (b) Matijaszewski, K. and Sawamoto, M. (1996) Cationic polymerization: mechanism, synthesis and application, in *Plastics Engineering*, vol. 35 (ed. K. Matyjaszewski), Marcel Dekker, New York, p. 343.

180 Liang, P.H., Ko, T.P., and Wang, A.H.J. (2002) Structure, mechanism and function of prenyltransferases. *Eur. J. Biochem.*, **269**, 3339.

181 Poulter, C.D. and Rilling, H.C. (1978) The prenyl transfer reaction. Enzymatic and mechanistic studies of the 1–4 coupling reaction in the terpene biosynthetic pathway. *Acc. Chem. Res.*, **11**, 307.

182 Subramaniam, A. (1987) Natural rubber, in *Rubber Technology*, 3rd edn (ed. M. Morton), Springer.

183 DuPont Performance Elastomers (2003) *Technical Information – Neoprene*, October 2003.

184 Tangpakdee, J., Tanaka, Y., Shiba, K.I., Kawahara, S., Sakurai, K., and Suzuki, Y. (1997) Structure and biosynthesis of *trans*-polyisoprene from Eucommia ulmoides. *Phytochemistry*, **45** (1), 75–80.

4
Bacterial Biopolymers and Genetically Engineered Biopolymers for Gel Systems Application

Deepti Singh and Ashok Kumar

4.1
Introduction

Biopolymers are also known as biomacromolecules. These can be either produced via a variety of mechanisms through microorganism or plants or can be made up from biological building blocks [1]. Natural biopolymers present a suitable example of how all the properties displayed by the biological systems and materials are exclusively determined by the chemical and physical properties of the monomers and their respective sequences [2]. The well-defined molecular structure can result in a rich complexity of structure that can define specific functions on the mesoscale [3], where the structure flexibility, functional properties, and competing molecular interactions are all tailored by small string of monomeric units. Biopolymers have found a wide range of applications in medical, food, and textile industries as new fabric material, biosensors, and plastics and in the treatment of water polluted with various chemicals. In the last decade, there has been significant development in genetically engineered biopolymers being used as biomaterials that can self-assemble into semisolid matrices depending upon the physiological environment [4–6]. Potential applications of these genetically engineered protein-based biopolymers range from drug delivery and tissue engineering to various surgical applications such as bone repair and soft tissue augmentation. The major advantage that genetically engineered polymers have over the conventional synthetic equivalents is the ability and ease with which these biopolymers of uniform composition, amino acid sequence, and molecular weight can be synthesized repeatedly [7]. Since polymer structure dictates physicochemical properties of the matrix, genetically engineered biopolymers offer the unique ability to control this microstructure, thereby governing the function and ultimately the fate of biomaterials. One of the major requirements of tissue engineering is designing biomaterial that regulates the cell proliferation and is tuned to provide appropriate signals for cell differentiation since this role in the *in vivo* system is played by the extracellular matrix (ECM) [8]. The polymer-based biomaterials should be able to replicate the dynamic structural, biochemical, and mechanical properties of the naturally occurring ECM, along with being both biodegradable and biocompatible [9]. This property has been widely

Handbook of Biopolymer-Based Materials: From Blends and Composites to Gels and Complex Networks,
First Edition. Edited by Sabu Thomas, Dominique Durand, Christophe Chassenieux, and P. Jyotishkumar.
© 2013 Wiley-VCH Verlag GmbH & Co. KGaA. Published 2013 by Wiley-VCH Verlag GmbH & Co. KGaA.

explored in the field of tissue engineering, where biopolymer-based matrix is fabricated for neotissue regeneration. The polymeric extracellular matrix also called scaffold or gels are most important component for engineering any damage or injured organs for restoration or replacement [10]. It is difficult to obtain the ideal tunable polymeric matrix from natural source; however, this can be easily achieved using genetic engineering tools, for example, protein-engineered biomaterials that are entirely made out of recombinant proteins in which amino acids sequences act as monomers of the material. The desired amino acids sequence is encoded into a deoxyribo nucleic acid (DNA) plasmid using recombinant DNA technology, and the genetic message is translated by the host expression systems, resulting in template biomaterials synthesized with exact molecular-level precision [9]. Broad families of biosynthesized materials and natural living organisms, which have the ability to produce a large variety of polymers, can be divided into various classes depending upon their chemical structure [11,12].

1) Nucleic acid or polynucleotides, for example, DNA, RNA
2) Polysaccharides (carbohydrates), for example, starch
3) Polyamides, for example, proteins
4) Polythioesters, for example, poly(3-mercaptopropionate)
5) Polyesters, for example, polyhydroxyalkanoates (PHA) also known as aliphatic polyesters
6) Polyisoprenoids, for example, rubber and its derivatives
7) Polyanhydrides, for example, polyphosphate
8) Polyphenols, for example, lignin.

The common relation for these major classes is the chemistry of repeating units, which is covalently linked, but the arrangements (cyclic, linear, or branched) and linkage between these units determine the difference and distinguish the members of each class. Like common polymers, biopolymers are strings of monomers and these are mostly linear strings but sometimes can be circular and closed or branched and open. When these strings of monomers are cross-linked, they form gels [13,14]. The determining factors in biopolymer synthesis are these monomeric units along with environmental factors such as temperature, medium, pH, and so on. The unique feature of biopolymers is that they are mostly heteropolymers, that is, made up of more than one monomeric unit. Few examples of the biopolymers are shown in Figure 4.1, but the emergent property of biopolymer cannot be attributed to these monomeric units since the monomers as such are unable of providing functionality. This can be best understood as in the case of DNA, which is made up of codes that are basically carbon, oxygen, and nitrogen and it is the arrangement of these bases that form the genetic code and individually these moieties do not hold much significance [15]. The unique feature that biopolymer exhibits is that there is hierarchy in the molecular structure and their associated biological functions emerge naturally. How the arrangements of building blocks govern the functionality of moiety can be understood by imagining the importance of double-helical construct of DNA during the replication process [16].

4.1 Introduction

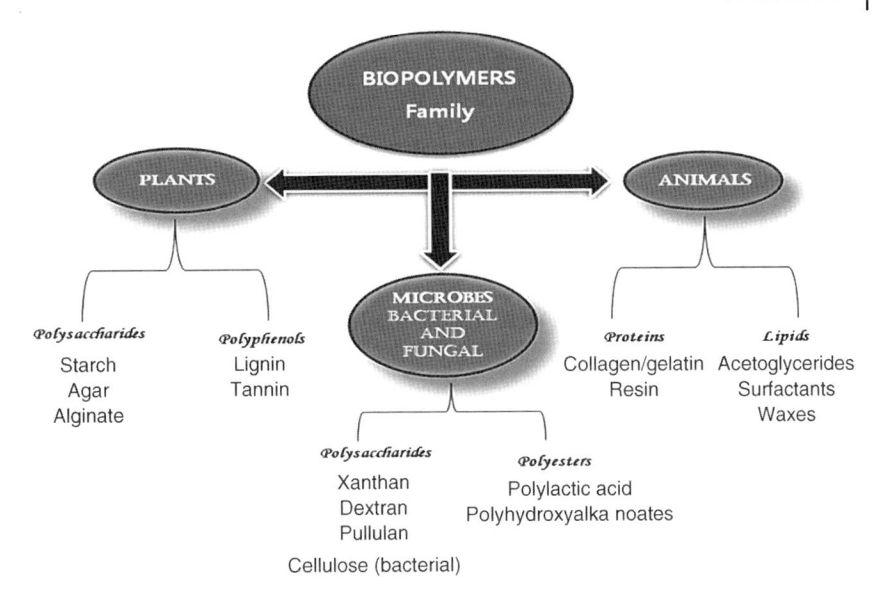

Figure 4.1 Naturally occurring biopolymer family from plants, animals, and microbes. Each of the broad family members are further divided mainly into proteins and polysaccharides, but have many other branches that are not shown here.

4.1.1
Nucleic Acid Biopolymers: Central Dogma

DNA and RNA are most important biopolymers that are constructed from nucleotides [17]. Single nucleotide is made up of a ribose, a sugar moiety as backbone, and phosphate group and nitrogen-containing base (Figure 4.2). Hydrogen group (H) in RNA differentiates it from the DNA that contains oxygen and hydrogen (>OH) group on the 2-position of the ribose, hence the name "deoxyribo" nucleic acid (DNA). The four bases in DNA are cytosine (C), guanine (G), adenine (A), and thymine (T); in RNA, thymine base is replaced by uracil (U) (Figure 4.2). Pairing rule is very specific in DNA where A = T and C = G and each strand of DNA has a precisely defined complementary strand [18]. Although DNA forms long, double-coiled strands, the RNA molecules are usually found as single stranded. For this reason, RNA is found to be more flexible in folding up in numerous ways that makes it more structurally versatile than the cumbersome counterpart DNA. Groundbreaking research led to the discovery of ribozymes giving the hypothesis that RNA could be the original molecule of life on earth. Identified as a biopolymer, which has the ability of self-replication, it can be the key for development from prebiotic organic molecules and is also known to store information and catalyze chemical reactions. Until the discovery of ribozymes, it was impossible to unveil the mystery of origin of life on earth just basing the evolution on DNA or proteins. DNA is a known carrier of all the genetic information acting as the blueprints for proteins, but without the help of enzymes the coding is impossible and for that proteins were needed. Proteins are the end product of genetic information.

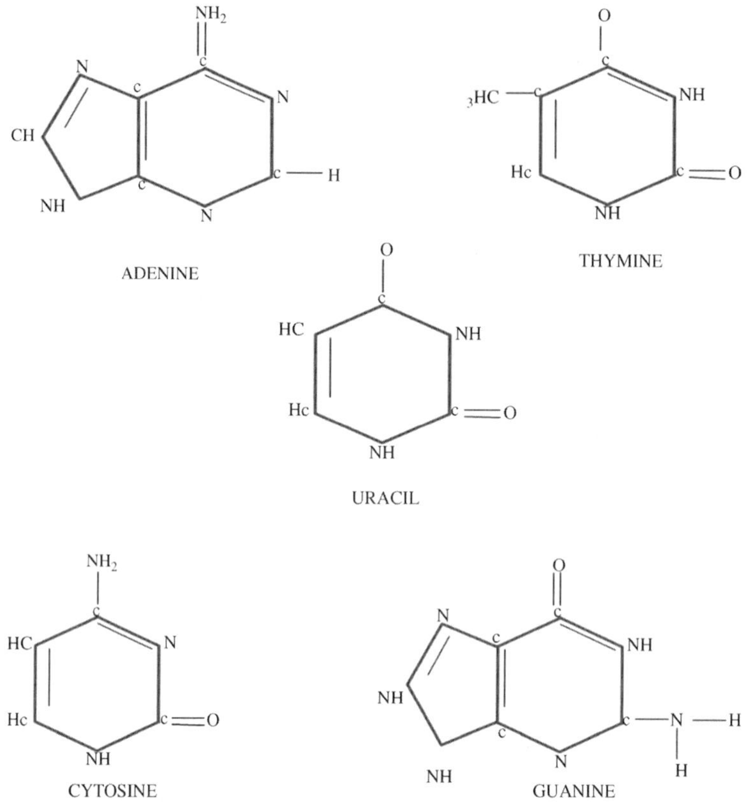

Figure 4.2 Chemical structure of the bases of DNA and RNA. In RNA, thymine is replaced by uracil and the remaining bases remain same. Nucleic acids are most common biopolymers found in the system and the understanding of its hierarchical discipline has revolutionized the concept of basic and applied biology.

DNA and protein were important partners, so both could not have been the reason for origin of life. RNA discovery therefore was the revolutionary research that answered the most important question. RNA was self-sufficient and can be the original molecule of life: information read and stored in itself and converts this information into biological functions. RNA is the most important and unique biopolymer and has constantly stirred the researcher's imagination worldwide [19].

4.2
Microbial Polysaccharides as Biopolymers

4.2.1
Synthesis and Applications

Bacterial cellulose is the most abundantly found biomass and feedstock in pulp and paper industries [20]. Traditionally, using certain bacterial species as fermenting

A. xylinum in Biomass fermentor   *Structure of cellulose synthase in bacteria*

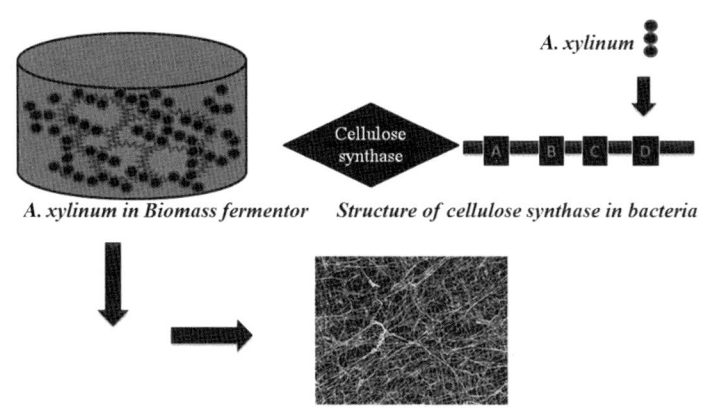

SEM image of engineered bacterial cellulose

Figure 4.3 Detailed explanation of bacterial cellulose biopolymer production. Steps involved in bacterial cellulose synthesis: first step is the production of cellulose synthase, which uses sugar precursor and consists of four subunits of which A and B form a glycosyltransferase that helps in catalyses of β-glycosyltransferase reaction, thereby joining two sugar molecules and creating chains of glucose gradually. It is assumed that the third subunit C governs the formation of pore for cellulose transportation and the fourth subunit D does the finishing role by crystallizing cellulose into larger fibers.

agent, this biomass can yield high amount of cellulose of pure quality with unique properties. Cellulose is a known polysaccharide composed of linear glucose chains (Figure 4.3). Best examples of cellulose-producing bacteria are *Acetobacter xylinum* and Gram-negative bacteria and are known to produce cellulose from a particular point on cell surface, thereby giving single linear network of interlocked fibers [21]. Steps involved in bacterial cellulose synthesis are first the production of cellulose synthase, which uses sugar precursor UDP-glucose for cellulose formation. Cellulose synthase enzyme consists of four subunits A–D. The subunits A and B are thought to form a glycosyltransferase that helps in catalyses of a β-glycosyltransferase reaction, thereby joining two sugar molecules and creating chains of glucose gradually. It is assumed that the third subunit C governs the formation of pore for cellulose transportation and the fourth subunit D does the finishing role by crystallizing cellulose into larger fibers. In addition to these, there are two complementary enzymes that appear to play an important role in bacterial cellulose production: a cellulose complementary protein (CcpAx) and an *endo*-β-1,4-glucanase (cellulose complementary protein 2 (Cmcax)), but the functions of this enzyme are not well elucidated [22]. The media for fermentation contain high salt, glucose, corn steep liquor, and iron chelators along with production enhancers. At present, around 0.2 g of cellulose has been obtained by using 1 g of glucose and an upscale production till 50 000 gallons fermentors has been achieved. In the final step of obtaining the biopolymer, the microbes are lysed using hot caustic treatment and as bacterial cellulose is water insoluble, it is easily removed from the fermentors. They have large surface area, which is almost 200 times more than wood pulp, and it is due to the large network of fibers, the surface of these fibers can be easily modulated to various applications [23]. Cellulose has found use in almost every industry right from clothing to the production of high-end speakers in audio system to coating agent. Industries have easily exploited the large surface area and water-absorbing ability of cellulose in food industries and also in the recovery of oil and gas. The fact that in the past 10 years almost 50 patents have been filed on production and applications of the bacterial cellulose explains the importance that the biopolymer has gained ever since being first used. Only bottleneck is that the production cost is high, thereby ultimately affecting the final price that is currently around $35 per pound; however, scientist and industries have been trying different methodology to reduce the initial production cost so that the use of cellulose could be enhanced to find its niche applications.

4.3
Microbial Biopolymers as Drug Delivery Vehicle

4.3.1
ε-Poly-L-Lysine (ε-PL) and Its Applications

Homo-poly amino acid is characterized by peptide linkage between the ε-amino group of lysine and carboxyl group [24]. At present, ε-PL in industries is produced by aerobic fermentation using the mutant strain of *Streptomyces albulus* and during the first fermentation process, the yield of ε-PL was 0.5 g/l. It was observed that the maximum ε-PL production was when the production was initiated at pH 6 and once cells reached the maximum growth, the broth pH had to be lowered to obtain the maximum ε-PL production. Few strains of *Streptomyces* like *S. albulus* can produce ε-PL and have been main subject of investigation as these biopolymers have shown antibacterial activity against wide range of microbes [24–26]. This antimicrobial activity is attributed to the electrostatic adsorption of ε-PL onto the microbial cell surface on the basis of polycationic properties [24]. Due to this property, ε-PL has been used in food preservative sector as food additives as antimicrobial component. It has been reported that ε-PL acts against both Gram-positive and Gram-negative bacteria with the minimal inhibitory concentration (MIC) found to be around 100 µg/ml [26]. Shima *et al.* [27] showed the correlation between lysine molecular size and MIC activity and found that more than 9-lysine residue showed lethal antimicrobial action; however, when the amino group of ε- PL was replaced by the aromatic carboxylic acid like 4-chlorobenzoic acid, the antimicrobial activity of this biopolymer completely diminished. The proposed mode of antimicrobial activity as discussed is based on the cationic properties of molecule and leads to the stripping of outer membrane that leads to the abnormal cytoplasm distribution [28].

4.3.2
Polyhydroxyalkanoates and Its Applications

The growing concern about the environmental issues and degradation of oil-based thermoplastics has given rise to new technology that uses the microbial source for

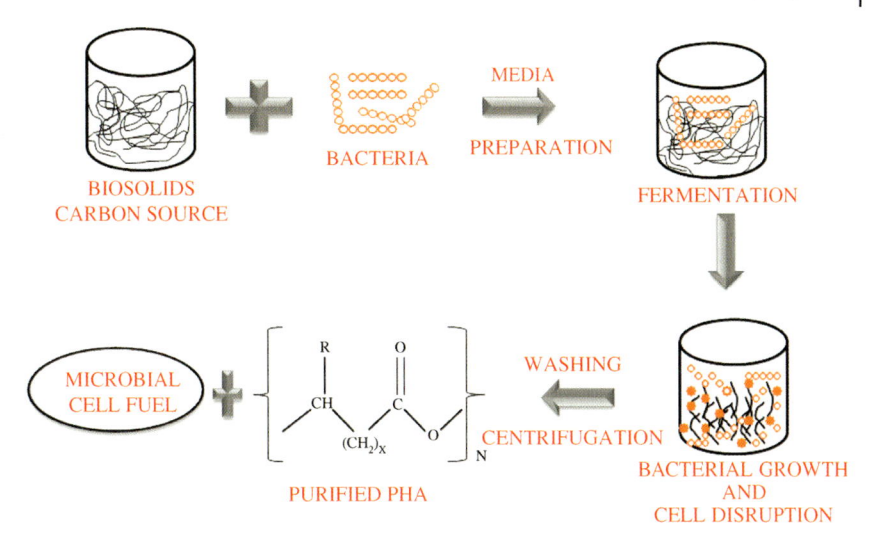

Figure 4.4 Schematic representation of microbial polyhydroxyalkanoates (PHA) production. PHA is known microbial energy reserve products that are mostly accumulated and stored as granules in the cytoplasm of cells and has found large commercial applications in the recent past.

generation of biodegradable and environment-friendly polymers that can easily replace thermoplastics and associated problems [29]. PHA is known microbial energy reserve products that are mostly accumulated and stored as granules in the cytoplasm of cells [30]. They have physical property similar to that of thermoplastics, that is, the melting temperature of these polyesters is between 50–180 °C, whereas the mechanical property can be easily modulated to resemble to that of crystalline plastic (hard) and rubber [31]. In 1927, the prototype of this family in Pasteur Institute in Paris was established and this polyester saw the daylight and was commercialized after 23 years of discovery. Production of PHA is illustrated in Figure 4.4.

Due to the biocompatibility nature of this polyester, it has found a wide range of application, in the biomedical field such as controlled drug release [28], bone plates, and surgical sutures and in the molding application in the personal care arena and monofilament fishing nets in Japan. PHA has been used in various sectors, and during the mid-1990s, a number of publications appeared on PHA synthesis, characterization, and applications [32,33].

4.4
Polyanhydrides

Polyanhydrides have emerged as a new class of biodegradable biopolymers and have been extensively used as drug delivery vehicle in a number of clinical trials [34]. The hydrophobic backbone of polyanhydride has hydrolytic labile anhydride linkage and the degradation can be easily controlled and manipulated by the composition of the polymers. Polyanhydrides have important advantages due to its nonimmunogenic/noninflammatory response once applied in the *in vivo* condition. Another advantage of using this polymer in the *in vivo* and *in vitro* system is that its rate of degradation is quick, which is due to unstable hydrolytic bond that breaks the anhydride bonds and results in two carboxylic moieties that are easily metabolized and the by-products of degradation is nonmutagenic and noncytotoxic [35]. Langer and coworkers [36] utilized the hydrolytic instability of these polymers for sustained drug release purposes. Polyanhydrides derived from the sebacic acid has been approved by Food and Drug Administration (FDA) for human use and its derivatives have been used to deliver chemotherapeutic agent in case of brain cancer [37]. Modification of the polyanhydrides by introducing imide group has provided the mechanical strength for widening the scope of its usage, such as in orthopedic applications [38].

During the period when microbial polymer synthesis became an established approach, the genetic engineering field was evolving rapidly. Engineering the genetics of bacterial species to produce biopolymer both in large quantity and of high quality led to the development of recombinant protein polymer products.

4.5
Recombinant Protein Polymer Production

The increasing ability to design a high structural organization in synthetic polymer materials offers the platform for powerful new technology that can open up new inroad toward the development of functional structure ranging from nanosize materials to the very advanced materials with a fine control. In addition to the differences in the chemical units forming the macromolecules, even the physical processes play critical role in structure formation of the molecules in diverse ways. This is where the main problem resides in our attempt to mimic nature by creating materials with a high degree of complexity and yet maintain the functionality of these molecules. The primary chemical composition along with monomer sequences that are much needed to achieve extraordinary macromolecular functionality cannot be obtained by classic chemical synthesis as these methods are not robust enough. Outstanding development in the field of molecular biology has allowed us to design DNA duplex that codes for any amino acid of interest [37]. Introducing the synthetic gene of interest into the genetic content of the microorganisms enables us to engineer the production of encoded protein-based polymers (Figure 4.5).

Exploiting this genetic engineering procedure, the use of modified cells as factory for producing polymers is highly desired. Owing to the current developments in molecular biology, scientists have the ability to create almost any duplex DNA strand with desired amino acid coding sequences. There are a number of tools that help in creating or introducing any synthetic gene in the genetic content of microbes or

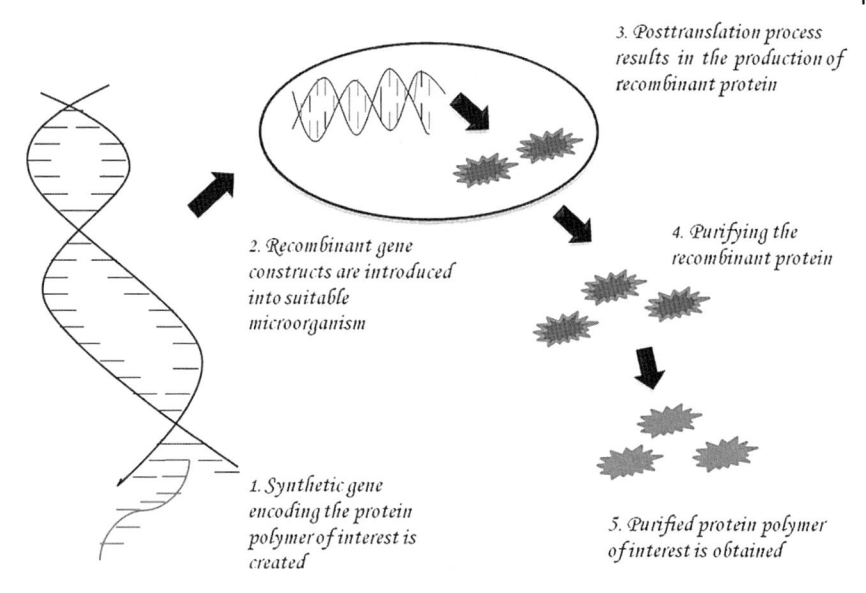

Figure 4.5 Schematic illustration of steps involved in the genetically engineered recombinant protein polymer production. Introducing the synthetic gene of interest into the genetic content of the microorganisms enables to engineer the production of encoded protein-based polymers.

1. Synthetic gene encoding the protein polymer of interest is created
2. Recombinant gene constructs are introduced into suitable microorganism
3. Posttranslation process results in the production of recombinant protein
4. Purifying the recombinant protein
5. Purified protein polymer of interest is obtained

higher organisms and induce the production of encoded protein-based polymers as recombinant proteins. The advantages of this approach are as follows:

i) In principle, it will be able to showcase the simple or complex properties present in natural polymers.
ii) Can exploit huge resources as proteins in living organisms ranging from prions to animals in ample sense.
iii) Stockpile and refinement of functionality to the extreme biology, especially during the long process of natural selection.
iv) Various different combinations can be achieved by clubbing the amino acids.
v) Production cost for both simple and complex polymers is nearly same, as the structural complexity does not change the genetic engineering protocol much.

4.6 Recombinant Genetically Engineered Biopolymer: Elastin

Still in the initial stage, genetically engineered biopolymers are yet to find some serious commercial investment. Reason of this could be attributed to the fact that this field is fairly new and selective groups use the methodology and techniques across the globe. However, elastin and spider silk are the two polymers that have caught the attention of the scientists worldwide. Elastin-like polymers (ELP) are the privileged family of the genetically engineered biopolymers (Figure 4.6) [38,39] and

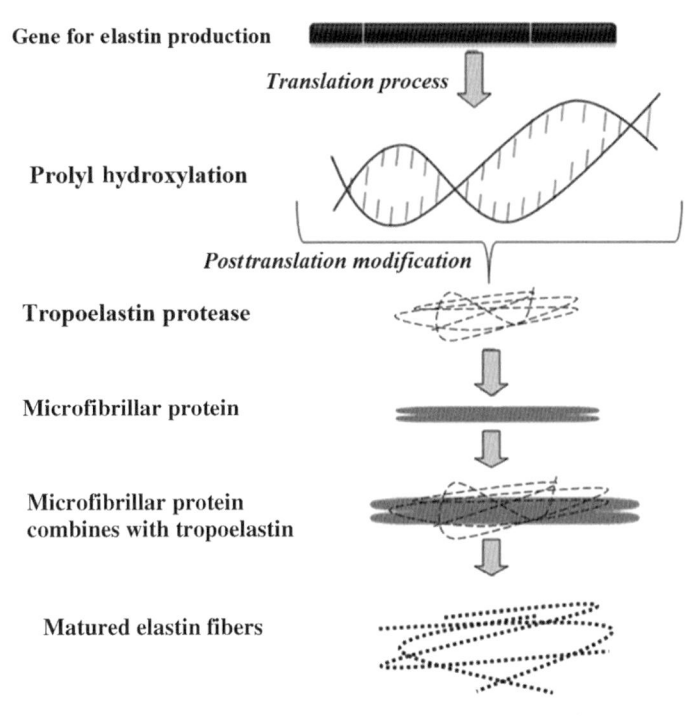

Figure 4.6 Schematic representation of steps involved in recombinant elastin production (ELPs). Genetically engineered ELPs are similar to mammalian elastic protein in repeated sequences of amino acids.

have been explored for a wide variety of properties unlike spider silk that is basically explored for its mechanical strength. ELPs have also helped scientists in understanding the basic protein folding and its importance in determining the functions of the proteins. Genetically engineered ELPs are similar to mammalian elastic protein in repeated sequences of amino acids. The most important property that the genetically engineered elastin resembles with their natural counterpart is when the matrix is synthesized, they retain the mechanical property of natural elastin, which is accompanied by extraordinary biocompatible nature along with the smart self-assembling property. This property is mainly attributed to the molecular changes or transition in presence of water. It is this transition known as inverse temperature transition (ITT) that holds the key in the development of the new peptide-based polymers as materials and molecular machines [40]. Although the phenomenon exhibited by the ELPs resembles to that exhibited by the amphiphilic polymers such as polyNIPAAM (N-isopropylacrylamide), it differs in the sense that the ordered state present in these ELPS above the transition temperature is not found in normal lower critical solution temperature (LCST) polymers, thus differentiating the use of ITT terminology and not the LCST for ELPs. This biopolymer has shown strong thermoresponsive changes, which is due to the existence of ITT. There are various subfamilies of ELPs that are based on pentapeptide VPGVG, where amino acid side chains are permutated and most of the ELP-based biopolymers are synthesized using this pentapeptide formula and VPGVG is replaced as VPGXG (X denoting modified amino acid sequence [37–39].

4.7
Collagen as an Ideal Biopolymer

Collagen is an important extracellular matrix protein in body that governs the main structural protein in mammals, invertebrates, and different multicellular organisms [40]. Collagen is triple helix fibrous protein that has repeated Gly-Pro-hydroxyproline subunits, but the last two subunits can sometimes change [41]. The sequence of these subunits are absolutely necessary for the collagen to assemble into fibrils that later organize to form fibers. Type I collagen is the best example of the prototype of fibrillar collagens and is most extensively found in connective tissue in body. The superfamily of collagen involves 30 different genes that provide codes for 20 genetically distinguishable collagen types, which are homo- or heteromeric types [42]. Fibers of collagen have exhibited unique structural properties, for example, they dissipate energy, provide biological signal, prevent premature aging that can lead to mechanical failure, and transmit forces along with low immunity and high affinity to water, and its ability to fasten regeneration of tissue has made collagen most ideal biopolymer for regenerative medicine [43]. The biodegradability along with its compatibility makes collagen also the most attractive polymer in tissue engineering and regenerative therapy. Since this is present in the connective tissue of the body, it is impossible to isolate this polymer from the natural source and the isolation process might not remove the risk of viral and other infectious agent that cadaver can harbor. The demand for this polymer had led to the use of genetically engineering tools, and in the tobacco mosaic plant, the triple-helical fully formed human collagen was expressed; however, its mechanical strength and thermal stability seem to have been compromised compared to that of the collagen from the animal counterpart. In-depth analysis of the collagen showed lack of hydroxyproline residues, which play an important role in the stability of the triple helix [44].

4.7.1
Microbial Recombinant Collagens: Production in *Pichia Pastoris*

P. pastoris has been engineered to coexpress P_4H for the production of full-strength hydroxylated triple-helical collagen of different type (recombinant human collagen (rhC)) at high levels (1–1.5 g/l) [38]. These microbial-derived fibrils of collagen are structurally, that is, the hydroxyproline content, similar to human tissue-derived collagens [44]. *P. pastoris*-derived recombinant collagen forms stable triple helices; a folding chaperone is thought to be involved in collagen assembly in animals (Figure 4.7) and helps in stabilizing the helical structure [45]. More than 90% of the nonhelical and helical procollagens is accumulated within the endoplasmic

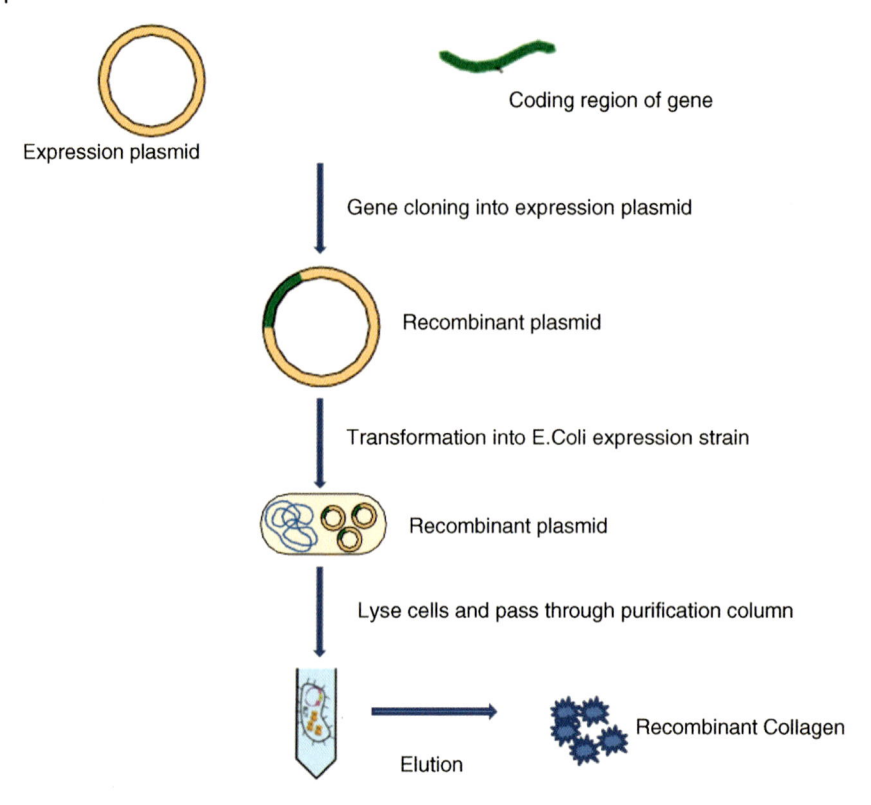

Figure 4.7 Steps involved in recombinant collagen production using genetic engineering technology and have been explored using the tobacco mosaic plant in which triple-helical fully formed human collagen was expressed successfully. However, the mechanical strength and thermal stability seems to have been compromised when compared to collagen from animal counterpart.

reticulum (ER) membrane of cells and these have not been found to proceed to secretory pathway (pathway that involves a number of steps via which cells move protein out of the cell) even after the authentic signal sequence of collagen is replaced with the *Saccharomyces cerevisiae* α-mating [44]. The accumulation levels of recombinant procollagen in fermentations of *P. pastoris* have been improved from 15 mg/l to 1.5 g/l by using genetic manipulations and optimization of the parameters to control fermentation [46]. Recombinant microbial technology has been anticipated for the production of different protein-based biopolymers that are important in medical applications like elastin, gelatin, and dragline spider silk. This technology also enables engineering of the biopolymers for improved performance of products containing these engineered biomaterials.

Modifying biopolymers using tools of genetic engineering has widened the scope of its applications as these modified biopolymers are used in gel system for bioengineering applications like drug delivery and tissue engineering scaffolds.

4.8
Biopolymers for Gel System

Due to their potential applications, biopolymers as biomaterials for matrix production in tissue engineering have opened new arena [47]. Biopolymer offers an important option of control over the structure, morphology, and chemistry of polymers as important functional units can be substituted to mimic extracellular matrix systems. These also provide control on the material property such as mechanical properties in case of gel, fiber, and porous scaffold formats [48]. The inherent rate of biodegradability of biopolymers is important as one can regulate the extent of cell and tissue regeneration or remodeling in both *in vitro* and *in vivo* systems. With the advent of genetically redesigning of these biopolymers, researchers have been able to incorporate features that regulate cell response to this matrix and also provide insight into the fundamentals of the chemistry of structure–function relationship between matrix and cells [49].

4.9
Hydrogels of Biopolymers for Regenerative Medicine

Hydrogels are chemically or physically cross-linked network of polymers that can absorb large quantity of water. Hydrogels can be classified into various classes depending upon the charges and mechanical and structural properties [50]. Depending upon the polymer used, they can be classified as follows.

4.9.1
Polysaccharide Hydrogels

The best example for this can be chondrotin sulfate (CS), which is glycosaminoglycan that contains *N*-acetyl-D-galactosamine and D-glucuronic acid as alternating units. CS is readily soluble in water due to which a chemical cross-linker is used during its application in gel system. They possess excellent bioproperty that helps in easy modulating and binding of growth factors. Kirker *et al.* [51] synthesized biocompatible hydrogel films by using adipic dihydrazide derivatives of CS (adipic dihydrazide derivatives of chondritin sulfate (CS-ADH)). In CS-ADH, a pendant hydrazide generated a gel using micro- or macromolecules of cross-linker such as poly(ethylene glycol)-propiondialdehyde. CS-based hydrogels have been applied immensely in tissue engineering, and clubbing CS with gelatin has been successfully used in drug delivery as it is easy to obtain the control release of the drug [52]. Bilayer wound dressing using same biopolymer combinations has yielded good results. Using bilayer dressing showed complete regeneration of skin with well-defined epithelial layer and adequately generated differentiated epithelial tissue [51]. Results also showed enhanced production of collagen type II and proteoglycans in the presence of CS hydrogel and when clubbed along with gelatin and hyaluronan, CS hydrogels were used in mimicking native cartilage environment and strength [53–55].

4.9.2
Cellulose-Derived Biopolymers-Based Hydrogels

Main difference between this and other biopolymers is the cellulose-derived methylcellulose (MC) and hydropropylmethylcellulose (HPMC) gelation occurs at high temperatures (upon heating), which are generally induced by the intermolecular hydrophobic interaction of the methoxy side chains leading to the hydration of the macromolecules at low temperatures [56]. The LCST of HPMC is 75–90 °C and for MC it is around 40–50 °C, and since the transition temperature of these polymers are above 37 °C, various chemical and physical methods have been employed to decrease the LSCT [57]. Incorporating sodium chloride or chemically reducing the hydroxypropyl concentration, this phenomenon could be achieved. Polymer–polymer interaction is high close to the transition temperature leading to formation of polymer network. Tate et al. [58] have applied this biopolymer in tissue engineering and showed the potential of this in brain cell support. These results demonstrated MC as potential matrix in treating multiple site injuries and in filling irregular defects [58]. Combined with hydroxyapatite, MC have shown positive effects in healing bone and hence applied in bone tissue engineering [59].

4.9.3
Protein Biopolymers-Based Hydrogels

Collagen, which is a major protein of connective tissue, can be used for synthesizing porous three-dimensional (3D) network using freeze-drying technique [60]. Collagen has already been used in bone tissue engineering in combination with calcium phosphate and microbeads of this protein has shown to modulate the stem cells differentiation ability by instructing the cells to follow adipogenic lineage [61,62]. To enhance the cell–matrix interaction, certain growth factors like, bone morphogenetic protein (BMP) and specific peptide sequences have been engineered onto the collagen biopolymers and have shown positive results. With the advancement of recombinant DNA technology, human recombinant collagen can find larger applications due to the safety issues in microbial-derived or chemically derived collagen.

4.10
Supermacroporous Cryogel Matrix from Biopolymers

Most recent development in scaffold synthesis is the use of cryogelation technique. Different biopolymers ranging from polysaccharides to proteins have been used to produce cryogel matrix that has found wide applications in bioengineering [10–63]. A well-defined "curtain"-like pore structure is obtained by inducing cryogenic treatment. Cryogelation takes place via cryogenic treatment of the systems containing aqueous solvent and monomer/polymer precursors. The crystallization of the solvent is the critical feature of cryogelation, which differentiates cryogelation technique from freeze-induced gelation. Depending upon the polymer precursor, the pore size ranges

4.10 Supermacroporous Cryogel Matrix from Biopolymers

from 0.1 to 200 μm. Cryogels are produced via gelation process at subzero temperatures where most of the solvents freeze and dissolved polymeric precursors concentrate in small nonfrozen regions (nonfrozen liquid microphase), where the chemical reaction and gel formation proceed over the set period of time. While all the reagents concentrate in nonfrozen liquid microphase, some portion of solvent remains unfrozen and this provides space for solute to concentrate into nonfrozen part with sufficient segmental or molecular mobility for reactions to perform [63]. After melting the solvent crystals (in the case of aqueous media, it will be ice), large continuous interconnected pores are formed. Shape and size of the pores generated directly depend upon the shape and size of the crystals formed during freezing (Figure 4.8). Other factors that contribute to the pore size are rate of freezing, freezing temperature, initial concentration of macromers/monomers in solution, amount of cross-linker, sample size, nucleation agent, and prehistory of the reaction [64]. Cryogels made from gelatin have been used as cell carriers, the sponges of agarose proved to be effective in culturing pancreatic islets, and the use of marcroporous beads and sheets have been employed in cell separation technology [65].

4.10.1
Protein Cryogel

The best example of this biopolymer is dragline silk [66]. Properties of silk such as hydrophobicity that influences cell attachment, tensile strength, easy in processing, biocompatible, thermal stability, and easy facilitation of chemical modifications make silk a promising biomaterial. Lee and Park had genetically engineered *Escherichia coli* to produce spider silk, which is believed to be five times stronger than steel. *E. coli* is the most feasible microbe generally cultured on industrial scale and is an ideal target for genetic engineering [60]. The DNA coding responsible for silk protein was first identified in spider *Nephilla clavipes* (known to produce silk web) and these sequences were engineered into *E. coli* [67–69]. After initial

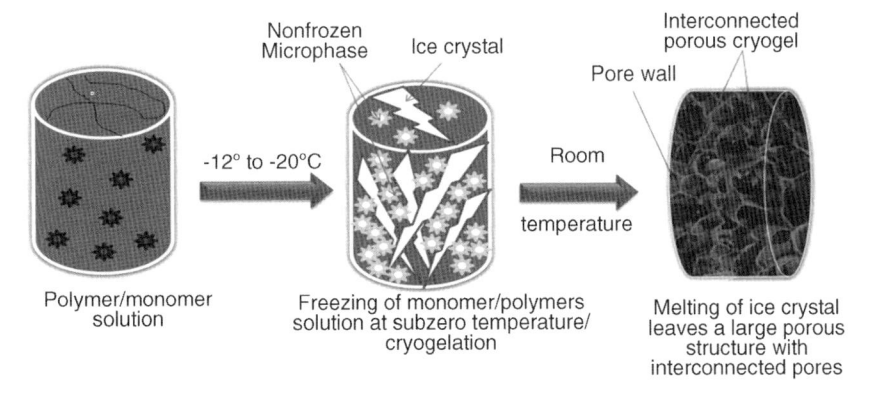

Figure 4.8 Schematic representation of three steps involved in the synthesis of cryogel biomaterial. Cryogelation takes place via cryogenic treatment of the systems containing aqueous solvent and monomer/polymer precursors.

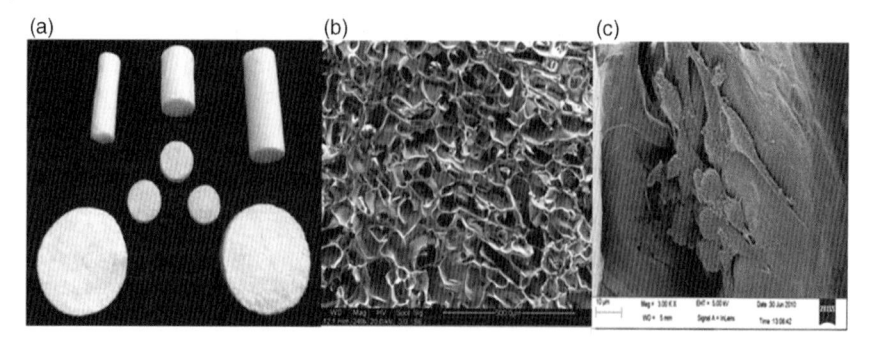

Figure 4.9 Different formats of silk cryogel, from monolith, disk to sheet (a). Scanning electron microscopy of the scaffold shows interconnected and evenly distributed pores (b). The biocompatibility of the synthesized silk cryogel was performed by seeding chondrocytes and scanning electron micrography shows well-adhered and proliferating cells at 96 h (c).

failures, scientists were able to obtain up to 250 kDa of silk protein in its native form. This was one of the biggest success stories in the field of recombinant technology [70]. The primary structure of the silk consists of glycine (Gly), serine, and alanine. Silk has been used as films, gels, nanofibers, and membrane, but no optimized technology has been found to aid in the development of macroporous scaffold of silk as biomaterial [70]. Using cryogel technique, silk was initially dissolved in lithium bromide and dialyzed before producing scaffolds. These scaffolds could be made into various formats ranging from disk to sheet to monoliths (Figure 4.9). Kumar and coworkers (unpublished data) also synthesized silk fibroin–agarose hybrid cryogel scaffolds and optimized their properties like swelling–deswelling behavior, mechanical properties, and so on. Silk cryogel was evaluated for *in vitro* and *in vivo* biocompatibility and there was no immunogenic reaction. The physical characterization using scanning electron microscopy of fibroin–agarose cryogels shows high porous internal architecture giving an ideal property to silk-based biocompatible and degradable scaffolds synthesized by cryogelation approach. Seeding chondrocytes isolated from the knee of 6-month-old goat checked the *in vitro* biocompatibility of synthesized fibroin–agarose cryogel. Cartilage was digested enzymatically and chondrocytes isolated were seeded into 3D fibrion–agarose cryogel (Figure 4.9). The *in vitro* testing showed effective cell attachment with enhanced proliferative rate confirming the biocompatible nature of the synthesized protein cryogel scaffold that can be further explored as an ideal 3D scaffold for tissue engineering and regenerative medicine.

4.11
Biopolymers Impact on Environment

Biopolymers can be sustainable and easily renewed. The basic materials come from the agricultural products (nonfood crop); thus, the use of biopolymers can help in creating sustainable industry in contrast to the chemical polymers that are derived from petrochemical and are feared to run out shortly. In addition to the benefits of using naturally occurring biopolymers, these have potential implications in cutting down the carbon emission that is known to cause serious damage to the earth atmospheric membrane. Most of the biopolymers are biodegradable and can be easily broken down by microbes to carbon dioxide and water. This CO_2 can be reused giving the end product carbon neutrals. Biopolymers that can degrade to 90% within short period, that is, within 6 months or less of time, in industries can be called as "compostable" under European Standard EN 13432 (2000). A simple example of a compostable polymer is polylactic acid (PLA) film under 20 µm thickness; films that are thicker than that does not qualify as compostable even if they are biodegradable.

4.12 Conclusion

This chapter shows the wide applications of biopolymers and the future this technology holds. Despite that recombinant genetic engineering has not completely matured and there are lots of key questions to be resolved, it has already been proven a promising technique. Using genetic engineering, most complex polymers can be tailor-made to obtain desired properties that natural polymers lack. The degree of complexity that can be achieved by this technique along with concurrent development of functions is incomparable to any other counter-techniques. Biopolymers including polysaccharides, lipids, and protein are promising materials in the field of tissue engineering and regenerative medicine. The thermoresponsive solubility behavior of biopolymers opens new perspectives to form gels or system that can polymerize at body temperature. Combining the biopolymers with genetic engineering approach gives the possibilities of synthesizing polymers with biomimetic properties comparable to natural polymers. In few years, the concept of polymers with small monomeric units might just vanish and we may have materials that are made of only desired functional subunits producing the most ideal materials for various biological and industrial applications. Protein-based materials that are genetically engineered can be the most interesting alternative for producing plastics, thereby saving gas and fuel. As this is an easy and cheap procedure, it will not be surprising to see genetically engineered products outlasting the petroleum-based polymers in near future.

Acknowledgments

We thank all the authors whose works are cited here. We also acknowledge Department of Biotechnology (DBT) and Department of Science and Technology (DST), Council of Scientific and Industrial Research (CSIR), Government of India organizations for financial support. Indian Institute of Technology Kanpur is duly acknowledged for all other support.

References

1 Bernad, H.A.R. (ed.) (2009) *Microbial Production of Biopolymers and Polymer Precursors*, Caister Academic Press.
2 Johan, R.C. and Van, D.M. (eds) (2007) *Introduction to Biopolymer Physics*, Wiley Scientific Publishing Co.
3 Carlos, R.C., Javer, R., Alessandar, G., Matilde, A., and Ana, M.T. (2005) Developing functionality in elastin-like polymers by increasing their molecular complexity: the power of genetic engineering approach. *Prog. Polym. Sci.*, **30**, 1119–1145.
4 Cappello, J., Crissman, J.W., Crissman, M., Ferrari, F.A., Textor, G., Wallis, O., Whitledge, J.R., Zhou, X., Burman, D., Aukerman, L., and Stedronsky, E.R. (1998) *In-situ* self-assembling protein polymer gel systems for administration, delivery, and release of drugs. *J. Control. Release*, **53**, 105–117.
5 Petka, W.A., Harden, J.L., McGrath, K.P., Wirtz, D., and Tirrell, D.A. (1998) Reversible hydrogels from self-assembling artificial proteins. *Science*, **281**, 389–392.
6 Urry, D.W., Harris, C.M., Luan, C.X., Luan, C.-H., Gowda, D.C., Parker, T.M., Peng, S.Q., and Xu, J. (1997) Transductional protein-based polymers as new controlled-release vehicles, in *Controlled Drug Delivery-Strategies and Challenges* (ed. K. Park), American Chemical Society, Washington, DC, pp. 405–436.
7 Nagarsekar, A. and Ghandehari, H. (1999) Genetically engineered polymers for drug delivery. *J. Drug Target.*, **7**, 11–32.
8 Mooney, D., Hansen, L., Vacanti, J., Langer, R., Farmer, S., and Ingber, D. (1992) Switching from differentiation to growth in hepatocytes: control by extracellular matrix. *J. Cell Physiol.*, **151**, 197.
9 Sengupta, D. and Heilshorn, S.C. (2010) Protein-engineered biomaterials: highly tunable tissue engineering scaffolds. *Tissue Eng. B*, **16**, 285–293.
10 Kumar, A. (2008) Designing new supermacroporous cryogel materials for bioengineering application. *Eur. Cell Mater.*, **16**, 11.
11 David, B. (ed.) (1991) *Biomaterials: Novel Materials from Biological Sources*, Stockton Press, New York, NY.
12 Tina, L.E., Klaus, B., Heinrich, L., and Alexander, S.C. (2001) Identification of a new class of biopolymer: bacterial synthesis of a sulfur-containing polymer with thioester linkages. *Microbiology*, **147**, 11–19.
13 Roger, R., Tor, S., and Ramani, N. (eds) (1992) *Emerging Technologies for Material and Chemicals from Biomass*, American Chemical Society, Washington, DC.
14 Kaplan, D.L. *et al.* (1994) Normally occurring biodegradable polymers, in *Polymer Systemics: Synthesis and Utility* (eds G. Swift and R. Narayan), Hanser Publishing, New York, NY.
15 Bernad, H.A.R. (2010) Bacterial biopolymers; biosynthesis, modification and application. *Nat. Rev. Microbiol*, **8**, 578–592.
16 Roger, C.H. (1993) *Biopolymers making material nature's way*. Background Paper, OTA-BP-E-102, U.S. Government Printing Office, Washington, DC.
17 Carlos, J., Crissman, J., Dorman, M., Mikolajczak, M., Textor, G., Marquet, M., and Ferrari, F. (1990) Genetic engineering of structural protein polymers. *Biotechnol. Prog.*, **6**, 198–202.
18 Crick, F. (1970) Central dogma of molecular biology. *Nature*, **227**, 561–563.
19 Pietzsch, J. (2001) Understanding the RNAissance. In Encyclopedia of Life Sciences. Nature Publishing Group, London, UK.
20 Ross, P., Mayer, R., and Benziman, M. (1991) Cellulose biosynthesis and function in bacteria. *Microbiol. Mol. Biol. Rev.*, **55**, 35–58.
21 Williams, S. and Cannon, R. (1998) Alternative environmental roles for cellulose produced by *Acetobacter xylinum*. *Appl. Environ. Microbiol.*, **55**, 2448–2452.
22 Yasutake, Y., Kawano, S., Tajima, K., Yao, M., Satoh, Y., Munekata, M., and Tanaka, I. (2006) Structural characterization of the *Acetobacter xylinum* endo-β-1,4-glucanase CMCax required for cellulose biosynthesis. *Proteins*, **64**, 1069–1077.
23 Shigeru, Y. and Junji, S. (2000) Structural modification of bacterial cellulose. *Cellulose*, **7**, 213–225.

24 Shima, S. and Sakai, H. (1981) Poly-L-lysine produced by *Streptomyces*: Part II. Taxonomy and fermentation studies. *Argric. Biol. Chem.*, **45**, 2497–2502.

25 Shima, S. and Sakai, H. (1981) Poly-L-lysine produced by *Streptomyces*: Part III. Chemical studies. *Argric. Biol. Chem.*, **45**, 2503–2508.

26 Shima, S. and Sakai, H. (1977) Polylysine produced by *Streptomyces*. *Agric. Biol. Chem.*, **41**, 1807–1809.

27 Shima, S., Matsuoka, H., Iwamoto, T., and Sakai, H. (1984) Antimicrobial action of ε-poly-L-lysine. *J. Antibiot.*, **37**, 1449–1455.

28 Shima, S., Fukuhara, Y., and Sakai, H. (1982) Inactivation of bacteriophages by ε-poly-L-lysine produced by *Streptomyces*. *Agric. Biol. Chem.*, **46**, 1917–1919.

29 Zinn, M., Witholt, B., and Egli, T. (2001) Occurrence, synthesis, medical application of bacterial polyhydroxyalkanoate. *Adv. Drug. Deliv. Rev.*, **53**, 5–21.

30 Jacquel, N. and Lo, C.W. (2008) Isolation and purification of bacterial poly(3-hydroxyalkanoates). *J. Biochem. Eng.*, **39**, 15–27.

31 Doi, Y. and Steinbuchel, A. (eds) (2002) *Biopolymers*, Wiley-VCH Verlag GmbH, Weinheim.

32 Husiman, G.W. and Madison, L. (1999) Metabolic engineering of poly(3-hydroxyalkanoates): from DNA to plastic. *Microbiol. Mol. Biol. Rev.*, **63**, 21–53.

33 Pouton, C.W. and Akhtar, S. (1996) Biosynthetic polyhydroxyalkanoates and their potential in drug delivery. *Adv. Drug. Deliv. Rev*, **18**, 132–162.

34 Kumar, N., Albertsson, A.C., Edlund, U., Teomim, D., Aliza, R., and Domb, A.J. (2005) Polyanhydrides. *Biopolymers*. doi: 10.1002/3527600035.bpol4007.

35 Abraham, J.D., Ludmila, T., and Raphael, N. (1994) Chemical interactions between drugs containing reactive amines with hydrolyzable insoluble biopolymers in aqueous solutions. *Pharm. Res.*, **11**, 865–868.

36 Rosen, H.B., Chang, J., Wnek, G.E., Linhardt, R.J., and Langer, R. (1983) Bioerodible polyanhydrides for controlled drug delivery. *Biomaterials*, **4**, 131–133.

37 Cappello, J. (1992) Genetic production of synthetic protein polymers. *MRS Bull.*, **17**, 48–53.

38 Gosline, J., Lillie, M., Carrington, E., Gurrette, P., Ortlepp, C., and Savage, K. (2002) Elastin proteins: biological roles and mechanical properties. *Philos. Trans. R. Soc. B*, **357**, 121–132.

39 Girrotti, A., Reguera, J., Arias, F.J., Alonso, M., Testera, A.M., and Cabbello, J.C. (2004) Influence of the molecular weight on the inverse temperature transition of model genetically engineered elastin-like pH responsive polymer. *Macromolecules*, **37**, 3396–3400.

40 Olsen, D., Yang, C., Bodo, M., Chang, R., Leigh, S., Baez, J., Carmichael, D., Perala, M., Hamalainen, E.R., Jarvinen, M., and Polarek, J. (2003) Recombinant collagen and gelatin for drug delivery. *Adv. Drug. Deliv. Rev.*, **55**, 1547–1567.

41 Myllyharju, J., Lamberg, A., Notbohm, H., Fietzek, P.P., Pihlajaniemi, T., and Kivirikko, K.I. (1997) Expression of wild-type and modified proalpha chains of human type I procollagen in insect cells leads to the formation of stable [α-1(I)2-α2 (I) collagen heterotrimers and [α-1(I)3 homotrimers but not [α-2(I)3 homotrimers. *J. Biol. Chem.*, **272**, 21824–21830.

42 Bateman, J.F., Lamande, S.R., and Ramshaw, J.A.M. (1996) Collagen superfamily, in *Extracellular Matrix*, 2nd edn (ed. W.D. Comper), Harwood Academic, Amsterdam, pp. 22–67.

43 Merle, C., Perret, S., Lacour, T., Jonval, V., Hudaverdian, S., Garrone, R., Ruggiero, F., and Theisen, M. (2002) Hydroxylated human homotrimeric collagen I in *Agrobacterium tumefaciens*-mediated transient expression and in transgenic tobacco plant. *FEBS Lett.*, **515**, 114–118.

44 Kivirikko, K.I., Myllyla, R., and Pihlajaniemi, T. (1992) Hydroxylation of proline and lysine residues in collagens and other animal and plant proteins, in *Posttranslational Modifications of Proteins* (eds J.J. Harding and M.J.C. Crabbe), CRC Press Inc., Boca Raton, pp. 1–51.

45 Keizer, G.I., Vuorela, A., Myllyharju, J., Pihlajaniemi, T., Kivirikko, K.I., and

Veenhuis, M. (2000) Accumulation of properly folded human type III procollagen molecules in specific intracellular membranous compartments in the yeast *Pichia pastoris*. *Matrix Biol.*, **19**, 29–36.

46 Bodo, B., Chang, R., Hamalainen, E., Leigh, S., McMullin, H., Olsen, D., Revak, T., Yang, C., and Polarek, J. (2004) *Production of triple-helical recombinant human collagen in P. pastoris*. Annual Meeting of the Society for Industrial Microbiology and Biotechnology, Anaheim, CA.

47 Urry, D.W. (2005) *What Sustains Life? Consilient Mechanisms for Protein-Based Machines and Materials*, Spinger, New York.

48 Van, V., Dubruel, P., and Schacht, E. (2011) Biopolymer-based hydrogel as scaffolds for tissue engineering applications: a review. *Biomacromolecules*. doi: 10.1021/bm200083n

49 James, V. and David, K. (2006) Biopolymer-based biomaterials as scaffolds for tissue engineering. *Adv. Biochem. Eng. Biotechnol.*, **102**, 187–238.

50 Peppas, N.A., Bures, P., Leobandung, W., and Ichikawa, H. (2002) Hydrogels in pharmaceutical formulations. *Eur. Pharm. Biopharm.*, **50**, 27–46.

51 Kirker, K.R., Luo, Y., Neilson, J.H., Shelby, J., and Prestwich, G.D. (2002) Glycosaminoglycan hydrogel film as biointeractive dressings for wound healing. *Biomaterials*, **23**, 3661–3671.

52 Kuijpers, A.J., Engbers, G.H.M., Meyvis, T.K.L., Smedt, S.S.C., Demeester, J., Krijgsveld, J., Zaat, S.A.J., Dankert, J., and Feijen, J. (2000) Combined gelatin–chondroitin sulfate hydrogels for controlled release of cationic antibacterial proteins. *Macromolecules*, **33**, 3705–3713.

53 Chang, C.H., Liu, H.C., Lin, C.C., Chou, C.H., and Lin, F.H. (2003) Gelatin–chondroitin–hyaluronan tri-copolymer scaffold for cartilage tissue engineering. *Biomaterials*, **24**, 4853–4858.

54 Muller, F.A., Muller, L., Hofmann, I., Greil, P., Wenzel, M.M., and Staudenmaier, R. (2006) Cellulose-based scaffold materials for cartilage tissue engineering. *Biomaterials*, **27**, 3955–3963.

55 Chang, C.H., Kuo, T.F., Lin, C.C., Chou, C.H., Chen, K.H., Lin, F.H., and Liu, H.C. (2006) Tissue engineering-based cartilage repair with allogenous chondrocytes and gelatin–chondroitin–hyaluronan tri-copolymer scaffold: a porcine model assessed at 18, 24, and 36 weeks. *Biomaterials*, **27**, 1876–1888.

56 Sarkar, N. (1979) Thermal gelation properties of methyl and hydroxypropyl methylcellulose. *J. Appl. Polym. Sci.*, **24**, 1073–1087.

57 Donhowe, I.G. and Fennema, O. (1993) The effects of solution composition and drying temperature on crystallinity, permeability and mechanical properties of methylcellulose films. *J. Food Process. Preserv.*, **17**, 231–246.

58 Tate, M.C., Shear, D.A., Hoffman, S.W., Stein, D.G., and LaPlaca, M.C. (2001) Biocompatibility of methylcellulose-based constructs designed for intracerebral gelation following experimental traumatic brain injury. *Biomaterials*, **22**, 1113–1123.

59 Chen, Y.M., Xi, T.F., Zheng, Y.D., and Wan, Y.Z. (2009) *In vitro* cytotoxicity study of the nano-hydroxyapatite/bacterial cellulose nanocomposites. *Mater. Res.*, **610–613**, 1011–1016.

60 Lee, J.E., Park, J.C., Hwang, Y.S., Kim, J.K., Kim, J.G., and Suh, H. (2001) Characterization of UV-irradiated dense/porous collagen membranes: morphology, enzymatic degradation, and mechanical properties. *Yonsei Med. J.*, **42**, 172–179.

61 Munajjed, A.A. and Brien'O, F.J. (2009) Influence of a novel calcium phosphate coating on the mechanical properties of highly porous collagen scaffolds for bone repair. *J. Mech. Behav. Biomed. Mater.*, **2**, 138–146.

62 Jungreuthmayer, C., Donahue, S.W., Jaasma, M.J., Al Munajjed, A.A., Zanghellini, J., Kelly, D.J., and Brien'O, F.J. (2009) A comparative study of shear stresses in collagen–glycosaminoglycan and calcium phosphate scaffolds in bone tissue-engineering bioreactors. *Tissue Eng. A*, **15**, 1141–1149.

63 Singh, D., Nayak, V., and Kumar, A. (2010) Proliferation of myoblast skeletal cell on three-dimensional supermacroporous cryogels. *Int. J. Biol. Sci.*, **6**, 371–381.

64 Singh, D., Tripathi, A., Nayak, V., and Kumar, A. (2010) Proliferation of chondrocytes on a 3-D modelled macroporous poly(hydroxyethyl methacrylate)–gelatin cryogel. *J. Biomater. Sci. Poly. Ed.* doi: 10.1163/092050610X522486

65 Kumar, A. and Srivastava, A. (2010) Cell separation using cryogel-based affinity chromatography. *Nat. Protocol.*, **5**, 1737–1747.

66 Rammensee, U.S., Slotta, T., Scheibel, A., and Bausch, R. (2008) Assembly mechanism of recombinant spider silk proteins. *Proc. Natl. Acad. Sci.*, **105**, 6590–6595.

67 Hinman, M.B. and Lewis, R.V. (1992) Isolation of a clone encoding a second dragline silk fibroin: *Nephila clavipes* dragline silk is a two-protein fiber. *J. Biol. Chem.*, **267**, 320–324.

68 Lewis, R.V. (1992) Spider silk: the unraveling of a mystery. *Acc. Chem. Res.*, **25**, 392–398.

69 Xu, M. and Lewis, R.V. (1990) Structure of a protein super fiber: spider dragline silk. *Proc. Natl. Acad. Sci. USA*, **87**, 7120–7124.

70 Xiao, X.X., Zhi, G.Q., Chang, S.K., Young, H.P., David, K., and Sang, Y.L. (2010) Native-sized recombinant spider silk protein produced in metabolically engineered *Escherichia coli* results in a strong fiber. *Proc. Natl. Acad. Sci.*, **61**, 14059–14063.

5
Biopolymers from Animals
Khaleelulla Saheb Shaik and Bernard Moussian

5.1
Introduction

Cells, tissues, and the whole body of an organism need to be protected against environmental harm and desiccation for full functionality and survival. Indeed, the invention of desiccation resistance was one major step in evolution when plants and animals invaded land. Most organisms, including plants, fungi, and the majority of animals, face the outer world with a more or less hard covering tunic that protects them against pathogens and predators and prevents their dehydration at the same time. In animals, it additionally serves as an exoskeleton allowing locomotion. In plants and fungi, the protective shield shelters every single cell, while in animals it usually covers the whole body. The cover of these organisms is an ordered extracellular matrix that is produced and organized by the underlying epithelial cells. The central components of these extracellular matrices are commonly polysaccharides, cellulose in plants, chitin and β-1,3-glucan in fungi, and chitin and hyaluronic acid in animals. Interestingly, the plant polysaccharides lack nitrogen residues reflecting the fact that nitrogen is a limiting factor in plant biology. Common to all these polysaccharides is that they are produced by evolutionary conserved membrane-bound glycosyltransferases. Especially, the amino acid sequence in the active center of these enzymes illustrates their close relationship (Figure 5.1) (cellulose synthase (EC 2.4.1.12); chitin synthase (EC 2.4.1.16); and hyaluronan synthase (EC 2.4.1.212)). In addition to these quasi-free polysaccharides, organisms possess protein-bound polysaccharides that cover the exposed surfaces of single cells. These polysaccharides are mainly heparan sulfate and chondroitin sulfate that are linked to specific proteins such as syndecans and perlecans. They are not produced at the plasma membrane, but in the endoplasmatic reticulum (ER) and the Golgi apparatus by specific glycosyltransferases that add different sugar monomers to the growing polysaccharide chain and by modifying enzymes that, for example, add sulfate groups to the sugar residues [1–3].

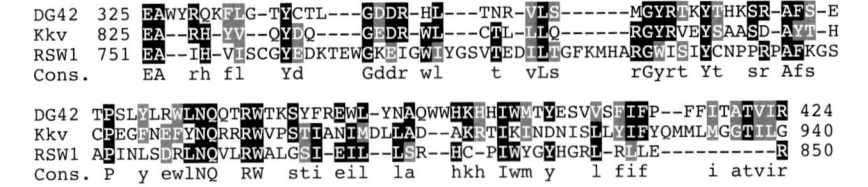

```
DG42  325 EAWYRQKFLG-TYCTL---GDDR-HL---TNR-VLS------MGYRLKYTHKSR-AFS-E
Kkv   825 EA--RH-YV--QYDQ----GEDR-WL---CTL-LLQ------RGYRVEYSAASD-AYT-H
RSW1  751 EA--IH-VLSCGYEDKTEWGKEIGWIYGSVTEDLL IGFKMHARGWISIYCNPPRPAFKGS
Cons.     EA  rh fl    Yd       Gddr wl    t  vLs         rGyrt Yt   sr Afs

DG42  TPSLYLRWLNQQTRWTKSYFREWL-YNAQWWHKHHIWMTYESVVSFIFP--FFITATVIR 424
Kkv   CPEGENEFYNQRRRWVPSTIANIMDLLAD--AKRTIKINDNISLLYIFYQMMLVGGTILG 940
RSW1  APINLSDRLNQVLRWALGSI-EIL--LSR--HC-PIWYGYHGRL-RLLE---------R 850
Cons. P  y ewlNQ  RW    sti eil   la      hkh Iwm y   l fif      i atvir
```

Figure 5.1 Chitin synthase, hyaluronic synthase, and cellulose synthase are related to each other. Chitin synthase, hyaluronic acid synthase, and cellulose synthase belong to the family of glycosyltransferase 2 (GT-2). Here, this relationship is illustrated by the alignment of the active centers of the hyaluronic acid synthase DG42 from *Xenopus laevis*, of the chitin synthase Kkv from *Drosophila melanogaster*, and of the cellulose synthase RSW1 from *Arabidopsis thaliana*.

5.2
Chitin and Hyaluronic Acid in the Living World

Chitin (which means tunic in Greek, χιτών) is after cellulose, the second most abundant polysaccharide in the living world. It is a defining trait of arthropods and fungi. Besides these taxa, it is also found in almost all invertebrates studied to date, namely, sponges, nematodes, polychaetes, and molluscs [4–8]. The abundance of chitin is also reflected by the presence of chitin-degrading enzymes, named chitinases in many different taxa, which use this weapon against chitin-producing pathogens and invaders without producing chitin themselves. Chitinases in insects have been studied extensively [9–11]. An important function of chitinases is to facilitate molting and to degrade the tracheal luminal chitin that is needed to establish tracheal tube diameter. These functions however strictly taken have not been demonstrated.

Hyaluronic acid or hyaluronan is present in vertebrates as a central element of cartilage and in some bacteria such as *Streptococcus pyogenes* [12]. In cartilage, hyaluronic acid binds to the extracellular glycosaminoglycan aggrecan, thereby immobilizing and retaining it at high concentrations required for compressive resilience [13]. Hyaluronic acid also plays an important role during development of vertebrates facilitating the migration of cells during gastrulation that leads to the separation of the three germ layers [14]. Hence, in brief, in contrast to chitin that seems to be important only as a structural element, hyaluronic acid assumes two different roles: like chitin as a structural component of certain extracellular matrices and as an essential factor in guiding development.

5.3
Milestones in Chitin History

Chitin was first discovered as a natural molecule occurring in the cell wall of fungi by the French botanist and chemist Henri Braconnot in 1811 (Figure 5.2) [15,16]. He named the substance that he isolated "fungine" after cooking different species

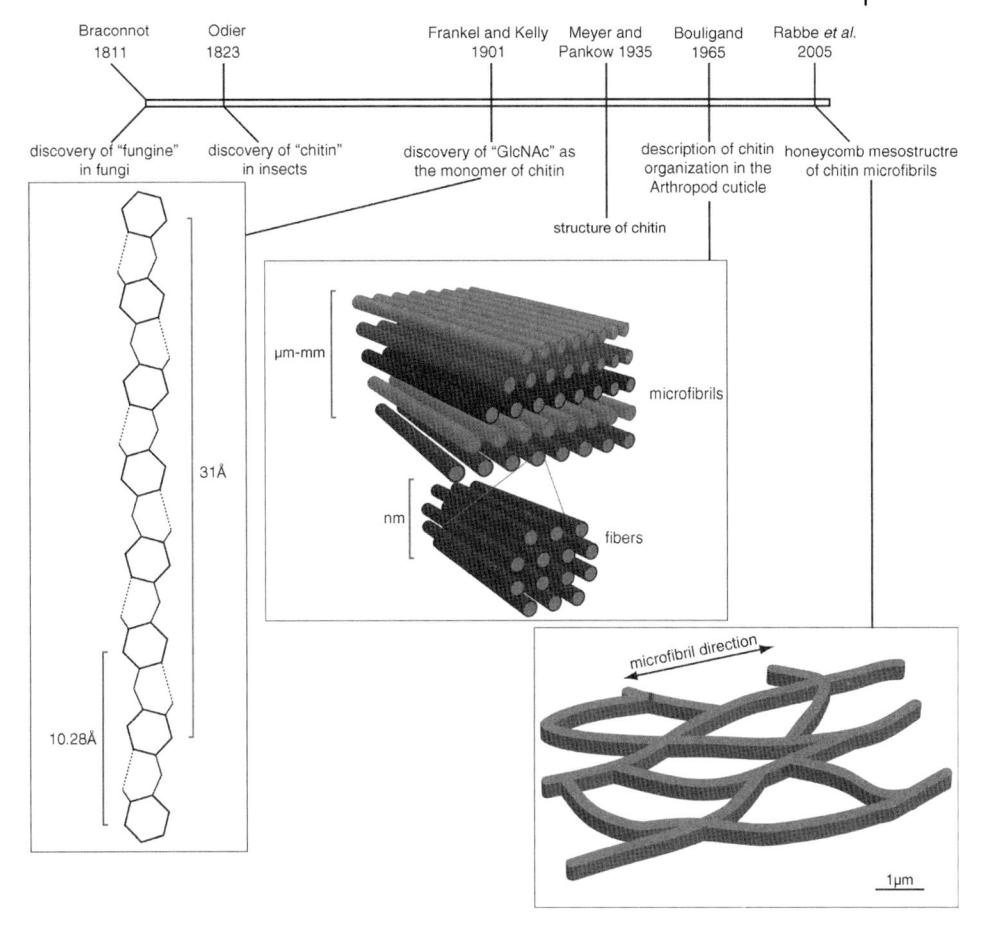

Figure 5.2 Chitin history. Chitin (fungine) was discovered as a component of fungal cell walls in 1811 by Henri Braconnot. Independently, Auguste Odier found in 1823 that a main component of the arthropod cuticle is a substance he called "chitine." The next milestone was in 1902 when Fränkel and Kelly identified an amino sugar to be the monomer of chitin. It was in 1935 when Meyer and Pankow [17] described for the first time the structure of chitin by X-ray diffraction analyses. Subsequent measurements have refined the model for chitin structure. The distance between two residues with the same orientation of the N-acetyl position is 10.28 Å. Chitin can therefore be considered as the polymer of two GlcNAc residues, chitobiose. The measured repeated unit is 31 Å suggesting that some acetyl groups are absent. In 1965, Bouligand [19] convincingly described the organization of chitin microfibrils (which are bundles of chitin fibers) in arthropod cuticles as twisted plywood. Finally, in 2005, the group of Raabe discovered that chitin microfibrils in the cuticle of the lobster *Humarus americanus* are not strictly arranged in parallel but rather form a honeycomb-like structure (adopted Ref. 24) that may contribute to the mechanical strength of the cuticle.

of fungi in alkali water. Interestingly, he noted that fungine contains considerable amounts of nitrogen as opposed to "wood." In 1823, the French naturalist Odier could demonstrate that what he named "chitine" is a component of arthropod cuticles, particularly in insects [17], without realizing that he had discovered the same substance as Braconnot in 1811. As Braconnot, Odier used simple stepwise chemical extraction methods. It took several decades until the monomer of chitin was characterized by Fränkel and Kelly [18], who probably knew about the presence of chitin in both arthropods and fungi. The structure of chitin was first described in X-ray diffraction experiments by Meyer and Pankow in 1935 [19]. Chitin structure has repeatedly been studied since then, the last study being published in 2011 [20]. This reflects the importance of this polysaccharide in material science. Another milestone in chitin research was the discovery by Bouligand in 1965 that chitin adopts a stereotypic arrangement in arthropod cuticles [21]. The organization of the arthropod cuticle chitin is described in Section 5.8. Finally, recently, the group of Raabe in Düsseldorf (Germany) found through extensive analyses of the cuticle ultrastructure by electron microscopy that chitin microfibrils in the lobster cuticle are arranged in a way to give rise to a honeycomb mesostructure (as shown in the following) that is thought to confer even more stability to the exoskeleton [22–26]. Whether this level of organization also exists in other arthropods remains to be shown.

The molecular constitution of hyaluronic acid or hyaluronan was found in 1934 by Meyer and Palmer [27] to be a polymer of "uronic acid and an amino sugar." Two decades later, in 1954, the chemical identity of hyaluronic acid was deciphered to be glucuronic acid alternating with N-acetylglucosamine [28].

5.4
From Trehalose to Chitin

Chitin $(C_8H_{13}O_5N)_n)$ is a polymer of N-acetylglucosamine (GlcNAc) that is produced by chitin synthases residing in the plasma membrane. Chitin synthases are inserted into the plasma membrane via 12–18 transmembrane domains that are organized such that the active center of the enzyme faces the cytoplasm, where GlcNAc residues activated by UDP (UDP-GlcNAc) are polymerized by the formation of β-1,4 linkages between GlcNAc residues before being extruded to the extracellular space. About 17–19 chitin fibers are assembled to microfibrils that are 3 nm in diameter and on average about 0.3 μm long [29,30].

The source of chitin is ultimately trehalose, which is the main storage sugar in the hemolymph (blood) of insects [31,32]. Trehalose is a disaccharide composed of two glucose molecules that are linked to each other through an α,α-1,1-glycosidic linkage abolishing the reducing nature of the anomeric carbon. The nonreducing properties of trehalose enable insects to store this sugar at high concentrations (1–2%) in the hemolymph instead of glucose, which is toxic at these concentrations due to its anomeric carbon-reducing properties. Thanks to its ability to make hydrogen bonds as substitute to the water, trehalose may also serve as a cryoprotectant preventing denaturation proteins and membranes [33]. Prior to chitin synthesis, trehalose is taken up by chitin-producing cells at their basal side through the function of membrane-inserted trehalose transporters [34]. It is then hydrolyzed to two glucose residues that eventually are polymerized to glycogen, which is used as an energy source and feeds at the same time the Leloir pathway that produces UDP-GlcNAc (Figure 5.3) [1,35]. In the fruit fly *D. melanogaster*, the supply of production of glycogen in the epidermal cells is under hormonal control. The nuclear receptor DHR38 controls the amounts of glycogen through the regulation of the expression of several genes coding for enzymes of the glucose metabolism such as glucose-1-phosphate mutase, which also acts in the Leloir pathway [36].

In the Leloir pathway (Figure 5.3), UDP-GlcNAc production runs through four cytoplasmic enzymes: glucosamine-6-phosphate synthase, also called glutamine: fructose-6-phosphate aminotransferase 1 (Gfat1, EC 2.6.1.16), glucosamine-6-phosphate acetyltransferase (EC 2.3.1.4), phosphoacetylglucosamine mutase (EC 5.4.2.3), and UDP-GlcNAc pyrophosphorylase (EC 2.7.7.23). This pathway is conserved in all organisms ranging from yeast to man.

UDP-GlcNAc is the substrate not only of chitin synthases but also of enzymes involved in *N*-glycosylation and GPI-anchor formation (Figure 5.3) [1]. Hence, chitin synthases compete for their substrates with several other enzymes that have essential functions during the production of the extracellular matrix. How is this competition controlled? A conceivable possibility is that UDP-GlcNAc is available in excess accommodating all different reactions. Another means to provide sufficient amounts of GlcNAc is to separate its supply spatially. Indeed, many animals possess at least two versions of the last enzyme of the Leloir pathway, UDP-GlcNAc pyrophosphorylase, either through alternative splicing of one gene or through the presence of two physically separate genes coding for this enzyme. In the fruit fly *D. melanogaster*, one splice variant of this enzyme harbors a longer N-terminal tail that contains features of ER associating signals [37]. This isoform could specifically produce UDP-GlcNAc for the reactions occurring in the ER, that is, glycosylation and GPI-anchor formation, while the other isoform could produce UDP-GlcNAc solely destined for chitin synthesis.

As a model organism, *D. melanogaster* has been very useful to study the biological importance of the Leloir pathway in multicellular organisms. Mutations in the gene coding for the UDP-GlcNAc pyrophosphorylase called *mummy* (*mmy*) are embryonic lethal and affect the correct development of different organs such as the nervous system, the respiratory organ (tracheae), and the skin [37–39]. Some defects observed in the tracheae and the epidermis are associated with reduced chitin. For instance, the cuticle of *mmy* mutant larvae is devoid of chitin and collapses. As a consequence, these animals do not possess a functional exoskeleton and are therefore unable to hatch. Moreover, presumably due to the disorganized cuticle, the *mmy* mutant cuticle is unable to prevent water loss, which may also contribute to the lethality of the mutation. Hypomorph mutations in *mmy*, interestingly, have no effect on glycosylation, but impair chitin organization, which may be caused by attenuated chitin synthesis. This genetic argument supports the idea that different GlcNAc consuming biochemical reactions have a different affinity to GlcNAc.

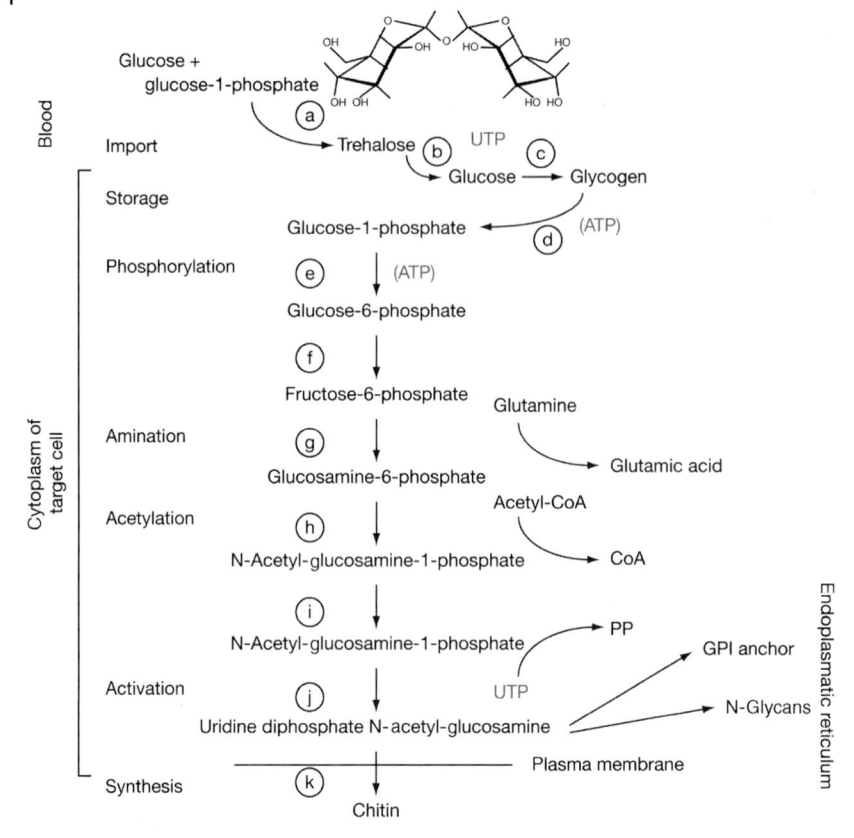

Figure 5.3 Biochemistry of chitin synthesis. The entire flowchart of chitin synthesis starting with glucose as a food source. Glucose is dimerized to form trehalose by the trehalose synthase (a). Trehalose is taken up at the basal side of the cells by the trehalose transporter and subsequently hydorlyzed to glucose and glucose-1-phosphate by the trehalose (EC 3.2.1.28) (b). Glucose is then used to build up glycogen by the glycogen synthase (EC 2.4.1.11) (c). Through the glycogen phosphorylase (EC 2.4.1.1) (d), glucose is again released as glucose-1-phosphate, which is isomerized to glucose-6-phosphate by the glucose-1-phosphate mutase (EC 5.4.2.2) (e). Through the activity of the glucose-6-phosphate isomerase (EC 5.3.1.9) (f), glucose-6-phosphate is isomerized to fructose-6-phosphate, the substrate of glucosamine-6-phosphate synthase, also called glutamine:fructose-6-phosphate aminotransferase 1 (Gfat1, EC 2.6.1.16) (g) that produces glucosamine-6-phosphate. Next, glucosamine-6-phosphate is acetylated by the glucosamine-6-phosphate acetyltransferase (EC 2.3.1.4) at the C2 position (h), which in turn is converted to glucosamine-6-phosphate by phospho-GlcNAc mutase (EC 5.4.2.3) (i). Finally, UDP-GlcNAc is formed by the UDP-GlcNAc pyrophosphorylase (EC 2.7.7.23) (j) using UTP and GlcNAc as substrates. GlcNAc is polymerized by chitin synthases (k) using UDP-GlcNAc as a substrate. GlcNAc is also used to form N-glycans and is an intermediate component of GPI anchors. In summary, the entire reaction consumes at least two UTPs and some amounts of ATP. Taking into account that glucose stores the energy equivalent of around 37 ATPs, the entire chitin-producing pathway is worth around 40 energy units per GlcNAc residue illustrating that chitin synthesis is an energy-consuming reaction.

This may have simple biochemical reasons. The kinetics of enzymatic reactions reflecting either the specific Michaelis constants of GlcNAc consuming enzymes or the actual amount of enzymes acting in the respective pathway may be different. Hence, to ensure stereotypic and robust functions, the cell has to overcome these differences by producing an excess of GlcNAc. This view does not *per se* contradict the compartmentalization of UDP-GlcNAc pyrophosphorylase activity.

In conclusion, chitin synthesis starting with trehalose is an energy-consuming biochemical reaction that requires tight control to ensure stereotypic development of the organism. Amazingly, despite the high amount of energy stored in chitin, arthropods shed of their cuticle after each molting with little if at all any recovery of the substances incorporated in the cuticle.

The Leloir pathway also supplies hyaluronic acid synthesis that besides UDP-GlcNAc requires equal amounts of UDP-GlcA. UDP-GlcA is produced by the UDP-Glc 6-dehydrogenase (EC 1.1.1.22). A chain of hyaluronic acid may contain 2500–25 000 repeating disaccharides. Predominantly, hyaluronic acid is found in vertebrates and plays a role as a structural component of extracellular matrices and as a milieu in which cells can migrate during development. Prominent extracellular matrices that contain hyaluronic acid are cartilage and the developing bone. Like chitin, hyaluronic acid is a linear polysaccharide that in contrast to chitin, however, has a high water retention capacity. Indeed, hyaluronic acid may retain high amounts of water constituting a gel-like matrix. Hyaluronic acid is therefore an excellent resilient and robust substance.

5.5
Chitin Synthase

Chitin synthases are family 2 glycosyltransferases that have multiple transmembrane domains inserting the enzyme into the plasma membrane (Figure 5.4). Insects possess two chitin synthases, while crustaceans have three [40]. One insect chitin synthase acts in the ectoderm-derived tissues such as epidermis, tracheae, and hindgut, while the other one produces chitin in the midgut where it is required to protect the midgut epithelium against pathogens. The function of the third chitin synthase in crustaceans is unknown. The two insect chitin synthases are highly similar over the entire sequence of the midgut chitin synthase. The sequence of the ectodermal chitin synthases, however, is longer. At the C-terminal or the catalytic center of the enzyme, there is an additional coiled-coil domain and three transmembrane domains. Due to these differences, although it has not been demonstrated yet, it can be expected that besides their enzymatic activity they are unable to replace the function of each other. The coiled-coil domain of the ectodermal enzyme suggests that this isoform may interact with partners that do not interact with the midgut enzyme. The differences in protein sequences moreover reflect the organization of the extracellular matrix they are contributing to. While the cuticle of the epidermis and the tracheae is highly organized, the peritrophic matrix that covers the midgut epithelium is rather unorganized. Mutations in the epidermal and

(a) β-chitin (b) α-chitin (c) γ-chitin

(d) chitin / chitin synthase dimer / plasma membrane / monomers

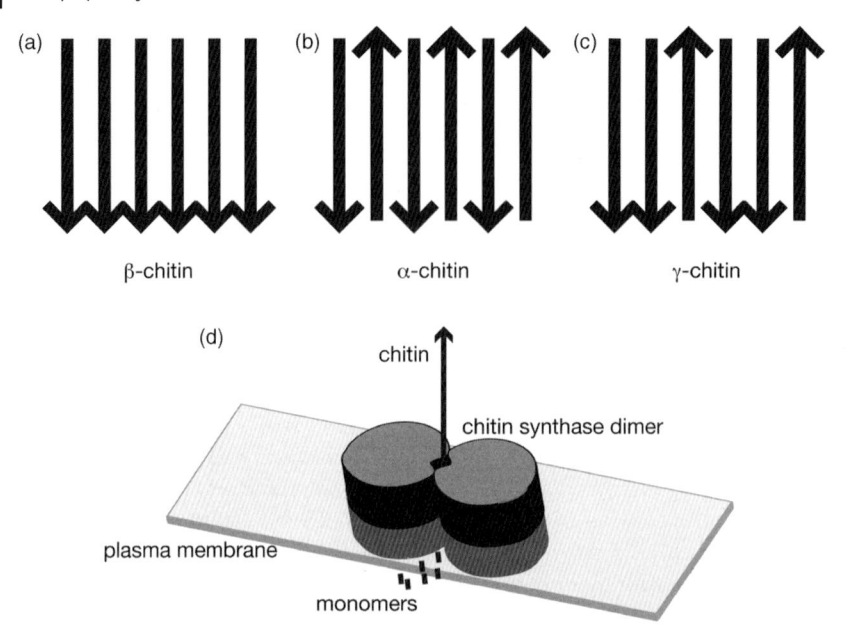

Figure 5.4 Chitin orientation. Three types of second-order chitin fiber orientation can be distinguished. The arrows point from the nonreducing to the reducing end. In the β-chitin arrangement, all fibers point to the same direction. In the α-chitin arrangement, fibers run antiparallel. In the γ-chitin arrangement, two fibers run in the same direction, while the third in the opposite direction. The organization found in the arthropod cuticle is α-chitin. As shown in part (d), chitin synthases, which are membrane-inserted glycosyltransferases, do not function alone but rather dimerize or multimerize. The activated monomers of chitin are supplied at the cytosolic side of the enzymes that subsequently polymerize them forming chitin fibers, which are extruded into the extracellular space.

tracheal chitin synthases in Drosophila cause a complete loss of chitin and detachment of the cuticle from the underlying epidermal and tracheal cells and excessive protein secretion suggesting that cuticle differentiation involves cross talk between chitin synthesis and protein secretion [41]. No mutations in the midgut chitin synthase are known. However, using immunohistochemical detection of chitin synthase in the midgut of the tobacco hornworm Manduca sexta, Zimoch and Merzendorfer have demonstrated that chitin synthase resides at the tip of the brush border microvilli [42]. Moreover, they showed that the activation of the midgut chitin synthase is under proteolytic control [43,44]. The chemotrypsin-like protease CTLP1 interacts with the midgut chitin synthase CHS2 at the tip of microvilli. The authors propose that this interaction takes place in response to the nutritional state of the animal.

Can we learn more about chitin synthases by looking at the situation in fungi? Fungi, for example, possess 3 chitin synthases in the yeast Saccharomyces cerevisiae and 20 chitin synthases in certain mucoromycete species. The main finding, regardless of which chitin synthase has been investigated, is that chitin synthases do not function alone. They rather require the assistance of an arsenal of cofactors and their anchoring to the cytoskeletal elements [45]. For instance, in the yeast *S. cerevisiae*, the CHSIII complex associates with the chain of Chs4p–Bni4p–Cdc10p, Cdc10p being a septin. These interactions are not required for Chs3p activity but for its correct localization. Deletion of Bni4p, for instance, causes mislocalization of Chs3p, which in turn results in aberrant cell shape. Concomitantly, Chs3p localization involves Chs5p and Myo2p, linking Chs3p to the actin cytoskeleton [46]. Like in the *M. sexta* midgut, the yeast Chs3p interacts with a protease, namely, the CaaX protease Ste24 [47] and requires this interaction for full function. An interesting aspect of chitin microfibril length regulation is encountered in *Candida albicans* [48]. The length of chitin microfibrils depends on the chitin synthase in action. For instance, Chs8p synthesizes long chitin microfibrils, whereas Chs3p synthesizes rather short ones at the same site. As many fungi such as *C. albicans* are human pathogens, the lack of chitin in vertebrates destines chitin synthesis as a target in medication of fungal infections. Chitin synthase inhibitors are discussed in Section 5.12.

It seems to be a paradigm that synthesis of quasi-free polysaccharides by glycosyltransferase takes place at the plasma membrane. Vertebrate hyaluronic acid synthases (HAS) are like chitin synthases membrane-inserted glycosyltransferases. The monomers of hyaluronic acid UDP-GlcNAc and UDP-GluA enter by turns the active center of the enzyme where they are linked together and extruded into the extracellular space. Hyaluronic acid chains can be up to 25 μm long and weigh 10^3–10^4 kDa. Interestingly, HAS activity is sufficient to induce plasma membrane protrusions [49]. Hyaluronic acid synthesis that occurs at tips of microvilli thus somehow regulates the formation of its own platform. It is not known to what extent the production of hyaluronic acid at the tips of microvilli is important for the organization of the extracellular matrix. Whether chitin synthases are able to influence plasma membrane dynamics as well remains to be investigated. In cartilage that is produced by chondrocytes, hyaluronic acid is synthesized by HAS2 and undergoes tight interaction with aggrecan forming large aggregates that account for cartilage structure and confer compressive resilience [13]. In the extracellular space, the major hyaluronic acid-interacting protein is the hyaluronic acid receptor CD44, a GPI-linked protein inserted into the plasma membrane [50]. CD44 mediates the linkage of hyaluronic acid with the cytoskeleton by recruiting ankyrin and ezrin/radixin/moesin, which in turn interacts with the actin cytoskeleton. Moreover, through Rho, Ras, and the c-Src tyrosine kinase, it activates different signaling pathways that change cell behavior. A linkage of chitin with the cytoskeleton occurs probably only at muscle attachment sites (apodemes). Here, a fortified connection between the specialized epidermal cells and the cuticle is necessary to withstand locomotion forces [51].

5.6
Regulation of Chitin Synthesis in Fungi

The reactions leading to chitin deposition into the extracellular space do not proceed automatically, but need to be controlled during the life cycle of the organism. Control of chitin synthesis is extensively studied in fungi [52–54]. In the yeast *S. cerevisiae*, chitin synthesis is regulated at the level of positioning the chitin synthase enzyme (CHS) into the plasma membrane [55]. Indeed, several proteins acting in the secretory pathways are required for this process. Chs4p, Chs6p, and Chs7p are Golgi-associated proteins, which upon a life cycle-specific signal interact to deliver the chitin synthesis complex consisting of the enzyme Chs3p to the plasma membrane. Although these factors are specific to fungi and are not found in other taxa, some general factors such as the G-protein Rho1p and protein kinase Pkc1p that are involved in intracellular transport and cytoskeleton remodeling are also required for Chs3p localization from cytoplasmic membrane stores to the plasma membrane [56]. These vesicles that contain the chitin synthesis complex are distinct from other kinds of vesicles. They were discovered by Bartnicki-Garcia and coworkers in 1975 in the fungus *Mucor rouxii* and because of their uniqueness they called them chitosomes [57–59]. Chitosomes are spherical membrane structures and have a diameter of around 60 nm. Since their discovery, they have also been found in other fungi, although their existence is still under debate [59]. In conclusion, some of the chitin synthase localization proteins do not have any orthologues in other organisms than fungi suggesting that besides the basic biochemical reaction of chitin synthesis, the control mechanisms have evolved completely independently.

Indeed, in *D. melanogaster* and *Tribolium castaneum*, several factors that are absent in fungi have been identified and characterized that play an essential role in chitin production and organization. These factors are presented in Section 5.8.

5.7
Organization of Chitin in the Fungal Cell Wall

To evaluate the organization of chitin in animals, it may be useful to first consider its organization in the cell wall of fungi. The fungal cell wall is, however, diverse and differences of its composition and structure reflect the evolution of fungi [60]. A scheme depicting the prototype of a fungal cell wall is therefore not meaningful. In any case, chitin is an important component of all fungal cell walls, the amount of chitin varying between fungal species [60–62]. To fulfill its function in cell wall integrity, chitin microfibrils, for instance, in *Conidiobolus obscurus*, do not seem to adopt a preferred orientation in the fungal cell wall [61]. Originally, chitin seems to be the first polysaccharide of fungal cell walls, with the abundant and membrane-bound polysaccharides β-1,3 and β-1,6 glucans being a secondary acquisition. Chitin covalently interacts with β-1,3 and β-1,6 glucans constituting a polysaccharide matrix with embedded and interacting proteins that withstands the pressure of the cell [63]. Linkage of β-1,3 glucan and chitin occurs in the extracellular space and is catalyzed in yeast by Crh1p and Crh2p, two relatives of glycosyltransferases [64]. β-1,3 Glucan-like chitin is produced by a membrane-inserted glycosyltransferase that utilizes UDP-glucose as a substrate. Thus, one can consider the fungal cell wall as lattice of mainly two linked polysaccharides. These polysaccharides are, however, not naked, but interact with proteins that bind to them and modify them. One of these proteins is Gelp in *Aspergillus fumigatus* that elongates β-1,3 glucan chains after the main polymerization reaction [65]. Cell wall formation also involves chitin deacetylases converting chitin partially into chitosan; these enzymes are also believed to be ancient players of fungal cell wall construction [66].

Several classes of secreted proteins have been proposed to be essential for cell wall physiology, integrity, and polysaccharide organization [54,60,67]. These proteins that are specific to fungal families are generally characterized by sequence repeats and may therefore play a structural rather than a regulatory role. In ascomycetes-like yeast, GPI-cell wall proteins (GPI-CWPs) and protein-with-internal-repeat (Pir)-cell wall proteins (Pir-CWPs) bind to β-glucans [67]. The group of GPI-CWP comprises both factors: those that act at the plasma membrane being anchored via their GPI moiety and those that are released from their anchor to function further away from the plasma membrane. Another class of cell wall proteins, named Flo, is characterized by the requirement of its members during yeast flocculation [68,69]. The exact molecular importance of these cell wall proteins is still obscure.

Taken together, the construction of the fungal cell wall implies the deployment of multiple mechanisms acting at the plasma membrane and within the cell wall. The result is an armor harboring a polysaccharide network and associated modifying or structural proteins. Interactions between polysaccharides and CWPs do not occur randomly, but give rise to a quasi-stratified extracellular matrix. These cell wall proteins predominantly occupy the outermost layer, while chitin fibers contact the plasma membrane [67,70]. The plasma membrane of fungal cells, besides being the interface of secretion and the platform for glycosyltransferases, has not been reported to play any active role in cell wall organization as it does during chitin synthesis in arthropods and hyaluronic acid synthesis in vertebrates (see the following section). The construction of the fungal cell wall is not a one-way process, but involves feedback controls. Indeed, one interesting property of the fungal cell wall is that the reduction of certain cell wall-organizing activities initiates a compensatory response resulting in overproduction of chitin [71,72]. Despite our knowledge on details of fungal cell wall assembly, a full picture of coordination of the underlying mechanisms remains to be drawn.

5.8
Organization of Chitin in the Arthropod Cuticle

With some exceptions, chitin biology is quite different in arthropods. In 1965, based on extensive electron microscopic work, the French biophysicist Bouligand proposed a model for the arrangement of chitin in the cuticle of crustaceans that is now widely accepted [21]. In principle, he analyzed consecutive sections of a specimen from different angles. Basically, microfibrils, which are bundles of 17–19 α-chitin fibers, are arranged in parallel forming horizontal sheets, called laminae, that are stacked helicoidally and cover the animal (Figures 5.2, 5.4, and 5.6). Subsequently, it was shown that chitin organization is also similar in the insect cuticle [73].

This arrangement is neither found in the fungal cell wall nor in the beaks of squids [7]. In other words, chitin organization as described by Bouligand is an arthropod-specific trait; in consequence, the mechanisms responsible for this organization have been evolved only in arthropods. This means that arthropod chitin synthesis complexes produce and extrude chitin fibers nonrandomly into the extracellular matrix. An intriguing role in chitin organization has been classically attributed to regular protrusions of the apical plasma membrane that are called microvilli, which like microvilli of gut epithelia contain a core of actin fibers. The tips of these structures carry the chitin synthesis complex, which is visible in electron micrographs as electron-dense plaques. Periodic and synchronized swinging of the microvilli has been proposed to be the cellular mechanism of chitin orientation, that is, organization (Figure 5.5).

In the *Drosophila* embryo, the situation at the apical plasma membrane during chitin synthesis and organization may be different [74]. Instead of the islands of microvilli, the apical plasma membrane forms longitudinal microtubuli-containing corrugations that run parallel to each other. These corrugations that are called *apical undulae* are ordered perpendicular to the anterior–posterior axis of the cuticle-producing embryo. Chitin microfibrils that are produced at the crests of these corrugations are extruded at an angle of 90° to them. This strict architecture suggests that it is reflecting the cellular mechanism of chitin synthesis and organization. In other words, these structures have two functions: supporting chitin synthesis and allowing chitin orientation at the same time. The actual mechanism, however, is unclear as the movement of chitin microfibrils across the apical corrugations in a static view is not self-evident. In plants, that is, in the thale cress *A. thaliana*, the cellulose synthase complex CESA of epidermal cells moves along microtubule tracks while synthesizing cellulose, thereby pulling cellulose fibers in one direction, that is, parallel to the microtubules [75,76]. Is a similar dynamic mechanism imaginable in arthropods? Movement of the chitin synthase complex along the microtubules would pull chitin microfibrils into the same direction and would therefore not explain the observed pattern in *Drosophila*. Movement of the chitin synthase complex from crest to crest is also unlikely as there is no ultrastructural evidence for such behavior. An alternative static view would imply a more active mode of chitin microfibril orientation by specialized factors that occupy

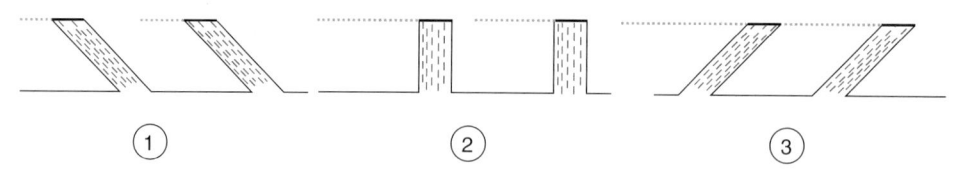

Figure 5.5 Chitin orientation by microvilli. Classically, microvilli that protrude at the apical plasma membrane are considered as the cellular structure, which is essential for chitin microfibril orientation. These actin-filled (dashed lines) structures oscillate back and forth from 1 to 3, while they are synthesizing chitin (dotted lines) at their tips that carry the chitin synthesis plaques (thick black line). The molecular mechanisms for this massive oscillation are absolutely obscure.

distinct positions at the crest of the immobile apical undulae or within the extracellular space. Besides in *Drosophila*, apical undulae have not yet been discovered in other arthropods.

How is chitin synthesis related to the construction of other parts of the arthropod cuticle? Chitin synthesis takes place during cuticle differentiation, which in the embryo runs through a sequence of events seemingly conserved in different arthropod species [40,74,77]. First, at mid-embryogenesis, the outermost envelope that is composed of lipids and proteins is deposited at the tip of irregular membrane protrusions. The envelope fragments eventually fuse to form a continuous film covering the animal. Next, the epicuticle that mainly harbors proteins and is devoid of chitin is assembled beneath the envelope. The stereotypic architecture of the arthropod cuticle is shown in Figure 5.6. At the same time, chitin is synthesized that associates with chitin binding proteins to constitute the procuticle. Often, the procuticle is bipartite with an upper exocuticle with densely packed laminae and a lower endocuticle with less densely packed laminae. The maturation of the envelope and the epicuticle occurs while the epidermal cells are producing chitin. Hence, the construction of the cuticle necessitates mechanisms to transport envelope and epicuticle material through the forming laminae. Indeed, as presented below (Section 5.10), secretion is not an automated process during cuticle differentiation, but must be tightly regulated to ensure a stereotypic cuticle structure. Since the crests of the apical undulae carry the chitin synthesis and probably also the

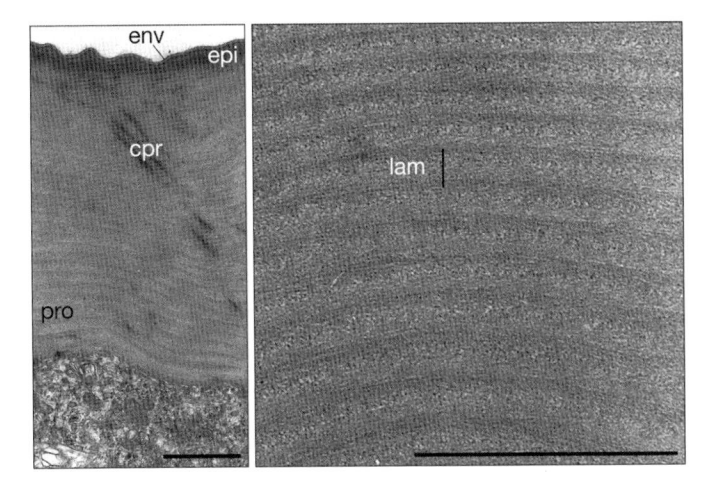

Figure 5.6 The prototype of the arthropod cuticle consists of three biochemically distinct composite layers. The outermost layer is the envelope (env) that is made up of proteins and lipids. The epicuticle (epi) beneath contains proteins that may be either directly cross-linked to each other or bridged by catecholamines. The thickest layer that contacts the apical surface of the cell is the procuticle (pro) that harbors a chitin–protein matrix. Chitin is organized in sheets called laminae (lam) that are stacked horizontally. Especially at muscle attachment sites (apodemes), procuticle proteins (cpr) become apparent. The identity of these proteins in not known. A good candidate is Dumpy (Dp) (see Section 5.9).

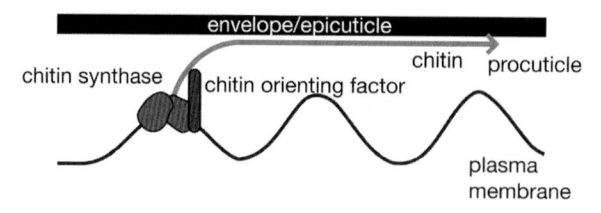

Figure 5.7 Model for chitin microfibril assembly and orientation. Chitin is synthesized at the crest of regular protrusions of the apical plasma membrane of epidermal cells called the apical undulae. The crest of these structures also harbors chitin-orienting factors that allow the polymerization of chitin fibers in a preferred direction. Chitin orientation could also involve the passive or active contribution of the cuticle layers formed above the chitin layer, the envelope, and the epicuticle. Due to spatial constraints, chitin fibers are forced to run in one direction.

orientation complex, secretion of cuticle material runs at the valley between the crests.

The presence of a stable layer above the chitin harboring one adds a new notion in speculating on mechanisms coupling chitin synthesis and orientation. Besides the chitin synthase, other factors may be involved acting either at the crest of the apical undulae or in the extracellular matrix guiding microfibrils in one direction (Figure 5.7). A potential extracellular factor could reside in the layer above the chitin harboring procuticle that is produced at least partially before the chitin synthesis starts. The pushing force of chitin synthesis and the binding of the putative extracellular factor to chitin are insufficient to explain directionality. The elucidation of this issue will be the most challenging task in the future.

Another major problem in chitin synthesis is the regulation of chitin fiber length and the cleavage, that is, release of chitin fibers from the synthesis complex. This problem becomes more complex considering that chitin fiber length might be different in different tissues and stages. A specific chitinase activity for this purpose has not yet been discovered. Random release is improbable, as this would not ensure a minimal length required for microfibril and lamina assembly. There is no evidence that this is an intrinsic function of (insect) chitin synthases. A solution of the problem as occurring in *C. albicans* already described, that is, the employment of different chitin synthases synthesizing chitin fibers of distinct length, is only of limited significance for arthropods since they possess only two chitin synthases expressed in cuticle or peritrophic matrix-producing tissues. Nevertheless, the *Candida* example suggests that chitin synthases may have an intrinsic property to regulate the length of chitin fibers they produce. Alternatively, different chitin synthases recruit different types of cofactors, which in turn determine chitin fiber length – an unsolved situation.

One major problem in understanding the mode of function of the arthropod chitin synthase is the antiparallel arrangement of chitin fibers in α-chitin bundles (Figure 5.4). How are fibers with opposed directions assembled? The classical model of chitin synthesis at the tip of microvilli-like structures would nicely explain how α-chitin is produced (Figure 5.5). At each end of an oscillation, a new chitin fiber is initiated. Considering the crest of the apical undulae as the site of chitin synthesis

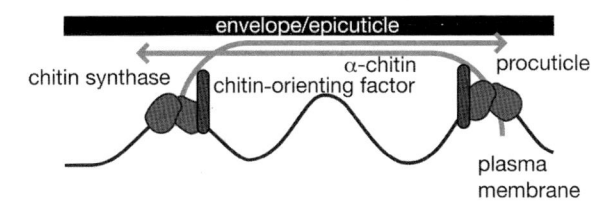

Figure 5.8 Combined model for α-chitin assembly and chitin microfibril formation. Chitin fibers in the alpha conformation run antiparallel to each other (see Figure 5.4). To produce this arrangement, chitin fibers may be synthesized by chitin synthesis complexes located at different apical undulate crests. By an unknown mechanism, the adjacent chitin fibers are fixed probably by chitin binding proteins, which thereby stabilize this arrangement forming microfibrils that contain around 17 chitin fibers.

and orientation, the situation becomes difficult. Oscillation of an entire longitudinal structure does not seem a probable explanation. In this model, conceivably, at least two chitin synthesis complexes at a certain distance from each other have to act together to order chitin fibers in such a manner (Figure 5.8). About the interaction between chitin synthases we know that in the midgut of the tobacco hornworm caterpillar, the chitin synthase complex seems to consist of three chitin synthases, which are located at the tips of the microvilli of the brush border. This is different from the rosettes of six cellulose synthase complexes that assemble cellulose bundles at the center of the hexamer. The composition of chitin synthase complexes in the epidermis awaits thorough analyses.

5.9
Chitin-Organizing Factors

Chitin organization may either be an intrinsic property of chitin or depend on the interaction between chitin and chitin binding and organizing proteins. From the first assumption, it would follow that unordered chitin in the fungal cell wall presupposes interaction with proteins that prevent the intrinsic forces in chitin to self-assemble. In arthropods, there is ample evidence that chitin organization such as observed in the cuticle very much involves the participation of structural as well as enzymatic active proteins. In the past 5 years, using *D. melanogaster* as a model insect, several groups have identified and characterized a number of factors that are essential for correct chitin organization [78]. Some of these factors are found only in arthropods, some others have counterparts in other phyla.

Some of the findings in *D. melanogaster* have been substantially extended using the red flour beetle *T. castaneum* as a model insect [78]. A central factor required for chitin organization is Knickkopf (Knk) [79]. The 689 amino acids of *D. melanogaster* Knk harbor three different types of domains: two N-terminal DM13 domain, a middle DOMON domain, and *n* C-terminal plastocyanin domain. The plastocyanin domain is known from plants where it is involved in electron transport during photosynthesis domains. The DM13 domain and the DOMON domain have been proposed to be implicated in electron transfer [80,81]. Hence, this notion suggests that Knk may be involved in electron transfer during chitin synthesis, a chemical reaction that is not intuitively understandable as there is no step during chitin synthesis that obviously presupposes electron transfer. The release of chitin chains from the chitin synthase complex may however be such a step, similar to a reaction that occurs during cellulose degradation. In phytopathogenic and saprotropic fungi, cellulose is degraded through the extracellular flavocytochrome dehydrogenase that oxidizes cellulose [82]. Indeed, it is not known how the length of chitin fibers in the arthropod cuticle is regulated. Random polymerization of chitin is certainly not desirable, when a stereotypic architecture with several layers of free chitin is the objective. Knk could represent the factor that regulates the length of chitin fibers by cleaving them at a certain frequency. The control of fiber length through oxidation or reduction presupposes that the whole arsenal of cofactors including the electron donor has to be present in the extracellular space. Chitin oxidation or reduction would not be the only redox reaction occurring within the extracellular space. Hardening and tanning of the cuticle, called sclerotization and melanization, involve several oxidizing enzymes such as phenol oxidases, laccases, and tyrosine oxidases [83]. These enzymes cross-link proteins, catecholamines, and chitin together to enhance cuticle impermeability and stability. Hence, the extracellular space is a suitable environment for various oxidation processes needed to structure the differentiating cuticle. Some of the enzymes involved in extracellular matrix oxidation are known. Phenol oxidases, for instance, cross-link catecholamines with proteins and maybe also chitin [83,84].

Another factor that is required for correct chitin arrangement in the cuticle of *D. melanogaster* is the protein retroactive (Rtv) [79,85]. Rtv belongs to the Ly6 group of proteins, most of which seem to be GPI-anchored factors directing the protein to membranes. Ly6 proteins in vertebrates are involved in multiple cell processes. However, their implication in extracellular matrix production and organization is not demonstrated. Turning back to Rtv, Rtv is the only example that a Ly6 protein is needed for extracellular matrix organization. In *rtv* null mutants, chitin microfibrils are misarranged. This is a very specific phenotype. The phenotype, although resulting from a null allele, is less severe than the phenotype caused by null mutations in *knk*. This difference indicates that Knk has more functions than Rtv. This notion in turn allows formulating two alternative scenarios. First, one may conclude that there are at least two different biochemical and cellular processes acting to arrange chitin. Second, Rtv is needed to assist the function of Knk, for instance, by facilitating its sorting or localization. Knk function would be diminished but not annihilated in lack of Rtv function. This view is somehow consistent with the common role of Ly6 proteins emerging in a recent work. Four other Ly6 proteins – Boudin, Crooked, Crimpled, and Coiled – that have been genetically identified and characterized promote the formation of the lateral septate junctions, which are a multiprotein complex connecting neighboring cells [86,87]. Boudin, Crooked, and Coiled proteins localize to intracellular compartments, probably to vesicles that drive correct septate junction assembly. In a similar way, Rtv could aid a Knk-containing complex to assemble in intracellular vesicles before being delivered to the plasma membrane where it supports chitin fiber organization. This hypothesis remains to be tested.

Besides the membrane-associated factors Knk and Rtv, chitin microfibril organization also requires the extracellular enzymes vermiform (Verm) and serpentine (Serp). These two enzymes represent two out of three bona fide chitin deacetylases (EC 3.5.1.41) in *D. melanogaster*. Mutations in the respective genes cause a phenotype that resembles the *rtv* mutant phenotype [88]. Interestingly again, the *serp* and *verm* double-loss of function situation does not result in a *knk* mutant phenotype. What is the significance of chitin deacetylation? Chitin deacetylation converts chitin to chitosan. Chitin in *D. melanogaster* as probably in other arthropods is only partially deacetylated, and it is proposed that the removal of the acetyl group from some GlcNAc residues enables chitin binding proteins to dock to chitin [29]. What is the relationship between Knk/Rtv and Verm/Serp? Verm and Serp are secreted proteins, while Rtv and Knk are membrane bound. Hence, it is conceivable that Knk and Rtv act before Verm and Serp suggesting that the function of Knk and Rtv may be a prerequisite for chitin deacetylase activity. A simple experiment to demonstrate any hierarchy between these factors is to determine the deacetylation grade of chitin in *knk* and *rtv* mutant animals.

Which proteins bind to chitin, be it deacetylated or not? In the last few years, a plethora of cuticle proteins in insects including mosquitoes (*Anopheles gambiae*) and *D. melanogaster* have been identified [89–94]. Some of these proteins harbor a chitin binding domain called the Riddiford and Rebers (R&R) domain. Three types of R&R domains have been described: R&R1, R&R2, and R&R3. They all consist of a common core sequence and additional specific flanking sequences. The R&R proteins from the mosquito *A. gambiae* and from *D. melanogaster* have been assessed in genomic approaches [94,95]. *A. gambiae* has been predicted to have 156, and *D. melanogaster* probably has 108 R&R proteins. Hence, despite their close relationship as both being dipterans, the difference in R&R content reflects probably the rapid evolution of these family of proteins. This difference also suggests that some of R&R proteins may have redundant functions. The high evolutionary activity of R&R proteins is also illustrated by their clustered localization in the respective genomes facilitating duplications and deletions of particular sequences that may account for the differences between these species. A prominent R&R protein is elastomeric resilin that has been discovered in the 1950s and published in 1960 by Weis-Fogh [96,97] before its protein sequence was "tentatively" discovered in 2001 by Ardell and Anderson in *D. melanogaster* and subsequently further characterized [98–103]. There are two isoforms of *Drosophila* resilin with 575 and 620 amino acids. The 45 extra amino acids of the 620-residue isoform derive from an additional exon. The information of this sequence is not obvious and deserves close investigation. Besides interacting with chitin, resilin molecules are probably linked to each other by dityrosine bridges forming a stable and high-molecular polysaccharide and protein network. In the nematode *C. elegans*, dityrosines are introduced to the cuticular matrix via the membrane-inserted dual oxidase protein (Duox), which using NADPH in the cytosol generates oxygen radicals, which are then consumed by the extracellular peroxidase domain to produce di- and trityrosines [104]. Whether the interaction between the R&R domain and chitin necessitates deacetylation of chitin residues remains to be shown.

Another class of proteins that have an essential function in cuticle organization are the zona pellucida (ZP) proteins. ZP proteins were first isolated from fertilized oocytes preventing double penetration by sperm cells [105]. In arthropods, the ZP proteins Papillote (Pot) and Piopio (Pio) are involved in attaching the cuticle to the epidermal surface [106]. Their molecular modes of action, that is, their binding partners within the extracellular space, are unknown. An interesting ZP protein is Dumpy (Dp), which is a huge extracellular protein that can be around 0.8 μm long [107], which is longer than chitin microfibrils (0.3 μm). Since the cuticle in the *Drosophila* larva is 0.5 μm thick, this protein has to adopt an oblique or horizontal position to fit in. Its sequence is modular with specific Dpy domains partially flanked by EGF-like repeats. The mRNA of *dp* is predominantly expressed at muscle attachment sites that are specialized epidermal cells producing a fortified cuticle to withstand mechanical work by muscles during locomotion. Again, the binding partners of Dp are unknown. An attractive model is that Dp through its 185 Dpy modules may bind to chitin microfibrils, thereby rendering the cuticle suitable for its specialized function (Figure 5.6).

More detailed molecular data are available for hyaluronic acid binding and organizing factors. The role of CD44 has already been briefly described. Some additional important hyaluronic acid binding proteins called hyaladherins are found in the extracellular matrix of, for example, the cumulus–oocyte complex (COC). HAS2-produced hyaluronic acid chains interact with the hyaladherins versican and TSG-6 (tumor necrosis factor-stimulated gene-6) and are anchored to the plasma membrane via CD44 (see above) where they play an essential role in fertility [108]. During bone differentiation, hyaluronic acid is produced by HAS2 and HAS3 in osteoblasts and is thought to control mineralization as it binds to hydroxyapatite [13]. These examples and those presented above illustrate the versatility of hyaluronic acid as a component of diverse extracellular matrices. Chitin, in contrast, seems to play solely a structural role. This issue is discussed in Section 5.14.

5.10
Secretion and Cuticle Formation

Naturally, the formation of an extracellular matrix employs the secretory machinery to localize the different types of factors acting within the plasma membrane and in the extracellular matrix itself. The secretory pathway starts with the endoplasmatic reticulum that is cotranslationally loaded with the proteins that ought to be sorted. The Sec61 complex is the gate to the ER, and indeed a subunit of the Sec61 complex Sec61β has been shown to be required for correct cuticle formation without, however, providing any details on the impact of the respective mutation in *sec61β* [109]. Supposedly, mutation in the other components of the Sec61 complex will result in a similar phenotype. The *sec61β* mutant phenotype is rather weak considering that it affects a major function of the cell. Most if not all cuticle differentiation factors have to travel through this gate. It is thus thinkable that maternal contribution masks the full requirement of Sec61β. Hence, a closer genetic study of the role of the Sec61 complex would teach us whether there may be routes running parallel to the canonical secretory pathway to establish the cuticle.

During ER entry, most proteins are N-glycosylated, that is, an oligosaccharide is added to certain asparagine residues. This process involves several cytoplasmic and ER-resident proteins. The last enzyme of the Leloir pathway, UDP-GlcNAc pyrophosphorylase, called Mummy (Mmy), in D. melanogaster has been shown to be essential for N-glycosylation [37–39]. However, as UDP-GlcNAc is used in several essential biochemical pathways, the analyses of the *mmy* mutant phenotype are of limiting value. Both chitin synthesis and N-glycosylation of, for example, Knk are affected in the respective mutant embryos and larvae resulting in a complex phenotype that does not allow major conclusions. Investigations on the D. melanogaster *wollknäuel* (*wol*) gene that codes for the Alg5 orthologue (EC 2.4.1.117) catalyzing the addition of glucose to dolichol, which in turn is used to garnish the oligosaccharide with glucose before it is attached to a protein, were somehow more fruitful [110]. The major outcome of this work was that coordination of cell functions comprising apical plasma membrane dynamics and guarantee of correct amounts of cuticle differentiation factors depends on N-glycosylation. In particular, the membrane determinants Crumbs (Crb) and aPKC fail to be removed from the apical plasma membrane, whereas Knk amounts are diminished in *wol* mutant animals. It is however not clear whether the proper supply of cuticle factors is a direct consequence of N-glycosylation. An alternative explanation is that effects of ER stress provoked by mutations in *wol* may account for the observed imbalance.

The assumed importance of general secretion for chitin synthesis and organization is illustrated by the phenotype of larvae suffering mutations in the *syntaxin1A* (*syx1A*) gene. Syx1A is a plasma membrane t-SNARE that mediates the fusion of secretory vesicles with the apical plasma membrane. Interestingly, mutations in *syx1A* do not affect the localization of chitin-organizing factors such as Knk and chitin synthesis is not blocked *per se* [111]. Hence, another apical plasma membrane syntaxin is needed to accommodate localization of chitin synthesis and organizing factors. The coordination of secretion hence reflects the topology of the apical plasma membrane during chitin production and cuticle differentiation (Figure 5.9). This second syntaxin remains, however, to be identified.

The separation of chitin synthesis and organization from general protein secretion might also suggest that chitin synthase and chitin organizing-factors follow a nonclassical secretion route. This view is supported by the observation that a signal peptide is missing at the N-terminus of insect chitin synthases [41]. This speculation turned out to be wrong. Indeed, *Drosophila* genetics allowed investigating the implication of the ER-to-Golgi vesicle transport on cuticle differentiation and chitin synthesis. These COPII vesicles carry factors destined for virtually all membrane compartments, including the Golgi apparatus and the plasma membrane. Mutations in *haunted* (*hau*) and *ghost* (*gho*) coding for the *Drosophila* COPII components Sec23 and Sec24, respectively, cause an almost complete lack of epidermal cuticle arguing that secretion starting at the ER is the major if not only secretory route

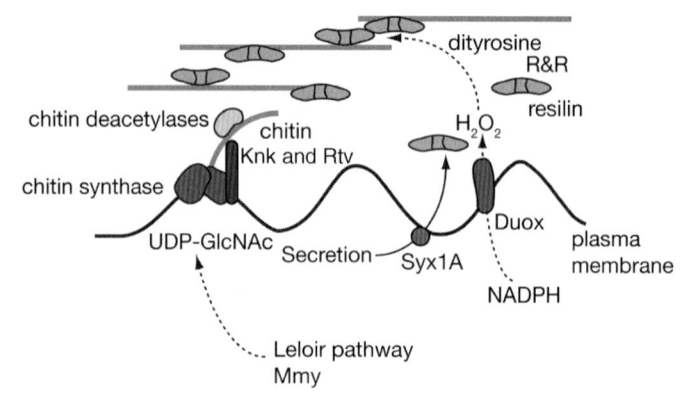

Figure 5.9 Simplified scheme on the assembly of the chitin–protein matrix in the procuticle of arthropods. Chitin synthesis, deposition, and organization imply the function of intracellular, membrane-located, and extracellular factors. Most of these factors have been identified and characterized genetically using *D. melanogaster* as a model insect. In brief, the active form of the chitin monomer UDP-GlcNAc is produced in the cytoplasm by the Leloir pathway, the last enzyme of which is Mmy, the UDP-GlcNAc pyrophosphorylase. The membrane-inserted chitin synthase here depicted as a dimer competes with different other essential enzymes for UDP-GlcNAc (see text) in order to synthesize chitin, which is extruded to the extracellular space. Here, chitin is organized involving Knk and Rtv, two membrane-associated factors that have yet been characterized biochemically. Chitin organization also involves chitin deacetylases and interaction with cuticle proteins such as the R&R protein resilin, which in this model are secreted via the tSNARE Syx1A. Of note, positioning of chitin synthases into the plasma membrane does not depend on Syx1A. Resilin molecules are probably covalently cross-linked to each other via dityrosine bridges. The formation of these bridges required the function of the membrane-inserted Duox enzyme that transports NADPH-derived electrons within the cytosol to produce H_2O_2 in the extracellular space where it used to oxidize tyrosine.

responsible for full cuticle differentiation. In addition, lack of COPII function perturbs cell polarity, a prerequisite for directed secretion and hence ordered cuticle deposition. In conclusion, the COPII vesicles are central carriers for essential determinants of cuticle formation and cell polarity. Consistently, as shown for tracheal cells, the function of COPII components is largely cell autonomous [112]. This means that each cuticle-producing cell coordinates differentiation in its own account with respect to cell polarity and directed secretion. At the end of differentiation, the cuticle has a uniform thickness and organization. Hence, the sum of self-organizing cells is sufficient to establish a fully functional cuticle that covers the animal as one piece.

5.11
Transcriptional Regulation of Cuticle Production

Chitin and cuticle production are also controlled and coordinated at the transcriptional level involving several evolutionary conserved transcription factors that are

5.11 Transcriptional Regulation of Cuticle Production

also essential in vertebrates, for instance, for skin development and function. One of these is Granyhead (Grh), an ETS-type transcription factor that has been shown to be important for cuticle differentiation in *D. melanogaster* [113,114]. This phenotypic indication is the only one arguing that Grh regulates cuticle differentiation. Indeed, we do not have any direct evidence for this assumption. In other words, the targets of Grh that would be implicated in cuticle differentiation and chitin synthesis and organization are not known. However, the cuticle defects of larvae mutant for *grh* and *knk* are very similar [114]; thus, *knk* could be one target of Grh. Moreover, Grh has been demonstrated to control reconstitution of the cuticle after wounding [115]. However, in this work only two targets of Grh were identified and analyzed: *pale* and *Ddc*. These two genes code for the Dopa decarboxylase (EC 4.1.28) and tyrosine dehydrogenase (EC 1.14.16.2), respectively. They act in the pathway producing Dopa and derivatives that are secreted to the extracellular space where they interact with proteins and chitin to cross-link the cuticle and thereby render it stiff [83].

The evolutionary conserved transcription factor CrebA (Cre binding A) has been repeatedly reported to drive the expression of factors that act in the secretory pathway [116,117]. In *Drosophila*, CrebA transcribes a battery of 34 putative and true targets that function at the different levels of secretion. The cuticle of *CrebA* mutant larvae is in consequence reduced but not absent. This observation indicates that either another factor is required for full expression of these genes or the maternal CrebA masks the zygotic mutant phenotype. An impact of CrebA on chitin synthesis and organization has not been demonstrated and can therefore not be ruled out. It is indeed conceivable to assume that at least the lack of secretion of cuticle proteins may affect chitin organization according to Bouligand. Of course, this remains to be shown.

Taken together, a master regulator of cuticle differentiation seems not to exist. Rather, differentiation relies on multiple regulators that function in parallel to ensure correct establishment of the functional extracellular matrix. This means that coordination of cuticle differentiation already occurs at the transcriptional level. The trigger of differentiation is hence a complex event involving different pathways that are not understood. One molecule that is classically viewed as a main trigger of cuticle differentiation is the steroid hormone ecdysone. Ecdysone acts over a nuclear receptor that consists of the ecdysone receptor (EcR) and its partner ultraspiracle (Usp) [118]. Once this complex receives the ecdysone signal, it activates or represses the transcription of other transcription factors, including E75 and FtzF1. These factors in turn drive the development of organs in a tissue autonomous manner [119]. The cascade leading to cuticle differentiation is only partly investigated. For instance, in *M. sexta*, ecdysone signaling controls the melanization process in the cuticle [120]. Again, the molecular impact of ecdysone on cuticle differentiation is not investigated in detail. However, recently, RNA interference (RNAi) against the lepitopteran *Spodoptera exigua* EcR results in decreased expression of *SeTre-1*, *SeG6PI*, *SeUAP*, and *SeCHSA* that act in the pathway catalyzing the conversion of glucose to UDP-GlcNAc (*SeTre-1*, *SeG6PI*, and *SeUAP*) or chitin synthesis (*SeCHSA*) [121]. This kind of broad approaches may fast open a new field in investigating the role of ecdysone in chitin synthesis and organization. However, the tissue-specific response of ecdysone on cuticle differentiation including chitin synthesis and organization may be more complex and require more elaborate experiments in different species to conclude on this issue.

Beyond this scenario, ecdysone has also been shown to be needed for the establishment of the secretory pathway in regulating the formation of the Golgi apparatus in *Drosophila* imaginal disks [122]. Hence, ecdysone may constitute a partner pathway of the CrebA defined pathway. Together, they may regulate secretion and localization of cuticle factors to the extracellular space or to the apical plasma membrane. The expression of these factors depends probably on the function of other transcription factors such as Grh.

5.12
Chitin Synthesis Inhibitors

Chitin is virtually absent in vertebrates; chitin synthesis is therefore an excellent reaction to be controlled by chitin synthesis inhibitors in both medicine and agriculture. The central target for chitin synthesis inhibition is the chitin synthase itself. A potent inhibitor of this enzyme is nikkomycin (e.g., Nikkomycin Z, $C_{20}H_{25}N_5O_{10}$), which is produced by *Streptomyces* species. It is a competitive nucleoside analogue of UDP-GlcNAc and binds irreversibly to the catalytic center of chitin synthases. Interestingly, it seems not to affect the enzymatic activity of other UDP-GlcNAc using enzymes. Indeed, application of Nikkomycin Z to *D. melanogaster* causes an embryonic lethal phenotype, which is identical to the one provoked by mutations in the chitin synthase encoding gene (already discussed) [123]. The strength of Nikkomycin Z defects is dosage dependent. Lower amounts of Nikkomycin Z induce phenotypes reminiscent of *knk*- and *rtv*-mutant larvae. This finding strongly indicated that Knk and Rtv play an essential role in chitin synthesis. Nikkomycins have been considered as antifungal drugs. However, several pathogenic fungi such as *Candida* species have been shown to be insensitive to nikkomycins [124,125]. This has been speculated to relate to the amount of chitin in the cell wall of these fungi.

Another group of chitin inhibitors consists of derivatives of benzoylphenyl urea. They are used in veterinary medicine as well as in agriculture as they are significantly less expensive than nikkomycins. Some recent investigations suggest the use of these substances also against mosquitoes that act as vectors of tropical diseases such as malaria and dengue fever [126–128]. Benzoylphenyl urea has been demonstrated to interact with ABC transporters, a family of membrane-inserted proteins that mediate the cross-membrane transport of a variety of substances ranging from peptides, lipids, and sugars to inorganic ions. The mode of action with respect to chitin synthesis inhibition has not yet been determined.

Some attempts to elucidate the impact of benzoylphenyl urea have been undertaken using the fruit fly *Drosophila* as a model insect [123,129]. For instance, the benzoylphenyl urea receptor SUR (sulfonylurea receptor, an ABC transporter) has been proposed to mediate chitin synthesis inhibition in intact cells but not in cell lysates suggesting that the activity of this receptor requires functional membrane compartments. At the same time, these results underline that the target of benzoylphenyl urea is not the chitin synthase itself. However, Abo-Elghar et al. [129] did not give direct evidence that SUR is the receptor of the applied benzoylphenyl urea (lufenuron and glibenclamide). Indeed, SUR itself is unlikely to be the ABC transporter that is needed for chitin synthesis, as it is not expressed in those tissues that produce chitin [130,131]. D. melanogaster possesses around 100 ABC transporters, some of which have been characterized; none of them, however, is a candidate to be necessary during chitin synthesis. In the future, it should therefore be a major task in the field to identify and characterize the responsible ABC transporter. An understanding of these processes will allow studying and designing new benzoylphenyl urea-based insecticides that specifically affect particular pests while beneficial insects would be spared.

How can we then imagine the mode a SUR-like ABC transporter is involved in chitin synthesis? There is no evidence that chitin synthesis in fungi involves ABC transporters. Hence, there are fundamental differences between arthropod and fungal chitin syntheses, in spite of the similar biochemical reaction. This indicates that the chitin synthesis-related ABC transporter in arthropods may be needed to control the enzymatic activity of the arthropod chitin synthase rather than being directly implicated in the chitin polymerization reaction. In line with this notion, ABC transporters modify the milieu in which they transport their cargo. Thus, a substance delivered by an ABC transporter into the extracellular space that is filled with chitin is apparently required for the activity of the chitin synthase present in the plasma membrane. The identification of this substance remains an important task.

An interesting notion that came up in the inhibitor research field is that attenuation of chitin synthesis does not stall animal development but leads to severe problems and ultimately to death. Of course, this observation may be trivial as often if not always reduction of the activity of an essential factors results in the establishment of a distinct phenotype, underlining that there is probably no controlled communication between organs and tissues during development.

5.13
Noncuticular Chitin in Insects

In 2005, the group of Anne Uv from the Gothenburg University in Sweden discovered an unexpected role of chitin in insect development. During embryogenesis, the radial expansion of the *Drosophila* larval tracheal tubes, which constitute the respiratory organ, involves a luminal chitin rod (Figure 5.10) that is produced and organized by largely the same factors that are essential for cuticular chitin production and organization [132]. This finding was confirmed by other

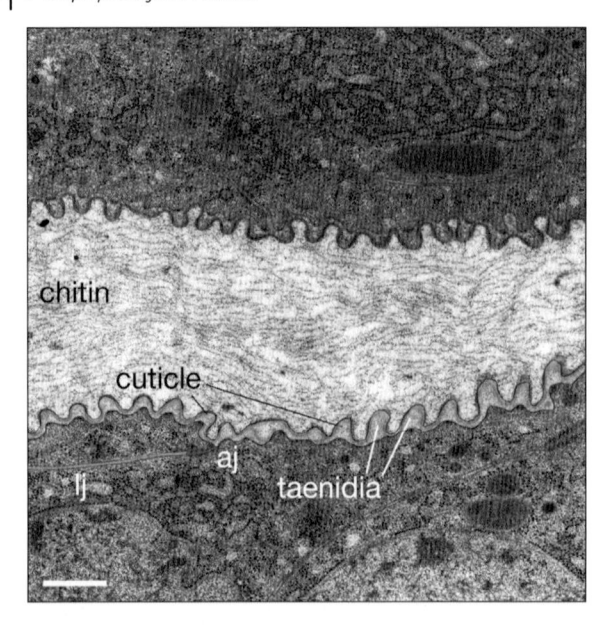

Figure 5.10 The tracheal luminal chitin. During development, the tracheae of the D. melanogaster embryo produce a luminal chitin rod that is proposed to be involved in tube diameter regulation, while at plasma membrane protrusions, the taenidia, the spiral apical tracheal cuticle is formed. In the absence of the luminal chitin rod, the tracheal lumen is irregular and the tubes eventually fail to air fill. This happens, for instance, in embryos mutant for the chitin synthase gene. Likewise, disorganization of the luminal chitin rod causes a similar phenotype as observed in knk and rtv mutant animals. This indicates that the presence of chitin as such is not sufficient to ensure correct diameter regulation. aj: adherens junction; lj: lateral junction.

laboratories [133]. Virtually, the whole complement of cell biological processes is involved in this process. The lateral cell contacts, for instance, are reported to direct apical secretion of chitin-organizing factors such as Knk and the chitin deacetylases Verm/Serp [134]. In larvae that are homozygous mutant for the underlying genes, the tracheal tubes are elongated. This suggests that chitin is also needed for tube length regulation. Secretion itself does not simply deliver the components to the extracellular space, but is believed to be the driving force of tube expansion [112]. The situation becomes complicated when endocytosis of luminal components is introduced [135]. Indeed, pulses of endocytosis have been shown to be essential for tracheal development. To date, the luminal chitin has not been shown to play a role in organisms other than *Drosophila*; however, a survey of earlier publications reveals that it may be common. In the caterpillar of the butterfly *Calpodes ethlius*, for example, it was published already in 1966 by Locke that "the (tracheal) lumen contains fibrous material arranged longitudinally" [136]. He did not know nor speculate that this fibrous material was chitin. The nature of chitin packing is not known to date. It is however tempting to consider that it may to some extent be organized as in the cuticle as many cuticle-organizing factors are also required for the luminal chitin organization. The length of luminal chitin microfibrils or any other state of chitin bundling is not quantified.

Besides being a component of the tracheal luminal matrix, noncuticular chitin is also found in the eye of the adult animal [137]. It is part of the extracellular optical system that collects light. The eye chitin is organized as chitin in the procuticle, that is, it is laminar and adopts the organization pattern according to Bouligand's model already described. Interestingly, the basal surface of these chitin layers does not contact the apical surface of epithelial cells but another extracellular matrix that is produced by the same cells. This indicates that the organization of laminar chitin does not strictly imply stabilization of the chitin layer by a plasma membrane. The eye chitin extracellular matrix may constitute an exception of the structural function of chitin.

5.14
Chitin as a Structural Element

Chitin is a structural element of arthropod cuticle. In particular, elasticity and stiffness have been attributed to chitin. How is this role assessed? Especially, chitin stiffness has been extensively studied by the laboratory of Raabe in Düsseldorf in Germany, with interesting results. Using nanoindentation, they determined the stiffness of cuticle layers of the American lobster *H. americanus*. In the upper portion of the procuticle called exocuticle, the stiffness is around 9 GPa. This is a similar value as measured for wood. Penetrating further into the lower part of the procuticle called endocuticle, the stiffness drops to around 4 GPa. Likewise, hardness in the exocuticle is around 250 MPa and decreases to around 40 MPa in the endocuticle. Correlating with these courses of stiffness and hardness along the different zones of the procuticle, package of chitin laminae is denser in the exocuticle than in the endocuticle. Beyond correlation, this stacking difference may explain the difference in stiffness. However, the molecular basis of packaging of chitin laminae is not clear. Conceivably, different chitin binding proteins are responsible for the differences in packaging. The process of cuticle hardening is called sclerotization already described and involves besides proteins the phenolic substances that cross-link proteins and chitin [83]. Whether denser packing of chitin laminae indeed is accompanied by enhanced sclerotization is a matter of speculation.

Raabe et al. [26] also demonstrated that besides the tendency of parallel orientation of chitin microfibrils, they draw a planar honeycomb-like mesostructure (see Figure 5.2). This organization has a central implication as proposed by the authors. By introducing discontinuities, the honeycomb organization confers resistance to crack initiation and propagation.

Insights into the hardness of chitin-based extracellular matrices come from studies on the beak of the Humboldt squid *Dosidicus gigas* [7]. Beak stiffness gradually increases from 0.05 to 5 GPa measured by nanoindentation with the amount of histidyl-Dopa cross-links as assessed by mass spectrometry.

Unexpectedly, beak stiffness is inversely proportional to chitin content. Apparently, the authors conclude that the degree of hydration of the beak has a greater influence on stiffness than chitin alone. Chitin microfibrils do not have a preferred orientation in the squid's beak but form a fibrous scaffold. Interestingly, chitin is nonoriented in *Drosophila* larval denticles that are stiff cuticular structures required for locomotion [78]. Thus, one may hypothesize that chitin orientation according to Bouligand seems to be necessary for elasticity, whereas unoriented chitin appears to confer stiffness to extracellular matrices.

5.15
Application of Chitin

Chitosan that is produced industrially by alkaline deacetylation of chitin obtained mainly from marine crustaceans such as shrimps is used as a positively charged matrix in various pharmaceutical application systems [143]. Chitosan is suitable in this context as it is an acid-soluble, nontoxic, nonallergenic, and biodegradable biopolymer. Indeed, humans are able to degrade chitosan but not chitin. Chitosan itself has been demonstrated to have antitumor properties [138,139]. Cells are able to take up chitosan nanoparticles via endocytosis, and chitosan may also enter the nucleus. Chitosan endocytosis is enhanced when chitosan is modified by arginine moieties [140]. The ease of chitosan uptake by the cell underlines its use as an excellent drug delivery material. As a carrier of antigens and drugs, chitosan is often structurally remodeled in order to encapsulate and orderly deliver the bioactive agent to its destination. For instance, porous chitosan microspheres are loaded with appropriate antigen to vaccinate poultry against Newcastle disease virus infection. In addition, chitosan function as a matrix can be optimized by chemical modification such as succinylation. Succinyl-chitosan microparticles were used to deliver 5-aminosalicylic acid (5-ASA) to the colon of rats suffering experimentally induced colitis [141].

In cancer chemotherapy, a number of modified types of chitosan particles are used to apply the anticancer agents mitomycin C (MMC) and doxorubicin, for example, against human bladder tumors [142,143]. In modern medical research, chitosan nanoparticles are used as carriers of short interfering (si) RNA to prevent tumor growth [144,145]. DNAzymes and siRNA act on the availability and expression of tumor-specific factors. Chitosan-conjugated DNAzyme nanoparticles were, for example, successfully tested against the expression of the oncogene c-Jun in SaOS2 osteosarcoma cells [146,147]. And chitosan-bound siRNA was directed against myeloid leukemia K562 cells [148]. Finally, chitosan in combination with other polymers is used as hydrating dressing material supporting wound healing [149].

The works cited in this chapter are only single examples out of the long list of chitosan use in pharmacology. Excellent reviews have been published recently [150,151], and the dynamics of the field will certainly continue to urge more reviews in the future.

5.16
Conclusion

Chitin is an abundant polysaccharide that is an important component in fungal cell walls and arthropod cuticle. In both cases, it adopts a characteristic organization that is essential for the function of the respective structure. It is considered since decades that at least in arthropods chitin organization relies on chitin binding proteins that associate with chitin. However, there is no causal evidence for this view. The genetic approach using *D. melanogaster* and *T. castaneum* will in the future contribute to elucidating this vividly discussed issue. An ultimate goal of this work will be to establish chitin and chitin-based biopolymers as a highly versatile biomaterial.

References

1. Moussian, B. (2008) The role of GlcNAc in formation and function of extracellular matrices. *Comp. Biochem. Physiol. B Biochem. Mol. Biol.*, **149**, 215–226.
2. Pinto, D.O., Ferreira, P.L., Andrade, L.R., Petrs-Silva, H., Linden, R., Abdelhay, E., Araujo, H.M., Alonso, C.E., and Pavao, M.S. (2004) Biosynthesis and metabolism of sulfated glycosaminoglycans during *Drosophila melanogaster* development. *Glycobiology*, **14**, 529–536.
3. Silbert, J.E. and Sugumaran, G. (2002) Biosynthesis of chondroitin/dermatan sulfate. *IUBMB Life*, **54**, 177–186.
4. Brunner, E., Ehrlich, H., Schupp, P., Hedrich, R., Hunoldt, S., Kammer, M., Machill, S., Paasch, S., Bazhenov, V.V., Kurek, D.V., Arnold, T., Brockmann, S., Ruhnow, M., and Born, R. (2009) Chitin-based scaffolds are an integral part of the skeleton of the marine demosponge *Ianthella basta*. *J. Struct. Biol.*, **168**, 539–547.
5. Michels, J. and Buntzow, M. (2010) Assessment of Congo red as a fluorescence marker for the exoskeleton of small crustaceans and the cuticle of polychaetes. *J. Microsc.*, **238**, 95–101.
6. Martin, R., Hild, S., Walther, P., Ploss, K., Boland, W., and Tomaschko, K.H. (2007) Granular chitin in the epidermis of nudibranch molluscs. *Biol. Bull.*, **213**, 307–315.
7. Miserez, A., Schneberk, T., Sun, C., Zok, F.W., and Waite, J.H. (2008) The transition from stiff to compliant materials in squid beaks. *Science*, **319**, 1816–1819.
8. Zhang, Y., Foster, J.M., Nelson, L.S., Ma, D., and Carlow, C.K. (2005) The chitin synthase genes chs-1 and chs-2 are essential for *C. elegans* development and responsible for chitin deposition in the eggshell and pharynx, respectively. *Dev. Biol.*, **285**, 330–339.
9. Zhu, Q., Arakane, Y., Beeman, R.W., Kramer, K.J., and Muthukrishnan, S. (2008) Characterization of recombinant chitinase-like proteins of *Drosophila melanogaster* and *Tribolium castaneum*. *Insect Biochem. Mol. Biol.*, **38**, 467–477.
10. Zhu, Q., Arakane, Y., Beeman, R.W., Kramer, K.J., and Muthukrishnan, S. (2008) Functional specialization among insect chitinase family genes revealed by RNA interference. *Proc. Natl. Acad. Sci. USA*, **105**, 6650–6655.
11. Zhu, Q., Arakane, Y., Banerjee, D., Beeman, R.W., Kramer, K.J., and Muthukrishnan, S. (2008) Domain organization and phylogenetic analysis of the chitinase-like family of proteins in three species of insects. *Insect Biochem. Mol. Biol.*, **38**, 452–466.
12. Toole, B.P. (2000) Hyaluronan is not just a goo! *J. Clin. Invest.*, **106**, 335–336.
13. Bastow, E.R., Byers, S., Golub, S.B., Clarkin, C.E., Pitsillides, A.A., and Fosang, A.J. (2008) Hyaluronan synthesis and degradation in cartilage and bone. *Cell Mol. Life Sci.*, **65**, 395–413.

14 Bakkers, J., Kramer, C., Pothof, J., Quaedvlieg, N.E., Spaink, H.P., and Hammerschmidt, M. (2004) Has2 is required upstream of Rac1 to govern dorsal migration of lateral cells during zebrafish gastrulation. *Development*, **131**, 525–537.

15 Braconnot, H. (1811) Sur la nature des champignons. *Annales de chimie ou recueil de mémoires concernant la chimie et les arts qui en dépendent et spécialement la pharmacie*, **79**, 265–304.

16 Labrude, P. and Becq, C (2003) Pharmacist and chemist Henri Braconnot. *Rev. Hist. Pharm. (Paris)*, **51**, 61–78.

17 Odier, A. (1823) Mémoires sur la composition chimique des parties cornées des insectes. *Mém. Soc. Hist. Naturelle de Paris*, **1**, 29–42.

18 Fränkel, S. and Kelly, A. (1901) Beiträge zur constitution des chitins. *Monatsh. Chem.*, **23**, 123–132.

19 Meyer, K.H. and Pankow, G. (1935) Sur la constitution et la structure de la chitine. *Helv. Chim. Acta*, **18**, 589–598.

20 Beckham, G.T. and Crowley, M.F. (2011) Examination of the alpha-chitin structure and decrystallization thermodynamics at the nanoscale. *J. Phys. Chem. B*, **115**, 4516–4522.

21 Bouligand, Y. (1965) On a twisted fibrillar arrangement common to several biologic structures. *C. R. Acad. Sci. Hebd Seances Acad. Sci. D*, **261**, 4864–4867.

22 Raabe, D., Al-Sawalmih, A., Yi, S.B., and Fabritius, H. (2007) Preferred crystallographic texture of alpha-chitin as a microscopic and macroscopic design principle of the exoskeleton of the lobster *Homarus americanus*. *Acta Biomater.*, **3**, 882–895.

23 Raabe, D., Romano, P., Al-Sawalmih, A., Sachs, C., Servos, G., and Hartwig, H.G. (2005) Mesostructure of the exoskeleton of the lobster *Homarus americanus*. *Struct. Mech. Behav. Biol. Mater.*, **874**, 155–160.

24 Raabe, D., Romano, P., Sachs, C., Fabritius, H., Al-Sawalmih, A., Yi, S., Servos, G., and Hartwig, H.G. (2006) Microstructure and crystallographic texture of the chitin-protein network in the biological composite material of the exoskeleton of the lobster *Homarus americanus*. *Mater. Sci. Eng. A*, **421**, 143–153.

25 Raabe, D., Sachs, C., and Romano, P. (2005) The crustacean exoskeleton as an example of a structurally and mechanically graded biological nanocomposite material. *Acta Mater.*, **53**, 4281–4292.

26 Raabe, D., Romano, P., Sachs, C., Al-Sawalmih, A., Brokmeier, H.-G., Yi, S.-B., Servos, G., and Hartwig, H.G. (2005) Discovery of a honeycomb structure in the twisted plywood patterns of fibrous biological nanocomposite tissue. *J. Cryst. Growth*, **283**, 1–7.

27 Meyer, K. and Palmer, J.W. (1934) The polysaccharide of the vitreous humor. *J. Biol. Chem.*, **107**, 629–634.

28 Weissman, B. and Meyer, K. (1954) The structure of hyalobiuronic acid and hyaluronic acid from umbilical cord. *J. Am. Chem. Soc.*, **76**, 1753–1757.

29 Neville, A.C. (1975) *Biology of the Arthropod Cuticle*, Springer, Berlin.

30 Vincent, J.F. and Wegst, U.G. (2004) Design and mechanical properties of insect cuticle. *Arthropod. Struct. Dev.*, **33**, 187–199.

31 Merzendorfer, H. and Zimoch, L. (2003) Chitin metabolism in insects: structure, function and regulation of chitin synthases and chitinases. *J. Exp. Biol.*, **206**, 4393–4412.

32 Becker, A., Schloder, P., Steele, J.E., and Wegener, G. (1996) The regulation of trehalose metabolism in insects. *Experientia*, **52**, 433–439.

33 Paul, M.J., Primavesi, L.F., Jhurreea, D., and Zhang, Y. (2008) Trehalose metabolism and signaling. *Annu. Rev. Plant Biol.*, **59**, 417–441.

34 Kanamori, Y., Saito, A., Hagiwara-Komoda, Y., Tanaka, D., Mitsumasu, K., Kikuta, S., Watanabe, M., Cornette, R., Kikawada, T., and Okuda, T. (2010) The trehalose transporter 1 gene sequence is conserved in insects and encodes proteins with different kinetic properties involved in trehalose import into peripheral tissues. *Insect Biochem. Mol. Biol.*, **40**, 30–37.

35 Milewski, S., Gabriel, I., and Olchowy, J. (2006) Enzymes of UDP-GlcNAc biosynthesis in yeast. *Yeast*, **23**, 1–14.

36 Ruaud, A.F., Lam, G., and Thummel, C.S. (2011) The *Drosophila* NR4A nuclear receptor DHR38 regulates carbohydrate metabolism and glycogen storage. *Mol. Endocrinol.*, **25**, 83–91.

37 Tonning, A., Helms, S., Schwarz, H., Uv, A.E., and Moussian, B. (2006) Hormonal regulation of mummy is needed for apical extracellular matrix formation and epithelial morphogenesis in *Drosophila*. *Development*, **133**, 331–341.

38 Schimmelpfeng, K., Strunk, M., Stork, T., and Klambt, C. (2006) Mummy encodes an UDP-*N*-acetylglucosamine-dipohosphorylase and is required during *Drosophila* dorsal closure and nervous system development. *Mech. Dev.*, **123**, 487–499.

39 Araujo, S.J., Aslam, H., Tear, G., and Casanova, J. (2005) mummy/cystic encodes an enzyme required for chitin synthesis, involved in trachea, embryonic cuticle and CNS development: analysis of its role in *Drosophila* tracheal morphogenesis. *Dev. Biol.*, **288**, 179–193.

40 Havemann, J., Muller, U., Berger, J., Schwarz, H., Gerberding, M., and Moussian, B. (2008) Cuticle differentiation in the embryo of the amphipod crustacean *Parhyale hawaiensis*. *Cell Tissue Res.*, **332**, 359–370.

41 Moussian, B., Schwarz, H., Bartoszewski, S., and Nusslein-Volhard, C. (2005) Involvement of chitin in exoskeleton morphogenesis in *Drosophila melanogaster*. *J. Morphol.*, **264**, 117–130.

42 Zimoch, L. and Merzendorfer, H. (2002) Immunolocalization of chitin synthase in the tobacco hornworm. *Cell Tissue Res.*, **308**, 287–297.

43 Broehan, G., Kemper, M., Driemeier, D., Vogelpohl, I., and Merzendorfer, H. (2008) Cloning and expression analysis of midgut chymotrypsin-like proteinases in the tobacco hornworm. *J. Insect Physiol.*, **54**, 1243–1252.

44 Broehan, G., Zimoch, L., Wessels, A., Ertas, B., and Merzendorfer, H. (2007) A chymotrypsin-like serine protease interacts with the chitin synthase from the midgut of the tobacco hornworm. *J. Exp. Biol.*, **210**, 3636–3643.

45 DeMarini, D.J., Adams, A.E., Fares, H., De Virgilio, C., Valle, G., Chuang, J.S., and Pringle, J.R. (1997) A septin-based hierarchy of proteins required for localized deposition of chitin in the *Saccharomyces cerevisiae* cell wall. *J. Cell Biol.*, **139**, 75–93.

46 Santos, B. and Snyder, M. (1997) Targeting of chitin synthase 3 to polarized growth sites in yeast requires Chs5p and Myo2p. *J. Cell Biol.*, **136**, 95–110.

47 Meissner, D., Odman-Naresh, J., Vogelpohl, I., and Merzendorfer, H. (2010) A novel role of the yeast CaaX protease Ste24 in chitin synthesis. *Mol. Biol. Cell*, **21**, 2425–2433.

48 Lenardon, M.D., Whitton, R.K., Munro, C.A., Marshall, D., and Gow, N.A. (2007) Individual chitin synthase enzymes synthesize microfibrils of differing structure at specific locations in the *Candida albicans* cell wall. *Mol. Microbiol.*, **66**, 1164–1173.

49 Kultti, A., Rilla, K., Tiihonen, R., Spicer, A.P., Tammi, R.H., and Tammi, M.I. (2006) Hyaluronan synthesis induces microvillus-like cell surface protrusions. *J. Biol. Chem.*, **281**, 15821–15828.

50 Camenisch, T.D. and McDonald, J.A. (2000) Hyaluronan: is bigger better? *Am. J. Respir. Cell Mol. Biol.*, **23**, 431–433.

51 Tepass, U. and Hartenstein, V. (1994) The development of cellular junctions in the *Drosophila* embryo. *Dev. Biol.*, **161**, 563–596.

52 Roncero, C. (2002) The genetic complexity of chitin synthesis in fungi. *Curr. Genet.*, **41**, 367–378.

53 Xie, X. and Lipke, P.N. (2010) On the evolution of fungal and yeast cell walls. *Yeast*, **27**, 479–488.

54 Lipke, P.N. and Ovalle, R. (1998) Cell wall architecture in yeast: new structure and new challenges. *J. Bacteriol.*, **180**, 3735–3740.

55 Cabib, E. (2004) The septation apparatus, a chitin-requiring machine in budding yeast. *Arch. Biochem. Biophys.*, **426**, 201–207.

56 Valdivia, R.H. and Schekman, R. (2003) The yeasts Rho1p and Pkc1p regulate the transport of chitin synthase III (Chs3p) from internal stores to the plasma membrane. *Proc. Natl. Acad. Sci. USA*, **100**, 10287–10292.

57 Bracker, C.E., Ruiz-Herrera, J., and Bartnicki-Garcia, S. (1976) Structure and transformation of chitin synthetase particles (chitosomes) during microfibril synthesis *in vitro*. *Proc. Natl. Acad. Sci. USA*, **73**, 4570–4574.

58 Ruiz-Herrera, J., Sing, V.O., Van der Woude, W.J., and Bartnicki-Garcia, S. (1975) Microfibril assembly by granules of chitin synthetase. *Proc. Natl. Acad. Sci. USA*, **72**, 2706–2710.

59 Bartnicki-Garcia, S. (2006) Chitosomes: past, present and future. *FEMS Yeast Res.*, **6**, 957–965.

60 Ruiz-Herrera, J. and Ortiz-Castellanos, L. (2010) Analysis of the phylogenetic relationships and evolution of the cell walls from yeasts and fungi. *FEMS Yeast Res.*, **10**, 225–243.

61 Latge, J.P. (2007) The cell wall: a carbohydrate armour for the fungal cell. *Mol. Microbiol.*, **66**, 279–290.

62 Latge, J.P. (2010) Tasting the fungal cell wall. *Cell Microbiol.*, **12**, 863–872.

63 Kollar, R., Petrakova, E., Ashwell, G., Robbins, P.W., and Cabib, E. (1995) Architecture of the yeast cell wall: the linkage between chitin and beta(1–3)-glucan. *J. Biol. Chem.*, **270**, 1170–1178.

64 Cabib, E., Blanco, N., Grau, C., Rodriguez-Pena, J.M., and Arroyo, J. (2007) Crh1p and Crh2p are required for the cross-linking of chitin to beta(1–6) glucan in the *Saccharomyces cerevisiae* cell wall. *Mol. Microbiol.*, **63**, 921–935.

65 Costachel, C., Coddeville, B., Latge, J.P., and Fontaine, T. (2005) Glycosylphosphatidylinositol-anchored fungal polysaccharide in *Aspergillus fumigatus*. *J. Biol. Chem.*, **280**, 39835–39842.

66 Ruiz-Herrera, J., Ortiz-Castellanos, L., Martinez, A. I., Leon-Ramirez, C., and Sentandreu, R. (2008) Analysis of the proteins involved in the structure and synthesis of the cell wall of Ustilago maydis. *Fungal. Genet. Biol.*, **45**, (Suppl 1), S71–S76.

67 Kapteyn, J.C., Van Den Ende, H., and Klis, F.M. (1999) The contribution of cell wall proteins to the organization of the yeast cell wall. *Biochim. Biophys. Acta*, **1426**, 373–383.

68 Van Mulders, S.E., Ghequire, M., Daenen, L., Verbelen, P.J., Verstrepen, K.J., and Delvaux, F.R. (2010) Flocculation gene variability in industrial brewer's yeast strains. *Appl. Microbiol. Biotechnol.*, **88**, 1321–1331.

69 Smukalla, S., Caldara, M., Pochet, N., Beauvais, A., Guadagnini, S., Yan, C., Vinces, M.D., Jansen, A., Prevost, M.C., Latge, J.P., Fink, G.R., Foster, K.R., and Verstrepen, K.J. (2008) FLO1 is a variable green beard gene that drives biofilm-like cooperation in budding yeast. *Cell*, **135**, 726–737.

70 Smits, G.J., Kapteyn, J.C., van den Ende, H., and Klis, F.M. (1999) Cell wall dynamics in yeast. *Curr. Opin. Microbiol.*, **2**, 348–352.

71 Machi, K., Azuma, M., Igarashi, K., Matsumoto, T., Fukuda, H., Kondo, A., and Ooshima, H. (2004) Rot1p of *Saccharomyces cerevisiae* is a putative membrane protein required for normal levels of the cell wall 1,6-beta-glucan. *Microbiology*, **150**, 3163–3173.

72 Kapteyn, J.C., Ram, A.F., Groos, E.M., Kollar, R., Montijn, R.C., Van Den Ende, H., Llobell, A., Cabib, E., and Klis, F.M. (1997) Altered extent of cross-linking of beta1,6-glucosylated mannoproteins to chitin in *Saccharomyces cerevisiae* mutants with reduced cell wall beta1,3-glucan content. *J. Bacteriol.*, **179**, 6279–6284.

73 Neville, A.C. and Luke, B.M. (1969) Molecular architecture of adult locust cuticle at the electron microscope level. *Tissue Cell*, **1**, 355–366.

74 Moussian, B., Seifarth, C., Muller, U., Berger, J., and Schwarz, H. (2006) Cuticle differentiation during *Drosophila* embryogenesis. *Arthropod. Struct. Dev.*, **35**, 137–152.

75 Gutierrez, R., Lindeboom, J.J., Paredez, A.R., Emons, A.M., and Ehrhardt, D.W. (2009) Arabidopsis cortical microtubules position cellulose synthase delivery to the plasma membrane and interact with cellulose synthase trafficking compartments. *Nat. Cell Biol.*, **11**, 797–806.

76 Paredez, A.R., Somerville, C.R., and Ehrhardt, D.W. (2006) Visualization of cellulose synthase demonstrates functional association with microtubules. *Science*, **312**, 1491–1495.

77 Ziese, S. and Dorn, A. (2003) Embryonic integument and "molts" in *Manduca sexta* (Insecta, Lepidoptera). *J. Morphol.*, **255**, 146–161.

78 Moussian, B. (2010) Recent advances in understanding mechanisms of insect cuticle differentiation. *Insect Biochem. Mol. Biol.*, **40**, 363–375.

79 Moussian, B., Tang, E., Tonning, A., Helms, S., Schwarz, H., Nusslein-Volhard, C., and Uv, A.E. (2006) *Drosophila* knickkopf and retroactive are needed for epithelial tube growth and cuticle differentiation through their specific requirement for chitin filament organization. *Development*, **133**, 163–171.

80 Aravind, L. (2001) DOMON: an ancient extracellular domain in dopamine beta-monooxygenase and other proteins. *Trends Biochem. Sci.*, **26**, 524–526.

81 Iyer, L.M., Anantharaman, V., and Aravind, L. (2007) The DOMON domains are involved in heme and sugar recognition. *Bioinformatics*, **23**, 2660–2664.

82 Henriksson, G., Johansson, G., and Pettersson, G. (2000) A critical review of cellobiose dehydrogenases. *J. Biotechnol.*, **78**, 93–113.

83 Andersen, S.O. (2010) Insect cuticular sclerotization: a review. *Insect Biochem. Mol. Biol.*, **40**, 166–178.

84 Schaefer, J., Kramer, K.J., Garbow, J.R., Jacob, G.S., Stejskal, E.O., Hopkins, T.L., and Speirs, R.D. (1987) Aromatic cross-links in insect cuticle: detection by solid-state 13C and 15N NMR. *Science*, **235**, 1200–1204.

85 Moussian, B., Soding, J., Schwarz, H., and Nusslein-Volhard, C. (2005) Retroactive, a membrane-anchored extracellular protein related to vertebrate snake neurotoxin-like proteins, is required for cuticle organization in the larva of *Drosophila melanogaster*. *Dev. Dyn.*, **233**, 1056–1063.

86 Hijazi, A., Masson, W., Auge, B., Waltzer, L., Haenlin, M., and Roch, F. (2009) Boudin is required for septate junction organisation in *Drosophila* and codes for a diffusible protein of the Ly6 superfamily. *Development*, **136**, 2199–2209.

87 Nilton, A., Oshima, K., Zare, F., Byri, S., Nannmark, U., Nyberg, K.G., Fehon, R.G., and Uv, A.E. (2010) Crooked, coiled and crimpled are three Ly6-like proteins required for proper localization of septate junction components. *Development*, **137**, 2427–2437.

88 Luschnig, S., Batz, T., Armbruster, K., and Krasnow, M.A. (2006) Serpentine and vermiform encode matrix proteins with chitin binding and deacetylation domains that limit tracheal tube length in *Drosophila*. *Curr. Biol.*, **16**, 186–194.

89 Cornman, R.S., Togawa, T., Dunn, W.A., He, N., Emmons, A.C., and Willis, J.H. (2008) Annotation and analysis of a large cuticular protein family with the R&R Consensus in *Anopheles gambiae*. *BMC Genomics*, **9**, 22.

90 Cornman, R.S. and Willis, J.H. (2008) Extensive gene amplification and concerted evolution within the CPR family of cuticular proteins in mosquitoes. *Insect Biochem. Mol. Biol.*, **38**, 661–676.

91 Cornman, R.S. and Willis, J.H. (2009) Annotation and analysis of low-complexity protein families of *Anopheles gambiae* that are associated with cuticle. *Insect Mol. Biol.*, **18**, 607–622.

92 Togawa, T., Augustine Dunn, W., Emmons, A.C., and Willis, J.H. (2007) CPF and CPFL, two related gene families encoding cuticular proteins of *Anopheles gambiae* and other insects. *Insect Biochem. Mol. Biol.*, **37**, 675–688.

93 Rebers, J.E. and Willis, J.H. (2001) A conserved domain in arthropod cuticular proteins binds chitin. *Insect Biochem. Mol. Biol.*, **31**, 1083–1093.

94 Karouzou, M.V., Spyropoulos, Y., Iconomidou, V.A., Cornman, R.S., Hamodrakas, S.J., and Willis, J.H. (2007) *Drosophila* cuticular proteins with the R&R Consensus: annotation and classification with a new tool for discriminating RR-1 and RR-2 sequences. *Insect Biochem. Mol. Biol.*, **37**, 754–760.

95 Togawa, T., Dunn, W.A., Emmons, A.C., Nagao, J., and Willis, J.H. (2008) Developmental expression patterns of cuticular protein genes with the R&R Consensus from *Anopheles gambiae*. *Insect Biochem. Mol. Biol.*, **38**, 508–519.

96 Bennet-Clark, H. (2007) The first description of resilin. *J. Exp. Biol.*, **210**, 3879–3881.

97 Weis-Fogh, T. (1960) A rubber-like protein in insect cuticle. *J. Exp. Biol.*, **37**, 889–907.

98 Qin, G., Lapidot, S., Numata, K., Hu, X., Meirovitch, S., Dekel, M., Podoler, I., Shoseyov, O., and Kaplan, D.L. (2009) Expression, cross-linking, and

characterization of recombinant chitin binding resilin. *Biomacromolecules*, **10**, 3227–3234.

99 Lyons, R.E., Nairn, K.M., Huson, M.G., Kim, M., Dumsday, G., and Elvin, C.M. (2009) Comparisons of recombinant resilin-like proteins: repetitive domains are sufficient to confer resilin-like properties. *Biomacromolecules*, **10**, 3009–3014.

100 Dutta, N.K., Choudhury, N.R., Truong, M. Y., Kim, M., Elvin, C.M., and Hill, A.J. (2009) Physical approaches for fabrication of organized nanostructure of resilin-mimetic elastic protein rec1-resilin. *Biomaterials*, **30**, 4868–4876.

101 Burrows, M., Shaw, S.R., and Sutton, G.P. (2008) Resilin and chitinous cuticle form a composite structure for energy storage in jumping by froghopper insects. *BMC Biol.*, **6**, 41.

102 Ardell, D.H. and Andersen, S.O. (2001) Tentative identification of a resilin gene in *Drosophila melanogaster*. *Insect Biochem. Mol. Biol.*, **31**, 965–970.

103 Elvin, C.M., Carr, A.G., Huson, M.G., Maxwell, J.M., Pearson, R.D., Vuocolo, T., Liyou, N.E., Wong, D.C., Merritt, D.J., and Dixon, N.E. (2005) Synthesis and properties of crosslinked recombinant pro-resilin. *Nature*, **437**, 999–1002.

104 Edens, W.A., Sharling, L., Cheng, G., Shapira, R., Kinkade, J.M., Lee, T., Edens, H.A., Tang, X., Sullards, C., Flaherty, D. B., Benian, G.M., and Lambeth, J.D. (2001) Tyrosine cross-linking of extracellular matrix is catalyzed by Duox, a multidomain oxidase/peroxidase with homology to the phagocyte oxidase subunit gp91phox. *J. Cell Biol.*, **154**, 879–891.

105 Clark, G.F. (2011) Molecular models for mouse sperm-oocyte binding. *Glycobiology*, **21**, 3–5.

106 Bokel, C., Prokop, A., and Brown, N.H. (2005) Papillote and Piopio: *Drosophila* ZP-domain proteins required for cell adhesion to the apical extracellular matrix and microtubule organization. *J. Cell Sci.*, **118**, 633–642.

107 Wilkin, M.B., Becker, M.N., Mulvey, D., Phan, I., Chao, A., Cooper, K., Chung, H.J., Campbell, I.D., Baron, M., and MacIntyre, R. (2000) *Drosophila* dumpy is a gigantic extracellular protein required to maintain tension at epidermal-cuticle attachment sites. *Curr. Biol.*, **10**, 559–567.

108 Rodgers, R.J., Irving-Rodgers, H.F., and Russell, D.L. (2003) Extracellular matrix of the developing ovarian follicle. *Reproduction*, **126**, 415–424.

109 Valcarcel, R., Weber, U., Jackson, D.B., Benes, V., Ansorge, W., Bohmann, D., and Mlodzik, M. (1999) Sec61beta, a subunit of the protein translocation channel, is required during *Drosophila* development. *J. Cell Sci.*, **112** (Part 23), 4389–4396.

110 Shaik, K.S., Pabst, M., Schwarz, H., Altmann, F., and Moussian, B. (2011) The Alg5 ortholog Wollknauel is essential for correct epidermal differentiation during *Drosophila* late embryogenesis. *Glycobiology*, **21**: 743–756.

111 Moussian, B., Veerkamp, J., Muller, U., and Schwarz, H. (2007) Assembly of the *Drosophila* larval exoskeleton requires controlled secretion and shaping of the apical plasma membrane. *Matrix Biol.*, **26**, 337–347.

112 Forster, D., Armbruster, K., and Luschnig, S. (2010) Sec24-dependent secretion drives cell-autonomous expansion of tracheal tubes in *Drosophila*. *Curr. Biol.*, **20**, 62–68.

113 Bray, S.J. and Kafatos, F.C. (1991) Developmental function of Elf-1: an essential transcription factor during embryogenesis in *Drosophila*. *Genes Dev.*, **5**, 1672–1683.

114 Ostrowski, S., Dierick, H.A., and Bejsovec, A. (2002) Genetic control of cuticle formation during embryonic development of *Drosophila melanogaster*. *Genetics*, **161**, 171–182.

115 Mace, K.A., Pearson, J.C., and McGinnis, W. (2005) An epidermal barrier wound repair pathway in *Drosophila* is mediated by grainy head. *Science*, **308**, 381–385.

116 Fox, R.M., Hanlon, C.D., and Andrew, D.J. (2011) The CrebA/Creb3-like transcription factors are major and direct regulators of secretory capacity. *J. Cell Biol.*, **191**, 479–492.

117 Abrams, E.W. and Andrew, D.J. (2005) CrebA regulates secretory activity in the *Drosophila* salivary gland and epidermis. *Development*, **132**, 2743–2758.

118 Spindler, K.D., Honl, C., Tremmel, C., Braun, S., Ruff, H., and Spindler-Barth, M. (2009) Ecdysteroid hormone action. *Cell Mol. Life Sci.*, **66**, 3837–3850.

119 Chavoshi, T.M., Moussian, B., and Uv, A. (2010) Tissue-autonomous EcR functions are required for concurrent organ morphogenesis in the *Drosophila* embryo. *Mech. Dev.*, **127**, 308–319.

120 Hiruma, K. and Riddiford, L.M. (2009) The molecular mechanisms of cuticular melanization: the ecdysone cascade leading to dopa decarboxylase expression in *Manduca sexta*. *Insect Biochem. Mol. Biol.*, **39**, 245–253.

121 Yao, Q., Zhang, D., Tang, B., Chen, J., Lu, L., and Zhang, W. (2011) Identification of 20-hydroxyecdysone late-response genes in the chitin biosynthesis pathway. *PLoS One*, **5**, e14058.

122 Dunne, J.C., Kondylis, V., and Rabouille, C. (2002) Ecdysone triggers the expression of Golgi genes in *Drosophila* imaginal discs via broad-complex. *Dev. Biol.*, **245**, 172–186.

123 Gangishetti, U., Breitenbach, S., Zander, M., Saheb, S.K., Muller, U., Schwarz, H., and Moussian, B. (2009) Effects of benzoylphenylurea on chitin synthesis and orientation in the cuticle of the *Drosophila* larva. *Eur. J. Cell Biol.*, **88**, 167–180.

124 Goldberg, J., Connolly, P., Schnizlein-Bick, C., Durkin, M., Kohler, S., Smedema, M., Brizendine, E., Hector, R., and Wheat, J. (2000) Comparison of nikkomycin Z with amphotericin B and itraconazole for treatment of histoplasmosis in a murine model. *Antimicrob. Agents Chemother.*, **44**, 1624–1629.

125 Tariq, V.N. and Devlin, P.L. (1996) Sensitivity of fungi to nikkomycin Z. *Fungal Genet. Biol.*, **20**, 4–11.

126 Belinato, T.A., Martins, A.J., Lima, J.B., Lima-Camara, T.N., Peixoto, A.A., and Valle, D. (2009) Effect of the chitin synthesis inhibitor triflumuron on the development, viability and reproduction of *Aedes aegypti*. *Mem. Inst. Oswaldo Cruz*, **104**, 43–47.

127 Martins, A.J., Belinato, T.A., Lima, J.B., and Valle, D. (2008) Chitin synthesis inhibitor effect on *Aedes aegypti* populations susceptible and resistant to organophosphate temephos. *Pest Manag. Sci.*, **64**, 676–680.

128 Arredondo-Jimenez, J.I. and Valdez-Delgado, K.M. (2006) Effect of Novaluron (Rimon 10 EC) on the mosquitoes *Anopheles albimanus*, *Anopheles pseudopunctipennis*, *Aedes aegypti*, *Aedes albopictus* and *Culex quinquefasciatus* from Chiapas. *Mexico Med. Vet. Entomol.*, **20**, 377–387.

129 Abo-Elghar, G.E., Fujiyoshi, P., and Matsumura, F. (2004) Significance of the sulfonylurea receptor (SUR) as the target of diflubenzuron in chitin synthesis inhibition in *Drosophila melanogaster* and *Blattella germanica*. *Insect Biochem. Mol. Biol.*, **34**, 743–752.

130 Akasaka, T., Klinedinst, S., Ocorr, K., Bustamante, E.L., Kim, S.K., and Bodmer, R. (2006) The ATP-sensitive potassium (KATP) channel-encoded dSUR gene is required for *Drosophila* heart function and is regulated by tinman. *Proc. Natl. Acad. Sci. USA*, **103**, 11999–12004.

131 Nasonkin, I., Alikasifoglu, A., Ambrose, C., Cahill, P., Cheng, M., Sarniak, A., Egan, M., and Thomas, P.M. (1999) A novel sulfonylurea receptor family member expressed in the embryonic *Drosophila* dorsal vessel and tracheal system. *J. Biol. Chem.*, **274**, 29420–29425.

132 Tonning, A., Hemphala, J., Tang, E., Nannmark, U., Samakovlis, C., and Uv, A. (2005) A transient luminal chitinous matrix is required to model epithelial tube diameter in the *Drosophila* trachea. *Dev. Cell*, **9**, 423–430.

133 Devine, W.P., Lubarsky, B., Shaw, K., Luschnig, S., Messina, L., and Krasnow, M.A. (2005) Requirement for chitin biosynthesis in epithelial tube morphogenesis. *Proc. Natl. Acad. Sci. USA*, **102**, 17014–17019.

134 Wang, S., Jayaram, S.A., Hemphala, J., Senti, K.A., Tsarouhas, V., Jin, H., and Samakovlis, C. (2006) Septate-junction-

dependent luminal deposition of chitin deacetylases restricts tube elongation in the *Drosophila* trachea. *Curr. Biol.*, **16**, 180–185.

135 Tsarouhas, V., Senti, K.A., Jayaram, S.A., Tiklova, K., Hemphala, J., Adler, J., and Samakovlis, C. (2007) Sequential pulses of apical epithelial secretion and endocytosis drive airway maturation in *Drosophila*. *Dev. Cell*, **13**, 214–225.

136 Locke, M. (1966) The structure and formation of the cuticulin layer in the epicuticle of an insect, *Calpodes ethlius* (Lepidoptera, Hesperiidae). *J. Morphol.*, **118**, 461–494.

137 Yoon, C.S., Hirosawa, K., and Suzuki, E. (1997) Corneal lens secretion in newly emerged *Drosophila melanogaster* examined by electron microscope autoradiography. *J. Electron. Microsc. (Tokyo)*, **46**, 243–246.

138 Hasegawa, M., Yagi, K., Iwakawa, S., and Hirai, M. (2001) Chitosan induces apoptosis via caspase-3 activation in bladder tumor cells. *Jpn. J. Cancer Res.*, **92**, 459–466.

139 Qi, L., Xu, Z., and Chen, M. (2007) *In vitro* and *in vivo* suppression of hepatocellular carcinoma growth by chitosan nanoparticles. *Eur. J. Cancer*, **43**, 184–193.

140 Zhang, H., Zhu, D., Song, L., Liu, L., Dong, X., Liu, Z., and Leng, X. (2011) Arginine conjugation affects the endocytic pathways of chitosan/DNA nanoparticles. *J. Biomed. Mater. Res. A*, **98**, 296–302.

141 Mura, C., Nacher, A., Merino, V., Merino-Sanjuan, M., Carda, C., Ruiz, A., Manconi, M., Loy, G., Fadda, A.M., and Diez-Sales, O. (2011) N-succinyl-chitosan systems for 5-aminosalicylic acid colon delivery: *in vivo* study with TNBS-induced colitis model in rats. *Int. J. Pharm.*, **416**, 145–154.

142 Bilensoy, E., Sarisozen, C., Esendagli, G., Dogan, A.L., Aktas, Y., Sen, M., and Mungan, N.A. (2009) Intravesical cationic nanoparticles of chitosan and polycaprolactone for the delivery of Mitomycin C to bladder tumors. *Int. J. Pharm.*, **371**, 170–176.

143 Grégorio Crini, Pierre-Marie Badot, èric Guibal (2009) Chitin et chitosane. Du biopolymère à l'application. *Presses universitaires de Franche-Comté. Besançon.*

144 Tan, M.L., Choong, P.F., and Dass, C.R. (2009) DNAzyme delivery systems: Getting past first base. *Expert Opin. Drug. Deliv.*, **6**, 127–138.

145 Andersen, M.O., Howard, K.A., and Kjems, J. (2009) RNAi using a chitosan/siRNA nanoparticle system: *in vitro* and *in vivo* applications. *Methods Mol. Biol.*, **555**, 77–86.

146 Tan, M.L., Dunstan, D.E., Friedhuber, A.M., Choong, P.F., and Dass, C.R. (2010) A nanoparticulate system that enhances the efficacy of the tumoricide Dz13 when administered proximal to the lesion site. *J. Control. Release*, **144**, 196–202.

147 Dass, C.R., Friedhuber, A.M., Khachigian, L.M., Dunstan, D.E., and Choong, P.F. (2008) Biocompatible chitosan-DNAzyme nanoparticle exhibits enhanced biological activity. *J. Microencapsul.*, **25**, 421–425.

148 Howard, K.A., Rahbek, U.L., Liu, X., Damgaard, C.K., Glud, S.Z., Andersen, M.O., Hovgaard, M.B., Schmitz, A., Nyengaard, J.R., Besenbacher, F., and Kjems, J. (2006) RNA interference *in vitro* and *in vivo* using a novel chitosan/siRNA nanoparticle system. *Mol. Ther.*, **14**, 476–484.

149 Murakami, K., Ishihara, M., Aoki, H., Nakamura, S., Yanagibayashi, H., Takikawa, M., Kishimoto, S., Yokoe, H., Kiyosawa, T., and Sato, Y. (2010) Enhanced healing of mitomycin C-treated healing-impaired wounds in rats with hydrosheets composed of chitin/chitosan, fucoidan, and alginate as wound dressings. *Wound Repair Regen.*, **18**, 478–485.

150 Tan, M.L., Choong, P.F., and Dass, C.R. (2009) Cancer, chitosan nanoparticles and catalytic nucleic acids. *J. Pharm. Pharmacol.*, **61**, 3–12.

151 Kato, Y., Onishi, H., and Machida, Y. (2005) Contribution of chitosan and its derivatives to cancer chemotherapy. *In Vivo*, **19**, 301–310.

6
Polymeric Blends with Biopolymers

Hero Jan Heeres, Frank van Mastrigt, and Francesco Picchioni

6.1
Introduction

The future scarcity of oil sources and the current strong awareness of sustainability issues in society are two of the main drivers behind the interest, at both academic and industrial levels, in the use of biopolymers (defined here as "polymers that involve living organisms in their synthesis process") in a variety of consumer products [1]. Biopolymers are in general characterized by relatively low costs and a large spread in geographic availability. However, they usually display (when taken alone) rather unsatisfactory mechanical properties (e.g., tensile properties in thermoplastic starch) and the variability of the feed on a (macro)molecular level (e.g., different amino acid compositions in proteins) is also a serious issue. In this respect, blending of biopolymers with commercial ones (e.g., polyesters) is the most common route for the production of bioplastics. Such blending processes are often aimed at overcoming the disadvantages outlined above while at the same time exploiting production technologies (e.g., extrusion) that are well established in the plastic industry [2].

Possible markets for bioplastics [3], as envisioned by the European Commission in 1998, include mainly packaging applications and the use as plastic bags. The total production levels were estimated to be 1 145 000 ton in the first decade of the new century. Almost 15 years later, these expectations are fulfilled and bioplastics have found applications in the foreseen application areas. In addition, the total volume is even considerably higher (1 145 000 ton/year as predicted in 1998 versus 1 500 000 ton/year estimated in 2009) [4]. By looking at these numbers, one might be tempted to consider the bioplastic industry a large one indeed. However, when comparing the bioplastic volumes with those of fossil-derived plastics (more than 30 000 000 ton/year in Europe only), it is clear that the bioplastics industry is in reality only in a state of infancy [1]. This is probably a consequence of the fact that many scientific/technological issues concerning the use of biopolymers in bioplastics have been only partially addressed and solved. Among these, the selection of a given biopolymer for a certain application is still a major issue in the design of new chemical products. Blends of commercial polymers with alginate

Handbook of Biopolymer-Based Materials: From Blends and Composites to Gels and Complex Networks,
First Edition. Edited by Sabu Thomas, Dominique Durand, Christophe Chassenieux, and P. Jyotishkumar.
© 2013 Wiley-VCH Verlag GmbH & Co. KGaA. Published 2013 by Wiley-VCH Verlag GmbH & Co. KGaA.

[5], starch [6], gluten [7], carboxymethyl cellulose [8–10], soya proteins [11,12], wood flour [13,14], and natural fibers [15,16] have been extensively studied and reported in the open literature. Generally, a plasticizer must be added to the biopolymer to be able to process it using conventional processing equipments such as extruders [17]. However, in some cases biopolymers have been used as simple solid fillers [18–20]. The situation is further complicated by the fact that within every class of biopolymers (e.g., soya proteins) further variations in the (macro)molecular structure as a function of the (botanical) origin (and in some cases even of the harvested region) are possible.

In this chapter, we will not consider all possible blends of biopolymers and synthetic plastics but focus on starch (St) and chitosan (Cht). These two materials were selected as they have already a broad application range, are produced in large volumes, and are considered as good examples of the advantages and disadvantages associated with the use of biopolymers in bioplastic materials.

Starch is considered one of the most promising candidates for use in bioplastics because of its wide availability (although from different sources) and relatively low cost. The monomeric unit of this biopolymer consists of D-glucose, which is arranged in a simple linear (amylose) or branched fashion (amylopectin) [21]. Starch is generally a semicrystalline polymer where crystallinity is the result of organization of amylopectin in the granules while, amylose constitutes the main part of the amorphous phase. Starches from different sources are in principle characterized by a different molecular weight as well as amylopectin/amylose ratio.

The large availability of starch makes this material a popular choice for a wide variety of products [22,23]. Moreover, besides commercial polymers, starch can be blended easily with other biopolymers such as chitosan [24,25], gluten [26,27], and lignin derivatives [28]. In general, starch blends and composites have found applications for packaging purposes, for foam production [29–34], and for tissue engineering [35,36] and biomedical applications in general [37–42]. In many cases, the main objective of starch addition to other polymers is the necessity to reduce feedstock costs while at the same time preserving/conferring a biodegradable character to the end product [43–47]. Furthermore, in some cases St is simply added to other polymeric systems as a filler [48–54].

Besides simple melt mixing processes, other routes to starch blends have been explored. Blending in solution is a widely studied possibility [55–59]; however, the use of less environment-friendly solvents is a serious drawback. *In situ* blends can also be prepared by chemically grafting a polymeric chain on the starch [60–62] or vice versa [63]. However, also in this case, the use of organic solvents renders the process less attractive from an industrial point of view and actually is only convenient when the product has a high-value specialty type of application, for example, in the biomedical industry [64].

Starch is, generally speaking, a hydrophilic polymer in which hydrogen bonding is mainly responsible for the intermacromolecular interactions. The latter must be overcome to render starch processable, usually by addition of a plasticizer, for example, glycerol or other polyols [65]. The same intermacromolecular interactions are actually also responsible for the low miscibility of starch with many commercial polymers, for example, polyesters and hydrophobic ones in general [66,67]. This incompatibility between starch and other polymers may be overcome by two strategies: either by the addition of a compatibilizer [68] or by the use of a (chemically) modified starch [69]. In both cases, the employed strategy has often consequences for the biodegradability of the blends. The system is further complicated when considering that often additional components are employed to the blends to fine-tune the mechanical properties. Such component could be another (commercial) polymer [70–73] or even a filler in the form of fibers, for example, natural ones such as cotton [74–76]. The influence of the additional components on the biodegradability should be carefully assessed [75,77]. This property is determined using standardized procedures involving assessment of the mechanical properties of the blends as a function of time [78,79] upon exposure to typical degradation conditions, for example, soil burial. In this chapter, we will limit the discussion to the synthesis and mechanical/rheological properties of starch- and chitosan-based blends and not to biodegradability as this topic has been recently reviewed [80].

From the above discussion, it is clear that starch may indeed represent a paradigmatic example for the scientific/technological issues relevant to biopolymer blends. The incompatibility at molecular level (thus the necessity of modification or compatibilization), the variability in the macromolecular structure (linear versus branched chains), the necessity to use a plasticizer (e.g., glycerol), and sensitivity to moisture and temperature are all factors that render starch an excellent representative of a biopolymer. However, due to the lack of variation in the chemical structure at monomer level in starch (the only functional groups being the hydroxyl ones), we decided to also include chitosan-based blends in this chapter. Chitosan is the second most abundant biopolymer in nature consisting of repeating 1,4-linked 2-amino-2-deoxy-β-D-glucan units [81]. As such, it is the only naturally occurring carbohydrate source with an amine functionality.

As seen for starch, chitosan needs to be used in combination with a plasticizer for processability [82] and is mostly incompatible with commercial polymers [83]. It finds application, in its pure form as well as in blends with other biopolymers (such as St, cellulose and derivatives, proteins, etc.), mainly in the food (packaging) industry [84–93]. It is very similar to St and it is not surprising if one takes into account the very similar chemical structure of these two polymers, which differ only by the presence on the C_2 of an –OH group for starch and an –NH_2 group for chitosan. Such slight variation in the chemical structure of the monomeric unit is, however, responsible for relevant differences in properties. Indeed, the presence of an easily ionizable (e.g., by protonation) amino group (responsible also for the antibacterial activity of this material [94,95]) along the backbone renders Cht-based blends particularly interesting for application in biomedical products (e.g., in tissue engineering and drug delivery) [96–105], in conductive materials [81,95,106,107], and in metal complexation resins [108–113].

In Sections 6.2 and 6.3, we will discuss starch- and chitosan-based blends by critically reviewing the scientific literature on the subject published in the past 15 years. In Section 6.4, we will provide a short summary of the more general concepts and a short outlook to future possibilities for both biopolymers.

6.2
Starch-Based Blends

In this section, we start by providing an overview of the most studied starch-based blends with synthetic polymers. The choice of the polymer to be blended with the starch and the physical form (e.g., as solid or as thermoplastic (TPS) material) and structural properties (e.g., amylose intake) of the latter are then discussed. Finally, general trends in terms of mechanical behavior for uncompatibilized and compatibilized blends are presented.

Blends of starch with a variety of polymeric materials have been widely studied and reported in the open literature. Figures 6.1 and 6.2 report the chemical

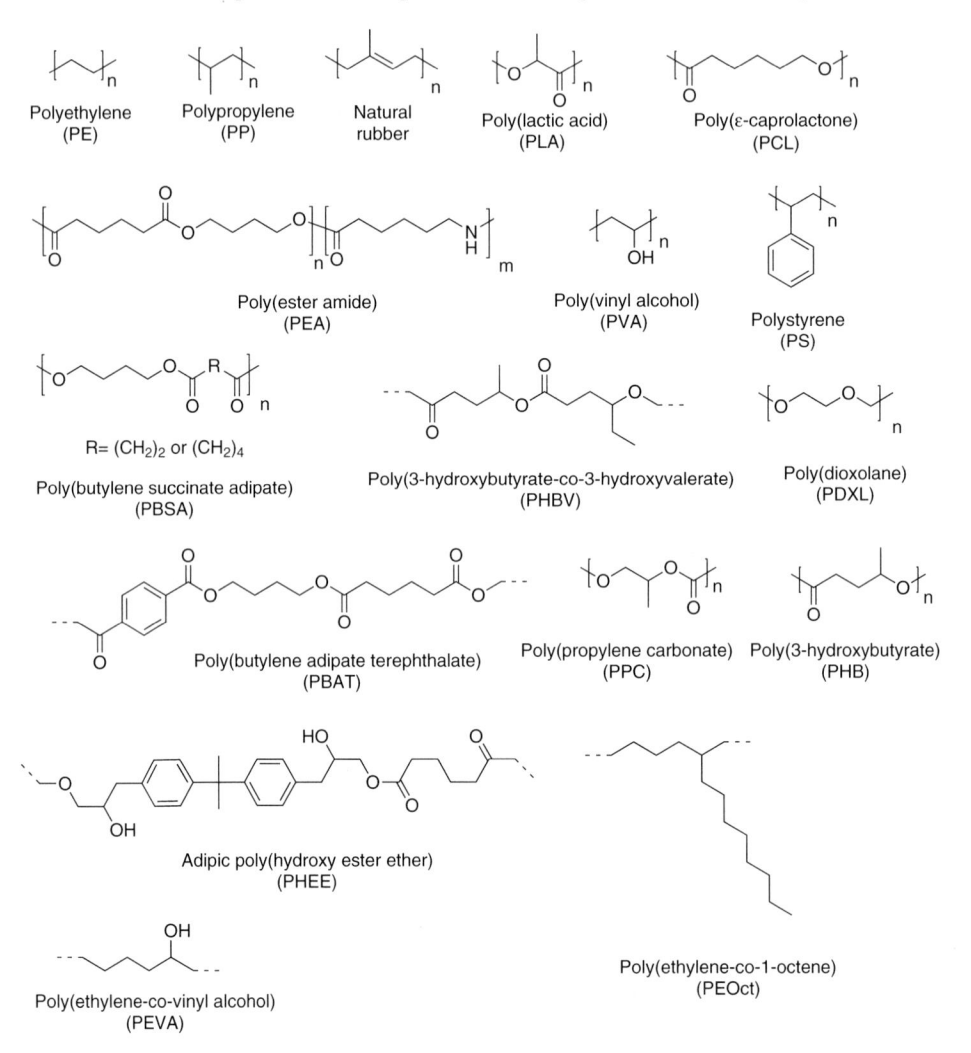

Figure 6.1 Chemical structures and full names of the most common polymers blended with starch.

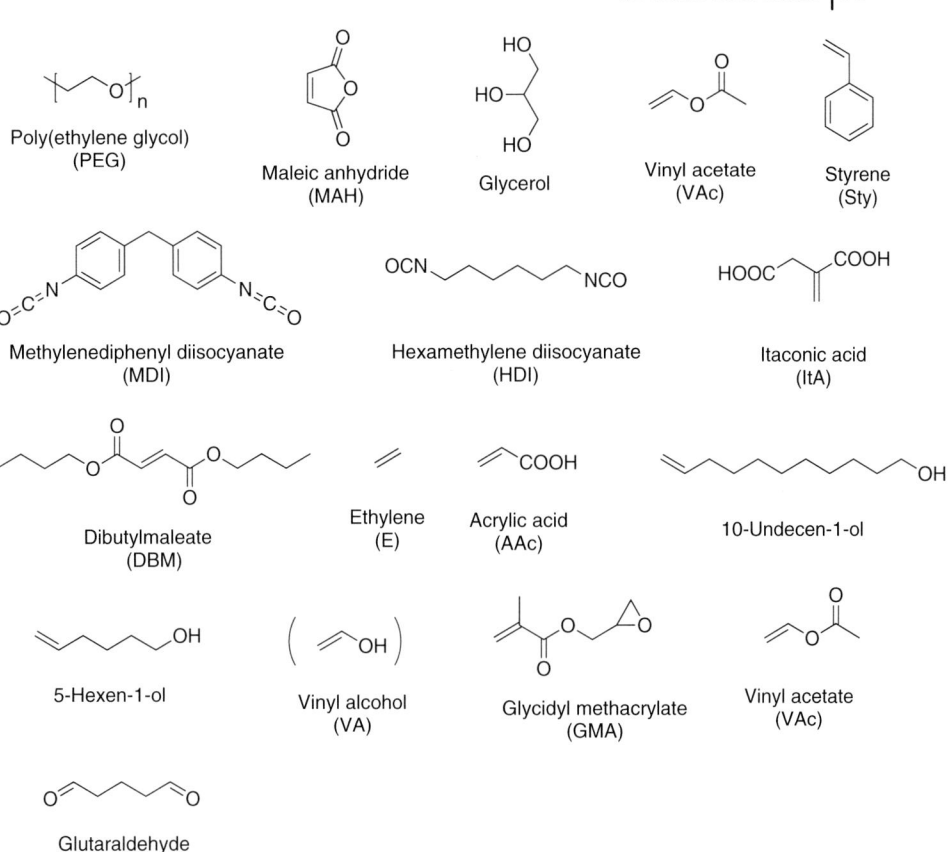

Figure 6.2 Chemical structures of the most common monomeric units and low molecular weight compounds used as additives for St-based blends.

structures and full names of the most important polymers and additives used in St-based blends.

An overview of the most popular starch-based blends together with the type of starch, its physical form (i.e., as solid or as thermoplastic material, with the corresponding plasticizer), and eventually the compatibilizer is reported in Table 6.1.

6.2.1
Polymer Selection for Starch Blending

The choice of the polymer to be blended with starch depends on many factors: mechanical and thermal behavior, biodegradability, and compatibility. By taking a general look at the most popular polymers (Table 6.1), it is quite difficult to define a generic framework for polymer selection. A suitable methodology may be based on

Table 6.1 Overview of St blends with synthetic polymers.

Polymer	Starch	Additives	Reference
LDPE	Maize (S)	PEG	[114]
LDPE	(Modified) sago (S)	—	[115]
LDPE	Sago (TPS, glycerol)	PE-g-MAH	[116]
LDPE	(Modified) potato (TPS, glycerol)	—	[117]
LDPE	Rice and potato (TPS, water)	—	[118]
LDPE	Corn (TPS, glycerol)	PE-g-MAH	[119]
LDPE	Corn (S)	PE-g-MAH	[120]
LDPE	Not specified	PE-co-10-undecen-1-ol, PE-co-5-hexen-1-ol	[121]
LDPE	Tapioca (TPS, glycerol, and water)	PE-g-DBM	[122]
LDPE	Potato (S)	PE-g-MAH	[123]
LDPE	Tapioca (TPS, glycerol)	PE-g-MAH, PE-g-AAc	[124]
LDPE	Corn (S)	PE-g-MAH	[125]
LDPE	Corn (S)	PE-g-(sty-co-MAH)	[126]
LDPE	Corn (S)	PE-g-GMA	[127]
LDPE	Wheat (S) and (TPS, glycerol)	PEVAc	[128]
LDPE	Corn (TPS, glycerol)	—	[129]
LDPE	Not specified	—	[130]
LDPE	Banana (S)	PE-g-MAH	[131]
LDPE	Wheat (TPS, water, and glycerol)	—	[132]
LDPE	Tapioca (TPS, water, and glycerol)	PEVA	[133]
LDPE	Corn (TPS, glycerol)	—	[129]
LDPE	Corn and rice (S)	—	[134]
LDPE	Rice (TPS, glycerol)	PE-g-MAH	[135]
LDPE	Corn (S)	PEAAc	[136]
HDPE	Tapioca (TPS, water, and glycerol)	HDPE-g-MAH	[137]
PE	Corn (TPS, glycerol)	PE-g-ItA	[138]
PEOct	Corn (S)	PEOct-g-MAH	[139]
PEOct	Corn (S)	PEOct-g-AAc	[140]
PLA, PHEE	Corn (S)	—	[141]
PP-g-MAH	Corn (TPS, glycerol)	—	[142]
PLA	Wheat (S)	MDI	[143]
PLA	Corn (S)	St-g-PLA	[144]
PLA	Corn (TPS, glycerol)	PLA-g-MAH	[145]
PLA	Corn (S)	—	[146]
PLA	Corn (S)	PVA	[147]
PLA	Wheat (TPS, glycerol, and sorbitol)	PLA-g-MAH	[148]
PLA	Corn (TPS, glycerol)	—	[149]
PLA	Maize (S, amylopectin only)	PEVA	[150]
PLA	Corn (TPS, glycerol)		[151]
PLA	Corn and tapioca (TPS, water, and glycerol)	—	[152]
PLA	Wheat (TPS, glycerol)	PCL	[153]
PLA	Wheat (TPS, glycerol)	Several compatibilizers	[154]
PLA	Wheat	MDI	[155]

PLA	Corn (S)	PLA-g-AAc	[156]
PLA	Corn (S)	PLA-g-AAc	[157]
PHB	Potato (water solution)	St-g-VAc	[158]
PHB	Corn (S)	—	[159]
PHB	Maize (TPS, water, and glycerol)	—	[160]
PHBV	Not specified	—	[161]
PHBV	Corn (TPS, water, and glycerol)	—	[162]
PHBV	Maize (S)	—	[163]
PHBV	Corn (TPS, acetyl tributyl citrate)	St-g-GMA	[164]
PEA	Wheat (TPS, glycerol)	—	[165]
PCL	Corn	HDI	[166]
PCL	Wheat and potato (TPS, glycerol)	—	[167]
PCL	Starch formate	—	[168]
PCL	Corn (S)	St-g-PCL	[169]
PCL	Sago (S) and (TPS, water, and glycerol)	—	[170]
PCL	Corn (S) and (TPS, glycerol)	—	[171]
PCL	Corn (S) and (TPS, glycerol)	PEG	[172]
PCL	Corn (TPS, water)	PCL-g-AAc	[173]
PCL	Wheat (TPS, glycerol, and/or water)	PCL-g-MAH	[174]
PCL	Tapioca (S)	PDXL	[175]
PCL	Corn (S) and (TPS, glycerol)	—	[21]
PCL	Corn	—	[176]
PCL	Corn (TPS, glycerol)	—	[177]
PCL	Corn (S)	PCL-g-GMA, PCL-g-DEM	[178]
PCL	Potato (TPS, glycerol)	—	[179]
PCL	Corn (S)	PCL-g-AAc	[180]
PCL	Not specified	PCL-g-MAH	[181]
PCL	Corn (S)	—	[182]
PP	Amylose	Modified amylose	[183]
NR	Corn (TPS, glycerol)	—	[184]
NR	Cassava (S)	NR-g-MAH	[185]
PS	Not specified	St-g-PS	[186]
PBS	Corn (TPS, glycerol)	—	[187]
PBSA	High-amylose starch	—	[188]
PBSA	Corn (S)	—	[189]
PEVA	Acetylated tapioca (TPS, glycerol)	—	[190]
PEVA	Corn (S)	—	[191]
PVA	Sago (S)	—	[192]
PVA	Corn (TPS, glycerol)	—	[193]
PVA	Potato	Glutaraldehyde	[194]
PVA	Cassava	—	[195]
PPC	Corn (S)	—	[196]
PPC	Corn (S)	—	[197]
PDXL	Corn (S)		[198]

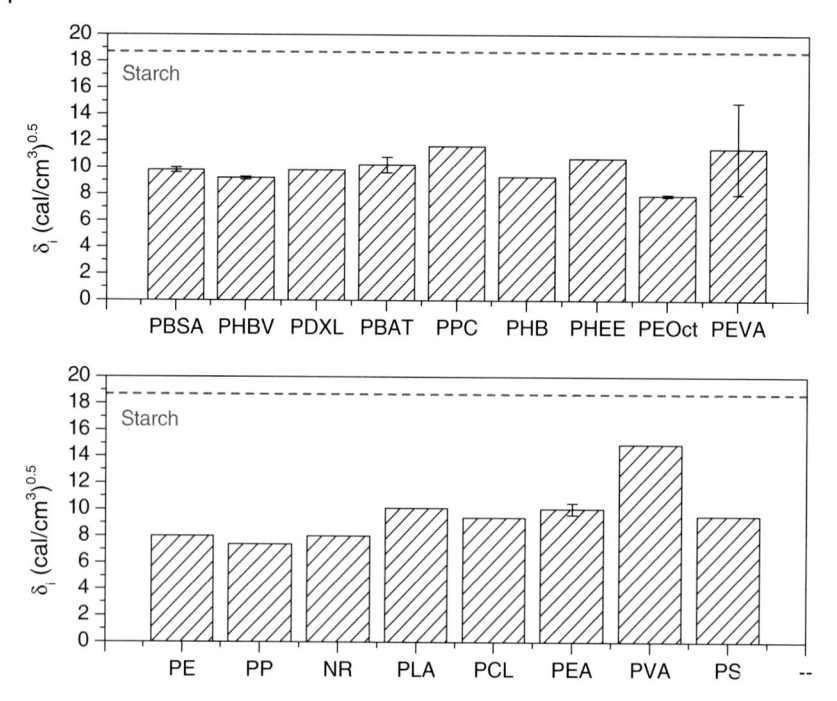

Figure 6.3 Solubility parameters for all polymeric materials in Table 6.1. Error bars take into account, for copolymers, changes in the chemical composition.

differences in solubility parameters between starch and the second polymer in the blend. The solubility parameter (δ), calculated using a group contribution approach [199], for all reported systems is given in Figure 6.3.

For all polymers, the $\delta_{polymer}$ is smaller than δ_{starch}, thus clearly indicating that the main idea behind blending is actually to attenuate the hydrophilic character of the starch component. However, in some cases (e.g., LDPE or PS), the selected polymer is not biologically degradable. This is not necessarily a major issue since, even when using commercial polymers that are in principle poorly biodegradable, the starch component is easily degraded and this also has a positive effect on the subsequent degradation rate of the second polymer [200]. In some cases, the final blend needs to be biodegradable rapidly and this puts constraints on the choice of the second polymer. In general terms, the higher the amount of St in the blend, the faster the degradation process [201,202].

6.2.2
Starch Structure

The starch source is also a variable and allows tuning of the properties of the starch–polymer blend. The amylose/amylopectin ratio, the moisture content, and the kind and amount of plasticizer used are known to affect the mechanical behavior of the blends.

The botanical origin of the starch, resulting in a.o. differences in the amylose/amylopectin ratio, has a strong influence on the properties of pure TPS [203]. This is also, although slightly, reflected in St-based blends with several different polyesters [204]. Inspection of blend morphology indicates that the starch phase becomes more finely dispersed as the amylopectin content in the blend increases. This leads to changes in tensile strengths, though a clear trend is absent. The same authors, working at a fixed starch intake of 70 wt% in blends with polyolefins, observed that the morphology is a clear function of the amylopectin/amylose ratio. This is not surprising if one takes into account the fact that the same ratio results generally in different viscosities of the St phase. Because of differences in the morphology, one would expect related differences in the mechanical behavior for blends containing starch at different amylose intakes. This has been only partially confirmed [205] and it is still a point of debate in the open literature.

The physical nature of the St phase (either as solid or as TPS) also has a clear influence on the final properties and rheological behavior [21]. Virgin starch gives plastic behavior in blends with PCL, while gelatinized starch results in brittle behavior with relatively high stress [137]. Ishiaku *et al.* [170] studied PCL blends with sago starch and found that the ultimate strength and elongation at break decrease with the starch intake; however, TPS performs better than normal starch. The overall inferior performance of TPS is explained by the formation of water (and thus voids after evaporation) in the molding stage. This does, however, not constitute a general concept since in other cases no differences are observed between solid St and TPS [128]. It must be stressed here that these discrepancies are more rule than exception, thus strongly suggesting that the influence of the plasticizer in general terms is strongly dependent on the system under examination.

The amount of water initially present in the starch source seems to have little effect on the final properties for blends with PLA, the only exception being the water uptake of the blends [146]. This has been confirmed by other researchers [206–209] and in particular by a systematic study on blends of sago starch with PCL [210]. Here, St is used in various states: native, predried, as TPS (20 wt% glycerol), and granules obtained by "powderizing" TPS. Elongation at break of the blends comprising of native and thermoplastic starches decreases almost linearly with the St volume fraction, whereas nonlinear dependences were observed for predried and thermoplastic starch granules (Figure 6.4). Except for blends containing native starch, the tensile strength was found to decrease linearly with the St volume fraction. One may conclude that in all cases, the tensile properties decrease almost linearly with the St volume fraction up to a maximum of around 0.6.

In successive research, the authors showed that predrying of the starch has a positive effect on properties and the drop rate as function of starch intake is reduced [211].

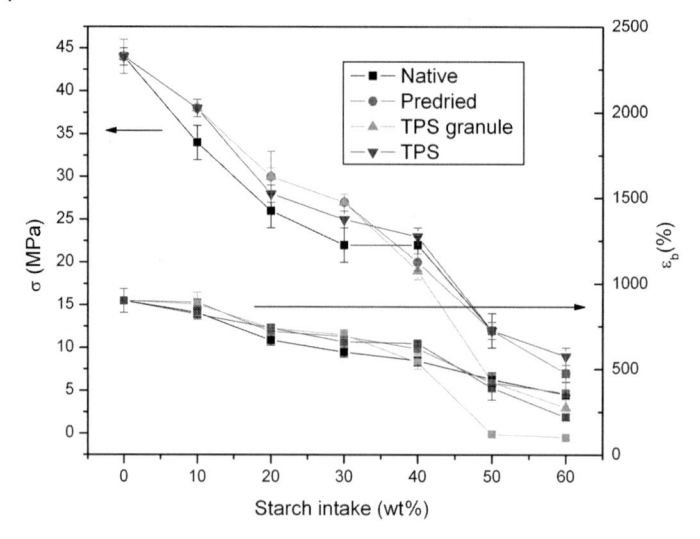

Figure 6.4 Mechanical properties versus ST intake as a function of the physical form of the St phase.

6.2.3
Uncompatibilized Blends

For blends in which St is the main component (i.e., the matrix), the addition of a second polymer (dispersed phase, for example, PEA) often results in an improvement in the mechanical properties (tensile strength, modulus, and elongation [165]). The opposite trend is generally observed when starch represents the minor component. For blends in which St is the dispersed phase, for example, with PPC [197], generally an increase in modulus [212] and a decrease in tensile strength with the St intake are observed. This is in agreement with semiempirical equations for composites with uniformly distributed (also in size) spherical particles [213]. Here, the decrease in tensile strength when the starch volume fraction (ϕ) increases can be described theoretically by

$$\sigma_C = \sigma_0(1 - 1.21\phi^{2/3}), \tag{6.1}$$

where σ_C and σ_0 are the tensile strength of the blend and the matrix, respectively, and ϕ is the volume fraction of the filler (starch in this case). For the modulus, a theoretical equation may be derived and this was shown to be a good model for the experimental trends:

$$E_C = E_0\left[1 + \left(\frac{\phi}{1-\phi}\right)\left(\frac{15(1-\nu)}{8-10\nu}\right)\right], \tag{6.2}$$

where E_C and E_0 are the modulus of the composite and the matrix, respectively, ϕ is the volume fraction of the filler (starch in this case), and ν is the Poisson ratio for the matrix [214]. These trends clearly indicate that starch acts as rigid component in the blends (increase in modulus, and decrease in tensile strength and elongation). This decrease in tensile strength and elongation at break (with respect to the pure components) as a function of the St intake is often perceived as a serious issue from an application point of view and hampers the use of larger amount of St in the blend without significant reductions in the mechanical properties.

It is clear that the morphology (i.e., the average particle size and particle size distribution of the minor component in the blend) and the chemical composition determine the final properties of the material. The first attempt to relate the blend morphology to the properties of the individual components has been proposed and is based on surface energy considerations [215]. For example, Biresaw and Carriere [216] reported surface energy measurements on St blends with PS, PCL, PHBV, PLA, PBAT, and PHEE. The surface energy of the solids and subsequently the interfacial adhesion was calculated. The surface tension of a liquid or solid is expressed as

$$\gamma_S^{TOT} = \gamma_S^D + \gamma_S^P = \gamma_S^D + 2(\gamma_S^+ \gamma_S^-), \tag{6.3}$$

where S is the solid, either starch (St) or polymer (Po), γ^{TOT} is the total surface energy, and γ^D is the contribution due to dispersive forces and γ^P due to polar ones, the latter being split into contributions for electron/H bonding donor (γ^+) and acceptor (γ^-) ability. The interfacial tension between starch and the polymer in the liquid phase ($\gamma_{St/Po}$), as determined from contact angle measurements for the individual components, is estimated by

$$\gamma_{St/Po}^{TOT} = (\sqrt{\gamma_{St}^D} - \sqrt{\gamma_{Po}^D})^2 + 2(\sqrt{\gamma_{St}^+ \gamma_{St}^-} + \sqrt{\gamma_{Po}^+ \gamma_{Po}^-} - \sqrt{\gamma_{St}^+ \gamma_{Po}^-} - \sqrt{\gamma_{Po}^+ \gamma_{St}^-}). \tag{6.4}$$

The results [216] show the absence of a clear correlation between the estimated interfacial tensions and the mechanical properties. This suggests that other factors, besides interfacial properties, also determine the mechanical behavior. Indeed, the morphology is also influenced by the processing conditions. A typical example is given for PCL/St blends. The rheological behavior of the individual components [167] as a function of the shear rate differs significantly. For instance, the viscosity of PCL follows the Carreau–Yasuda model:

$$\eta = \frac{\eta_0}{[1 + (\lambda\dot{\gamma})^a]^{(1-n)/a}}, \tag{6.5}$$

with η_0 the viscosity at zero shear rate, λ the relaxation time, $\dot{\gamma}$ the shear rate, n the pseudoplasticity index, and a the Carreau–Yasuda fitting parameter. On the other hand, the viscosity of TPS generally follows a power law model:

$$\eta = K\dot{\gamma}^{n-1}, \tag{6.6}$$

with K being the consistency index. In the above-mentioned example [167], such differences in viscosity behavior have direct consequences for the processing of the blend. Between 1 and $100\,s^{-1}$, the viscosity of TPS decreases with the shear rate while it is at a Newtonian plateau for PCL. As a result of the nonmiscibility and differences in viscosities ($\eta_{TPS} \gg \eta_{PCL}$), different morphologies (i.e., different average St particle sizes) are accessible by simply controlling the shear rate inside the processing equipment.

The fact that both interfacial tension between starch and the other polymeric material and the rheological properties of both have a clear influence on the morphology of the blends (and thus on the final properties) is not surprising when taking into account the general theories for morphology development in the melt [217]. Indeed, when mixing a given polymer with TPS (taken here as example of dispersed phase), the presence of shear causes breakup of the TPS droplets, thus in principle leading to a finer dispersion of the latter in the matrix (Figure 6.5b), while coalescence of the TPS droplets leads to a higher average particle size (thus to a coarse dispersion).

The balance between these two phenomena is the governing factor for morphology formation and it usually expressed in terms of the Weber number (We):

$$We = \frac{\eta_d G r}{\gamma}, \tag{6.7}$$

where G is the velocity gradient in the system (a function of the kind of mixing equipment used and the kind of flow), r is the droplet radius, and γ is the interfacial tension between the two liquid polymers. For a given droplet to break up, the Weber number must be higher than a critical value (We_{cr}), which is in turn a function of the kind of flow during mixing (e.g., shear or elongational) and of the viscosity ratio (Figure 6.5a) between dispersed and continuous phases (η_d/η_c). From these theoretical considerations, it is clear that both the interfacial tension between the polymers (directly affecting the We values) and the rheological behavior (affecting the We_{cr} values) must be taken into account when trying to gain a more fundamental understanding of the relationship between the blend morphology and the properties of the individual components.

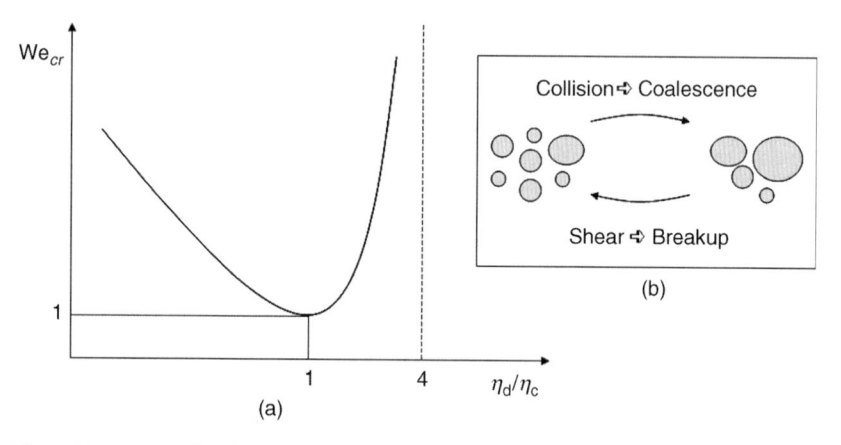

Figure 6.5 (a) Critical Weber number (We_{cr}) as a function of the viscosity ratio between dispersed phase (η_d) and continuous one (η_c) for shear flows. (b) Schematic representation of droplet breakup and coalescence during melt mixing.

6.2.4
Compatibilization

To overcome the trends in mechanical properties discussed above (namely, a decrease in tensile strength and elongation at break especially at higher St intakes in the blend), compatibilization of the blends is often perceived as a necessity. A compatibilized blend is characterized in general by a lower interfacial tension between the components (thus resulting in higher W_e values, see above) and also better interfacial adhesion. In this respect, two main strategies for compatibilization have been developed for starch-based blends. The first consists of starch modification with hydrophobic chains, while the second involves the use of a functionalized polymer (e.g., PE-g-MAH) to be used in combination with the virgin one (e.g., PE). In this case, the intention is to graft the starch on the compatibilizer *in situ*, that is, during processing. In both cases, the general idea is to improve the affinity of the St with the other polymer and in particular the interfacial adhesion between the St particles and the matrix.

The first strategy comprises the use of modified starch, usually with hydrophobic chains (see above) [218–223] in combination with native St. Starch can be modified before blending with hydrophobic polymers such as LDPE. In particular, the reaction with (long-chain) anhydrides (Figure 6.6) should theoretically result in improved compatibility with apolar polymers [115].

Indeed, the presence of hydrophobic chains (even for a one-carbon chain as in starch formate [168,224]) grafted on St results in general in a better compatibility [190,225] and better mechanical properties (particularly a higher modulus). Tensile strength and elongation at break still decrease with the St intake but to a lesser extent with respect to blends containing unmodified starch. It is postulated that the presence of aliphatic chains on the St increases the interfacial adhesion with the other polymer and ultimately favors the stress transfer mechanism between the two phases [115,117]. This approach has one drawback, besides the necessity of an extra processing step for the St modification, and this involves biodegradability. Modified starch displays usually a lower biodegradation rate (the effect being more relevant as the length of the grafted chains increases) with respect to the unmodified one [226].

The second strategy involves a chemical reaction between one of the two components (usually St) with a compatibilizer precursor (e.g., polymers grafted with MAH). In some special cases, the compatibilizer precursor is also generated

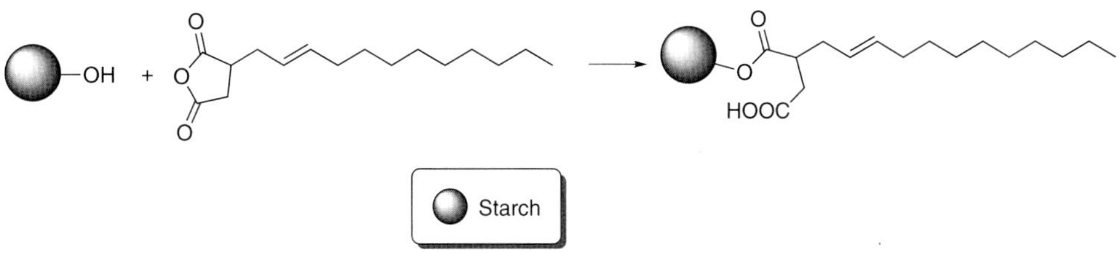

Figure 6.6 Modification of starch with dodecen-1-yl-succinic anhydride.

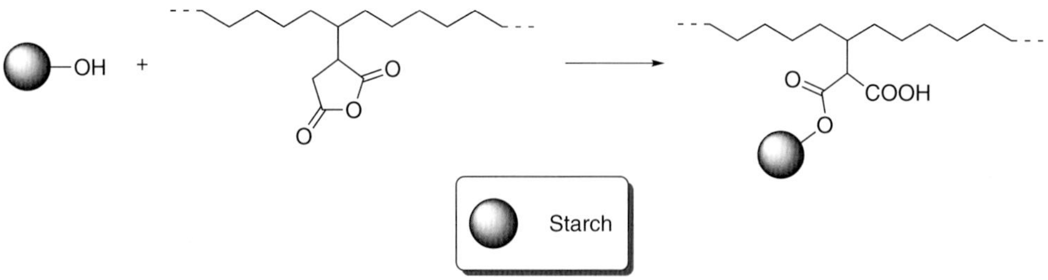

Figure 6.7 Schematic reaction between –OH groups on the surface of the ST particles (S) or droplets (TPS) and the MAH groups on PE-g-MAH (taken here as example).

in situ directly by addition of a peroxide and MAH to the St blend [227–231]. Independently of the way in which the maleated polymer is added (either directly or generated in situ), a chemical reaction is supposed to take place between the –OH groups on the starch and the anhydride group on the compatibilizer (Figure 6.7).

Confirmation of the occurrence of this reaction has been obtained in many studies, mainly by spectroscopic methods (e.g., FTIR) [119,232–234]. The disappearance of the peaks assigned to the anhydride (typically around 1850, 1780, and 1720 cm^{-1}) in FTIR spectra of the blend and the appearance of those typical of esters and acid groups (at 1730 and 1710 cm^{-1}, respectively) is often considered as proof of the reaction. This is, however, not entirely correct when using plasticized starch in the blend. Typical plasticizers for starch are polyols (e.g., glycerol and sorbitol) as well as water, that is, molecules containing –OH groups as in the starch. As a consequence, the possibility that the observed trends in the FTIR spectra are actually due to the reaction of the plasticizer with MAH may not be excluded. The occurrence of such side reaction has been demonstrated in binary blends of functionalized polymers with St. Kim et al. [235] studied PCL-g-GMA blends and observed a decrease in gel content (PCL-g-GMA acts as cross-linker for St) as the glycerol intake increases. The possible competition of St and the plasticizer with the reactive groups of the compatibilizer represents a very important factor and determines the properties of the ultimate blends. This was also illustrated by Taguet et al. [236], when studying blends of TPS (wheat, glycerol) with HDPE compatibilized by PE-g-MAH. The average particle size for uncompatibilized blends decreased with the glycerol content, while an increase with the PE-g-MAH intake at relatively high glycerol content was observed for compatibilized blends. The authors attributed the first effect to the differences in TPS viscosity as a function of the glycerol intake. The second effect is explained by the formation of two different TPS phases during mixing: a glycerol-rich one on the outside and a starch-rich one on the inside. This is governed by the spreading coefficient ($S_{St/Gly}$):

$$S_{St/Gly} = \gamma_{St/HDPE} - \gamma_{Gly/HDPE} - \gamma_{St/Gly}, \quad (6.8)$$

with $\gamma_{i/j}$ being the interfacial tension between component i and j. St and glycerol have about the same surface energy but starch has a much higher average molecular weight. Thus, it can be readily assumed that $\gamma_{Gly/HDPE}$ will be significantly smaller than $\gamma_{St/HDPE}$ and that $\gamma_{St/Gly}$ will be very low, as typical for a partially miscible mixture. Thus, the spreading coefficient of glycerol/starch is most probably a positive number. This would lead to the spontaneous formation of a thin glycerol-rich layer during melt mixing at the TPS/polyethylene interface to reduce the overall surface free energy of the system. This layer is expected to hinder the interaction between St and PE-g-MAH, probably through reaction of the glycerol itself with the compatibilizer precursor. As a result, at relatively high glycerol content, the reaction between St and PE-g-MAH is hindered, the compatibilizer (PE-g-St) is not formed, and thus the particle size, as observed experimentally, does not decrease with respect to the uncompatibilized blend.

A critical comparison, besides empirical ones [154], of the two compatibilization strategies (see above) is very difficult, not in the last place due to the rather long and time-consuming synthetic steps needed for the preparation of well-defined compatibilizers [237,238]. Moreover, some authors preferred a combined approach to the problem, for example, by using modified starch as the main component together with a compatibilizer precursor [239–241] or the use of modified starch as the compatibilizer precursor [242]. This renders the rationalization of the observed effects very difficult to achieve.

When selecting a compatibilization strategy, not only the chemistry of the system should be taken into account, but also the effect on the melt viscosity (crucial in determining the blend morphology) should be considered. When aiming for relatively low melt viscosities, the use of compatibilizer precursors (as in a maleated polymer) is an advantage with respect to premade compatibilizers, since the latter cause a significant increase in the melt viscosity [243].

Generally, both compatibilization strategies are effective. The decrease in tensile strength and elongation at break at higher St intakes is attenuated when the blend is compatibilized. However, in almost all studied systems, such attenuation is only partial and the mechanical properties (e.g., tensile strength and elongation) of the pure polymer (e.g., PCL) remain in almost all cases unattainable.

6.2.5
Composites

Compatibilization of blends is generally not sufficient to improve (see above) the mechanical properties to the desired values, especially at relatively high starch contents. The use of inorganic fillers is a very attractive route to further improve product properties. Among all possible fillers, clays in general [244–247] and montmorillonite in particular are the most popular choices [248,249]. This is likely due to the "nano" size of the filler particles, which ultimately results in a large increase in the stiffness of the end product [250,251]. Arroyo et al. [252] recently reported nanocomposites of TPS (wheat, water, and glycerol) with PLA (possibly grafted with MAH, PLAg) and montmorillonite. The authors found that TPS can intercalate the clay, the latter being mostly present in the starch phase. Clay composites with TPS, PLA, and/or PLA-g-MAH show very similar mechanical behavior, with rather similar values of E, σ, and ε (see Figure 6.8 for modulus values).

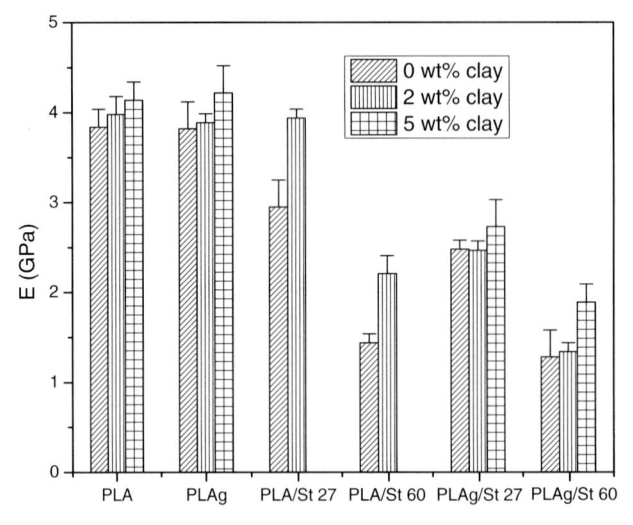

Figure 6.8 Tensile modulus as a function of composition. The number in the sample code indicates the TPS wt% in the blend.

Besides clays, carbon nanotubes were also used as nanofiller in St-based blends. In this case, besides improvements in mechanical behavior, a less pronounced moisture sensitivity of the final product is usually observed [253]. Also, inorganic salts such as $CaCO_3$ may be used to prevent swelling of St-based blends [254].

6.3
Blends with Chitosan (One Amino Group Too Much . . .)

Based on the close resemblance in chemical structure between chitosan and starch, the only difference being an amino group instead of a hydroxyl one in the chemical structure of the monomeric unit, one might anticipate similar blending behavior. However, this is actually not the case and the presence of amino groups results in specific interactions between the Cht chains. These must be overcome upon blending to obtain good dispersions [255]. However, as generally observed for polymeric systems, fully miscible blends are more exception than rule [199]. In the case of chitosan, full miscibility has been reported with hydroxypropyl cellulose and a few other polymers [256,257]. In most cases, as for St-based blends, immiscibility remains a common issue. Despite the general immiscibility, Cht has often been blended with commercial polymers to combine its positive properties (e.g., conductibility and antibacterial activity) with favorable properties of the other component. Correlo et al. [258] studied blends of Cht with several different polyesters (PBS, PCL, PLA, PBSA, and PBTA) and determined relevant mechanical properties as a function of the chemical composition. The mechanical properties of Cht/PBS blends over a wide range of compositions from pure PBS up to 70 wt% Cht are given in Figure 6.9.

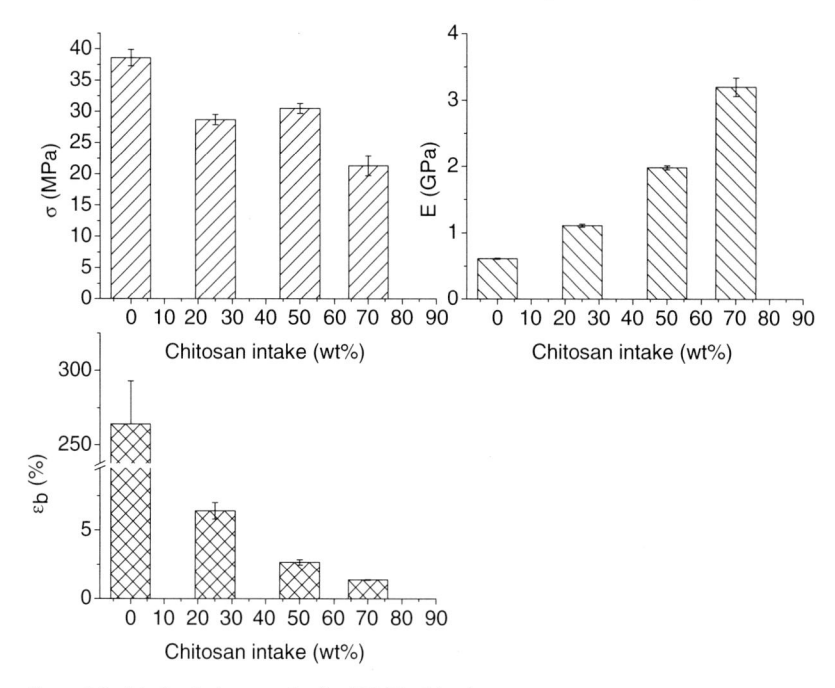

Figure 6.9 Mechanical properties for PBS/Cht blends.

The addition of Cht results in a reduction in tensile strength and elongation at break and a higher modulus. As for St (see above), these trends are easily explained by the lack of compatibility between the components and are in agreement with semiempirical relations. At relatively high Cht intakes (>50 wt%), aggregates can be formed, which further lower the stress value at which the materials fails. These trends are also valid for blends with different polyesters (Figure 6.10).

The elongation at break decreases dramatically for all blends except the one with PLA. This may be rationalized when considering that PLA is the only polyester with a T_g above room temperature, thus showing brittle behavior.

The results discussed above clearly point out the necessity for Cht-based blends for compatibilization. The use of diisocyanates is a promising option [259]. The relatively higher reactivity of the –NH$_2$ groups with –NCO groups [260] compared to hydroxyl groups renders this possibility even more attractive for Cht than for St. However, the difficulties associated with diisocyanate synthesis, mainly based on the use of phosgene, as well as the necessity for a controlled reaction (isocyanates being extremely reactive), make this strategy not widely popular. The use of a modified polymer as compatibilizer is a more convenient route. Wu [261] studied blends of PEOct compatibilized by PEOct-g-AAc and found similar effects as described for starch (see above). The main action of the compatibilizer is an attenuation in the decrease of tensile strength and elongation at break at higher biopolymer intake. Deviations from this trend have, however, already been reported. For example, Johns and Rao [262] used MAH (as monomer) for the compatibilization of Cht/NR blends.

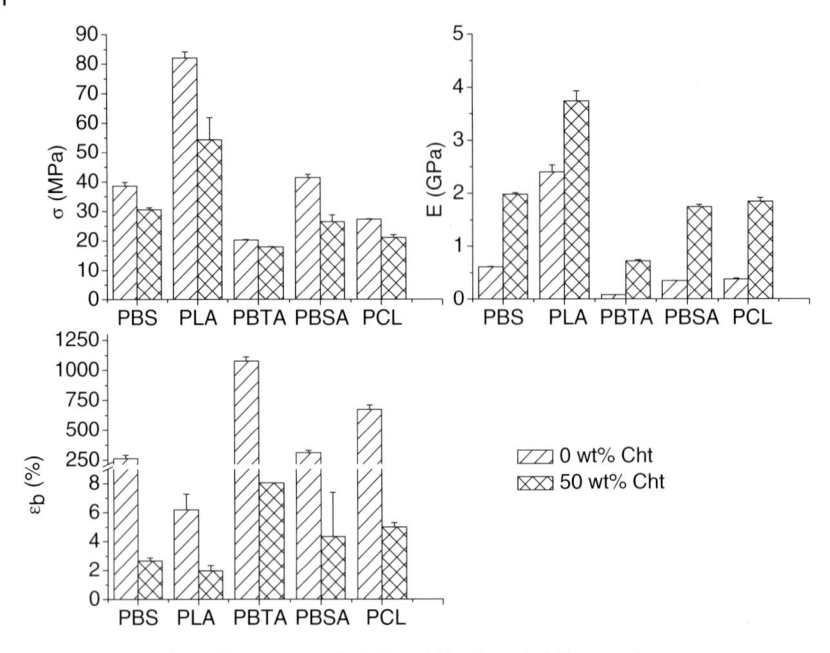

Figure 6.10 Mechanical properties of Cht-based blends with different polyesters.

The authors assumed that MAH is grafted on the NR chains and that the corresponding NR-g-MAH chains react successively with the amino groups of Cht to yield the desired block copolymer (Figure 6.11), the effective compatibilizer for this system.

Figure 6.11 Grafting of MAH on NR and reaction of NR-g-MAH with chitosan.

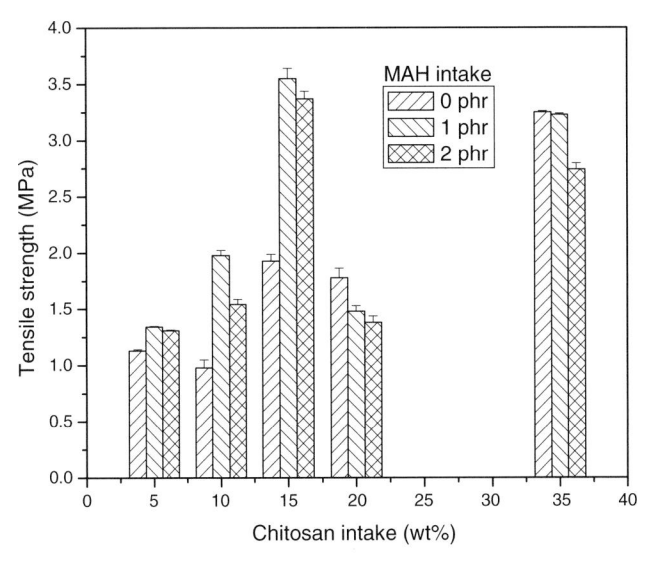

Figure 6.12 Tensile strength as a function of Cht intake for blends with NR.

When looking at the tensile strength as a function of composition (Figure 6.12), the modulus and elongation are not shown for brevity since the trends closely resemble the one of σ, and a strong negative effect of this compatibilization strategy is observed.

Both the σ and ε_b decrease with the MAH and Cht intake in the blends, the only exception being the tensile strength of the blend containing 15 wt% Cht. The authors attributed this lack of effect to a delicate balance between compatibilization and the plasticization of the blend, the latter due to unreacted MAH. The observed trends remain, however, at least peculiar when considering the general behavior of compatibilized blends based on St and Cht, that is, a decrease in tensile strength at higher biopolymer intake [83].

6.4
Future Perspectives

The above discussion clearly points out the existence of a number of general strategies for the preparation of biopolymer-based blends with good product properties. These can be extrapolated to improved routes for these materials.

6.4.1
Biopolymer Plasticization

The use of a plasticizer (like polyols) is in most cases an absolute necessity for processing of biopolymers and biopolymer-based blends. This is a direct result of the specific interactions in the materials as well as their sensitivity to relatively high temperatures. The plasticizer (both structure/functionality and intake) has a clear influence on the rheological properties of the biopolymer and in turn on the morphology and end properties of the blends. The use of mixed plasticizer systems, as shown for St [263], allows fine-tuning of the rheological behavior and can be seen as a tool for the design of improved processes for these materials. The use of new plasticizers is also a possibility. In particular, the use of supercritical CO_2 (scCO2) represents a "green" option in this respect. Indeed, it has already been demonstrated that $scCO_2$ can induce starch gelatinization [264] in combination with water. Besides the necessity to work at relatively high pressure (>80 bar), the inert nature of $scCO_2$ and the possibility to remove it by simple degassing of the system constitute clear advantages for this system over more classical ones. Furthermore, the possibility to recycle the CO_2 stream, to use relatively low processing temperatures, and to integrate the plasticization process with, for example, a foaming one renders this approach even more attractive. The addition of "plasticizer enhancers" [265], such as citric acid for starch [266], is also a viable option to modify the product properties of the blends. Citric acid aids rupture of the St granules and was shown to improve the TPS dispersion in blends of corn St with LDPE. The mechanical properties were better than St alone and in some cases similar to those of pure LDPE [267].

6.4.2
Blend Morphology and Compatibilization

The morphology of a polymeric blend is in general a function of the composition (volume fractions) and the viscosity and surface energies of the individual components. Blends of biopolymers do not constitute in this respect an exception to the rule. Process and product design must therefore take into account and when necessary comprise all of these aspects. From a scientific point of view, this requires a multidisciplinary approach. However, to the best of our knowledge, such studies are not known in the open literature.

The strong differences in polarity of many biopolymers with respect to commercial ones render the blends almost always immiscible and not compatible. The use of a compatibilizer is often needed to obtain the desired thermal and mechanical behavior. The use of diisocyanates represents a popular choice even if this is not completely in line with the "green" and "sustainable" character of these materials. The use of compatibilizers' precursors (e.g., maleated polymers) represents a viable option. The relatively lower reactivity of the anhydride groups with respect to the isocyanates is in this case compensated by the commercial availability of the polymers (e.g., PE-g-MAH) or in any case by the easiness of their production process. From a purely scientific point of view, the use of premade block or graft copolymers is most useful to gain a better understanding of the compatibilization mechanism as well as of compatibilizer effects on the thermal and mechanical behavior of the blend. This means that synthetic routes should be available for well-characterized grafted polymers (e.g., St-g-PCL). However, the use of a biopolymer (e.g., starch) together with a monomer (e.g., styrene) and initiator generally results in grafting efficiencies on the order of 30% [268] because of the fact that the reaction

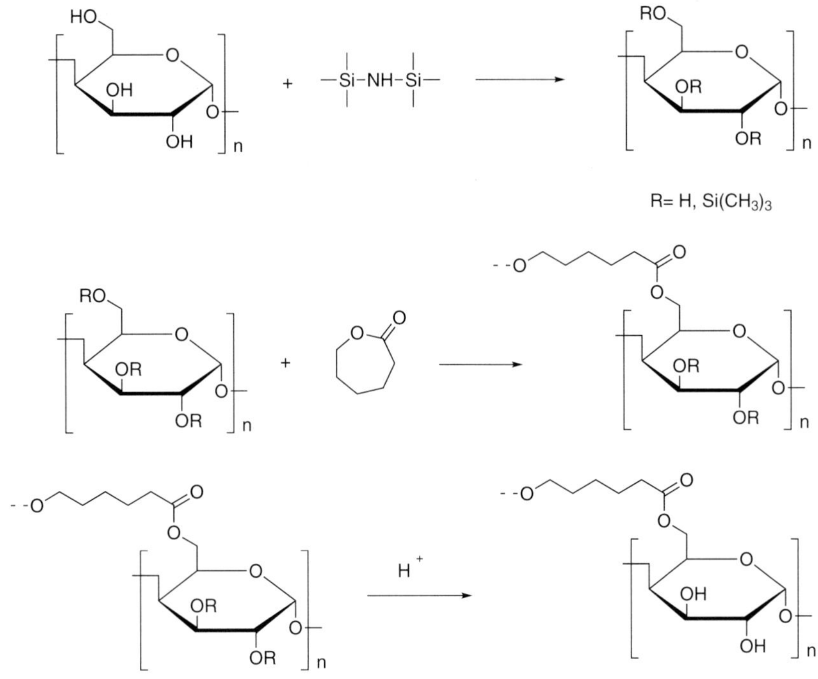

Figure 6.13 General strategy for starch silylation, grafting of PCL, and desilylation.

is generally heterogeneous. This makes the systems not well characterized and thus in principle unsuitable for a better understanding of the compatibilization mechanism. As suggested by Sugih et al. [269] (Figure 6.13), silylation of the St represents a viable route (at academic level) for the preparation of well-characterized systems.

Silylation of the starch is a crucial step since the resulting product is soluble in common organic solvent, thus allowing the grafting reaction to proceed in relatively homogeneous conditions. Upscaling of such processes at industrial level is at the moment strongly hindered by the use of organic solvents. The possibility to carry out such "grafting from" processes in alternative solvents such as ionic liquids [270] or even in scCO$_2$ [271–273] has already been reported and is a popular research topic at the moment.

6.4.3
Blend Processing: Technological Aspects

Improvement of the mechanical and thermal behavior of blends can also be achieved by proper selection of the processing technology. A typical example is the use of a one-step extrusion system for ST-based blends [274]. As in the case of TPS/LDPE blends [132], the general idea is to feed the polymer (LDPE) via a single-screw extruder to a double-screw containing starch and the plasticizers (in this case glycerol). Water is used as processing aid but is removed (volatilization) before St is mixed with LDPE. The connection between the two extruders contains efficient mixing elements, thus allowing accurate control of the blend morphology. The mechanical properties of the corresponding blends are comparable with the ones of compatibilized blends.

New blending technologies, such as solid-state shear pulverization, have been proposed recently [275]. However, simple modification of existing processing tools still remains preferable in terms of industrial applicability. In this respect, the formation of fibers in the biopolymer matrix (in this case St) during extrusion is an interesting opportunity [276]. Blends produced via this new concept display significantly higher tensile strengths and modulus compared to simple extruded blends.

References

1 Queiroz, A.U.B. and Collares-Queiroz, F. P. (2009) *Polym. Rev.*, **49** (2), 65–78.
2 Raquez, J., Narayan, R., and Dubois, P. (2008) *Macromol. Mater. Eng.*, **293** (6), 447–470.
3 Lorcks, J. (1998) *Polym. Degrad. Stabil.*, **59** (1–3), 245–249.
4 Markarian, J. (2008) *Plast. Addit. Compd.*, **10** (3) 22.
5 Basavaraja, C., Jo, E.A., Kim, B.S., Kim, D. G., and Huh, D.S. (2010) *Macromol. Res.*, **18** (11), 1037–1044.
6 Pielichowska, K. and Pielichowski, K. (2010) *J. Appl. Polym. Sci.*, **116** (3), 1725–1731.
7 Mohamed, A., Finkenstadt, V.L., Gordon, S.H., and Palmquist, D.E. (2010) *J. Appl. Polym. Sci.*, **118** (5), 2778–2790.
8 Ghanbarzadeh, B., Almasi, H., and Entezami, A.A. (2011) *Ind. Crops Prod.*, **33** (1), 229–235.
9 Ghanbarzadeh, B., Almasi, H., and Entezami, A.A. (2010) *Innovat. Food Sci. Emerg. Technol.*, **11** (4), 697–702.
10 Landreau, E., Tighzert, L., Bliard, C., Berzin, F., and Lacoste, C. (2009) *Eur. Polym. J.*, **45** (9), 2609–2618.
11 Fang, K., Wang, B., Sheng, K., and Sun, X.S. (2009) *J. Appl. Polym. Sci.*, **114** (2), 754–759.
12 Sam, S.T., Ismail, H., and Ahmad, Z. (2010) *J. Vinyl Addit. Technol.*, **16** (4), 238–245.
13 Liu, X., Khor, S., Petinakis, E., Yu, L., Simon, G., Dean, K., and Bateman, S. (2010) *Thermochim. Acta*, **509** (1–2), 147–151.
14 Yao, F. and Wu, Q. (2010) *J. Appl. Polym. Sci.*, **118** (6), 3594–3601.
15 Galicia-Garcia, T., Martinez-Bustos, F., Jimenez-Arevalo, O., Martinez, A.B., Ibarra-Gomez, R., Gaytan-Martinez, M., and Mendoza-Duarte, M. (2011) *Carbohydr. Polym.*, **83** (2), 354–361.
16 Guimaraes, J.L., Wypych, F., Saul, C.K., Ramos, L.P., and Satyanarayana, K.G. (2010) *Carbohydr. Polym.*, **80** (1), 130–138.
17 Dai, H., Chang, P.R., Peng, F., Yu, J., and Ma, X. (2009) *J. Polym. Res.*, **16** (5), 529–535.
18 Kramarova, Z., Alexy, P., Chodak, I., Spirk, E., Hudec, I., Kosikova, B., Gregorova, A., Suri, P., Feranc, J., Bugaj, P., and Duracka, M. (2007) *Polym. Adv. Technol.*, **18** (2), 135–140.
19 Tao, L., Jing-Xin, L., and Xin-Yuan, L. (2007) *Polym. Plast. Technol. Eng.*, **46** (6), 569–573.
20 Lan, C., Yu, L., Chen, P., Chen, L., Zou, W., Simon, G., and Zhang, X. (2010) *Macromol. Mater. Eng.*, **295** (11), 1025–1030.
21 Rosa, D.S., Guedes, C.G.F., Pedroso, A. G., and Calil, M.R. (2004) *Mater. Sci. Eng. C: Biomimetic Supramol. Syst.*, **24** (5), 663–670.
22 Liu, H., Xie, F., Yu, L., Chen, L., and Li, L. (2009) *Prog. Polym. Sci.*, **34** (12), 1348–1368.
23 Parulekar, Y. and Mohanty, A.K. (2007) *Macromol. Mater. Eng.*, **292** (12), 1218–1228.
24 Abd El Wahab, M. and Abdou, E.S. (2010) *J. Appl. Polym. Sci.*, **116** (5), 2874–2883.
25 Serrero, A., Trombotto, S., Cassagnau, P., Bayon, Y., Gravagna, P., Montanari, S.,

and David, L. (2010) *Biomacromolecules*, **11** (6), 1534–1543.

26 Chen, J.-S., Deng, Z.-Y., Wu, P., Tian, J.-C., and Xie, Q.-G. (2010) *Agric. Sci. China*, **9** (12), 1836–1844.

27 Corradini, E., Marconcini, J.M., Agnelli, J.A.M., and Mattoso, L.H.C. (2011) *Carbohydr. Polym.*, **83** (2), 959–965.

28 Morais, L.C., Consolin-Filho, N., Sartori, R.A., Cadaxo-Sobrinho, E.S., Souza, T.M.H., and Regiani, A.M. (2010) *J. Thermoplast. Compos. Mater.*, **23** (5), 699–716.

29 Lee, S.Y. and Hanna, M.A. (2009) *Polym. Compos.*, **30** (5), 665–672.

30 Mihai, M., Huneault, M.A., and Favis, B.D. (2007) *J. Cell. Plast.*, **43** (3), 215–236.

31 Mihai, M., Huneault, M.A., Favis, B.D., and Li, H. (2007) *Macromol. Biosci.*, **7** (7), 907–920.

32 Preechawong, D., Peesan, M., Supaphol, P., and Rujiravanit, R. (2004) *Polym. Test.*, **23** (6), 651–657.

33 Rosa, D.S., Guedes, C.G.F., and Casarin, F. (2005) *Polym. Bull.*, **54** (4–5), 321–333.

34 Sjoqvist, M., Boldizar, A., and Rigdahl, M. (2010) *J. Cell. Plast.*, **46** (6), 497–517.

35 Ciardelli, G., Chiono, V., Vozzi, G., Pracella, M., Ahluwalia, A., Barbani, N., Cristallini, C., and Giusti, P. (2005) *Biomacromolecules*, **6** (4), 1961–1976.

36 Duarte, A.R.C., Mano, J.F., and Reis, R.L. (2010) *Polym. Degrad. Stabil.*, **95** (10), 2110–2117.

37 Mano, J.F., Vaz, C.M., Mendes, S.C., Reis, R.L., and Cunha, A.M. (1999) *J. Mater. Sci.: Mater. Med.*, **10** (12), 857–862.

38 Mano, J.F., Reis, R.L., and Cunha, A.M. (2000) *J. Appl. Polym. Sci.*, **78** (13), 2345–2357.

39 Neves, N.M., Kouyumdzhiev, A., and Reis, R.L. (2005) *Mater. Sci. Eng. C: Biomimetic Supramol. Syst.*, **25** (2), 195–200.

40 Mano, J.F., Koniarova, D., and Reis, R.L. (2003) *J. Mater. Sci.: Mater. Med.*, **14** (2), 127–135.

41 Pashkuleva, I., Azevedo, H.S., and Reis, R.L. (2008) *Macromol. Biosci.*, **8** (2), 210–219.

42 Bagri, L.P., Bajpai, J., and Bajpai, A.K. (2009) *J. Macromol. Sci. Part A*, **46** (11), 1060–1068.

43 Sanchez-Garcia, M.D., Nordqvist, D., Hedenqvist, M., and Lagaron, J.M. (2011) *J. Appl. Polym. Sci.*, **119** (6), 3708–3716.

44 Innocentini-Mei, L.H., Bartoli, J.R., and Ballieri, R.C. (2003) *Macromol. Symp.*, **197**, 77–87.

45 Petinakis, E., Liu, X., Yu, L., Way, C., Sangwan, P., Dean, K., Bateman, S., and Edward, G. (2010) *Polym. Degrad. Stabil.*, **95** (9), 1704–1707.

46 Rahman, W.A.W.A., Sin, L.T., Rahmat, A.R., and Samad, A.A. (2010) *Carbohydr. Polym.*, **81** (4), 805–810.

47 Ramis, X., Cadenato, A., Salla, J.M., Morancho, J.M., Valles, A., Contat, L., and Ribes, A. (2004) *Polym. Degrad. Stabil.*, **86** (3), 483–491.

48 Dufresne, A. and Cavaille, J.Y. (1998) *J. Polym. Sci. Polym. Phys.*, **36** (12), 2211–2224.

49 Li, Y., Shoemaker, C.F., Ma, J., Shen, X., and Zhong, F. (2008) *Food Chem.*, **109** (3), 616–623.

50 Tang, H.G., Qi, Q., Wu, Y.P., Liang, G.H., Zhang, L.Q., and Ma, J. (2006) *Macromol. Mater. Eng.*, **291** (6), 629–637.

51 Valodkar, M. and Thakore, S. (2010) *Int. J. Polym. Anal. Charact.*, **15** (6), 387–395.

52 Wang, Z., Li, S., Fu, X., Lin, H., She, X., and Huang, J. (2010) *e-Polymers*, No. 115.

53 Wu, Y.P., Ji, M.Q., Qi, Q., Wang, Y.Q., and Zhang, L.Q. (2004) *Macromol. Rapid Commun.*, **25** (4), 565–570.

54 Wu, Y., Qi, Q., Liang, G., and Zhang, L. (2006) *Carbohydr. Polym.*, **65** (1), 109–113.

55 Nakashima, T., Xu, C., Bin, Y., and Matsuo, M. (2001) *Colloid Polym. Sci.*, **279** (7), 646–654.

56 Pereira, A.G.B., Gollveia, R.F., de Carvalho, G.M., Rubira, A.F., and Muniz, E.C. (2009) *Mater. Sci. Eng. C: Biomimetic Supramol. Syst.*, **29** (2), 499–504.

57 Pereira, A.G.B., Paulino, A.T., Rubira, A.F., and Muniz, E.C. (2010) *Express Polym. Lett.*, **4** (8), 488–499.

58 Ray, D., Roy, P., Sengupta, S., Sengupta, S.P., Mohanty, A.K., and Misra, M. (2009) *J. Polym. Environ.*, **17** (1), 49–55.

59 Ray, D., Roy, P., Sengupta, S., Sengupta, S.P., Mohanty, A.K., and Misra, M. (2009) *J. Polym. Environ.*, **17** (1), 56–63.

60 Rui-He, Wang, X., Wang, Y., Yang, K., Zeng, H., and Ding, S. (2006) *Carbohydr. Polym.*, **65** (1), 28–34.

61 Najemi, L., Jeanmaire, T., Zerroukhi, A., and Raihane, M. (2010) *Starch/Starke*, **62** (3–4), 147–154.
62 Chang, P.R., Zhou, Z., Xu, P., Chen, Y., Zhou, S., and Huang, J. (2009) *J. Appl. Polym. Sci.*, **113** (5), 2973–2979.
63 Maharana, T. and Singh, B.C. (2006) *J. Appl. Polym. Sci.*, **100** (4), 3229–3239.
64 Shaikh, M.M. and Lonikar, S.V. (2009) *J. Appl. Polym. Sci.*, **114** (5), 2893–2900.
65 Qiao, X., Tang, Z., and Sun, K. (2011) *Carbohydr. Polym.*, **83** (2), 659–664.
66 Moghaddam, L., Rintoul, L., Halley, P.J., and Frederick, P.M. (2006) *Polym. Test.*, **25** (1), 16–21.
67 Simoes, R.D., Rodriguez-Perez, M.A., de Saja, J.A., and Constantino, C.J.L. (2010) *J. Therm. Anal. Calorim.*, **99** (2), 621–629.
68 Moura, I., Machado, A.V., Duarte, F.M., and Nogueira, R. (2011) *J. Appl. Polym. Sci.*, **119** (6), 3338–3346.
69 Monnet, D., Joly, C., Dole, P., and Bliard, C. (2010) *Carbohydr. Polym.*, **80** (3), 747–752.
70 Liao, H. and Wu, C. (2009) *Mater. Sci. Eng. A: Struct. Mater.*, **515** (1–2), 207–214.
71 Kim, G.Y., Park, E.S., Kim, K., Chin, I.J., and Yoon, J.S. (2005) *Macromol. Symp.*, **224**, 333–341.
72 Averous, L., Moro, L., Dole, P., and Fringant, C. (2000) *Polymer*, **41** (11), 4157–4167.
73 Tanrattanakul, V. and Chumeka, W. (2010) *J. Appl. Polym. Sci.*, **116** (1), 93–105.
74 Majid, R.A., Ismail, H., and Taib, R.M. (2010) *Iran. Polym. J.*, **19** (7), 501–510.
75 Moriana, R., Vilaplana, F., Karlsson, S., and Ribes-Greus, A. (2011) *Compos. Part A: Appl. Sci. Manuf.*, **42** (1), 30–40.
76 Prachayawarakorn, J., Hommanee, L., Phosee, D., and Chairapaksatien, P. (2010) *Starch/Starke*, **62** (8), 435–443.
77 Moriana, R., Karlsson, S., and Ribes-Greus, A. (2010) *Polym. Compos.*, **31** (12), 2102–2111.
78 Lawton, J.W., Doane, W.M., and Willett, J.L. (2006) *J. Appl. Polym. Sci.*, **100** (4), 3332–3339.
79 Majid, R.A., Ismail, H., and Taib, R.M. (2010) *Polym. Plast. Technol. Eng.*, **49** (11), 1142–1149.
80 Leja, K. and Lewandowicz, G. (2010) *Pol. J. Environ. Stud.*, **19** (2), 255–266.
81 Park, W.H., Jeong, L., Yoo, D.I., and Hudson, S. (2004) *Polymer*, **45** (21), 7151–7157.
82 Epure, V., Griffon, M., Pollet, E., and Averous, L. (2011) *Carbohydr. Polym.*, **83** (2), 947–952.
83 Chiono, V., Vozzi, G., D'Acunto, M., Brinzi, S., Domenici, C., Vozzi, F., Ahluwalia, A., Barbani, N., Giusti, P., and Ciardelli, G. (2009) *Mater. Sci. Eng. C: Mater. Biol. Appl.*, **29** (7), 2174–2187.
84 Lopez-Caballero, M.E., Gomez-Guillen, M.C., Perez-Mateos, M., and Montero, P. (2005) *Food Hydrocolloids*, **19** (2), 303–311.
85 Ferreira, C.O., Nunes, C.A., Delgadillo, I., and Lopes-da-Silva, J.A. (2009) *Food Res. Int.*, **42** (7), 807–813.
86 Almeida, E.V.R., Frollini, E., Castellan, A., and Coma, V. (2010) *Carbohydr. Polym.*, **80** (3), 655–664.
87 Mao, L. and Wu, T. (2007) *J. Food Eng.*, **82** (2), 128–134.
88 Baruk Zamudio-Flores, P., Vargas Torres, A., Salgado-Delgado, R., and Arturo Bello-Perez, L. (2010) *J. Appl. Polym. Sci.*, **115** (2), 991–998.
89 Shih, C., Shieh, Y., and Twu, Y. (2009) *Carbohydr. Polym.*, **78** (1), 169–174.
90 Tripathi, S., Mehrotra, G.K., and Dutta, P.K. (2010) *Carbohydr. Polym.*, **79** (3), 711–716.
91 Zhang, B., Wang, D., Li, H., Xu, Y., and Zhang, L. (2009) *Ind. Crops Prod.*, **29** (2–3), 541–548.
92 Bourtoom, T. and Chinnan, M.S. (2008) *Food Sci. Technol.*, **41** (9), 1633–1641.
93 Kachanechai, T., Jantawat, P., and Pichyangkura, R. (2008) *Food Hydrocolloids*, **22** (1), 74–83.
94 Torres-Giner, S., Jose Ocio, M., and Maria Lagaron, J. (2009) *Carbohydr. Polym.*, **77** (2), 261–266.
95 Abou-Aiad, T.H.M., Abd-El-Nour, K.N., Hakim, I.K., and Elsabee, M.Z. (2006) *Polymer*, **47** (1), 379–389.
96 Lin, S., Hsiao, W., Jee, S., Yu, H., Tsai, T., Lai, J., and Young, T. (2006) *Biomaterials*, **27** (29), 5079–5088.
97 Simi, C.K. and Abraham, T.E. (2007) *Bioprocess Biosyst. Eng.*, **30** (3), 173–180.
98 Argin-Soysal, S., Kofinas, P., and Lo, Y.M. (2009) *Food Hydrocolloids*, **23** (1), 202–209.

99 Chen, Z.G., Wang, P.W., Wei, B., Mo, X.M., and Cui, F.Z. (2010) *Acta Biomater.*, **6** (2), 372–382.
100 Osugi, N., Dong, T., Hexig, B., and Inoue, Y. (2007) *J. Appl. Polym. Sci.*, **104** (5), 2939–2946.
101 Jin, J., Song, M., and Hourston, D.J. (2004) *Biomacromolecules*, **5** (1), 162–168.
102 Zuo, D., Tao, Y., Chen, Y., and Xu, W. (2009) *Polym. Bull.*, **62** (5), 713–725.
103 Lee, D.W., Yun, K., Ban, H., Choe, W., Lee, S.K., and Lee, K.Y. (2009) *J. Control. Release*, **139** (2), 146–152.
104 Kim, I.Y., Yoo, M.K., Kim, B.C., Kim, S.K., Lee, H.C., and Cho, C.S. (2006) *Int. J. Biol. Macromol.*, **38** (1), 51–58.
105 Grant, J., Tomba, J.P., Lee, H., and Allen, C. (2007) *J. Appl. Polym. Sci.*, **103** (6), 3453–3460.
106 Smitha, B., Sridhar, S., and Khan, A.A. (2005) *Eur. Polym. J.*, **41** (8), 1859–1866.
107 Smitha, B., Sridhar, S., and Khan, A.A. (2006) *J. Power Sources*, **159** (2), 846–854.
108 Boricha, A.G. and Murthy, Z.V.P. (2009) *J. Membr. Sci.*, **339** (1–2), 239–249.
109 Hiroki, A., Tran, H.T., Nagasawa, N., Yagi, T., and Tamada, M. (2009) *Radiat. Phys. Chem.*, **78** (12), 1076–1080.
110 Silva, S.S., Goodfellow, B.J., Benesch, J., Rocha, J., Mano, J.F., and Reis, R.L. (2007) *Carbohydr. Polym.*, **70** (1), 25–31.
111 Saxena, A., Kumar, A., and Shahi, V.K. (2006) *J. Colloid Interface Sci.*, **303** (2), 484–493.
112 Liu, C. and Bai, R. (2006) *J. Membr. Sci.*, **284** (1–2), 313–322.
113 Rutnakornpituk, M. and Ngamdee, P. (2006) *Polymer*, **47** (23), 7909–7917.
114 El-Rehim, H.A.A., Hegazy, E.S.A., Ali, A.M., and Rabie, A.M. (2004) *J. Photochem. Photobiol. A*, **163** (3), 547–556.
115 Khalil, H.P.S.A., Chow, W.C., Rozman, H.D., Ismail, H., Ahmad, M.N., and Kumar, R.N. (2001) *Polym. Plast. Technol. Eng.*, **40** (3), 249–263.
116 Majid, R.A., Ismail, H., and Taib, R.M. (2009) *Polym. Plast. Technol. Eng.*, **48** (9), 919–924.
117 Aburto, J., Thiebaud, S., Alric, I., Borredon, E., Bikiaris, D., Prinos, J., and Panayiotou, C. (1997) *Carbohydr. Polym.*, **34** (1–2), 101–112.
118 Arvanitoyannis, I., Biliaderis, C.G., Ogawa, H., and Kawasaki, N. (1998) *Carbohydr. Polym.*, **36** (2–3), 89–104.
119 Bikiaris, D. and Panayiotou, C. (1998) *J. Appl. Polym. Sci.*, **70** (8), 1503–1521.
120 Chandra, R. and Rustgi, R. (1997) *Polym. Degrad. Stabil.*, **56** (2), 185–202.
121 Dominguez, A.M., Quijada, R., and Yazdani-Pedram, M. (2006) *Macromol. Mater. Eng.*, **291** (8), 962–971.
122 Girija, B.G. and Sailaja, R.R.N. (2006) *J. Appl. Polym. Sci.*, **101** (2), 1109–1120.
123 Gupta, A.P., Sharma, M., and Kumar, V. (2008) *Polym. Plast. Technol. Eng.*, **47** (9), 953–959.
124 Huang, C.Y., Roan, M.L., Kuo, M.C., and Lu, W.L. (2005) *Polym. Degrad. Stabil.*, **90** (1), 95–105.
125 Liu, W., Wang, Y.J., and Sun, Z. (2003) *J. Appl. Polym. Sci.*, **88** (13), 2904–2911.
126 Park, E.S. and Yoon, J.S. (2003) *J. Appl. Polym. Sci.*, **88** (10), 2434–2438.
127 Pedroso, A.G. and Rosa, D.S. (2005) *Polym. Adv. Technol.*, **16** (4), 310–317.
128 Prinos, J., Bikiaris, D., Theologidis, S., and Panayiotou, C. (1998) *Polym. Eng. Sci.*, **38** (6), 954–964.
129 Pushpadass, H.A., Bhandari, P., and Hanna, M.A. (2010) *Carbohydr. Polym.*, **82** (4), 1082–1089.
130 Raj, B., Annadurai, V., Somashekar, R., Raj, M., and Siddaramaiah, S. (2001) *Eur. Polym. J.*, **37** (5), 943–948.
131 Ratanakamnuan, U. and Aht-Ong, D. (2006) *J. Appl. Polym. Sci.*, **100** (4), 2717–2724.
132 Rodriguez-Gonzalez, F.J., Ramsay, B.A., and Favis, B.D. (2003) *Polymer*, **44** (5), 1517–1526.
133 Sailaja, R.R.N. and Chanda, M. (2002) *J. Appl. Polym. Sci.*, **86** (12), 3126–3134.
134 Wang, Y.J., Liu, W.J., and Sun, Z.H. (2003) *J. Mater. Sci. Lett.*, **22** (1), 57–59.
135 Wang, Y.J., Liu, W., and Sun, Z. (2004) *J. Appl. Polym. Sci.*, **92** (1), 344–350.
136 Yin, Q., Dong, A., Wang, J., and Yin, Y. (2008) *Polym. Compos.*, **29** (7), 745–749.
137 Sailaja, R.R.N. and Chanda, M. (2001) *J. Appl. Polym. Sci.*, **80** (6), 863–872.
138 Ermolovich, O.A. and Makarevich, A.V. (2006) *Russ. J. Appl. Chem.*, **79** (9), 1526–1531.

139 Fu, X., Chen, X., Wen, R., He, X., Shang, X., Liao, Z., and Yang, L. (2007) *J. Polym. Res.*, **14** (4), 297–304.
140 Wu, C.S. and Liao, H.T. (2002) *J. Appl. Polym. Sci.*, **86** (7), 1792–1798.
141 Garlotta, D., Doane, W., Shogren, R., Lawton, J., and Willett, J.L. (2003) *J. Appl. Polym. Sci.*, **88** (7), 1775–1786.
142 DeLeo, C., Goetz, J., Young, B., and Velankar, S.S. (2010) *J. Appl. Polym. Sci.*, **116** (3), 1775–1781.
143 Acioli-Moura, R. and Sun, X.S. (2008) *Polym. Eng. Sci.*, **48** (4), 829–836.
144 Chen, L., Qiu, X., Xie, Z., Hong, Z., Sun, J., Chen, X., and Jing, X. (2006) *Carbohydr. Polym.*, **65** (1), 75–80.
145 Huneault, M.A. and Li, H. (2007) *Polymer*, **48** (1), 270–280.
146 Ke, T.Y. and Sun, X.Z. (2001) *J. Appl. Polym. Sci.*, **81** (12), 3069–3082.
147 Ke, T.Y. and Sun, X.Z.S. (2003) *J. Polym. Environ.*, **11** (1), 7–14.
148 Li, H. and Huneault, M.A. (2011) *J. Appl. Polym. Sci.*, **119** (4), 2439–2448.
149 Ning, W., Jiugao, Y., and Xiaofei, M. (2008) *Polym. Compos.*, **29** (5), 551–559.
150 Nordqvist, D., Sanchez-Garcia, M.D., Hedenqvist, M.S., and Lagaron, J.M. (2010) *J. Appl. Polym. Sci.*, **115** (3), 1315–1324.
151 Park, J.W., Im, S.S., Kim, S.H., and Kim, Y.H. (2000) *Polym. Eng. Sci.*, **40** (12), 2539–2550.
152 Phetwarotai, W., Potiyaraj, P., and Aht-Ong, D. (2010) *J. Appl. Polym. Sci.*, **116** (4), 2305–2311.
153 Sarazin, P., Li, G., Orts, W.J., and Favis, B.D. (2008) *Polymer*, **49** (2), 599–609.
154 Schwach, E., Six, J., and Averous, L. (2008) *J. Polym. Environ.*, **16** (4), 286–297.
155 Wang, H., Sun, X.Z., and Seib, P. (2001) *J. Appl. Polym. Sci.*, **82** (7), 1761–1767.
156 Wu, C.S. (2005) *Macromol. Biosci.*, **5** (4), 352–361.
157 Xiao, Y., Che, J., Bergeret, A., Mao, C., and Shen, J. (2010) *e-Polymers*, No. 008.
158 Don, T., Chung, C., Lai, S., and Chiu, H. (2010) *Polym. Eng. Sci.*, **50** (4), 709–718.
159 Ismail, A.M. and Gamal, M.A.B. (2010) *J. Appl. Polym. Sci.*, **115** (5), 2813–2819.
160 Thire, R.M.S.M., Ribeiro, T.A.A., and Andrade, C.T. (2006) *J. Appl. Polym. Sci.*, **100** (6), 4338–4347.
161 Avella, M. and Errico, M.E. (2000) *J. Appl. Polym. Sci.*, **77** (1), 232–236.
162 Hong, F., Peng, J., and Lui, W. (2011) *J. Appl. Polym. Sci.*, **119** (3), 1797–1804.
163 Reis, K.C., Pereira, J., Smith, A.C., Carvalho, C.W.P., Wellner, N., and Yakimets, I. (2008) *J. Food Eng.*, **89** (4), 361–369.
164 Willett, J.L., Kotnis, M.A., O'Brien, G.S., Fanta, G.F., and Gordon, S.H. (1998) *J. Appl. Polym. Sci.*, **70** (6), 1121–1127.
165 Averous, L., Fauconnier, N., Moro, L., and Fringant, C. (2000) *J. Appl. Polym. Sci.*, **76** (7), 1117–1128.
166 Barikani, M. and Mohammadi, M. (2007) *Carbohydr. Polym.*, **68** (4), 773–780.
167 Belard, L., Dole, P., and Averous, L. (2009) *Polym. Eng. Sci.*, **49** (6), 1177–1186.
168 Bossard, F., Pillin, I., Aubry, T., and Grohens, Y. (2008) *Polym. Eng. Sci.*, **48** (9), 1862–1870.
169 Choi, E.J., Kim, C.H., and Park, J.K. (1999) *J. Polym. Sci. Polym. Phys.*, **37** (17), 2430–2438.
170 Ishiaku, U.S., Pang, K.W., Lee, W.S., and Ishak, Z.A.M. (2002) *Eur. Polym. J.*, **38** (2), 393–401.
171 Kim, C.H., Choi, E.J., and Park, J.K. (2000) *J. Appl. Polym. Sci.*, **77** (9), 2049–2056.
172 Kim, C., Kim, D., and Cho, K.Y. (2009) *Polym. Bull.*, **63** (1), 91–99.
173 Kim, C.H., Cho, K.Y., and Park, J.K. (2001) *Polym. Eng. Sci.*, **41** (3), 542–553.
174 Li, G., Sarazin, P., and Favis, B.D. (2008) *Macromol. Chem. Phys.*, **209** (10), 991–1002.
175 Noomhorm, C. and Tokiwa, Y. (2006) *J. Polym. Environ.*, **14** (2), 149–156.
176 Rosa, D.S., Lopes, D.R., and Calil, M.R. (2005) *Polym. Test.*, **24** (6), 756–761.
177 Shin, B.Y., Lee, S.I., Shin, Y.S., Balakrishnan, S., and Narayan, R. (2004) *Polym. Eng. Sci.*, **44** (8), 1429–1438.
178 Sugih, A.K., Drijfhout, J.P., Picchioni, F., Janssen, L.P.B.M., and Heeres, H.J. (2009) *J. Appl. Polym. Sci.*, **114** (4), 2315–2326.
179 Vikman, M., Hulleman, S.H.D., Van der Zee, M., Myllarinen, P., and Feil, H. (1999) *J. Appl. Polym. Sci.*, **74** (11), 2594–2604.
180 Wu, C.S. (2003) *J. Appl. Polym. Sci.*, **89** (11), 2888–2895.

181 Wu, C.S. (2003) *Polym. Degrad. Stabil.*, **80** (1), 127–134.
182 Yavuz, H. and Babac, C. (2003) *J. Polym. Environ.*, **11** (3), 107–113.
183 Basu, D., Datta, C., and Banerjee, A. (2002) *J. Appl. Polym. Sci.*, **85** (7), 1434–1442.
184 Carvalho, A.J.F., Job, A.E., Alves, N., Curvelo, A.A.S., and Gandini, A. (2003) *Carbohydr. Polym.*, **53** (1), 95–99.
185 Nakason, C., Kaesman, A., Homsin, S., and Kiatkamjornwong, S. (2001) *J. Appl. Polym. Sci.*, **81** (11), 2803–2813.
186 De Graaf, R.A. and Janssen, L.P.B.M. (2001) *Polym. Eng. Sci.*, **41** (3), 584–594.
187 Lai, S.M., Huang, C.K., and Shen, H.F. (2005) *J. Appl. Polym. Sci.*, **97** (1), 257–264.
188 Dean, K., Yu, L., Bateman, S., and Wu, D.Y. (2007) *J. Appl. Polym. Sci.*, **103** (2), 802–811.
189 Ratto, J.A., Stenhouse, P.J., Auerbach, M., Mitchell, J., and Farrell, R. (1999) *Polymer*, **40** (24), 6777–6788.
190 Jiang, W., Qiao, X., and Sun, K. (2006) *Carbohydr. Polym.*, **65** (2), 139–143.
191 Reis, R.L., Mendes, S.C., Cunha, A.M., and Bevis, M.J. (1997) *Polym. Int.*, **43** (4), 347–352.
192 Khan, M.A., Bhattacharia, S.K., Kader, M.A., and Bahari, K. (2006) *Carbohydr. Polym.*, **63** (4), 500–506.
193 Mao, L.J., Imam, S., Gordon, S., Cinelli, P., and Chiellini, E. (2000) *J. Polym. Environ.*, **8** (4), 205–211.
194 Ramaraj, B. (2007) *J. Appl. Polym. Sci.*, **103** (2), 909–916.
195 Sin, L.T., Rahman, W.A.W.A., Rahmat, A.R., and Mokhtar, M. (2011) *Carbohydr. Polym.*, **83** (1), 303–305.
196 Ma, X., Yu, H., and Zhao, A. (2006) *Compos. Sci. Technol.*, **66** (13), 2360–2366.
197 Peng, S.W., Wang, X.Y., and Dong, L.S. (2005) *Polym. Compos.*, **26** (1), 37–41.
198 Wang, X.L., Yang, K.K., Wang, Y.Z., Wang, D.Y., and Yang, Z. (2004) *Acta Mater.*, **52** (16), 4899–4905.
199 Coleman, M.M., Graf, J.F., and Painter, P.C. (1991) *Specific Interactions and the Miscibility of Polymer Blends*, Technomic.
200 Hakkarainen, M., Albertsson, A.C., and Karlsson, S. (1997) *J. Appl. Polym. Sci.*, **66** (5), 959–967.
201 Araujo, M.A., Cunha, A.M., and Mota, M. (2010) *J. Biomed. Mater. Res. A*, **94** (3), 720–729.
202 Bikiaris, D., Prinos, J., and Panayiotou, C. (1997) *Polym. Degrad. Stabil.*, **58** (1–2), 215–228.
203 Chaudhary, A.L., Torley, P.J., Halley, P.J., McCaffery, N., and Chaudhary, D.S. (2009) *Carbohydr. Polym.*, **78** (4), 917–925.
204 Mani, R. and Bhattacharya, M. (1998) *Eur. Polym. J.*, **34** (10), 1477–1487.
205 Ke, T.Y., Sun, S.X.Z., and Seib, P. (2003) *J. Appl. Polym. Sci.*, **89** (13), 3639–3646.
206 Orts, W.J., Nobes, G.A.R., Glenn, G.M., Gray, G.M., Imam, S., and Chiou, B. (2007) *Polym. Adv. Technol.*, **18** (8), 629–635.
207 Walia, P.S., Lawton, J.W., Shogren, R.L., and Felker, F.C. (2000) *Polymer*, **41** (22), 8083–8093.
208 Walia, P.S., Lawton, J.W., and Shogren, R.L. (2002) *J. Appl. Polym. Sci.*, **84** (1), 121–131.
209 Willett, J.L. and Doane, W.M. (2002) *Polymer*, **43** (16), 4413–4420.
210 Odusanya, O.S., Ishiaku, U.S., Azemi, B.M.N., Manan, B.D.M., and Kammer, H.W. (2000) *Polym. Eng. Sci.*, **40** (6), 1298–1305.
211 Odusanya, O.S., Manan, D.M.A., Ishiaku, U.S., and Azemi, B.M.N. (2003) *J. Appl. Polym. Sci.*, **87** (6), 877–884.
212 Rosa, D.d.S., Volponi, J.E., and Guedes, C.d.G.F. (2006) *J. Appl. Polym. Sci.*, **102** (1), 825–832.
213 Nicolais, L. and Narkis, M. (1971) *Polym. Eng. Sci.*, **11** (3), 194-&.
214 Willett, J.L. (1994) *J. Appl. Polym. Sci.*, **54** (11), 1685–1695.
215 Schwach, E. and Averous, L. (2004) *Polym. Int.*, **53** (12), 2115–2124.
216 Biresaw, G. and Carriere, C.J. (2001) *J. Polym. Sci. Polym. Phys.*, **39** (9), 920–930.
217 Grace, H.P. (1982) *Chem. Eng. Commun.*, **14** (3–6), 225–277.
218 Seves, A., Beltrame, P.L., Selli, E., and Bergamasco, L. (1998) *Angew. Makromol. Chem.*, **260**, 65–70.
219 Thakore, I.M., Iyer, S., Desai, A., Lele, A., and Devi, S. (1999) *J. Appl. Polym. Sci.*, **74** (12), 2791–2802.
220 Thakore, I.M., Desai, S., Sarawade, B.D., and Devi, S. (2001) *Eur. Polym. J.*, **37** (1), 151–160.

221 Thiebaud, S., Aburto, J., Alric, I., Borredon, E., Bikiaris, D., Prinos, J., and Panayiotou, C. (1997) *J. Appl. Polym. Sci.*, **65** (4), 705–721.

222 Wootthikanokkhan, J. and Santikunakorn, S. (2005) *J. Appl. Polym. Sci.*, **96** (6), 2154–2162.

223 Rivero, I.E., Balsamo, V., and Muller, A.J. (2009) *Carbohydr. Polym.*, **75** (2), 343–350.

224 Pillin, I., Divers, T., Feller, J.F., and Grohens, Y. (2005) *Macromol. Symp.*, **222**, 233–238.

225 Raquez, J., Nabar, Y., Narayan, R., and Dubois, P. (2008) *Polym. Eng. Sci.*, **48** (9), 1747–1754.

226 Bikiaris, D., Aburto, J., Alric, I., Borredon, E., Botev, M., Betchev, C., and Panayiotou, C. (1999) *J. Appl. Polym. Sci.*, **71** (7), 1089–1100.

227 Maliger, R.B., McGlashan, S.A., Halley, P.J., and Matthew, L.G. (2006) *Polym. Eng. Sci.*, **46** (3), 248–263.

228 Wang, S.J., Yu, J.G., and Yu, J.L. (2005) *Polym. Degrad. Stabil.*, **87** (3), 395–401.

229 Wang, S.J., Yu, J.G., and Yu, J.L. (2004) *J. Appl. Polym. Sci.*, **93** (2), 686–695.

230 Wang, S.J., Yu, J.G., and Yu, J.L. (2006) *J. Polym. Environ.*, **14** (1), 65–70.

231 Wang, S.J., Yu, J.G., and Yu, J.L. (2005) *Polym. Int.*, **54** (2), 279–285.

232 Gupta, A.P., Kumar, V., Sharma, M., and Shukla, S.K. (2009) *Polym. Plast. Technol. Eng.*, **48** (6), 587–594.

233 Gupta, A.P., Kumar, V., and Sharma, M. (2010) *J. Polym. Environ.*, **18** (4), 484–491.

234 Yoo, S.I., Lee, T.Y., Yoon, J.S., Lee, I.M., Kim, M.N., and Lee, H.S. (2002) *J. Appl. Polym. Sci.*, **83** (4), 767–776.

235 Kim, C.H., Cho, K.Y., and Park, J.K. (2001) *J. Appl. Polym. Sci.*, **81** (6), 1507–1516.

236 Taguet, A., Huneault, M.A., and Favis, B.D. (2009) *Polymer*, **50** (24), 5733–5743.

237 Duquesne, E., Rutot, D., Degee, P., and Dubois, P. (2001) *Macromol. Symp.*, **175**, 33–43.

238 Pascente, C., Marquez, L., Balsamo, V., and Mueller, A.J. (2008) *J. Appl. Polym. Sci.*, **109** (6), 4089–4098.

239 Sailaja, R.R.N., Reddy, A.P., and Chanda, M. (2001) *Polym. Int.*, **50** (12), 1352–1359.

240 Sailaja, R.R.N. (2005) *Polym. Int.*, **54** (2), 286–296.

241 Sailaja, R.R.N. and Seetharamu, S. (2008) *React. Funct. Polym.*, **68** (4), 831–841.

242 Kweon, D.K., Kawasaki, N., Nakayama, A., and Aiba, S. (2004) *J. Appl. Polym. Sci.*, **92** (3), 1716–1723.

243 Nakason, C., Kaesaman, A., and Eardrod, K. (2005) *Mater. Lett.*, **59** (29–30), 4020–4025.

244 Nayak, S.K. (2010) *Polym. Plast. Technol. Eng.*, **49** (14), 1406–1418.

245 Mondragon, M., Hernandez, E.M., Rivera-Armenta, J.L., and Rodriguez-Gonzalez, F.J. (2009) *Carbohydr. Polym.*, **77** (1), 80–86.

246 Namazi, H., Mosadegh, M., and Dadkhah, A. (2009) *Carbohydr. Polym.*, **75** (4), 665–669.

247 Majdzadeh-Ardakani, K., Navarchian, A.H., and Sadeghi, F. (2010) *Carbohydr. Polym.*, **79** (3), 547–554.

248 Bocchini, S., Battegazzore, D., and Frache, A. (2010) *Carbohydr. Polym.*, **82** (3), 802–808.

249 Vertuccio, L., Gorrasi, G., Sorrentino, A., and Vittoria, V. (2009) *Carbohydr. Polym.*, **75** (1), 172–179.

250 Kalambur, S.B. and Rizvi, S.S. (2004) *Polym. Int.*, **53** (10), 1413–1416.

251 Kalambur, S. and Rizvi, S.S.H. (2005) *J. Appl. Polym. Sci.*, **96** (4), 1072–1082.

252 Arroyo, O.H., Huneault, M.A., Favis, B.D., and Bureau, M.N. (2010) *Polym. Compos.*, **31** (1), 114–127.

253 Cao, X., Chen, Y., Chang, P.R., and Huneault, M.A. (2007) *J. Appl. Polym. Sci.*, **106** (2), 1431–1437.

254 Yang, J.H., Park, J., Kim, D., and Lee, D. (2004) *J. Appl. Polym. Sci.*, **93** (4), 1762–1768.

255 Safronov, A.P., Suvorova, A.I., Tyukova, I.S., and Smirnova, Y.A. (2007) *J. Polym. Sci. Polym. Phys.*, **45** (18), 2603–2613.

256 Jayaraju, J., Raviprakash, S.D., Keshavayya, J., and Rai, S.K. (2006) *J. Appl. Polym. Sci.*, **102** (3), 2738–2742.

257 Veerapur, R.S., Gudasi, K.B., and Aminabhavi, T.M. (2007) *J. Membr. Sci.*, **304** (1–2), 102–111.

258 Correlo, V.M., Boesel, L.F., Bhattacharya, M., Mano, J.F., Neves, N.M., and Reis, R.L. (2005) *Mater. Sci. Eng. A: Struct. Mater.*, **403** (1–2), 57–68.

259 Suyatma, N.E., Copinet, A., Coma, V., and Fricoteaux, F. (2010) *J. Appl. Polym. Sci.*, **117** (5), 3083–3091.

260 Ghosh, B. and Urban, M.W. (2009) *Science*, **323** (5920), 1458–1460.

261 Wu, C.S. (2003) *J. Polym. Sci. Polym. Chem.*, **41** (24), 3882–3891.

262 Johns, J. and Rao, V. (2009) *Fibers Polym.*, **10** (6), 761–767.

263 Zhou, X.Y., Cui, Y.F., Jia, D.M., and Xie, D. (2009) *Polym. Plast. Technol. Eng.*, **48** (5), 489–495.

264 Muljana, H., Picchioni, F., Heeres, H.J., and Janssen, L.P.B.M. (2009) *Carbohydr. Polym.*, **78** (3), 511–519.

265 Wang, N., Yu, J., Chang, P.R., and Ma, X. (2008) *Carbohydr. Polym.*, **71** (1), 109–118.

266 You, X.D., Li, L., Gao, J.P., Yu, J.G., and Zhao, Z.X. (2003) *J. Appl. Polym. Sci.*, **88** (3), 627–635.

267 Ning, W., Jiugao, Y., Xiaofei, M., and Chunmei, H. (2007) *Polym. Compos.*, **28** (1), 89–97.

268 Kaewtatip, K. and Tanrattanakul, V. (2008) *Carbohydr. Polym.*, **73** (4), 647–655.

269 Sugih, A.K., Picchioni, F., Janssen, L.P.B.M., and Heeres, H.J. (2009) *Carbohydr. Polym.*, **77** (2), 267–275.

270 Xu, Q., Wang, Q., and Liu, L. (2008) *J. Appl. Polym. Sci.*, **107** (4), 2704–2713.

271 Muljana, H., Picchioni, F., Heeres, H.J., and Janssen, L.P.B.M. (2010) *Starch/Starke*, **62** (11), 566–576.

272 Muljana, H., Picchioni, F., Heeres, H.J., and Janssen, L.P.B.M. (2010) *Carbohydr. Polym.*, **82** (3), 653–662.

273 Muljana, H., van der Knoop, S., Keijzer, D., Picchioni, F., Janssen, L.P.B.M., and Heeres, H.J. (2010) *Carbohydr. Polym.*, **82** (2), 346–354.

274 StPierre, N., Favis, B.D., Ramsay, B.A., Ramsay, J.A., and Verhoogt, H. (1997) *Polymer*, **38** (3), 647–655.

275 Walker, A.M., Tao, Y., and Torkelson, J.M. (2007) *Polymer*, **48** (4), 1066–1074.

276 Jiang, L., Liu, B., and Zhang, J. (2009) *Macromol. Mater. Eng.*, **294** (5), 301–305.

7
Macro-, Micro-, and Nanocomposites Based on Biodegradable Polymers

Luc Avérous and Eric Pollet

7.1
Introduction

During the past few decades, biodegradable polymers have attracted more and more interest due to the increasing environmental concern and the decreasing fossil resources. This recent evolution incites researchers and industrials to develop novel materials labeled as "environment-friendly," for example, materials produced from alternative resources, with low energy consumption, biodegradable, and nontoxic for the environment. Since these polymers are biodegradable with a large production obtained from renewable resources such as agro-resources from the biomass, they represent an interesting alternative route to common nondegradable polymers for short-life range applications (packaging, agriculture, leisure, hygiene, etc.). Nevertheless and till now, their properties are sometimes too weak for certain end uses. Therefore, it appears necessary to improve the properties and reduce the cost of these biodegradable polymers to make them fully competitive toward common thermoplastics.

Tailoring new composites within a perspective of ecodesign or sustainable development is a philosophy that is applied to more and more materials. Ecological concerns have resulted in a resumed interest in renewable resource-based and/or compostable products. It is the reason why material components such as natural fibers and biodegradable polymers can be considered as "interesting" – environmentally safe – alternatives. Biocomposites (biodegradable composites) are a special class of composite materials. They are obtained by blending biodegradable polymers with fillers (e.g., lignocellulose fibers). Since the final material is biodegradable (biodegradability ≥ 90 wt%, according to the definition given by the conventional standard EN 13432), the corresponding composite is called biocomposite.

Nanocomposites are novel materials with drastically improved properties due to the incorporation of small amounts (less than 10 wt%) of nanosized fillers into a polymer matrix. Nanofillers can be considered depending on the morphology such as (i) layered particles (e.g., clays), (ii) spherical ones (e.g., silica), or (iii) acicular ones (e.g., whiskers and carbon nanotubes). Their specific geometrical dimensions, and thus aspect ratios, partly affect the final material properties. Considering the layered silicate clays, they offer high surface area, up to 700 m^2/g. This huge interfacial surface, with the polymer matrix, governs the global material properties. The final behavior can be considerably improved, thanks to the strong and large polymer–nanofiller interactions, as well as good particle dispersion. Nanobiocomposites are based on biodegradable polymers, resulting in very promising materials since they show improved properties with a preservation of the final material biodegradability without ecotoxicity. Such materials are mainly destined to biomedical applications and different environmental and short-term applications (e.g., packaging, agriculture, leisure, catering, or hygiene devices), where long-lasting polymers are not entirely adequate. They thus represent a strong and emerging answer for improved and eco-friendly materials. Until the end of the twentieth century, only few articles have been published on this very specific topic, but during the past few years an exponential number of publications have appeared.

This chapter aims at reporting a partial state of the art in macro-, micro-, and nanocomposites from biodegradable polymers. This chapter is mainly focused on two main multiphase systems with different biodegradable polymers (agro-polymers and biopolyesters):

- (macro/micro)biocomposites also called biocomposites that can be elaborated, for example, with lignocellulose fibers from different botanical resources;
- (micro/nano)biocomposites also called nanobiocomposites, based on nanofillers, for example, nanoclays.

7.2
Biodegradable Polymers

7.2.1
Classification

A vast number of biodegradable polymers are chemically or biologically synthesized during the growth cycles of organisms. Some enzymes and microorganisms capable of degrading them have also been identified. Depending on the synthesis way, a classification of the biodegradable polymers into four different categories can be proposed [1]: (i) polymers from biomass such as the agro-polymers from agro-resources (e.g., starch and cellulose), (ii) polymers obtained by microbial production (e.g., the polyhydroxyalkanoates), (iii) polymers chemically synthesized using monomers obtained from agro-resources (e.g., the poly(lactic acid) (PLA)), and (iv) biodegradable polymers whose monomers and polymers are both obtained by chemical synthesis from fossil resources. Only three categories (i–iii) are obtained from renewable resources. We can sort these different biodegradable polymers into two main families: (a) the agro-polymers (category i) and (b) the biodegradable polyesters (categories ii–iv), also called biopolyesters.

7.2.2
Agro-Polymers: The Case of Starch

Starch is an important material as it is the main storage supply in botanical resources (cereals, legumes, and tubers), a widely available raw material, and, after processing, it can be useful for many different nonfood applications such as bioplastics, paper, textile, or adhesives. In Europe, less than 50% of starch production is used for food applications.

7.2.2.1 Native Starch Structure

Depending on the botanical origin of the plant, starch granules can have very different shapes (sphere, platelet, polygon, etc.) and sizes (from 0.5 to 175 µm). These granules are composed of two α-D-glucopyranose homopolymers, the amylose and the amylopectin. Their proportions into the granules depend directly on the botanical source. In addition, starch contains also in smaller proportion other compounds, such as proteins, lipids, and minerals, which can interfere with starch, for example, by the formation of lipid complexes or with the proteins by "Maillard reaction" during the process.

The amylose is mainly a linear polysaccharide composed of D-glucose units linked by α(1 → 4) linkages (Figure 7.1). These chains are partially ramified with some α(1 → 6) linkages. Their number is directly proportional to the amylose molecular weight (from 105 to 106 g/mol) [2]. The amylose chains show a single- or double-helix conformation with a rotation on the α(1 → 4) linkage [3]. The helix is composed of six glucose units per turn with a diameter of 4.5 Å.

The amylopectin is the main starch component and has the same monomeric unit as amylose. It shows 95% of α(1 → 4) and 5% of α(1 → 6) linkages. The latter are found every 24–79 glucose units [4] and bring to the amylopectin a highly branched structure. Depending on the botanical source, the molecular weight varies from 107 to 108 g/mol. Consequently, the amylopectin structure and organization, which have been elucidated for the first time by Hizukuri [5], can be seen as a grape with pending chains (Figure 7.2).

The starch granule consists of alternate crystalline and amorphous areas leading to a concentric structure from the hilum [6]. The amorphous areas are composed of the amylose chains and the amylopectin branching points. The semicrystalline areas are mainly composed of the amylopectin side chains. Some cocrystalline structures with the amylose chains have also been identified [7]. Depending on the botanical

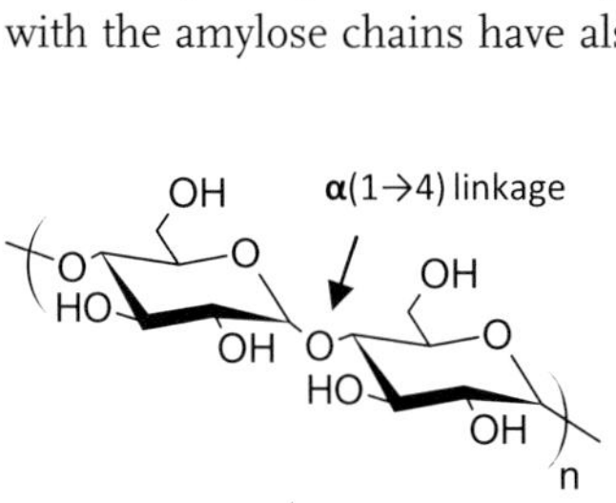

Figure 7.1 Chemical structure of amylose.

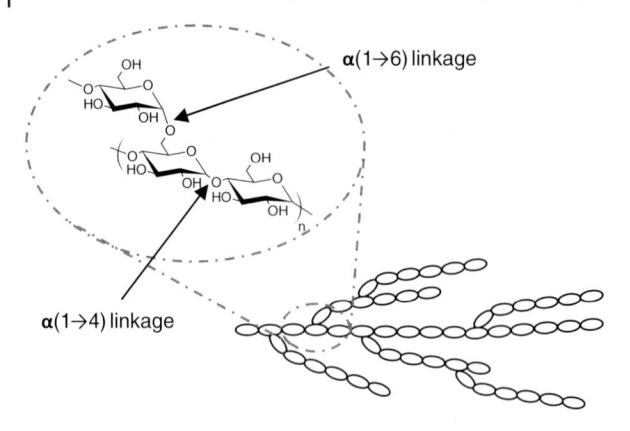

Figure 7.2 Chemical structure and grape representation of amylopectin.

origin, starch granules present a crystallinity varying from 20 to 45%. Four starch allomorphic structures exist [7].

7.2.2.2 Plasticized Starch

Because of the numerous intermolecular hydrogen bonds existing between the chains, starch melting temperature is higher than its degradation temperature [8]. Consequently, to elaborate a plastic-like material with conventional plastic processing techniques, it is necessary to introduce high water content or/and some nonvolatile plasticizers (glycerol, sorbitol, etc.), which will increase the free volume and thus decrease the glass transition and the melting temperature [9]. These plasticized materials are currently named "thermoplastic starch (or TPS)" or "plasticized starch." Since starch is a hydrophilic material, water is the best plasticizer [10–13]. Nevertheless, the water content and thus the plasticized starch properties strongly depend on the storage conditions (temperature and atmosphere relative humidity) through sorption–desorption exchanges. This drawback is partially solved with the use of less volatile plasticizers, which, however, present lower plasticization efficiency. These products possess hydroxyl groups (polyols) and thus can interact with the starch chains through hydrogen bonds. Glycerol is the most common plasticizer [14–16], but numerous other polyols, such as sorbitol [17] and xylitol [18], or plasticizers with amino groups, such as urea, can be used [19]. Nevertheless, these plasticizers are more hydrophilic than starch and are also sensitive to the relative humidity.

To be transformed, the starch granule structure has to be disrupted. The disruption can be obtained by two different processes, namely, casting and melting processes.

At ambient temperature, starch remains insoluble in water and keeps its granular structure. Water temperature increase induces an irreversible swelling called "gelatinization." This phenomenon occurs at a given temperature defined as "gelatinization temperature." During this gelatinization, the granule semicrystalline structure disappears and the granules swell rapidly. To obtain full starch

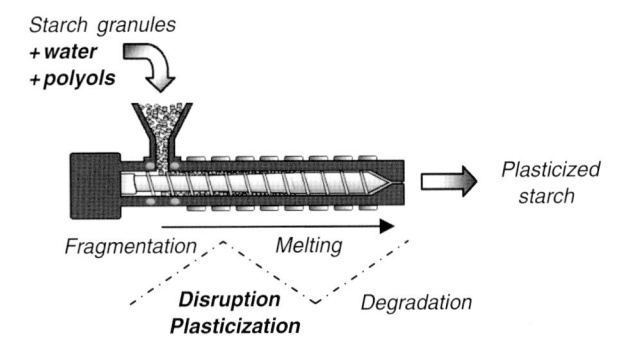

Figure 7.3 Schematic representation of the starch extrusion process.

solubilization, hot DMSO is often used as solvent. Then, to end the casting this solvent is volatilized under vacuum and heat.

The granules' melting is often carried out in association with plasticizers using a common thermomechanical mixing process, for example, extrusion, to obtain a homogeneous molten phase. During this transformation, different successive phenomena can occur, such as (i) the fragmentation of the starch granules, (ii) the disruption and the plasticization of the destructured granules, (iii) the material melting, and (iv) partial chain degradation, under the thermomechanical input (Figure 7.3) [20].

During processing, the starch granules lose their crystalline structure and become an amorphous material. This physical state is nonstable and the material will evolve with time. This evolution corresponds to molecular reorganizations, which depend on the process protocol and the storage conditions. When the samples are stored below the T_g, the samples will undergo physical aging with a material densification [21]. When $T > T_g$, the samples will retrograde with an increase in crystallinity [22].

7.2.3
Biodegradable Polyesters

Figure 7.4 shows the structures of main biodegradable polyesters (biopolyesters). One can see how large the range of possible chemical structures is.

7.2.3.1 Polyesters Based on Agro-Resources

Poly(lactic acid) Lactic acid is a chiral molecule existing as two enantiomers, L- and D-lactic acid, which can be produced by different ways, that is, biologically or chemically synthesized [23]. The cyclic dimer, L- or D-lactide, is usually formed as an intermediate step to the production of PLA. The ring-opening polymerization (ROP) of the lactide can lead to macromolecular chains with L- and D-lactic acid units. This ROP route has the advantage of reaching high molecular weight [24–27] and allows the control of the PLA final properties by adjusting the proportions and the sequencing of L- and D-lactic acid units.

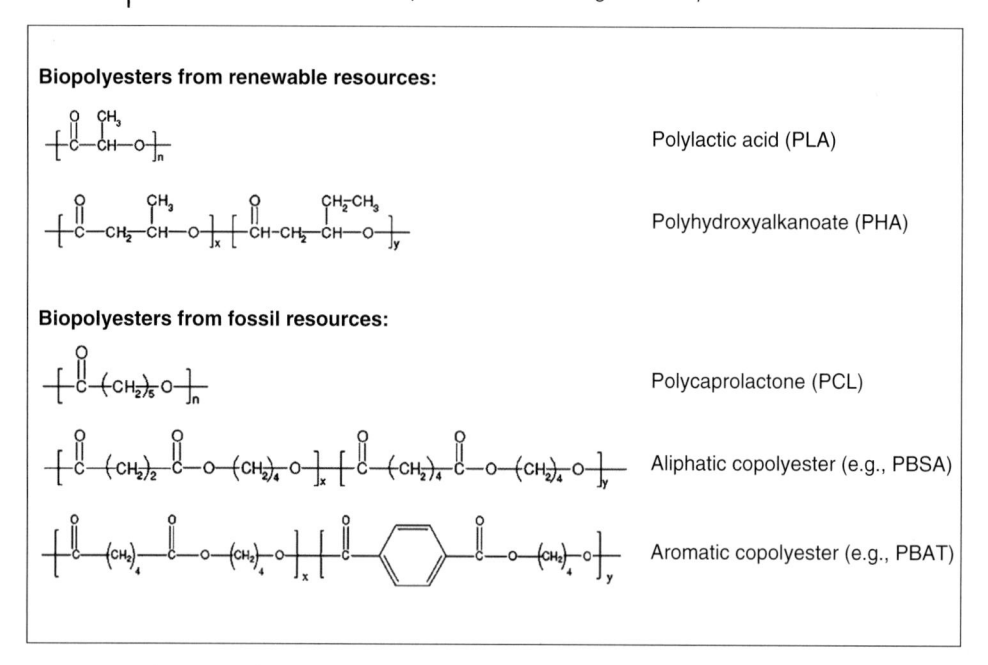

Figure 7.4 Chemical structures of main biopolyesters.

At present, due to its availability on the market and its low price [28–30], PLA has one of the highest potential among biopolyesters, particularly for packaging [30] and biomedical applications. For instance, Cargill has developed processes that use corn and other feedstock to produce different PLA grades (NatureWorks®) [29,31]. With about 100 kton/year, it is the highest worldwide production of biodegradable polyester and its price is around 2 €/kg. Different companies such as Futerro (Belgium), Purac (The Netherlands), Teijin (Japan), or Zhejiang Hisun (China) produce smaller PLA outputs with different D/L ratios. 100% poly(L-lactic acid) (PLLA) that presents a high crystallinity (C-PLA) and copolymers of PLLA and poly (D,L-lactic acid) (PDLLA) that are rather amorphous (A-PLA) are commercially available [31–33]. Furthermore, PLA can be plasticized using oligomeric lactic acid (o-LA) [34], citrate ester [35], or low molecular weight polyethylene glycol (PEG) [34,36–38]. The effect of plasticization increases the chain mobility and then favors the PLA organization and crystallization. PLA presents a medium water and oxygen permeability level [30,39] comparable to polystyrene [40]. Its different properties associated with its tunability and its availability favor its actual development in packaging (trays, cups, bottles, films, etc.) [28,30,31] or biomedical applications.

Polyhydroxyalkanoates Polyhydroxyalkanoate is naturally produced by microorganisms from various carbon substrates as a carbon or energy reserve. A wide variety of prokaryotic organisms [41] accumulate PHA from 30 to 80% of their cellular dry weight. Depending on the carbon substrates and the metabolism of the microorganism, different monomers, and thus (co)polymers, could be obtained [42]. The main polymer of the polyhydroxyalkanoate family is the polyhydroxybutyrate (PHB) homopolymer, but different poly(hydroxybutyrate-co-hydroxyalkanoate) copolyesters exist, such as poly(hydroxybutyrate-co-hydroxyvalerate) (PHBV), poly(hydroxybutyrate-co-hydroxyhexanoate) (PHBHx), poly(hydroxybutyrate-co-hydroxyoctanoate) (PHBO), and poly(hydroxybutyrate-co-hydroxyoctadecanoate) (PHBOd). The recovery process, that is, the extraction and purification steps, is decisive to obtain a highly pure PHA.

PHB is a highly crystalline polyester (above 50%) with a high melting point, $T_m = 175–180\,°C$, compared to the other biodegradable polyesters. Glass transition temperature (T_g) is around 5 °C. The homopolymer shows a narrow window for the processing conditions. To ease the transformation, PHB can be plasticized with citrate ester. But the PHBV copolymer is more adapted for the process. Figure 7.4 shows the chemical structure of PHBV. Material properties can be tailored by varying the HV content. An increase in the HV content induces an increase in the impact strength and a decrease in the melting and glass transition temperatures [43], the crystallinity [44], the water permeability [44], and the tensile strength [45]. A large range of bacterial copolymer grades had been industrially produced by Monsanto under the Biopol® trademark, with HV contents reaching 20%. The production was stopped at the end of 1999. Metabolix bought Biopol® assets in 2001. But recently, Telles™, the joint venture between Metabolix and Archer Daniels Midlands Company (ADM), marketing the Mirel™ product from corn sugar, has been stopped. Different small companies currently produce bacterial PHA, for example, PHB Industrial (Brazil) produces PHB and PHBV (HV = 12%) 45% crystalline from sugarcane molasses [46]. In 2004, Procter & Gamble (US) and Kaneka Corporation (Japan) announced the commercialization of Nodax®, a large range of polyhydroxybutyrate-co-hydroxyalkanoates (PHBHx, PHBO, and PHBOd) [47]. But the industrial development was stopped. Recently, different Chinese companies such as Tianjin or Tianan Biologic have launched new PHA productions.

PHA is intended to replace synthetic nondegradable polymers for a wide range of applications [48] such as packaging, agriculture, leisure, and medicine [42,49], since PHA is biocompatible.

7.2.3.2 Petroleum-Based Polyesters

A large number of biodegradable polyesters are based on petroleum resources and obtained chemically from synthetic monomers [26–29,31,32,50]. According to the chemical structures (see Figure 7.4), we can distinguish polycaprolactones, polyesteramides, and aliphatic or aromatic copolyesters. All these polyesters are soft at room temperature.

Polycaprolactone Poly(ε-caprolactone) (PCL) is usually obtained by ring-opening polymerization of ε-caprolactone in the presence of metal alkoxides (aluminum isopropoxide, tin octoate, etc.). The chemical structure of PCL is presented in Figure 7.4. PCL finds some applications based on its biodegradable character in domains such as biomedicine (e.g., controlled release drugs) and environment (e.g., soft compostable packaging). Different commercial grades were produced by Solvay

(CAPA®), which has sold this activity to Perstorp (UK), and by Dow Chemical (Tone®). PCL shows a very low T_g (−61 °C) and a low melting point (65 °C), which could be a handicap in some applications. Therefore, PCL is generally blended [51–54] or modified (e.g., copolymerization and cross-linking [55]).

Aliphatic Copolyesters Different aliphatic copolyesters are biodegradable and based on petroleum resources. They are obtained by the combination of diols such as 1,2-ethanediol, 1,3-propanediol, or 1,4-butadenediol and dicarboxylic acids such as adipic, sebacic, or succinic acid. Since the production of bio-based succinic acid and diols (1,3-propanediol and 1,4-butanediol) increases a lot with new plants throughout the world, the production of these copolyesters should increase in the next years. Showa Highpolymer (Japan) has developed a large range of polybutylene succinate (PBS) obtained by polycondensation of 1,4-butanediol and succinic acid. Polybutylene succinate/adipate (PBSA), presented in Figure 7.4, is obtained by the addition of adipic acid. These copolymers are commercialized under the Bionolle® trademark [31]. Table 7.1 shows some properties of such biopolyesters. Ire Chemical (Korea) commercialized exactly the same kind of copolyesters under EnPol® trademark. Skygreen®, a product from SK Chemicals (Korea), is obtained by polycondensation of 1,2-ethanediol and 1,4-butanediol with succinic and adipic acids [56]. The properties of these copolyesters depend on the structure [57], that is, the combination of diols and diacids used.

Aromatic Copolyesters Compared to totally aliphatic copolyesters, aromatic copolyesters are often based on terephthalic acid. Figure 7.4 shows the chemical structure of polybutylene adipate/terephthalate or PBAT (e.g., Eastar Bio® from Eastman). BASF and DuPont commercialized similar aromatic copolyesters under Ecoflex® [31] and Biomax® trademarks, respectively. Biomax® shows a high terephthalic acid content that modifies some properties such as the melting temperature (200 °C).

Table 7.1 Main TGA results of biocomposites based on aromatic copolyesters.

	Transition 1: filler		Transition 2: matrix		Loss of weight at 300 °C
	Onset 1	Degradation temperature (maximum 1 of DTG)	Onset 2	Degradation temperature (maximum 2 of DTG)	
PBAT	No visible "transition"		382 °C (15%)	410 °C (60%)	1%
PBAT-10%			384 °C (15%)	411 °C (59%)	3%
PBAT-30%	323 °C (6%)	364 °C (17%)	398 °C (61%)	413 °C (63%)	5%
PBAT-40%	324 °C (10%)	357 °C (22%)	401 °C (48%)	421 °C (70%)	7%

Source: Ref. [85].
Note: The total weight loss at each corresponding temperature is given in parentheses.

7.3
Biocomposites

7.3.1
Generalities

Biocomposites are obtained by the association of macro/microfillers (mainly lignocellulose fibers) into a biomatrix. Since the size of the fillers varies from the micrometer to the millimeter, with good or weak dispersion, we obtain a micro- or macrostructure. For short-term applications, biocomposites based on biodegradable polymers present strong advantages and a large number of papers have been published on this topic. Although few publications are based on agro-polymer matrices, mainly with plasticized starch, most of the published studies on biocomposites are based on biopolyesters (biodegradable polyesters). These different systems and some major results will be developed in the next sections.

7.3.2
The Case of Biocomposites Based on Agro-Polymers

Since most of the studies have been focused on a starchy matrix, this section will be dedicated to the biocomposites based on plasticized starch.

7.3.2.1 Cellulose Fiber Reinforcement

Various types of (ligno)cellulose fibers or microfibrils were tested in association with plasticized starch, such as microfibrils from potato pulp [58], bleached leafwood fibers [1,59–61], fibers from bleached eucalyptus pulp [62], and flax and jute fibers [63]. These different authors have shown high compatibility between starch and these fibers. For instance, Averous *et al.* have found a strong T_g increase by addition of cellulose fibers into a plasticized starch matrix [61]. This behavior is linked to the fiber–matrix interactions, which decrease starch chain mobility. For instance, Figure 7.5 shows

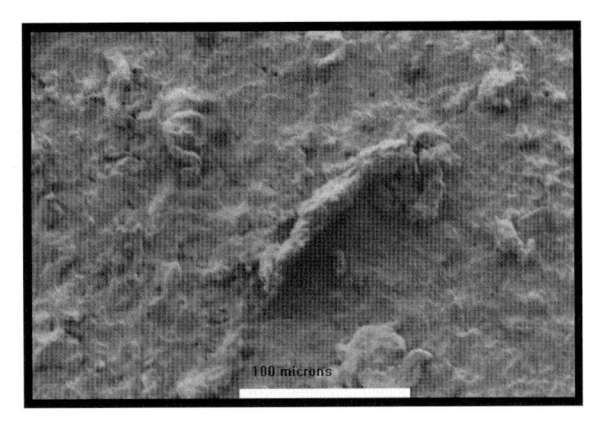

Figure 7.5 SEM observation. Cryogenic fracture of composites: plasticized starch–lignocellulose fibers (white scale = 100 μm). (*Source*: Ref. [61].)

a SEM image of a cryogenic fracture. The cellulose fibers are fully embedded into the starchy matrix. Similar results have also been found by Curvelo et al. [62].

After mixing, spectacular properties are observed with large improvement in the material performances. Some of these enhancements are linked to typical matrix reinforcement [60], and some others are linked to the fiber–matrix interface interactions. For instance, one can highlight the following unexpected improvements:

- higher mechanical properties [20]: compared to biopolyester-based biocomposites, starch-based biocomposites present superior properties linked to higher interactions between the matrix and the filler;
- higher thermal resistance [64], due to the transition shift of T_g and an increase in the rubber plateau;
- reduced water sensitivity, due to fiber–matrix interactions and the higher hydrophobic character of the cellulose, which is linked to its high crystallinity [58,59,61,62,64];
- reduced postprocessing aging, due to the formation of a 3D network between the different matrix–filler carbohydrates based on hydrogen bonds [1].

7.3.2.2 Lignin and Mineral Fillers

Different types of fractioned or modified lignins (i.e., kraft lignins and Acell® lignins) have been tested in association with plasticized starch. Biocomposites were obtained by film casting preparation and by extrusion. According to Braumberger et al. [65,66], the lignins act as fillers or as extenders for the starchy matrix, with the soluble lignin fractions interacting with the plasticized starch.

Mineral microfillers were tested into a plasticized starch matrix [67]. Kaolin particles, with micron size, were incorporated by extrusion. Due to a significant compatibility between matrix and filler, subsequent behavior such as a glass transition increase, a reduction in water uptake, and an increase in the stiffness can be observed.

7.3.3
The Case of Biocomposites Based on Biopolyesters

7.3.3.1 Generalities

One of the main advantages of the polyesters is their polar character. Then, they can found good compatibility with other polar materials (e.g., materials based on polysaccharides).

For instance, PHA shows a very good adhesion to lignocellulose fibers, compared to conventional polyolefins [68]. The addition of cellulose fibers and different fillers has often been proposed as a solution for increased mechanical performance and toughness of PHB and PHBV [68–74]. In terms of crystallization and thermal behavior, no significant effect of cellulose on PHB crystallinity was reported. A slight increase in T_g (glass transition temperature) and delay in the crystallization process were observed [69]. The presence of cellulose filler fibers increased the rate of PHBV crystallization, due to a nucleating effect, while the thermal parameters such as crystallinity content remained unchanged. Studies on the crystallization behavior of PHB/kenaf fiber biocomposites showed that nucleation from kenaf fibers affected the crystallization kinetics of the PHB [70]. Differences in the effect of cellulose fibers on the crystallization process have been attributed to the lignin content at the surface/interface of the cellulose fiber. The increase in the HV content, the addition of compatibilizers, and the rise in fiber content in PHA-based composites influenced the mechanical performance of the corresponding biocomposites. The addition of HV led to a reduction in the stiffness of the PHA but to increased elongation at break. In reinforced PHBV, a 50–150% enhancement in tensile strength, 30–50% in bending strength, and 90% in impact strength have been reported [71]. The addition of varying HV content to PHB polymers improved the toughness of the natural fiber composites and increased the ductility, but lowered the crystallization rate. It has been suggested, however, that the combination of coupling agents and HV improved the storage modulus and led to a reduction in the tan δ [68], due to an improvement in the interfacial bonding between PHB and the fibers and an increase in transcrystallinity near the fiber interfaces. The addition of cellulose fibers led to some improvement in tensile strength and stiffness, but the composites remained brittle [72]. At low content, the incorporation of cellulose fibers lowered the stiffness; however, higher amounts of cellulose fibers greatly improved the mechanical properties of PHB. For biocomposites based on cellulose fibers and PHB, the effect of fiber length and surface modification on the tensile and flexural properties has been investigated. Results on PHB reinforced with straw fibers have been published [69]. Fracture toughness values of composite materials containing 10–20 wt% straw fibers were higher than those of pure PHB, while biocomposites containing 30–50 wt% straw fibers presented almost the same values as neat PHB. With the addition of interface modifiers, the interfacial shear strength also improved [75]. PHB with wood flour and plasticizers presents modest increase in tensile strength, while some improvement in terms of thermal stability was demonstrated.

Polylactic acid has been associated with a great number of lignocellulose fillers such as paper waste fibers, wood flour, kenaf [76], bamboo [77], papyrus [78], jute [79], or flax fibers [80].

Some authors have also tested different lignocellulose fibers such as flax [80] or sisal [81] with polycaprolactone.

Aliphatic copolyesters have been used with several cellulosic fibers [63], bamboo fibers [82], and flax, oil palm, jute, or ramie fibers [63].

Aromatic copolyesters have been associated with wheat straw fillers [83–85]. The results obtained on these systems have shown a good compatibility between the lignocellulose fillers and the biodegradable matrix without using compatibilizers or special filler treatment.

7.3.3.2 The Case of Biocomposites Based on Aromatic Copolyesters

This section is focused on the presentation of results based on biocomposites that have been elaborated with different lignocellulose fillers displaying various lignin contents and different lengths and contents. These fillers are combined with biodegradable aromatic copolyester PBAT. Thermal and mechanical properties have been determined and analyzed [85].

Thermal Properties of Biocomposites Thermogravimetric analysis (TGA) has been carried out to evaluate the thermal behavior of biocomposites with different filler contents. Up to 10 wt%, no change directly linked to the filler was observed. At 300 °C, the variations in weight losses are due to the water uptake at equilibrium, which is higher for lignocellulosic fillers (LCFs) compared to PBAT. Then, the weight loss increases with the filler content. This result can be obtained by the addition of the matrix water uptake (1%) and the filler water uptake (13–14%) corrected with the corresponding contents. Table 7.1 shows that the matrix degradation temperature and the corresponding onset increase with the filler content. The latter results are in agreement with the data reported for studies based on PCL matrix [81]. In the same way, the filler degradation temperature (around 360 °C) is consistent with the values obtained by various authors on LCF [1,81]. The degradation behavior of these fillers is then compatible with the plastic processing temperatures.

Table 7.2 gives the main thermal characteristics of these PBAT biocomposites. Compared to most thermoplastics, T_g and T_f are rather low and thus the processing temperature is not so high, around 130 °C. We can notice that this copolymer presents a single transition for T_g, T_c, and T_m due to the repartition of the different sequences. These temperatures are intermediate between the data of both homopolymers, polybutylene adipate and polybutylene terephthalate [86]. At room temperature, PBAT is on the rubber plateau, that is, between T_g and T_f. Without knowing the theoretical enthalpy for 100% crystalline PBAT, we have used the approach presented by Herrera et al., where the theoretical enthalpy is calculated by the contribution of the different chain groups. The calculated value ($\Delta H_{100\%}$) is equal to 22.3 kJ/mol, that is, 114 J/g. Thus, the crystallinity of PBAT is rather low, around 12%. The different thermodynamic values are consistent with data obtained by other authors [87].

Table 7.2 presents the different values obtained by DSC determinations on PBAT with increasing filler contents. As can be seen, the addition of increasing amounts of LCF results in a slight but significant increase in T_g of PBAT, from -39.3 to -35.7 °C. According to Avella et al. [88], this trend may be explained by intermolecular interactions between the hydroxyl groups of the fillers and the carbonyl groups of the PBAT ester functions. These hydrogen bonds would probably reduce the polymer mobility and then increase T_g values. The PBAT/LCF biocomposites do not show any significant variation in T_f, in agreement with the data of Avella et al. [88].

We can notice that crystallization and fusion heats decrease. This is due to a dilution effect linked to the filler incorporation into the matrix. But when the enthalpy is corrected by the filler content (see the $\Delta H'_c$ data in Table 7.2), these values stay rather constant. The corrected heats of crystallization and fusion are equivalent; so, no significant crystallization occurs during the second scan. The heats of crystallization and fusion are equal to 13–14 J/g, that is, around

Table 7.2 Main DSC results of biocomposites based on aromatic copolyesters.

Sample	T_g (°C)	ΔC_p (W/g)	T_c (°C)	T_i (°C)	ΔH_c (J/g)	$\Delta H'_c$ (J/g)	T_f (°C)	ΔH_f (J/g)	X_c (%)
PBAT	−39.3 ± 0.3	0.038 ± 0.001	68.0 ± 1.0	84 ± 1	13.5 ± 0.2	13.5 ± 0.2	113.9 ± 0.5	13.9 ± 0.3	12
PBAT-10%	−38.2 ± 0.2	0.028 ± 0.002	68.0 ± 0.4	87 ± 1	11.3 ± 0.1	12.6 ± 0.1	113.2 ± 0.7	11.7 ± 0.2	11
PBAT-20%	−36.6 ± 0.2	0.020 ± 0.001	71.0 ± 0.3	90 ± 1	11.4 ± 0.2	14.2 ± 0.3	113.8 ± 0.5	11.2 ± 0.4	12
PBAT-30%	−35.7 ± 0.2	0.010 ± 0.001	70.7 ± 0.3	91 ± 1	9.3 ± 0.5	13.3 ± 0.7	114.2 ± 0.6	10.0 ± 0.3	12

Source: Ref. [85].

2.6 kJ/mol. The dilution effect also seems to affect ΔC_p, which tends to decrease with filler incorporation. We can notice that an increase in the amount of LCF does not affect the degree of PBAT crystallinity that stays constant, at around 12%. The incorporation of LCF induces a slight but significant increase in T_c. This is probably linked to the reduction in the polymer mobility. The beginning of the crystallization (T_i) during the cooling tends to increase with increasing filler content. The fillers modify the crystallization by increasing the number of nucleating sites.

Mechanical Properties of Biocomposites According to the stress–strain evolutions, PBAT is a ductile material at room temperature with a high elongation at break (ε_b), more than 200%. This is consistent with T_g and T_f values.

Evaluation of the filler modulus is always problematic. In such PBAT/LCF biocomposites, the filler modulus has been estimated by fitting a semiempirical Halpin–Tsai model on the evolution of the Young's modulus of composites as a function of filler volume fraction [85]. By extrapolation at 100% of fillers, the estimated value of filler modulus (E_f) is 6.7 GPa. This result is coherent with previous data reported in the literature [89,90]. These composites show a common behavior compared to equivalent reinforced thermoplastics. Addition of LCF fractions, acting as reinforcing materials, results in strong evolutions of the mechanical properties. For instance, the moduli are 3.3–6.4 times higher for the biocomposites compared to the neat matrix. Besides, by increasing the filler size, both modulus and yield stress (σ_Y) increase while a decrease in ε_Y and a decrease in the values at break (ε_b and σ_b) are observed.

Takayanagi's model is a phenomenological model obtained by a combination of serial and parallel models. The composite modulus ($E_{c(T)}$) is determined by Equation 7.1, with λ being an adjustment parameter:

$$E_{c(T)} = (1-\lambda)E_m + \frac{\lambda}{[(1-\phi_f)/E_m] + (\phi_f/E_f)}. \tag{7.1}$$

Takayanagi's equation seems to be an excellent model to predict the modulus evolution, in the range 0–30 wt% of filler. From this equation, the parameter λ has been determined by adjustment, at 4.5.

7.4
Nanobiocomposites

7.4.1
Generalities

Nanobiocomposites are organic/inorganic hybrid materials composed of nanosized fillers (nanofillers) incorporated into a biodegradable polymer (matrix) [91,92]. Depending on the chosen nanofiller, the nanocomposite materials can exhibit drastic modifications in their properties, such as improved mechanical properties, barrier properties, or change in their thermal and electrical conductivities [93].

Besides, the nanofiller often acts as fire retardant. Such properties' enhancements rely on the nanofiller geometry, on the nanofiller surface area (e.g., 700 m^2/g for the montmorillonite (MMT) when the nanoclay is fully exfoliated), and on the nanofiller surface chemistry [94].

During the past decade, the most intensive research on nanobiocomposites was focused on layered particles [91,92,95], and especially on nanoclays such as montmorillonite, due to their availability, versatility, and low environment and health concerns.

Phyllosilicates are a wide family in which smectites with different structures, textures, or morphologies can be found. For instance, montmorillonites are anisotropic flexible particles with a high aspect ratio, a width of hundreds of nanometers, and a thickness of around 1 nm.

The distance observed between two platelets of the primary particle, termed the interlayer spacing or d-spacing (d_{001}), depends on the silicate type. This value does not depend entirely on the layer crystal structure, but depends also on the type of the countercation and on the hydration state of the silicate. To increase the d-spacing and then enhance the nanostructure effects, a chemical modification of the clay surface, with the aim to match the polymer matrix polarity, is often carried out [93]. Cationic exchange is the most common technique for chemical surface modification, but other original techniques such as the organosilane grafting [96,97], the use of ionomers [98,99], or block copolymer adsorption [100] are also used.

The cationic exchange consists of substitution of the original inorganic cations by organic ones. These surfactants are often quaternary alkylammonium cations (Table 7.3). The ionic substitution is performed in water because of its ability to swell clay. This latter is then washed to remove the salt formed during the organomodifier adsorption and the surfactant excess and finally lyophilized to obtain organomodified clays. In addition to the modification of the clay surface polarity, organomodification increases the interlayer spacing, which further facilitates the polymer chain intercalation by crawling [101].

Most of the published studies on nanobiocomposites are based on biopolyesters (biodegradable polyesters). Only few publications are based on agro-polymer matrices, recently with plasticized starch. The two main systems based on polysaccharides and biopolyesters and some major results will be discussed in the next sections.

7.4.2
Nanobiocomposites Based on Agro-Polymers (Starch)

According to the literature, to obtain nanobiocomposites based on plasticized starch, three main types of nanoparticles have been used:

- whiskers obtained from cellulose;
- nanocrystals from native starch;
- nanoclays with or without organomodification.

Table 7.3 Montmorillonites, with and without organomodifications: types (trade name and code) and corresponding counterion chemical structures.

Code	Name	Countercation
MMT-Na	"Natural" sodium montmorillonite	Na^+
OMMT-Alk1	Cloisite® 15A – Southern Clay	$H_3C-N^+(CH_3)(HT)(HT)$ (Dimethyl-dihydrogenated tallow ammonium)
OMMT-Alk2	Cloisite® 6A – Southern Clay	
OMMT-Alk3	Cloisite® 20A – Southern Clay	
OMMT-Alk4	Cloisite® 25A – Southern Clay	(Dimethyl-hydrogenated tallow-2-ethylhexylammonium)
OMMT-Alk5	Nanomer® I.30E – Nanocor	$H_{35}C_{18}-NH_3^+$ (Octadecylammonium)
OMMT-Alk6	—	(Trimethyldodecylammonium)
OMMT-Alk7	Cloisite® 93A – Southern Clay	(Methyl-dihydrogenated tallow ammonium)

OMMT-Bz	Cloisite® 10A – Southern Clay	H$_3$C–N$^+$(CH$_3$)(HT)–CH$_2$–C$_6$H$_5$ (Dimethyl-benzyl-hydrogenated tallow ammonium)
OMMT-OH1	Cloisite® 30B – Southern Clay	H$_3$C–N$^+$(T)(CH$_2$CH$_2$OH)$_2$ (Methyl-tallow-bis-2-hydroxyethylammonium)
OMMT-OH2	Nanofil® 804 – Süd Chemie	
OMMT-NH$_4$	Bentone® 111 – Elementis Specialities	NH$_4^+$
OMMT-EtA	—	HO–CH$_2$CH$_2$–NH$_2$ (Ethanolamine)
OMMT-CitA	—	Citric acid

T = tallow (~65% C$_{18}$, ~30% C$_{16}$, and ~5% C$_{14}$); HT = hydrogenated tallow.

7.4.2.1 Whisker-Based Nanobiocomposites

Through a long pretreatment, whiskers can be isolated from their original biomass by acid hydrolysis with concentrated mineral acids under strictly controlled conditions of time and temperature [102,103]. Acid action results in a decrease in the amorphous parts by removing polysaccharide material closely bonded to the crystallite surface and breaks down portions of glucose chains in most accessible, noncrystalline regions. Although chitin can be used, whiskers are typically cellulose monocrystals. Some authors used tunicin (seafood cellulose) whiskers [104–107], which are slender parallelepiped rods of 500 nm to 1–2 µm length and 10 nm width, into starchy matrices. The whisker–matrix interactions are important and the high shape ratio of the nanoparticles (50–200) with the corresponding high specific area ($\approx 170 \, m^2/g$) increases the interfacial phenomena. The global behavior of nanowhisker-based material is primarily driven by the matrix/nanofiller interface, which in turn controls the subsequent performance properties (mechanical properties and permeability). For instance, tunicin whiskers favor starch crystallization due to the nucleating effect of the nanofiller [105].

7.4.2.2 Starch Nanocrystal-Based Nanobiocomposites

Starch nanocrystals were obtained by acid hydrolysis of native granules by strictly controlling the temperature of the process, the acid and starch concentrations, the hydrolysis duration, and the stirring speed. For instance, waxy maize starch nanocrystals consist of 5–7 nm thick platelet-like particles with a 20–40 nm length and a 15–30 nm width. They were used as a reinforcing agent in a waxy maize starch matrix plasticized with glycerol. Angellier *et al.* [108] have shown that the reinforcing effect of starch nanocrystals can be attributed to strong filler/filler and filler/matrix interactions due to the establishment of increased hydrogen bonding. The presence of starch nanocrystals leads to a slowing down of the recrystallization of the matrix during aging in high-humidity atmospheres [108].

7.4.2.3 Nanoclay-Based Nanobiocomposites

Main recent publications on plasticized starch-based nanobiocomposites are based on nanoclays (e.g., montmorillonite or sepiolite) with or without organomodification.

Elaboration and Structure of Nanobiocomposites Based on Nanoclay and Starch Different nanofillers and dispersion protocols have been tested to produce plasticized starch-based nanobiocomposites [92] in order to obtain good filler dispersion and an exfoliation (for the lamellar nanoclays). Hydrophobic nanofillers were incorporated into plasticized starch based on wheat [109], potato [110–112], or corn [110], with MMT content varying from 0 to 9 wt%. It was clearly demonstrated that the incorporation of OMMT-Alk1, OMMT-Alk2, or OMMT-Bz (see Table 7.3 for the OMMT designations) led to the formation of microbiocomposites [109–112], evidenced by the unchanged values of the d_{001}. Higher spacing results were obtained with OMMT-OH, which presents a more hydrophilic character, with a slight d_{001} shift and a strong decrease in the diffraction peak intensity [109,113,114], corresponding to a better dispersion. This state was likely achieved due to the hydrogen bonds established between the clay surfactant and the starch chains [113].

Nanobiocomposites were also elaborated with natural sodium MMT (MMT-Na) due to the starch hydrophilic characters and the Na-based nanofiller [109–120]. These materials were prepared with corn starch [110,118,119], wheat [109,117], or potato [111–113,115,116,120]. It was highlighted that glycerol content higher than 10 wt% led to the formation of an intercalated structure with a d_{001} increasing to 18 Å, a value which is commonly reported in the literature as generally attributed to glycerol intercalation [114,121]. The influence of the plasticizer on the MMT dispersion and on the exfoliation state was also highlighted by Dean et al. [121]. Corresponding results showed a homogeneous dispersion with an exfoliated structure in agreement with results of Cyras et al. [120], which have demonstrated that for glycerol content lower than 10 wt% exfoliation was achieved. Under certain conditions, MMT-Na seemed suitable to achieve exfoliation. These results were in agreement with some studies that have highlighted the formation of hydrogen bonds and deep interactions between glycerol and MMT platelets [118,119,122]. Huang et al. [123–126] have studied the effect of the plasticizer nature, showing that exfoliation could also be reached when using urea or urea/formamide. Nevertheless, these compounds generate ecotoxic residues after biodegradation or composting and cannot be used for safe biodegradable materials. Kampeerapappun et al. [127] have focused their attention on the use of a new eco-friendly compatibilizer, chitosan, to promote the MMT platelet exfoliation. But, only a small increase in the d_{001} was achieved. Nevertheless, this approach was successfully applied by Chivrac et al. with cationic starch (CS) as MMT and sepiolite organomodifier [128]. According to these authors, no diffraction peak was observed by X-ray diffraction, suggesting an exfoliated morphology, which was confirmed by TEM analysis.

Properties of Nanobiocomposites Based on Nanoclay and Starch Large improvements in the material performance were found with nanoclay-based nanobiocomposites: some of them are linked to traditional matrix reinforcement, and some others are brought by the high interface area between nanoclay and the matrix, and by the corresponding dispersion state.

As usual, these nanocomposites displayed substantial improvement in mechanical properties such as Young's modulus, which is correlated to the clay loading for MMT-Na (with corn and wheat starch) [110] or OMMT-CS (with wheat starch) [128]. The mechanical improvement depends on the nanobiocomposite structure. The modulus increase is highest in the case of exfoliation with modified MMT. For instance, Figure 7.6 shows the modulus (uniaxial tensile test) gaps between an exfoliated structure (based on OMMT-CS) and a nonexfoliated one (based on MMT-Na). Similar behavior is obtained on the elongation at break, which slightly increases with the clay content in the case of exfoliation and decreases in the case of a microcomposite structure. In the same way, the energy at break increases in the case of exfoliation and decreases in the case of a microcomposite structure [128].

From thermomechanical measurements based on DMTA characterization, the influence of the nanofillers on the local mobility of the chains can be determined and thus on their relaxation temperatures that could in turn be associated with the glass transitions. In the case of plasticized starch/MMT-Na nanobiocomposites, the

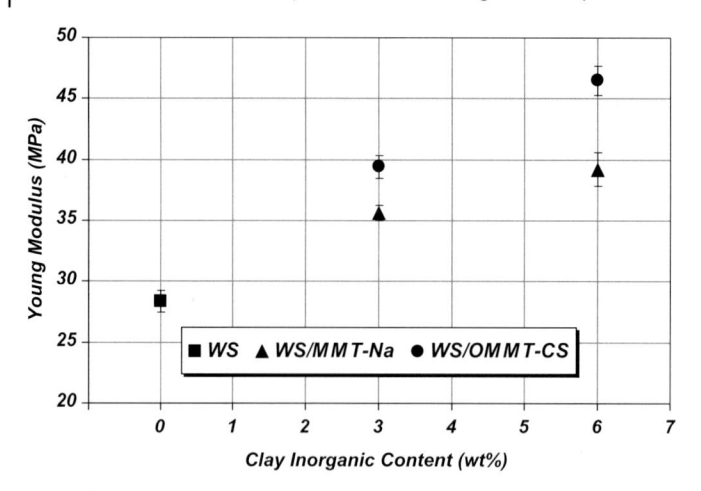

Figure 7.6 Variations of the Young's modulus versus clay content, for plasticized wheat starch with MMT-Na (microbiocomposites) and OMMT-CS (exfoliated nanobiocomposite), stabilized at 57% RH at 23 °C [128].

temperatures of two main relaxation peaks shifted toward higher temperatures indicating that the layered clays strongly restricted the starch chain mobility. This tendency [111] was attributed to the MMT-Na higher affinity with the starch chains. The same trends were observed by DSC [124], meaning that starch/clay hybrids were strongly affected by the clay surface polarity and the clay/matrix interactions.

Some authors studied in detail the thermal stability of nanobiocomposites by TGA. Park et al. [113] showed that the potato starch/MMT-Na and OMMT-OH hybrids have a higher degradation temperature than the neat matrix. The MMT-Na thermal stability was higher than that of OMMT-OH nanobiocomposites. Such a result highlighted a relationship between the clay dispersion state and the thermal stability. Such behavior is observed in most nanocomposite systems. It is linked to the clay aspect ratio and the dispersion state. The exfoliation of the MMT nanoplatelets into the starch matrix increases the tortuosity of the combustion gas diffusion pathway and promotes charring at the surface.

Nanoclays also impact the water vapor permeability of the corresponding nanocomposite materials. Park et al. [111] examined the water vapor permeabilities of potato starch nanobiocomposites with different types of clays. According to these authors, all the clay-based films showed lower water vapor permeabilities compared to the neat matrix. Best results are obtained with MMT-Na nanobiocomposites, which present the higher dispersion state according to these authors. The same trends are observed in different studies [112,129]. However, a recent study shows that with OMMT-CS results obtained with high glycerol content are rather poor [130]. The relatively high plasticizer content (23 wt% glycerol) induces a phase separation, with plasticizer-rich and carbohydrate-rich phases, resulting in the nanoclay being preferentially located in the carbohydrate-rich domains. As a consequence, a preferential way for water transfer was more likely created in the very hydrophilic glycerol-rich domains where the nanoclay platelets were almost totally absent. Thus, even if exfoliated morphology is achieved, the heterogeneous clay distribution and phase separation phenomena explain the lack of improvement and even the decline in the moisture barrier properties for these glycerol-plasticized starch nanobiocomposites. This behavior is induced by these two distinct phenomena, that is, (i) the silicate layer dispersion and (ii) the solubility of the gas into the material [93].

To conclude, the literature shows a very active field with a lot of recent publications on plasticized starch-based nanobiocomposites. Several answers and strategies are proposed to improve the material properties since starch is fully renewable, 100% bio-based, cheap, and largely available.

7.4.3
Nanobiocomposites Based on Biopolyesters

A lot of researchers have developed nanobiocomposites based on biopolyesters, mainly with nanoclays such as montmorillonites, with or without organomodification [91].

7.4.3.1 Poly(lactic acid)-Based Nanobiocomposites

PLA is a very promising material. During the past two decades, many attempts were carried out to reach the exfoliation state in corresponding nanobiocomposites. Various organoclays and several elaboration routes were tested. Ogata [131] first attempted to prepare PLA-based nanocomposites by the solvent intercalation method. Unfortunately, the layered silicates were not individually well dispersed but rather formed tactoids consisting of several stacked clay platelets. Finally, Krikorian and Pochan [132] successfully prepared exfoliated materials with randomly distributed clay platelets via solvent intercalation with OMMT-OH1. According to the authors, interactions between OH (from the clay organomodifier) and C=O groups (from the PLA backbone) favored the exfoliation. Consequently, the mechanical properties were improved. Exfoliated structures in PLA-based nanocomposites were also obtained by Wu et al. [133] using a solution mixing process. They increased the interactions between the filler and the matrix by treating MMT with n-hexadecyltrimethylammonium bromide cations and then modified the clay with chitosan.

The elaboration of PLA/clay nanobiocomposites by melt intercalation has also been widely described in the literature [134–148] and led to various material structures. Okamoto and his group at Toyota Technological Institute (Nagoya, Japan) tested a lot of PLA-based systems differing in the aspect ratio of the inorganic platelets, the nature of the organomodifier, and the clay content [136–146]. Depending on these parameters, intercalated, intercalated and flocculated, nearly exfoliated, and coexisting intercalated and exfoliated states were obtained. They even proposed an interpretation of the nanocomposite structure related to the aspect ratio and the organomodifier chain lengths [136]. Regarding the aspect ratio, it was demonstrated that the smaller the silicate layer size, the lower the physical jamming, restricting the conformation of organomodifier alkyl chains, and thus, the lower the coherency of the organoclay. As a consequence of the clay nanodispersion and despite an incomplete exfoliation, these nanobiocomposites exhibited dramatic enhancements of various properties. These improvements included tensile and flexural moduli, heat distortion temperature, and O_2 gas permeability. Therefore, by a judicious choice of the OMMT, it is possible to tune the material properties [143]. Furthermore, the process could also play a key role in the final nanobiocomposite structure and properties [149]. Finally, the melt intercalation process allows incorporation of additives such as compatibilizers, plasticizers, and so on. The addition of oligo-PCL (o-PCL) [139] or PCL [150] to PLA–clay systems did not give a beneficial effect on the intercalation extent. However, the o-PCL, used as compatibilizer, induced a flocculated state due to hydroxyl edge–edge interactions of layered silicates leading to great enhancement of mechanical properties. Hasook et al. [150] also obtained reinforced material properties when adding 5 wt% of PCL with rather short chain length (<40 000).

Some authors were also interested in developing plasticized PLA-based nanocomposites to reduce the brittleness and to improve the flowability during the process. Thus, PLAs plasticized with PEG [151–156] or diglycerine tetraacetate [157] were melt compounded with different organoclays leading to intercalated structures. Tanoue et al. [156] also studied the structure of such materials using PEG with different molecular weight. Mechanical properties can be improved [156,157] by selecting the right organoclay, plasticizer nature and content, and chain length. Paul et al. [151] and Pluta et al. [154] showed that there is a real competition between PEG and PLA for the intercalation into the clay interlayer. Therefore, Paul et al. [153] settled a protocol consisting in the *in situ* polymerization of lactide from end-hydroxylated PEG in the presence of OMMT-OH1 with tin octoate $(Sn(Oct)_2)$ as an activator/initiator. This polymerization method, called the "coordination–insertion" method, leads to PLA chains grafted onto the clay surface via the hydroxylated ammonium organomodifier and to PLA-*b*-PEG-*b*-PLA triblock copolymer intercalated into the clay gallery. The plasticizing effect is ensured by the PEG sequence of the triblock without phase separation. More generally, this mechanism allows complete exfoliation by ring-opening polymerization of lactide, after adequate activation [158].

To conclude, the literature shows a lot of publications on PLA-based nanobiocomposites. Several answers and strategies are proposed to improve the material properties of PLA, which is renewable, bio-based, and largely commercially available.

7.4.3.2 Polyhydroxyalkanoate-Based Nanobiocomposites

Some drawbacks of PHA, such as brittleness and poor thermal stability, restrict their development and uses. Thus, nanobiocomposites appear as a possible answer to overcome these issues and to improve PHA properties.

First, Maiti et al. [159] prepared PHB-based nanocomposites by melt extrusion. These authors have shown an enhanced degradation in the presence of OMMT, due to the presence of Al Lewis acid sites in the inorganic layers that catalyze the ester linkage hydrolysis. Moreover, Hablot et al. [160] reported that PHB-enhanced degradation can also be caused by decomposition products of clay organomodifiers that have a catalytic effect on the thermal or thermomechanical degradation of the biopolyester. Scientists were also interested in the development of PHBV-based nanocomposites since PHBV presents good properties and better processability than PHB. In 2003, Choi et al. [161] described the morphology as well as the thermal and mechanical properties of PHBV/MMT-OH1 nanocomposites with low clay content. These materials were prepared by melt intercalation. XRD and TEM analyses clearly confirmed that intercalated nanostructures were obtained. Such morphologies were formed thanks to the strong hydrogen bond interactions between PHBV and the hydroxyl groups of the organomodifier. They concluded that the well-dispersed and layered structure accounts for an efficient barrier to the permeation of oxygen and combustion gas. Besides, tensile tests showed that nanoclay can also act as an effective reinforcing agent since the Young's modulus significantly increases.

Wang et al. [162] and Zhang and coworkers [163,164] have investigated the structure and the properties of PHBV/OMMT nanocomposites. They studied PHBV with 3 and 6.6 mol% of HV units and a MMT organomodified with hexadecyltrimethylammonium bromide. The nanocomposites, prepared by solution intercalation, showed intercalated structures evidenced by XRD and clay aggregation occurred for clay content of 10 wt%.

The PHBV processing behavior could be improved with OMMT-based nanocomposites since the processing temperature range enlarged by lowering melting temperature with the increasing clay content. The tensile properties of the corresponding materials were improved by incorporation of 3 wt% of clay [164]. Above this clay content, aggregation of clay occurred and tensile strength and strain at break decrease. Different studies have confirmed that intercalated nanocomposites were formed.

To conclude, most of these recent studies reported the preparation of PHA-based nanocomposite by solvent intercalation, and irrespective of the elaboration route, full exfoliation state was neither obtained nor clearly demonstrated and only micro/nanocomposites were achieved. However, these studies have shown that several answers and strategies could be considered to improve the properties of materials based on PHA that are renewable and bio-based polyesters that should be more largely commercially available in a near future.

7.4.3.3 Polycaprolactone-Based Nanobiocomposites

PCL-based nanocomposite was probably the first studied nanobiocomposite. In the early 1990s, Giannelis' group (Cornell University, USA) started to work on the elaboration of PCL-based nanocomposite by intercalative polymerization [165–167]. This work was motivated by previous studies involving polymerization of ε-caprolactam in the presence of layered silicates, which suggested that lactone ROP can be catalyzed by layered silicates. Then, they decided to investigate the intercalation and polymerization of ε-caprolactone within the gallery of layered silicates. Some results were also obtained by Kiersnowski et al. [168] who prepared the PCL-based composites by *in situ* polymerization catalyzed by water. Afterward, Messersmith and Giannelis [165] attempted to prepare PCL-based nanocomposites by *in situ* polymerization thermally activated and initiated by organic acid [165]. More precisely, the protonated form of 12-aminododecanoic acid, which was used as the OMMT organomodifier, was thus present on the clay surface and initiated the ε-caprolactone ROP by a nucleophilic attack on the carbonyl function. The resulting PCL was therefore ionically bound to the silicate layers through the protonated amine chain end. XRD results suggested that individual clay layers were dispersed in the matrix. On the contrary, OMMT layers organomodified with a less polar ammonium (MMT-Alk1) [165] showed no dispersion in CL or PCL. Subsequently, the interactions occurring at the interface of a PCL/OMMT exfoliated nanocomposite were investigated [169] and the crystallinity, the permeability, and the rheological behaviors were examined [165,167,169,170]. Tortora et al. [170], who examined the water and dichloromethane permeability, assumed that the diffusion path of the polar water molecules is slowed down compared to dichloromethane vapor, not only because of the physical barrier of the clay layers but also because of the hydrophilic character of the platelets. Dubois' group (University of Mons, Belgium) has worked on PCL nanocomposites. They were interested in the *in situ* ROP of ε-CL and in the melt intercalation route. They demonstrated that the formation of PCL-based nanocomposites depends not only on the ammonium cation and related functionality but also on the elaboration route. Contrary to Messersmith and Giannelis [165,166], PCL-based nanocomposites were prepared by *in situ* ROP according to a "coordination–insertion" mechanism [171–177], as for PLA [153,158]. This reaction consists in swelling the OMMT organomodified by alkylammonium bearing hydroxyl groups (e.g., OMMT-OH1) and then adding an initiator/activator such as tin(II) octoate ($Sn(Oct)_2$), dibutyltin(IV) dimethoxide ($Bu_2Sn(OMe)_2$), or triethylaluminum ($AlEt_3$). The ammonium is thus activated and can yield surface-grafted PCL chains. Every hydroxyl function generates a PCL chain. Consequently, the higher the hydroxyl group content, the lower the PCL average molar mass. It is worth noting that, in the presence of tin(IV) catalysts, since they are more efficient toward ε-CL ROP, the preparation took place under milder conditions compared to $Sn(Oct)_2$ [171]. This *in situ* polymerization process led to well-exfoliated PCL-based nanocomposites with 3 wt% of clay while with higher content (10 wt%) partially exfoliated/partially intercalated structures were observed. The water permeability decreased since the well-dispersed fillers with high aspect ratio acted as barriers to oxygen [171,178]. In contrast, nanocomposites filled with non-hydroxyl functional clays exhibited only intercalated structures [172,173,176].

The "coordination–insertion" mechanism, that is, *in situ* intercalation catalyzed by initiators, was compared to the thermally activated *in situ* intercalation with various OMMTs [176]. First, they confirmed Giannelis' results [165] that large catalytic surface of montmorillonite can contribute to polymerization of ε-CL. Exchanged cations bearing protic functions such as NH_3^+, OH, and COOH significantly favored the polymerization and led to similar structures to those obtained by the "coordination–insertion" mechanism. Nevertheless, the PCL molecular weights remained low and the polydispersity index at high conversion reached values higher than 2, confirming that the *in situ* intercalation in the presence of OH groups and initiators allows a better polymerization control.

Eventually, the melt intercalation route led to intercalated or intercalated/exfoliated structures when PCL was associated with OMMT-Alk4, OMMT-OH1, and so on [171,179,180]. On the contrary, MMT-Na and MMT organomodified with ammonium bearing 12-aminododecanoic acid formed microcomposites since no change of interlayer gap was observed, whereas the *in situ* intercalation showed exfoliation in the case of MMT-NH_3^+(C_{11}COOH) [165,171,176]. Thus, contrary to the *in situ* intercalative process, complete exfoliation was not reached by the melt intercalation route whatever the OMMT considered. However, the tensile properties were improved with an almost twofold increase in the PCL rigidity [179,180]. Chen and Evans [181] demonstrated on similar systems that the elastic modulus trends according to clay volume fraction can be interpreted using well-established theory for conventional composites, namely, the Hashin–Shtrikman bounds. At OMMT content higher than 5 wt%, the elongation at break dropped off due to clay aggregation [179,180]. Only Di *et al.* [182] reached exfoliated state in the case of PCL/OMMT-OH1 systems prepared by direct melt intercalation. Obviously, they reported great enhancements of mechanical and thermal properties caused by the strong interactions between the organoclay layers and PCL, and by the good dispersion of exfoliated platelets.

Since the direct melt intercalation suffers a lack of efficiency toward clay dispersion, an elaboration route combining *in situ* ε-CL polymerization and material redispersion by melt intercalation was settled. This masterbatch process [183], or equivalent [184], allowed yielding intercalated/exfoliated structures that are rather difficult to reach by direct melt blending. This process also turned out to be a good way to compatibilize and thus to reinforce other thermoplastics such as conventional polymers or biodegradable polymers such as PBS or PBAT [185]. Conversely, PCL was blended by a reactive process with thermoplastic–clay systems [186,187] to improve the properties of the final material.

To conclude, PCL-based nanocomposites have been widely studied and several well-controlled routes have been settled to reach exfoliated state. To be precise, the solvent intercalation route is not one of those since no satisfactory results were obtained regarding the structure [188]. Except this, elaboration of such materials turned out to be very useful not only for PCL properties' enhancement but also for other thermoplastics ones.

7.4.3.4 Biodegradable Aliphatic Copolyester-Based Nanobiocomposites

Polybutylene Succinate PBS presents many interesting properties but low gas barrier properties and softness still limit its use. Therefore, particular attention has been paid to the elaboration of PBS-based nanobiocomposites to overcome these issues and to improve the material properties.

Sinha Ray *et al.* [189] first reported structure and properties of PBS/clay nanocomposites obtained by melt intercalation. Other studies [190–192] investigated the effect of the organoclay type on the composite structures and properties. High molecular weight PBS was synthesized by a coupling reaction with a chain extender, namely, hexamethylene diisocyanate (OCN–C_6H_{12}–NCO), resulting in urethane moieties and terminal hydroxyl groups. Different OMMTs were tested. Intercalated and extended flocculated nanocomposites were obtained with the PBS/MMT-$NH_3^+(C_{18})$ and PBS/MMT-$N^+(Me)_3(C_{18})$ systems. According to the authors, flocculation occurred due to the urethane moieties of PBS that make hydrogen bonds with the silicate hydroxyl edge groups leading to very strong interactions between matrix and silicate layers [191,192]. These structures were confirmed by mechanical, rheological, and barrier properties. Someya et al. [193] have tested different OMMTs to determine the effect of variations in hydrophobicity and polar/steric interactions on the nanostructures. Highly intercalated structures and homogeneous clay dispersion were observed with MMT-$NH_3^+(C_{12})$, MMT-$NH_3^+(C_{18})$, and MMT-$NH^+(EtOH)_2(C_{12})$. On the contrary, some clusters or agglomerated particles were observed with other types of organomodified clays and with MMT-Na.

Other authors improved PBS properties using the compatibilization approach. Chen et al. [194] proposed to graft epoxy groups onto MMT-Alk4 with (glycidoxypropyl)trimethoxysilane leading to twice functionalized organoclay as also done for PLA [195]. Higher exfoliation degrees and better properties were obtained. The intercalation/exfoliation coexistence has been demonstrated by TEM images, whereas only intercalated clay tactoids have been observed without grafting. Up to 10 wt% of filler, the mechanical properties are increased. Pollet et al. also used MMT-OH1 for the preparation of PBS-based nanocomposites by melt intercalation [185]. Since the d-spacing increased, intercalated structure was obtained resulting in PBS rigidity improvement (+25% in Young's modulus). Recently, Shih et al. [196] prepared, by solvent intercalation, PBS nanocomposites with different organoclays functionalized by various ammonium. Stiffness and toughness of PBS were simultaneously improved suggesting an intercalated structure and inhomogeneous dispersion due to clay aggregation as mentioned by Chen et al. [194].

Polybutylene Succinate-*co*-adipate   Sinha Ray et al. [197–200] reported PBSA-based nanocomposite studies. First, they investigated various types of organoclays (OMMT-Alk1, OMMT-Alk7, and OMMT-OH1) with different polarities [198]. XRD and TEM have shown that the polymer–clay compatibility plays a key role in achieving high d-spacing and high quality of dispersion. Finally, OMMT-OH1 appeared as the most suitable nanofiller for PBSA-based nanobiocomposites prepared by melt intercalation. Nevertheless, the dispersion quality was lowered with increasing clay content [197]. Consequently, the properties were affected by the clay dispersion according to the organoclay nature and content. Tensile modulus and elongation at break increased in the presence of organoclay but the improvement was strongly dependent on the degree of dispersion and the OMMT-OH1 content.

Lee et al. [56] also tested OMMT-OH1 dispersion in Skygreen® at various clay contents, ranging from 1 to 30 wt%, by melt intercalation. The highest content values are unusual for nanocomposites since high clay content promotes aggregation. Nevertheless, in this case, authors claimed that exfoliated states were achieved above 15 wt%, whereas intercalation was obtained with lower clay contents. Authors attributed this phenomenon to the combination of high shear rate reached during melt intercalation and the polymer–OMMT affinity. As seen before [147,161,201], strong interactions or miscibility exist between the polymer and the OMMT-OH1, due to strong hydrogen bonding between the carboxyl (from the biodegradable polyester) and the OH group (from OMMT-OH1). These results were confirmed by TEM since the nanocomposites showed ordered intercalated structures with expanded interlayer gap and good dispersion at 10 wt%. Tensile properties' enhancements were consistent with the structural results. OMMT-Alk4 was also tested to improve Skygreen® properties [202–205]. Regardless of the elaboration route, solvent [202,203] or melt intercalation [204], and the organoclay content studied (<15 wt%), intercalated structures were observed. Even when Skygreen® was blended with polyepichlorohydrin (PECH), no larger interlayer spacing was obtained [205,206], although the d_{001} of PECH/OMMT nanocomposites can attain high values (>55 Å). Moreover, the decrease and the broadening of d_{001} peak intensity with the decreasing clay content and TEM observations suggest that structures were much more inhomogeneous at low clay content showing exfoliated platelets at 3 wt% of OMMT-Alk4 [204]. The characterizations also led to define a critical volume fraction. Beyond this threshold, the tactoids and individual layers are prevented from relaxing completely when subjected to shear, due to physical jamming or percolation leading to the solid-like behavior observed in both intercalated and exfoliated nanocomposites. Thus, although the tensile modulus dramatically increases with the clay content, the elongation at break decreases [203] due to the OMMT aggregation.

The OMMT-Alk4 was also modified by Chen et al. [207] based on the protocol established for PLA-based [195] and PBS-based nanocomposites [194]. OMMT-Alk4 was functionalized by grafting epoxy group at the surface using different silanes, as coupling agents. According to the authors, the silane compounds are located mainly on the edge of the clay layers where the silanol concentration is higher than that on the plain surface. Larger increase in d-spacing and better clay layer dispersion were observed. Tensile modulus and strength at break were greatly improved.

To conclude, classical and original processes were investigated to reach exfoliated state and improved aliphatic copolyester-based nanobiocomposite properties with more or less success. Nevertheless, irrespective of the systems considered, the properties were well correlated to the material structures.

7.4.3.5 Aromatic Copolyester-Based Nanobiocomposites

PBAT is flexible and has a higher elongation at break than most biodegradable polyesters such as PLA and PBS. Only a few articles report the study of PBAT/clay nanobiocomposites.

Chivrac et al. [208,209] have tested various organoclays such as OMMT-Alk4 and OMMT-OH2. The results relating to the structure and properties were compared to neat PBAT and to PBAT/MMT-Na. They also compared the elaboration processes, that is, solvent or melt intercalation. This study revealed that higher intercalation degrees have been obtained from solvent intercalation, OMMT-Alk4 presenting better affinity with PBAT. Tensile tests have shown that the stiffness increases continuously with clay content. This was attributed to the existence of strong interactions between PBAT and nanofillers, particularly with OMMT-Alk4, since the crystallinity decreases with increasing clay content. Nevertheless, decreases in the strain at yield and at break have been observed due to more aggregated structures at higher clay contents.

To conclude, a large range of OMMTs was tested with such aromatic copolyesters and these studies highlighted "structure–property" relationships.

References

1 Averous, L. and Boquillon, N. (2004) Biocomposites based on plasticized starch: thermal and mechanical behaviours. *Carbohydr. Polym.*, **56** (2), 111–122.

2 Della Valle, G., Buleon, A., Carreau, P.J., Lavoie, P.A., and Vergnes, B. (1998) Relationship between structure and viscoelastic behavior of plasticized starch. *J. Rheol.*, **42** (3), 507–525.

3 Hayashi, A., Kinoshita, K., Miyake, Y., and Cho, C.H. (1981) Conformation of amylose in solution. *Polym. J.*, **13** (6), 537–541.

4 Zobel, H.F. (1988) Molecules to granules: a comprehensive starch review. *Starch/Starke*, **40** (2), 44–50.

5 Hizukuri, S. (1986) Polymodal distribution of the chain lengths of amylopectins, and its significance. *Carbohydr. Res.*, **147** (2), 342–347.

6 Jenkins, P.J. and Donald, A.M. (1995) The influence of amylose on starch granule structure. *Int. J. Biol. Macromol.*, **17** (6), 315–321.

7 Van Soest, J.J.G., Hulleman, S.H.D., De Wit, D., and Vliegenthart, J.F.G. (1996) Crystallinity in starch bioplastics. *Ind. Crops Prod.*, **5** (1), 11–22.

8 Shogren, R.L. (1992) Effect of moisture content on the melting and subsequent physical aging of cornstarch. *Carbohydr. Polym.*, **19** (2), 83–90.

9 Swanson, C.L., Shogren, R.L., Fanta, G.F., and Imam, S.H. (1993) Starch-plastic materials – preparation, physical properties, and biodegradability (a review of recent USDA research). *J. Environ. Polym. Degrad.*, **1** (2), 155–166.

10 Zeleznak, K.J. and Hoseney, R.C. (1987) The glass transition in starch. *Cereal Chem.*, **64** (2), 121–124.

11 Kalichevsky, M.T., Jaroszkiewicz, E.M., Ablett, S., Blanshard, J.M.V., and Lillford, P.J. (1992) The glass transition of amylopectin measured by DSC, DMTA and NMR. *Carbohydr. Polym.*, **18** (2), 77–88.

12 Van Soest, J.J.G. and Knooren, N. (1997) Influence of glycerol and water content on the structure and properties of extruded starch plastic sheets during aging. *J. Appl. Polym. Sci.*, **64** (7), 1411–1422.

13 Tomka, I. (1991) Thermoplastic starch. *Adv. Exp. Med. Biol.*, **302**, 627–637.

14 Forssell, P., Mikkila, J., Suortti, T., Seppala, J., and Poutanen, K. (1996) Plasticization of barley starch with glycerol and water. *J. Macromol. Sci. Pure Appl. Chem.*, **33** (5), 703–715.

15 Hulleman, S.H.D., Kalisvaart, M.G., Janssen, F.H.P., Feil, H., and Vliegenthart, J.F.G. (1999) Origins of B-type crystallinity in glycerol-plasticised, compression-moulded potato starches. *Carbohydr. Polym.*, **39** (4), 351–360.

16 Van Soest, J.J.G., De Wit, D., Tournois, H., and Vliegenthart, J.F.G. (1994) The influence of glycerol on structural changes in waxy maize starch as studied by Fourier transform infra-red spectroscopy. *Polymer*, **35** (22), 4722–4727.

17 Gaudin, S., Lourdin, D., Forssell, P.M., and Colonna, P. (2000) Antiplasticisation and oxygen permeability of starch–sorbitol films. *Carbohydr. Polym.*, **43** (1), 33–37.

18 Lourdin, D., Della Valle, G., and Colonna, P. (1995) Influence of amylose content on starch films and foams. *Carbohydr. Polym.*, **27** (4), 261–270.

19 Shogren, R.L., Swanson, C.L., and Thompson, A.R. (1992) Extrudates of cornstarch with urea and glycols: structure/mechanical property relations. *Starch/Starke*, **44** (9), 335–338.

20 Averous, L. (2004) Biodegradable multiphase systems based on plasticized starch: a review. *J. Macromol. Sci. Part C: Polym. Rev.*, **44** (3), 231–274.
21 Thiewes, H.J. and Steeneken, P.A.M. (1997) The glass transition and the sub-T_g endotherm of amorphous and native potato starch at low moisture content. *Carbohydr. Polym.*, **32** (2), 123–130.
22 Lu, T.J., Jane, J.L., and Keeling, P.L. (1997) Temperature effect on retrogradation rate and crystalline structure of amylose. *Carbohydr. Polym.*, **33** (1), 19–26.
23 Averous, L. (2008) Polylactic acid: synthesis, properties and applications, in *Monomers, Oligomers, Polymers and Composites from Renewable Resources* (eds N. Belgacem and A. Gandini), Elsevier, pp. 433–450.
24 Kaplan, D.L., Mayer, J.M., Ball, D., McCassie, J., Allen, A.L., and Stenhouse, P. (1993) Fundamentals of biodegradable polymers, in *Biodegradable Polymers and Packaging* (eds C Ching, D.L. Kaplan, and E.L. Thomas), Technomic Publishing Co., Lancaster, pp. 1–42.
25 Garlotta, D. (2001) A literature review of poly(lactic acid). *J. Polym. Environ.*, **9** (2), 63–84.
26 Okada, M. (2002) Chemical syntheses of biodegradable polymers. *Prog. Polym. Sci. (Oxford)*, **27** (1), 87–133.
27 Albertsson, A.-C. and Varma, I.K. (2002) Aliphatic polyesters: synthesis, properties and applications. *Adv. Polym. Sci.*, **157**, 1–40.
28 Sinclair, R.G. (1996) The case for polylactic acid as a commodity packaging plastic. *J. Macromol. Sci. Pure Appl. Chem.*, **33** (5), 585–597.
29 Lunt, J. (1998) Large-scale production, properties and commercial applications of polylactic acid polymers. *Polym. Degrad. Stabil.*, **59** (1–3), 145–152.
30 Auras, R., Harte, B., and Selke, S. (2004) An overview of polylactides as packaging materials. *Macromol. Biosci.*, **4** (9), 835–864.
31 Steinbuchel, A. and Doi, Y. (2002) *Polyesters III – Applications and Commercial Products*, Wiley-VCH Verlag GmbH, Weinheim.
32 Bigg, D.M. (1996) Effect of copolymer ratio on the crystallinity and properties of polylactic acid copolymers. *J. Eng. Appl. Sci.*, **2**, 2028–2039.
33 Perego, G., Cella, G.D., and Bastioli, C. (1996) Effect of molecular weight and crystallinity on poly(lactic acid) mechanical properties. *J. Appl. Polym. Sci.*, **59** (1), 37–43.
34 Martin, O. and Averous, L. (2001) Poly (lactic acid): plasticization and properties of biodegradable multiphase systems. *Polymer*, **42** (14), 6209–6219.
35 Labrecque, L.V., Kumar, R.A., Dave, V., Gross, R.A., and McCarthy, S.P. (1997) Citrate esters as plasticizers for poly(lactic acid). *J. Appl. Polym. Sci.*, **66** (8), 1507–1513.
36 Jacobsen, S. and Fritz, H.G. (1999) Plasticizing polylactide – the effect of different plasticizers on the mechanical properties. *Polym. Eng. Sci.*, **39** (7), 1303–1310.
37 Kranz, H., Ubrich, N., Maincent, P., and Bodmeier, R. (2000) Physicomechanical properties of biodegradable poly(D,L-lactide) and poly(D,L-lactide-*co*-glycolide) films in the dry and wet states. *J. Pharm. Sci.*, **89** (12), 1558–1566.
38 Ljungberg, N., Andersson, T., and Wesslen, B. (2003) Film extrusion and film weldability of poly(lactic acid) plasticized with triacetine and tributyl citrate. *J. Appl. Polym. Sci.*, **88** (14), 3239–3247.
39 Van Tuil, R., Fowler, P., Lawther, M., and Weber, C.J. (2000) Properties of biobased packaging materials, in *Biobased Packaging Materials for the Food Industry – Status and Perspectives*, KVL, Frederiksberg, Denmark, pp. 8–33.
40 Lehermeier, H.J., Dorgan, J.R., and Way, J.D. (2001) Gas permeation properties of poly(lactic acid). *J. Membr. Sci.*, **190** (2), 243–251.
41 Madison, L.L. and Huisman, G.W. (1999) Metabolic engineering of poly(3-hydroxyalkanoates): from DNA to plastic. *Microbiol. Mol. Biol. Rev.*, **63** (1), 21–53.
42 Zinn, M., Witholt, B., and Egli, T. (2001) Occurrence, synthesis and medical application of bacterial polyhydroxyalkanoate. *Adv. Drug Deliv. Rev.*, **53** (1), 5–21.

43 Amass, W., Amass, A., and Tighe, B. (1998) A review of biodegradable polymers: uses, current developments in the synthesis and characterization of biodegradable polyesters, blends of biodegradable polymers and recent advances in biodegradation studies. *Polym. Int.*, **47** (2), 89–144.

44 Shogren, R. (1997) Water vapor permeability of biodegradable polymers. *J. Environ. Polym. Degrad.*, **5** (2), 91–95.

45 Kotnis, M.A., O'Brien, G.S., and Willett, J.L. (1995) Processing and mechanical properties of biodegradable poly (hydroxybutyrate-*co*-valerate)–starch compositions. *J. Environ. Polym. Degrad.*, **3** (2), 97–105.

46 El-Hadi, A., Schnabel, R., Straube, E., Müller, G., and Henning, S. (2002) Correlation between degree of crystallinity, morphology, glass temperature, mechanical properties and biodegradation of poly(3-hydroxyalkanoate) PHAs and their blends. *Polym. Test.*, **21** (6), 665–674.

47 Noda, I., Green, P.R., Satkowski, M.M., and Schechtman, L.A. (2005) Preparation and properties of a novel class of polyhydroxyalkanoate copolymers. *Biomacromolecules*, **6** (2), 580–586.

48 Philip, S., Keshavarz, T., and Roy, I. (2007) Polyhydroxyalkanoates: biodegradable polymers with a range of applications. *J. Chem. Technol. Biotechnol.*, **82** (3), 233–247.

49 Williams, S.F., Martin, D.P., Horowitz, D.M., and Peoples, O.P. (1999) PHA applications: addressing the price performance issue. I. Tissue engineering. *Int. J. Biol. Macromol.*, **25** (1–3), 111–121.

50 Vert, M., Schwarch, G., and Coudane, J. (1995) Present and future of PLA polymers. *J. Macromol. Sci. Pure Appl. Chem.*, **A32** (4), 787–796.

51 Bastioli, C., Cerutti, A., Guanella, I., Romano, G.C., and Tosin, M. (1995) Physical state and biodegradation behavior of starch–polycaprolactone systems. *J. Environ. Polym. Degrad.*, **3** (2), 81–95.

52 Bastioli, C. (1998) Biodegradable materials – present situation and future perspectives. *Macromol. Symp.*, **135**, 193–204.

53 Bastioli, C. (1998) Properties and applications of Mater-Bi starch-based materials. *Polym. Degrad. Stabil.*, **59** (1–3), 263–272.

54 Averous, L., Moro, L., Dole, P., and Fringant, C. (2000) Properties of thermoplastic blends: starch–polycaprolactone. *Polymer*, **41** (11), 4157–4167.

55 Koenig, M.F. and Huang, S.J. (1994) Evaluation of crosslinked poly (caprolactone) as a biodegradable, hydrophobic coating. *Polym. Degrad. Stabil.*, **45** (1), 139–144.

56 Lee, S.-R., Park, H.-M., Lim, H., Kang, T., Li, X., Cho, W.-J., and Ha, C.-S. (2002) Microstructure, tensile properties, and biodegradability of aliphatic polyester/clay nanocomposites. *Polymer*, **43** (8), 2495–2500.

57 Muller, R.-J., Witt, U., Rantze, E., and Deckwer, W.-D. (1998) Architecture of biodegradable copolyesters containing aromatic constituents. *Polym. Degrad. Stabil.*, **59** (1–3), 203–208.

58 Dufresne, A., Dupeyre, D., and Vignon, M.R. (2000) Cellulose microfibrils from potato tuber cells: processing and characterization of starch–cellulose microfibril composites. *J. Appl. Polym. Sci.*, **76** (14), 2080–2092.

59 Funke, U., Bergthaller, W., and Lindhauer, M.G. (1998) Processing and characterization of biodegradable products based on starch. *Polym. Degrad. Stabil.*, **59** (1–3), 293–296.

60 Averous, L. (2007) Cellulose-based biocomposites: comparison of different multiphasic systems. *Compos. Interfaces*, **14**, 787–805.

61 Averous, L., Fringant, C., and Moro, L. (2001) Plasticized starch–cellulose interactions in polysaccharide composites. *Polymer*, **42** (15), 6565–6572.

62 Curvelo, A.A.S., de Carvalho, A.J.F., and Agnelli, J.A.M. (2001) Thermoplastic starch–cellulosic fibers composites: preliminary results. *Carbohydr. Polym.*, **45** (2), 183–188.

63 Wollerdorfer, M. and Bader, H. (1998) Influence of natural fibres on the mechanical properties of biodegradable polymers. *Ind. Crops Prod.*, **8** (2), 105–112.

64 Dufresne, A. and Vignon, M.R. (1998) Improvement of starch film performances using cellulose microfibrils. *Macromolecules*, **31** (8), 2693–2696.

65 Baumberger, S., Lapierre, C., Monties, B., and Della Valle, G. (1998) Use of kraft lignin as filler for starch films. *Polym. Degrad. Stabil.*, **59** (1–3), 273–277.

66 Baumberger, S., Lapierre, C., and Monties, B. (1998) Utilization of pine kraft lignin in starch composites: impact of structural heterogeneity. *J. Agric. Food Chem.*, **46** (6), 2234–2240.

67 de Carvalho, A.J.F., Curvelo, A.A.S., and Agnelli, J.A.M. (2001) A first insight on composites of thermoplastic starch and kaolin. *Carbohydr. Polym.*, **45** (2), 189–194.

68 Shanks, R.A., Hodzic, A., and Wong, S. (2004) Thermoplastic biopolyester natural fiber composites. *J. Appl. Polym. Sci.*, **91** (4), 2114–2121.

69 Avella, M., Martuscelli, E., Pascucci, B., Raimo, M., Focher, B., and Marzetti, A. (1993) A new class of biodegradable materials: poly-3-hydroxy-butyrate/steam exploded straw fiber composites. I. Thermal and impact behavior. *J. Appl. Polym. Sci.*, **49** (12), 2091–2103.

70 Avella, M., Bogoeva-Gaceva, G., Buzõarovska, A., Errico, M.E., Gentile, G., and Grozdanov, A. (2007) Poly(3-hydroxybutyrate-*co*-3-hydroxyvalerate)-based biocomposites reinforced with kenaf fibers. *J. Appl. Polym. Sci.*, **104** (5), 3192–3200.

71 Mohanty, A.K., Misra, M., and Hinrichsen, G. (2000) Biofibres, biodegradable polymers and biocomposites: an overview. *Macromol. Mater. Eng.*, **276–277** (1), 1–24.

72 Gatenholm, P., Kubát, J., and Mathiasson, A. (1992) Biodegradable natural composites. I. Processing and properties. *J. Appl. Polym. Sci.*, **45** (9), 1667–1677.

73 Cyras, V.P., Commisso, M.S., Mauri, A.N., and Vázquez, A. (2007) Biodegradable double-layer films based on biological resources: polyhydroxybutyrate and cellulose. *J. Appl. Polym. Sci.*, **106** (2), 749–756.

74 Barkoula, N.M., Garkhail, S.K., and Peijs, T. (2010) Biodegradable composites based on flax/polyhydroxybutyrate and its copolymer with hydroxyvalerate. *Ind. Crops Prod.*, **31** (1), 34–42.

75 Wong, S., Shanks, R.A., and Hodzic, A. (2007) Effect of additives on the interfacial strength of poly(L-lactic acid) and poly(3-hydroxy butyric acid)–flax fibre composites. *Compos. Sci. Technol.*, **67** (11–12), 2478–2484.

76 Nishino, T., Hirao, K., Kotera, M., Nakamae, K., and Inagaki, H. (2003) Kenaf reinforced biodegradable composite. *Compos. Sci. Technol.*, **63** (9), 1281–1286.

77 Lee, S.H., Ohkita, T., and Kitagawa, K. (2004) Eco-composite from poly(lactic acid) and bamboo fiber. *Holzforschung*, **58** (5), 529–536.

78 Nishino, T., Hirao, K., and Kotera, M. (2007) Papyrus reinforced poly(L-lactic acid) composite. *Adv. Compos. Mater.*, **16** (4), 259–267.

79 Plackett, D., Andersen, T.L., Pedersen, W.B., and Nielsen, L. (2003) Biodegradable composites based on L-polylactide and jute fibres. *Compos. Sci. Technol.*, **63** (9), 1287–1296.

80 Van de Velde, K. and Kiekens, P. (2002) Biopolymers: overview of several properties and consequences on their applications. *Polym. Test.*, **21** (4), 433–442.

81 Ruseckaite, R.A. and Jimenez, A. (2003) Thermal degradation of mixtures of polycaprolactone with cellulose derivatives. *Polym. Degrad. Stabil.*, **81** (2), 353–358.

82 Nakamura, M., Sahoo, S., Ishiaku, U.S., Kotaki, M., Nakai, A., Hamada, H., and Kitagawa, K. (2006) Interfacial adhesion in bamboo fiber/biodegradable polymer composites, in *Design, Manufacturing and Applications of Composites* (eds J. Lo, T. Nishino, S.V. Hoa, H. Hamada, A. Nakai, and C. Poon), DEStech Publications, Inc., Lancaster, pp. 52–58.

83 Le Digabel, F. and Averous, L. (2006) Effects of lignin content on the properties of lignocellulose-based biocomposites. *Carbohydr. Polym.*, **66** (4), 537–545.

84 Le Digabel, F., Boquillon, N., Dole, P., Monties, B., and Averous, L. (2004) Properties of thermoplastic composites based on wheat-straw lignocellulosic fillers. *J. Appl. Polym. Sci.*, **93** (1), 428–436.

85 Averous, L. and Le Digabel, F. (2006) Properties of biocomposites based on lignocellulosic fillers. *Carbohydr. Polym.*, **66** (4), 480–493.

86 Chang, S.J. and Tsai, H.B. (1994) Copolyesters. 7. Thermal transitions of poly(butylene terephthalate-*co*-isophthalate-*co*-adipate)s. *J. Appl. Polym. Sci.*, **51** (6), 999–1004.

87 Herrera, R., Franco, L., Rodriguez-Galan, A., and Puiggali, J. (2002) Characterization and degradation behavior of poly(butylene adipate-*co*-terephthalate)s. *J. Polym. Sci. Polym. Chem.*, **40** (23), 4141–4157.

88 Avella, M., Rota, G.L., Martuscelli, E., Raimo, M., Sadocco, P., Elegir, G., and Riva, R. (2000) Poly(3-hydroxybutyrate-*co*-3-hydroxyvalerate) and wheat straw fibre composites: thermal, mechanical properties and biodegradation behaviour. *J. Mater. Sci.*, **35** (4), 829–836.

89 Hornsby, P.R., Hinrichsen, E., and Tarverdi, K. (1997) Preparation and properties of polypropylene composites reinforced with wheat and flax straw fibres. 2. Analysis of composite microstructure and mechanical properties. *J. Mater. Sci.*, **32** (4), 1009–1015.

90 Kronbergs, E. (2000) Mechanical strength testing of stalk materials and compacting energy evaluation. *Ind. Crops Prod.*, **11** (2–3), 211–216.

91 Bordes, P., Pollet, E., and Avérous, L. (2009) Nano-biocomposites: biodegradable polyester/nanoclay systems. *Prog. Polym. Sci. (Oxford)*, **34** (2), 125–155.

92 Chivrac, F., Pollet, E., and Averous, L. (2009) Progress in nano-biocomposites based on polysaccharides and nanoclays. *Mater. Sci. Eng. R*, **67** (1), 1–17.

93 Alexandre, M. and Dubois, P. (2000) Polymer-layered silicate nanocomposites: preparation, properties and uses of a new class of materials. *Mater. Sci. Eng. R*, **28** (1), 1–63.

94 Sinha Ray, S. and Okamoto, M. (2003) Polymer/layered silicate nanocomposites: a review from preparation to processing. *Prog. Polym. Sci. (Oxford)*, **28** (11), 1539–1641.

95 Raquez, J.M., Nabar, Y., Narayan, R., and Dubois, P. (2007) New developments in biodegradable starch-based nanocomposites. *Int. Polym. Proc.*, **22**, 463–470.

96 Dai, J.C. and Huang, J.T. (1999) Surface modification of clays and clay–rubber composite. *Appl. Clay Sci.*, **15** (1–2), 51–65.

97 Ke, Y., Lü, J., Yi, X., Zhao, J., and Qi, Z. (2000) The effects of promoter and curing process on exfoliation behavior of epoxy/clay nanocomposites. *J. Appl. Polym. Sci.*, **78** (4), 808–815.

98 Lagaly, G. (1999) Introduction: from clay mineral–polymer interactions to clay mineral–polymer nanocomposites. *Appl. Clay Sci.*, **15** (1–2), 1–9.

99 Shen, Z., Simon, G.P., and Cheng, Y.B. (2002) Comparison of solution intercalation and melt intercalation of polymer–clay nanocomposites. *Polymer*, **43** (15), 4251–4260.

100 Fischer, H.R., Gielgens, L.H., and Koster, T.P.M. (1999) Nanocomposites from polymers and layered minerals. *Acta Polym.*, **50** (4), 122–126.

101 Lagaly, G. (1986) Interaction of alkylamines with different types of layered compounds. *Solid State Ion.*, **22** (1), 43–51.

102 Azizi Samir, M.A.S., Alloin, F., Sanchez, J.-Y., El Kissi, N., and Dufresne, A. (2004) Preparation of cellulose whiskers reinforced nanocomposites from an organic medium suspension. *Macromolecules*, **37** (4), 1386–1393.

103 Azizi Samir, M.A.S., Alloin, F., and Dufresne, A. (2005) Review of recent research into cellulosic whiskers, their properties and their application in nanocomposite field. *Biomacromolecules*, **6** (2), 612–626.

104 Mathew, A.P., Thielemans, W., and Dufresne, A. (2008) Mechanical properties of nanocomposites from sorbitol plasticized starch and tunicin whiskers. *J. Appl. Polym. Sci.*, **109** (6), 4065–4074.

105 Mathew, A.P. and Dufresne, A. (2002) Morphological investigation of nanocomposites from sorbitol plasticized starch and tunicin whiskers. *Biomacromolecules*, **3** (3), 609–617.

106 Angles, M.N. and Dufresne, A. (2001) Plasticized starch/tunicin whiskers nanocomposite materials. 2. Mechanical behavior. *Macromolecules*, **34** (9), 2921–2931.

107 Angles, M.N. and Dufresne, A. (2000) Plasticized starch/tunicin whiskers nanocomposites. 1. Structural analysis. *Macromolecules*, **33** (22), 8344–8353.

108 Angellier, H., Molina-Boisseau, S., Dole, P., and Dufresne, A. (2006) Thermoplastic starch–waxy maize starch nanocrystals nanocomposites. *Biomacromolecules*, **7** (2), 531–539.

109 Chiou, B.-S., Yee, E., Glenn, G.M., and Orts, W.J. (2005) Rheology of starch–clay nanocomposites. *Carbohydr. Polym.*, **59** (4), 467–475.

110 Zhang, Q.X., Yu, Z.Z., Xie, X.L., Naito, K., and Kagawa, Y. (2007) Preparation and crystalline morphology of biodegradable starch/clay nanocomposites. *Polymer*, **48** (24), 7193–7200.

111 Park, H.M., Li, X., Jin, C.Z., Park, C.Y., Cho, W.J., and Ha, C.S. (2002) Preparation and properties of biodegradable thermoplastic starch/clay hybrids. *Macromol. Mater. Eng.*, **287** (8), 553–558.

112 Chen, B. and Evans, J.R.G. (2005) Thermoplastic starch–clay nanocomposites and their characteristics. *Carbohydr. Polym.*, **61** (4), 455–463.

113 Park, H.M., Lee, W.K., Park, C.Y., Cho, W.J., and Ha, C.S. (2003) Environmentally friendly polymer hybrids. Part I. Mechanical, thermal, and barrier properties of thermoplastic starch/clay nanocomposites. *J. Mater. Sci.*, **38** (5), 909–915.

114 Chiou, B.-S., Yee, E., Wood, D., Shey, J., Glenn, G., and Orts, W. (2006) Effects of processing conditions on nanoclay dispersion in starch–clay nanocomposites. *Cereal Chem.*, **83** (3), 300–305.

115 Avella, M., De Vlieger, J.J., Errico, M.E., Fischer, S., Vacca, P., and Volpe, M.G. (2005) Biodegradable starch/clay nanocomposite films for food packaging applications. *Food Chem.*, **93** (3), 467–474.

116 Chen, M., Chen, B., and Evans, J.R.G. (2005) Novel thermoplastic starch–clay nanocomposite foams. *Nanotechnology*, **16** (10), 2334–2337.

117 Chiou, B.S., Wood, D., Yee, E., Imam, S.H., Glenn, G.M., and Orts, W.J. (2007) Extruded starch–nanoclay nanocomposites: effects of glycerol and nanoclay concentration. *Polym. Eng. Sci.*, **47** (11), 1898–1904.

118 Huang, M.F., Yu, J.G., and Ma, X.F. (2004) Studies on the properties of montmorillonite-reinforced thermoplastic starch composites. *Polymer*, **45** (20), 7017–7023.

119 Pandey, J.K. and Singh, R.P. (2005) Green nanocomposites from renewable resources: effect of plasticizer on the structure and material properties of clay-filled starch. *Starch/Starke*, **57** (1), 8–15.

120 Cyras, V.P., Manfredi, L.B., Ton-That, M.T., and Vazquez, A. (2008) Physical and mechanical properties of thermoplastic starch/montmorillonite nanocomposite films. *Carbohydr. Polym.*, **73** (1), 55–63.

121 Dean, K., Yu, L., and Wu, D.Y. (2007) Preparation and characterization of melt-extruded thermoplastic starch/clay nanocomposites. *Compos. Sci. Technol.*, **67** (3–4), 413–421.

122 Wilhelm, H.-M., Sierakowski, M.-R., Souza, G.P., and Wypych, F. (2003) Starch films reinforced with mineral clay. *Carbohydr. Polym.*, **52** (2), 101–110.

123 Huang, M., Yu, J., and Ma, X. (2005) Studies on properties of the thermoplastic starch/montmorillonite composites. *Acta Polym. Sin.*, **1** (6), 862–867.

124 Huang, M., Yu, J., and Ma, X. (2006) High mechanical performance MMT–urea and formamide–plasticized thermoplastic cornstarch biodegradable nanocomposites. *Carbohydr. Polym.*, **63** (3), 393–399.

125 Huang, M.F., Yu, J.G., and Ma, X.F. (2005) Preparation of the thermoplastic starch/montmorillonite nanocomposites by melt-intercalation. *Chin. Chem. Lett.*, **16** (4), 561–564.

126 Huang, M.F., Yu, J.G., Ma, X.F., and Jin, P. (2005) High performance biodegradable thermoplastic starch–EMMT nanoplastics. *Polymer*, **46** (9), 3157–3162.

127 Kampeerapappun, P., Aht-ong, D., Pentrakoon, D., and Srikulkit, K. (2007) Preparation of cassava starch/montmorillonite composite film. *Carbohydr. Polym.*, **67** (2), 155–163.

128 Chivrac, F., Pollet, E., and Averous, L. (2008) New approach to elaborate exfoliated starch-based nanobiocomposites. *Biomacromolecules*, **9** (3), 896–900.

129 Huang, M. and Yu, J. (2006) Structure and properties of thermoplastic corn starch/montmorillonite biodegradable composites. *J. Appl. Polym. Sci.*, **99** (1), 170–176.

130 Chivrac, F., Angellier-Coussy, H., Guillard, V., Pollet, E., and Avérous, L. (2010) How does water diffuse in starch/montmorillonite nano-biocomposite materials? *Carbohydr. Polym.*, **82** (1), 128–135.

131 Ogata, N., Jimenez, G., Kawai, H., and Ogihara, T. (1997) Structure and thermal/mechanical properties of poly(L-lactide)–clay blend. *J. Polym. Sci. Polym. Phys.*, **35** (2), 389–396.

132 Krikorian, V. and Pochan, D.J. (2003) Poly (L-lactic acid)/layered silicate nanocomposite: fabrication, characterization and properties. *Chem. Mater.*, **15** (22), 4317–4324.

133 Wu, T.-M. and Wu, C.-Y. (2006) Biodegradable poly(lactic acid)/chitosan-modified montmorillonite nanocomposites: preparation and characterization. *Polym. Degrad. Stabil.*, **91** (9), 2198–2204.

134 Pluta, M., Galeski, A., Alexandre, M., Paul, M.-A., and Dubois, P. (2002) Polylactide/montmorillonite nanocomposites and microcomposites prepared by melt blending: structure and some physical properties. *J. Appl. Polym. Sci.*, **86** (6), 1497–1506.

135 Maiti, P., Giannelis, E.P., and Batt, C.A. (2002) Biodegradable polyester/layered silicate nanocomposites. *Mater. Res. Soc. Symp. Proc.*, **740**, 141–145.

136 Maiti, P., Yamada, K., Okamoto, M., Ueda, K., and Okamoto, K. (2002) New polylactide/layered silicate nanocomposites: role of organoclays. *Chem. Mater.*, **14** (11), 4654–4661.

137 Sinha Ray, S., Yamada, K., Okamoto, M., and Ueda, K. (2002) Polylactide–layered silicate nanocomposite: a novel biodegradable material. *Nano Lett.*, **2** (10), 1093–1096.

138 Sinha Ray, S., Yamada, K., Ogami, A., Okamoto, M., and Ueda, K. (2002) New polylactide/layered silicate nanocomposite: nanoscale control over multiple properties. *Macromol. Rapid Commun.*, **23** (16), 943–947.

139 Sinha Ray, S., Maiti, P., Okamoto, M., Yamada, K., and Ueda, K. (2002) New polylactide/layered silicate nanocomposites. 1. Preparation, characterization and properties. *Macromolecules*, **35** (8), 3104–3110.

140 Sinha Ray, S., Yamada, K., Okamoto, M., and Ueda, K. (2003) New polylactide–layered silicate nanocomposites. 2. Concurrent improvements of material properties, biodegradability and melt rheology. *Polymer*, **44** (3), 857–866.

141 Sinha Ray, S., Yamada, K., Okamoto, M., Ogami, A., and Ueda, K. (2003) New polylactide/layered silicate nanocomposites. 3. High-performance biodegradable materials. *Chem. Mater.*, **15** (7), 1456–1465.

142 Sinha Ray, S., Yamada, K., Okamoto, M., Ogami, A., and Ueda, K. (2003) New polylactide/layered silicate nanocomposites. 4. Structure, properties and biodegradability. *Compos. Interfaces*, **10** (4–5), 435–450.

143 Sinha Ray, S., Yamada, K., Okamoto, M., Fujimoto, Y., Ogami, A., and Ueda, K. (2003) New polylactide/layered silicate nanocomposites. 5. Designing of materials with desired properties. *Polymer*, **44** (21), 6633–6646.

144 Sinha Ray, S. and Okamoto, M. (2003) New polylactide/layered silicate nanocomposites. 6. Melt rheology and foam processing. *Macromol. Mater. Eng.*, **288** (12), 936–944.

145 Sinha Ray, S. and Okamoto, M. (2003) Biodegradable polylactide and its nanocomposites: opening a new dimension for plastics and composites. *Macromol. Rapid Commun.*, **24** (14), 815–840.

146 Sinha Ray, S., Yamada, K., Okamoto, M., and Ueda, K. (2003) Biodegradable polylactide/montmorillonite nanocomposites. *J. Nanosci. Nanotechnol.*, **3** (6), 503–510.

147 Nam, P.H., Fujimori, A., and Masuko, T. (2004) The dispersion behavior of clay particles in poly(L-lactide)/organo-modified montmorillonite hybrid systems. *J. Appl. Polym. Sci.*, **93** (6), 2711–2720.

148 Paul, M.-A., Delcourt, C., Alexandre, M., Degée, P., Monteverde, F., and Dubois, P. (2005) Polylactide/montmorillonite nanocomposites: study of the hydrolytic degradation. *Polym. Degrad. Stabil.*, **87** (3), 535–542.

149 Lewitus, D., McCarthy, S., Ophir, A., and Kenig, S. (2006) The effect of nanoclays on the properties of PLLA-modified polymers. Part 1. Mechanical and thermal properties. *J. Polym. Environ.*, **14** (2), 171–177.

150 Hasook, A., Tanoue, S., Lemoto, Y., and Unryu, T. (2006) Characterization and mechanical properties of poly(lactic acid)/poly(ε-caprolactone)/organoclay nanocomposites prepared by melt compounding. *Polym. Eng. Sci.*, **46** (8), 1001–1007.

151 Paul, M.-A., Alexandre, M., Degée, P., Henrist, C., Rulmont, A., and Dubois, P. (2003) New nanocomposite materials based on plasticized poly(L-lactide) and organo-modified montmorillonites: thermal and morphological study. *Polymer*, **44** (2), 443–450.

152 Pluta, M. (2004) Morphology and properties of polylactide modified by thermal treatment, filling with layered silicates and plasticization. *Polymer*, **45** (24), 8239–8251.

153 Paul, M.-A., Delcourt, C., Alexandre, M., Degée, P., Monteverde, F., Rulmont, A., and Dubois, P. (2005) (Plasticized) polylactide/(organo-)clay nanocomposites by *in situ* intercalative polymerization. *Macromol. Chem. Phys.*, **206** (4), 484–498.

154 Pluta, M., Paul, M.-A., Alexandre, M., and Dubois, P. (2006) Plasticized polylactide/clay nanocomposites. I. The role of filler content and its surface organo-modification on the physico-chemical properties. *J. Polym. Sci. Polym. Phys.*, **44** (2), 299–311.

155 Pluta, M., Paul, M.-A., Alexandre, M., and Dubois, P. (2006) Plasticized polylactide/clay nanocomposites. II. The effect of aging on structure and properties in relation to the filler content and the nature of its organo-modification. *J. Polym. Sci. Polym. Phys.*, **44** (2), 312–325.

156 Tanoue, S., Hasook, A., Iemoto, Y., and Unryu, T. (2006) Preparation of poly(lactic acid)/poly(ethylene glycol)/organoclay nanocomposites by melt compounding. *Polym. Compos.*, **27** (3), 256–263.

157 Shibata, M., Someya, Y., Orihara, M., and Miyoshi, M. (2006) Thermal and mechanical properties of plasticized poly (L-lactide) nanocomposites with organo-modified montmorillonites. *J. Appl. Polym. Sci.*, **99** (5), 2594–2602.

158 Paul, M.-A., Alexandre, M., Degée, P., Calberg, C., Jerome, R., and Dubois, P. (2003) Exfoliated polylactide/clay nanocomposites by *in-situ* coordination–insertion polymerization. *Macromol. Rapid Commun.*, **24** (9), 561–566.

159 Maiti, P., Batt, C.A., and Giannelis, E.P. (2003) Renewable plastics: synthesis and properties of PHB nanocomposites. *Polym. Mater. Sci. Eng.*, **88**, 58–59.

160 Hablot, E., Bordes, P., Pollet, E., and Avérous, L. (2008) Thermal and thermo-mechanical degradation of PHB-based multiphase systems. *Polym. Degrad. Stabil.*, **93** (2), 413–421.

161 Choi, W.M., Kim, T.W., Park, O.O., Chang, Y.K., and Lee, J.W. (2003) Preparation and characterization of poly (hydroxybutyrate-*co*-hydroxyvalerate)–organoclay nanocomposites. *J. Appl. Polym. Sci.*, **90** (2), 525–529.

162 Wang, S., Song, C., Chen, G., Guo, T., Liu, J., Zhang, B., and Takeuchi, S. (2005) Characteristics and biodegradation properties of poly(3-hydroxybutyrate-*co*-3-hydroxyvalerate)/organophilic montmorillonite (PHBV/OMMT) nanocomposite. *Polym. Degrad. Stabil.*, **87** (1), 69–76.

163 Chen, G.X., Hao, G.J., Guo, T.Y., Song, M.D., and Zhang, B.H. (2004) Crystallization kinetics of poly(3-hydroxybutyrate-*co*-3-

hydroxyvalerate)/clay nanocomposites. *J. Appl. Polym. Sci.*, **93** (2), 655–661.

164 Chen, G.X., Hao, G.J., Guo, T.Y., Song, M.D., and Zhang, B.H. (2002) Structure and mechanical properties of poly(3-hydroxybutyrate-*co*-3-hydroxyvalerate) (PHBV)/clay nanocomposites. *J. Mater. Sci. Lett.*, **21** (20), 1587–1589.

165 Messersmith, P.B. and Giannelis, E.P. (1995) Synthesis and barrier properties of polycaprolactone–layered silicate nanocomposites. *J. Polym. Sci. Polym. Chem.*, **33** (7), 1047–1057.

166 Messersmith, P.B. and Giannelis, E.P. (1993) Polymer–layered silicate nanocomposites: *in situ* intercalative polymerization of ε-caprolactone in layered silicates. *Chem. Mater.*, **5** (8), 1064–1066.

167 Krishnamoorti, R. and Giannelis, E.P. (1997) Rheology of end-tethered polymer layered silicate nanocomposites. *Macromolecules*, **30** (14), 4097–4102.

168 Kiersnowski, A., Dabrowski, P., Budde, H., Kressler, J., and Piglowski, J. (2004) Synthesis and structure of poly (ε-caprolactone)/synthetic montmorillonite nano-intercalates. *Eur. Polym. J.*, **40** (11), 2591–2598.

169 Pucciariello, R., Villani, V., Belviso, S., Gorrasi, G., Tortora, M., and Vittoria, V. (2004) Phase behavior of modified montmorillonite–poly(ε-caprolactone) nanocomposites. *J. Polym. Sci. Polym. Phys.*, **42** (7), 1321–1332.

170 Tortora, M., Vittoria, V., Galli, G., Ritrovati, S., and Chiellini, E. (2002) Transport properties of modified montmorillonite–poly(ε-caprolactone) nanocomposites. *Macromol. Mater. Eng.*, **287** (4), 243–249.

171 Pantoustier, N., Lepoittevin, B., Alexandre, M., Kubies, D., Calberg, C., Jerome, R., and Dubois, P. (2002) Biodegradable polyester layered silicate nanocomposites based on poly (ε-caprolactone). *Polym. Eng. Sci.*, **42** (9), 1928–1937.

172 Kubies, D., Pantoustier, N., Dubois, P., Rulmont, A., and Jerome, R. (2002) Controlled ring-opening polymerization of ε-caprolactone in the presence of layered silicates and formation of nanocomposites. *Macromolecules*, **35** (9), 3318–3320.

173 Lepoittevin, B., Pantoustier, N., Devalckenaere, M., Alexandre, M., Kubies, D., Calberg, C., Jerome, R., and Dubois, P. (2002) Poly(ε-caprolactone)/ clay nanocomposites by *in-situ* intercalative polymerization catalyzed by dibutyltin dimethoxide. *Macromolecules*, **35** (22), 8385–8390.

174 Lepoittevin, B., Pantoustier, N., Alexandre, M., Calberg, C., Jerome, R., and Dubois, P. (2002) Polyester layered silicate nanohybrids by controlled grafting polymerization. *J. Mater. Chem.*, **12** (12), 3528–3532.

175 Lepoittevin, B., Pantoustier, N., Alexandre, M., Calberg, C., Jerome, R., and Dubois, P. (2002) Layered silicate/polyester nanohybrids by controlled ring-opening polymerization. *Macromol. Symp.*, **183**, 95–102.

176 Pantoustier, N., Alexandre, M., Degée, P., Kubies, D., Jerome, R., Henrist, C., Rulmont, A., and Dubois, P. (2003) Intercalative polymerization of cyclic esters in layered silicates: thermal vs. catalytic activation. *Compos. Interfaces*, **10** (4), 423–433.

177 Viville, P., Lazzaroni, R., Pollet, E., Alexandre, M., Dubois, P., Borcia, G., and Pireaux, J.-J. (2003) Surface characterization of poly(ε-caprolactone)-based nanocomposites. *Langmuir*, **19** (22), 9425–9433.

178 Gorrasi, G., Tortora, M., Vittoria, V., Pollet, E., Lepoittevin, B., Alexandre, M., and Dubois, P. (2003) Vapor barrier properties of polycaprolactone montmorillonite nanocomposites: effect of clay dispersion. *Polymer*, **44** (8), 2271–2279.

179 Pantoustier, N., Alexandre, M., Degée, P., Calberg, C., Jerome, R., Henrist, C., Cloots, R., Rulmont, A., and Dubois, P. (2001) Poly(ε-caprolactone) layered silicate nanocomposites: effect of clay surface modifiers on the melt intercalation process. *e-Polymers*, No. 009.

180 Lepoittevin, B., Devalckenaere, M., Pantoustier, N., Alexandre, M., Kubies, D., Calberg, C., Jerome, R., and Dubois, P. (2002) Poly(ε-caprolactone)/clay

nanocomposites prepared by melt intercalation: mechanical, thermal and rheological properties. *Polymer*, **43** (14), 4017–4023.

181 Chen, B. and Evans, J.R.G. (2006) Poly(ε-caprolactone)–clay nanocomposites: structure and mechanical properties. *Macromolecules*, **39** (2), 747–754.

182 Di, Y., Iannace, S., Di Maio, E., and Nicolais, L. (2003) Nanocomposites by melt intercalation based on polycaprolactone and organoclay. *J. Polym. Sci. Polym. Phys.*, **41** (7), 670–678.

183 Lepoittevin, B., Pantoustier, N., Devalckenaere, M., Alexandre, M., Calberg, C., Jerome, R., Henrist, C., Rulmont, A., and Dubois, P. (2003) Polymer/layered silicate nanocomposites by combined intercalative polymerization and melt intercalation: a masterbatch process. *Polymer*, **44** (7), 2033–2040.

184 Shibata, M., Teramoto, N., Someya, Y., and Tsukao, R. (2007) Nanocomposites based on poly(ε-caprolactone) and the montmorillonite treated with dibutylamine-terminated ε-caprolactone oligomer. *J. Appl. Polym. Sci.*, **104** (5), 3112–3119.

185 Pollet, E., Delcourt, C., Alexandre, M., and Dubois, P. (2006) Transesterification catalysts to improve clay exfoliation in synthetic biodegradable polyester nanocomposites. *Eur. Polym. J.*, **42** (6), 1330–1341.

186 Kalambur, S.B. and Rizvi, S.S. (2004) Starch-based nanocomposites by reactive extrusion processing. *Polym. Int.*, **53** (10), 1413–1416.

187 Yoshioka, M., Takabe, K., Sugiyama, J., and Nishio, Y. (2006) Newly developed nanocomposites from cellulose acetate/layered silicate/poly(ε-caprolactone): synthesis and morphological characterization. *J. Wood Sci.*, **52** (2), 121–127.

188 Jimenez, G., Ogata, N., Kawai, H., and Ogihara, T. (1997) Structure and thermal/mechanical properties of poly(ε-caprolactone)–clay blend. *J. Appl. Polym. Sci.*, **64** (11), 2211–2220.

189 Sinha Ray, S., Okamoto, K., Maiti, P., and Okamoto, M. (2002) New poly(butylene succinate)/layered silicate nanocomposites: preparation and mechanical properties. *J. Nanosci. Nanotechnol.*, **2** (2), 171–176.

190 Okamoto, K., Ray, S.S., and Okamoto, M. (2003) New poly(butylene succinate)/layered silicate nanocomposites. II. Effect of organically modified layered silicates on structure, properties, melt rheology, and biodegradability. *J. Polym. Sci. Polym. Phys.*, **41** (24), 3160–3172.

191 Sinha Ray, S., Okamoto, K., and Okamoto, M. (2003) Structure–property relationship in biodegradable poly(butylene succinate)/layered silicate nanocomposites. *Macromolecules*, **36** (7), 2355–2367.

192 Sinha Ray, S., Okamoto, K., and Okamoto, M. (2006) Structure and properties of nanocomposites based on poly(butylene succinate) and organically modified montmorillonite. *J. Appl. Polym. Sci.*, **102** (1), 777–785.

193 Someya, Y., Nakazato, T., Teramoto, N., and Shibata, M. (2004) Thermal and mechanical properties of poly(butylene succinate) nanocomposites with various organo-modified montmorillonites. *J. Appl. Polym. Sci.*, **91** (3), 1463–1475.

194 Chen, G.-X., Kim, E.-S., and Yoon, J.-S. (2005) Poly(butylene succinate)/twice functionalized organoclay nanocomposites: preparation, characterization, and properties. *J. Appl. Polym. Sci.*, **98** (4), 1727–1732.

195 Chen, G.-X., Kim, H.-S., Kim, E.-S., and Yoon, J.-S. (2005) Compatibilization-like effect of reactive organoclay on the poly(L-lactide)/poly(butylene succinate) blends. *Polymer*, **46** (25), 11829–11836.

196 Shih, Y.F., Wang, T.Y., Jeng, R.J., Wu, J.Y., and Teng, C.C. (2007) Biodegradable nanocomposites based on poly(butylene succinate)/organoclay. *J. Polym. Environ.*, **15** (2), 151–158.

197 Sinha Ray, S., Bousmina, M., and Okamoto, K. (2005) Structure and properties of nanocomposites based on poly(butylene succinate-*co*-adipate) and organically modified montmorillonite. *Macromol. Mater. Eng.*, **290** (8), 759–768.

198 Sinha Ray, S. and Bousmina, M. (2005) Poly(butylene succinate-*co*-adipate)/montmorillonite nanocomposites: effect

of organic modifier miscibility on structure, properties, and viscoelasticity. *Polymer*, **46** (26), 12430–12439.

199 Sinha Ray, S., Bandyopadhyay, J., and Bousmina, M. (2007) Effect of organoclay on the morphology and properties of poly (propylene)/poly[(butylene succinate)-*co*-adipate] blends. *Macromol. Mater. Eng.*, **292** (6), 729–747.

200 Sinha Ray, S., Bandyopadhyay, J., and Bousmina, M. (2007) Thermal and thermomechanical properties of poly [(butylene succinate)-*co*-adipate] nanocomposite. *Polym. Degrad. Stabil.*, **92** (5), 802–812.

201 Nam, P.H., Fujimori, A., and Masuko, T. (2003) Flocculation characteristics of OM clay particles in PLA–MMT hybrid systems. *e-Polymers*, No. 005.

202 Lee, C.H., Lim, S.T., Hyun, Y.H., Choi, H. J., and Jhon, M.S. (2003) Fabrication and viscoelastic properties of biodegradable polymer/organophilic clay nanocomposites. *J. Mater. Sci. Lett.*, **22** (1), 53–55.

203 Lim, S.T., Hyun, Y.H., Choi, H.J., and Jhon, M.S. (2002) Synthetic biodegradable aliphatic polyester/montmorillonite nanocomposites. *Chem. Mater.*, **14** (4), 1839–1844.

204 Lim, S.T., Lee, C.H., Choi, H.J., and Jhon, M.S. (2003) Solidlike transition of melt-intercalated biodegradable polymer/clay nanocomposites. *J. Polym. Sci. Polym. Phys.*, **41** (17), 2052–2061.

205 Lim, S.T., Lee, C.H., Kim, H.B., Choi, H. J., and Jhon, M.S. (2004) Polymer/organoclay nanocomposites with biodegradable aliphatic polyester and its blends: preparation and characterization. *e-Polymers*, No. 026.

206 Lee, C.H., Kim, H.B., Lim, S.T., Choi, H. J., and Jhon, M.S. (2005) Biodegradable aliphatic polyester–poly(epichlorohydrin) blend/organoclay nanocomposites: synthesis and rheological characterization. *J. Mater. Sci.*, **40** (15), 3981–3985.

207 Chen, G. and Yoon, J.-S. (2005) Nanocomposites of poly[(butylene succinate)-*co*-(butylene adipate)] (PBSA) and twice-functionalized organoclay. *Polym. Int.*, **54** (6), 939–945.

208 Chivrac, F., Kadlecova, Z., Pollet, E., and Averous, L. (2006) Aromatic copolyester-based nano-biocomposites: elaboration, structural characterization and properties. *J. Polym. Environ.*, **14** (4), 393–401.

209 Chivrac, F., Pollet, E., and Averous, L. (2007) Nonisothermal crystallization behavior of poly(butylene adipate-*co*-terephthalate)/clay nano-biocomposites. *J. Polym. Sci. Polym. Phys.*, **45** (13), 1503–1510.

8
IPNs Derived from Biopolymers

Fernando G. Torres, Omar Paul Troncoso, and Carlos Torres

8.1
Introduction

Heterogeneous materials have been developed in order to improve their properties for a wide range of industrial applications. Heterogeneous materials include composites, blends, and interpenetrating polymer networks (IPNs). Blends and IPNs are comprised of two or more polymers that are mixed together. Unlike blends, IPNs are formed by two different polymer networks that are cross-linked producing a typical morphology that will be studied later. In composites, typical polymers are reinforced by adding load-bearing reinforcing agents such as fibers into their matrices, whereas IPNs are formed by interlacements between polymer networks. One of the major advantages of IPN materials is that they can combine the advantages of their constituent polymers collectively in one system leading to the improvement of their properties.

Biopolymers have been intensively studied due to their remarkable properties. Their chemical and structural properties have been used to form IPN-based biomaterials with specific properties for a wide range of applications. Also, there are IPN materials that combine the advantages of natural and synthesized polymers collectively in one system leading to the improvement of their physical properties. This type of mixtures are called hybrid IPNs. Hybrids IPNs have been considered to study the influence of natural polymers in the enhancement of properties of synthetic polymers.

In this chapter, we review the properties and applications of natural polymer-based IPNs. First, we summarize recent studies with select natural polymers that have been used as IPN biomaterials. Then, the formation and variety of types of natural polymer-based IPNs at different scales are discussed. Different manufacturing techniques used for the production of natural polymer-based IPNs and the main techniques used to characterize the IPNs are presented in Sections 8.4 and 8.5, respectively. Finally, we give several examples of these IPNs for different applications.

8.2
Types of IPNs

Interpenetrating polymer networks are materials that combine two or more polymer networks that are interlaced at a molecular scale by physical cross-links [1]. Several criteria are used to identify the different types of IPNs (Table 8.1). For instance, full IPNs are formed by two different polymer networks, whereas semi-IPNs are formed by one polymer network and a linear or branched polymer (Figure 8.1). The same constituent polymers may be used to form semi- or full IPNs depending on the polymerization technique and the cross-linking agents used for the establishment of the inter- or intramolecular cross-linkages. For example, Pulat *et al.* [2] prepared semi- and full IPNs of acrylic acid (AA), acrylamide (AAm), and chitosan (CS). Semi-IPNs were prepared by cross-linking of AA and AAm chains using ethylene glycol dimethacrylate (EGDMA) as a cross-linker. Full IPNs were prepared by cross-linking of AA, AAm, and CS chains using EGDMA and glutaraldehyde (GA) as cross-linkers.

Table 8.1 Classification of IPN materials in connection with different criteria.

Criteria	Classification
Constituent polymer structure	Full IPN
	Homo-IPN
	Semi-IPN
Polymerization procedure	Sequential
	Simultaneous
Polymer type	Thermoplastic
	Gradient
	Latex
	Hydrogel

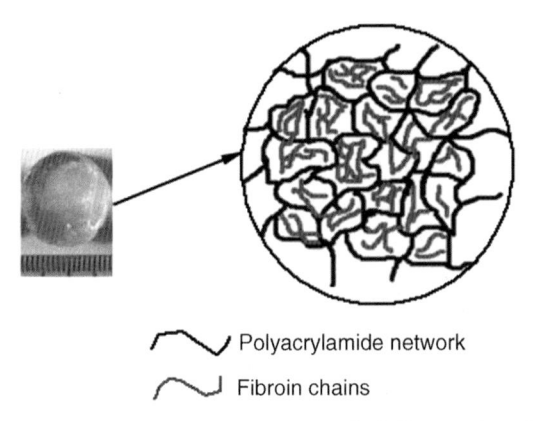

Figure 8.1 Schematic representation of silk fibroin/polyacrylamide semi-IPN hydrogels (right). Macroscopic sample of the IPN hydrogel (left) [3].

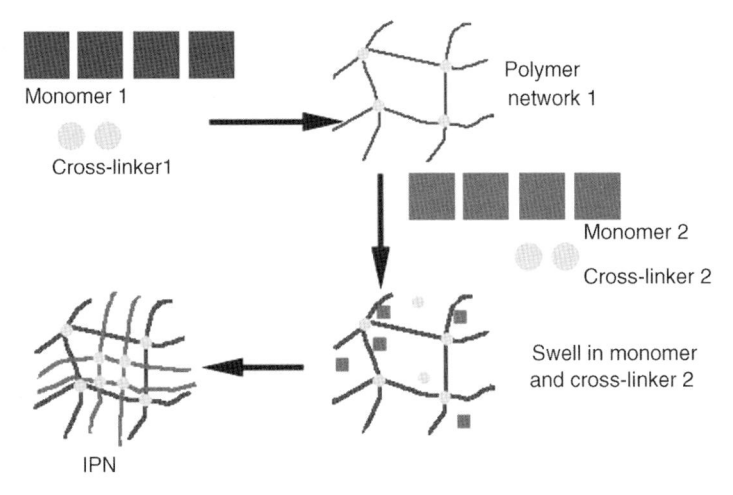

Figure 8.2 Schematic representation of the sequential process.

If the polymerization technique is taken into account, IPNs can be classified as sequential IPNs or simultaneous IPNs. Sequential IPNs are polymerized in two steps. In the first step, one constituent polymer and its cross-linking agent are polymerized to form the first polymer network. Then, *in situ* polymerization is used to polymerize the second polymer network within the first polymer network [4,5]. A schematic representation of the sequential process is depicted in Figure 8.2.

Simultaneous IPNs (Figure 8.3) are obtained by polymerizing the constituent polymer networks at the same time. Both constituents are polymerized by reactions that do not interfere with each other following independent routes [4,5]. Each polymer network is produced by a different polymerization process, or both constituent polymers are produced by the same polymerization process, but independently.

IPNs can also be classified according to the type of polymers used to form the constituent networks. Thermoplastic IPNs are formed by thermoplastic polymers. Each individual network is interlaced by physical cross-links rather than chemical cross-links. One constituent polymer is usually a block copolymer and the other is a semicrystalline or glassy polymer [4]. These IPNs behave as thermosets at their application temperature, but they can flow like thermoplastics at elevated temperatures.

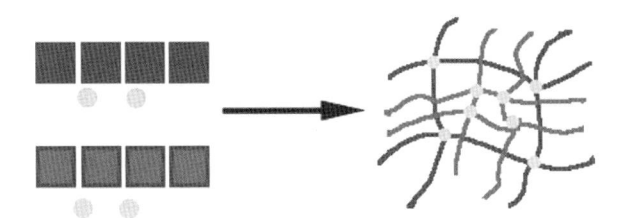

Figure 8.3 Schematic representation of the simultaneous process.

Latex is also used to obtain IPNs. In order to produce latex IPNs, the constituent polymers are synthesized by emulsion polymerization. The IPN morphology depends on how the polymerization is carried out. When using a sequential polymerization process, the diffusion rate of the second monomer into the first polymer latex determines if a homogeneous morphology or a core–shell morphology is obtained. If the second monomer diffuses fast in the first polymer latex, a homogeneous morphology is obtained. In contrast, if the second monomer diffuses slowly, most of it reacts near the surface of the first polymer latex and, as a consequence, a core–shell morphology is obtained.

Hydrogels are also made from IPNs. Hydrogels are able to absorb large amounts of water. Hydrogel IPNs are formed by the sequential or simultaneous polymerization process [6]. Several hydrogel IPNs exhibit high water sensitivity and thermoreversibility.

8.3
IPNs Derived from Biopolymers

Biopolymers such as polysaccharides, proteins, and polyhydroxyalkanoates (PHAs) have been used to produce IPNs with different applications. Biopolymer networks have also been interlaced with synthetic networks to obtain hybrid IPNs or semi-IPNs. Table 8.2 shows a list of several IPNs derived from biopolymers and their applications.

Table 8.2 Example of IPNs derived from biopolymers and their applications.

Composition		Applications	References
Konjac glucomannan/poly(acrylic acid)	Biodegradable, pH-sensitive KGM/PAA IPN hydrogels	Potential carriers for colon-specific drug delivery	[7]
Gelatin/poly(acrylic acid)	Semi-IPN and IPN of G/PAA nanogels	Potential targeted drug delivery system in the treatment of solid tumors	[8]
Fibrin (F)/alginate (A)	FA IPN hydrogel	Potential biomaterial for fertility preservation	[9]
Collagen/chitosan	C/C IPN scaffold	Suitable substrate for cancer cell culture	[10]
Partially hydrolyzed polyacrylamide/scleroglucan	Semi-IPN hydrogels of HPAM/Scl	Enhanced oil recovery applications	[11]
Poly(N-isopropylacrylamide)/chondroitin sulfate	Thermoresponsive semi-IPN hydrogels of PNIPAAm/ChS	Biocompatible hydrogels suitable for sensors, actuators, or artificial muscle applications. Potential hydrogel to load cationic drug for controlled delivery applications	[12]

8.3.1
Alginate

Alginate is a naturally derived polysaccharide that is produced by brown algae. As IPN biomaterial, this natural polymer has been used as a natural scaffold for tissue growth where alginate has been linked with fibrin to form interpenetrating matrices for *in vitro* ovarian follicle development [9]. Sodium alginate (SA) and poly(diallyldimethylammonium chloride) (PDADMAC) have been used to form IPN hydrogel with an electrical-sensitive behavior. When this swollen IPN hydrogel was placed between a pair of electrodes, the IPN exhibited bending behavior upon the application of an electric field [13]. New biomedical and pharmaceutical IPNs have been investigated by using calcium alginate (Ca(II)-Alg) hydrogel and dextran methacrylate (Dex-MA) derivative. The UV curing of this semi-IPN, by cross-linking of the methacrylate moieties, led to an IPN strong hydrogel used for a modulated delivery of bioactive molecules [14].

Other alginate-based IPNs were produced by Wang *et al.* [15]. They obtained a pH-sensitive semi-IPN superabsorbent hydrogel composed of sodium alginate-g-poly (sodium acrylate) (NaAlg-g-PNaA) network and linear polyvinylpyrrolidone (PVP). This hydrogel possesses sensitivity to external pH stimulus and shows reversible on–off switching swelling characteristic. Pescosolido *et al.* [16] have used calcium alginate and hydroxyethyl methacrylate-derivatized dextran (Dex-HEMA) in order to obtain semi-IPNs for protein release and cell encapsulation. They were able to use these semi-IPNs to release bovine serum albumin and to encapsulate chondrocytes.

8.3.2
Agarose

Agarose is a natural polysaccharide obtained from red algae [17]. As IPN biomaterial, this natural polymer is used to design hydrogels with improved mechanical properties suitable for tissue engineering. DeKosky *et al.* [18] have used agarose in the design of hydrogels of superior mechanical integrity for cell encapsulation. Agarose–poly(ethylene glycol) (PEG) diacrylate IPN hydrogels displayed a 4-fold increase in shear modulus relative to a pure PEG diacrylate network (39.9 kPa versus 9.9 kPa) and a 4.9-fold increase relative to a pure agarose network (8.2 kPa). Lomakin *et al.* [19] have prepared an IPN of PEG diacrylate in agarose. This material mimics the insect cuticle, which is composed of chitin fibers interpenetrated with a protein matrix. Agarose was selected to mimic the role of chitin, whereas PEG diacrylate was used to mimic the role of the cuticular proteins. The toughness of these agarose–PEG diacrylate IPNs was found to be 100 times greater than that of agarose and 5 times greater than that of a cross-linked PEG diacrylate network.

8.3.3
Chitosan

Chitosan, a natural biopolymer, is the deacetylated form of chitin that is a long-chain polymer of *N*-acetylglucosamine extracted from the exoskeleton of arthropods such as marine shellfish of crabs and shrimps. It has been widely studied due to its potential applications as a biomaterial in tissue engineering, separation membrane, and field responsive material. Peng et al. [20] reported a procedure for the preparation of semi-IPN hydrogel of chitosan and polyether. They studied the pH sensitivity, swelling and release kinetics, and structural changes of the gel in different pH solutions. They found that the physicochemical properties of the chitosan-based IPNs depend not only on the molecular structure, the gel structure, and the degree of cross-linking, but also on the content and state of the water in the hydrogel.

Superporous hydrogels have been prepared using chitosan and poly(vinyl alcohol) (PVA) IPNs [21]. In this case, poly(vinyl alcohol) contributes to the molecular weight of the strengthener. It was found that the introduction of a small amount of high molecular weight PVA significantly enhanced the mechanical strength but slightly reduced the swelling capacity. Other chitosan-based IPNs reported include the use of poly(ethylene glycol) [22], polyallylamine [23], polyvinylpyrrolidone [24], acrylated polylactide (PLA) [25], and glycine–glutamic acid [26].

8.3.4
Starch

Starch is a polymer composed of anhydroglucose units, which occurs widely in plants. Native starch granules are composed of two major polymers, namely, amylose and amylopectin. Amylose is essentially linear ($\alpha(1\rightarrow4)$-linked glucose unit) and soluble in boiling water, whereas amylopectin is branched ($\alpha(1\rightarrow4)$-linked glucose unit interlinked by $\alpha(1\rightarrow6)$-D-glucosidic linkage). Starch is used in various commercial applications due to its gelatin characteristic and it is used as thickener and sizing agent in textile and paper industries. Murthy et al. [27] prepared semi-IPN hydrogels composed of starch and random copolymer of poly(acrylamide-co-sodium methacrylate). The swelling pattern of these IPN hydrogels was studied in different physiological, pH, and ionic/salt solutions and showed great responsiveness due to their ionic character.

However, it has been reported that starch has some limitations such as poor processability and high brittleness [28]. Also, starch-based hydrogels exhibit some disadvantages such as poor mechanical strength, which limits their application. Li et al. [29,30] have used a new cationic starch-g-acrylic acid amphoteric hydrogel that showed better swelling capacity and reversible behavior. These semi-IPN hydrogels were found to possess appreciable compatibility, good swellability, and mechanical strength.

Bajpai and Saxena [31] have used soluble starch and poly(acrylic acid) (PAA) to obtain semi-IPN gels with drug delivery applications. They used riboflavin in order to determine their drug releasing capacity. The amount of drug released was found to increase with the starch content in the gels and with the initial water content. Other chemically modified starch is benzyl starch (BS). Benzyl starch has been proposed for use as an acrylic resin additive, detergent component, adhesive, and emulsifier for textile finishing, paper coatings, soil additives, and other applications [32]. Cao and Zhang [33] have prepared semi-IPNs from castor oil-based polyurethane (PU) and BS films. Their results showed that these semi-IPN films had good or certain miscibility with BS concentrations of 5–70 wt%. The tensile strength and Young's modulus of such semi-IPNs increased with an increase in the concentration of BS.

8.3.5
Dextran

Dextran is a natural hydrophilic degradable polysaccharide that has been used as an alternative instead of PEG in the formation of hydrogels [34,35]. Dextran-based IPNs have been studied as promising materials for 3D scaffolds for vascular tissue engineering and regeneration due to its biocompatibility with vascular cells [35].

8.3.6
Gum Arabic

Gum arabic (GA) is a natural polysaccharide derived from exudates of *Acacia senegal* and *Acacia seyal* trees [36] Among the remarkable properties are its high solubility, pH stability, nontoxicity, and antioxidant activity [37–39]. GA has been used in a wide range of applications such as food industry, textile, pottery, lithography, cosmetics, and pharmaceutical industry [40].

8.3.7
Fibrinogen

Fibrinogen is a soluble protein that is polymerized into fibrin through the action of thrombin in the presence of calcium [9]. Fibrin has been used as a scaffold for tissue growth.

8.3.8
Collagen and Gelatin

Collagen and gelatin are biopolymers formed by proteins. Collagen is part of the extracellular matrices of animal tissues and is formed by a triple-helical molecule of polypeptide chains. It has been used due to its capacity to form physical gels that are thermoreversible. Collagen chemical gels can also been obtained by using cross-linking agents such as glutaraldehyde and diphenylphosphoryl azide. Gelatin is obtained from denatured collagen isolated from bovine or porcine skin. Unlike the triple-helix structure of collagen, gelatin is composed of single-strand molecules. As IPN biomaterial, gelatin has been used in the form of IPN microgels to be assessed as a potential drug delivery system [41].

8.3.9
Cellulose and Cellulose Derivatives

Cellulose is the most abundant polymer on Earth. This natural polymer is the major constituent of cotton and wood that are used for the production of paper, textiles, and cellulose derivatives as cellophane and rayon. Williamson et al. [42] prepared semi-IPNs of cellulose and N,N-dimethylacrylamide (DMAm) utilizing a 9% LiCl/N,N-dimethylacetamide solvent system. The resulting IPNs were optically transparent and exhibited a sixfold higher modulus than the DMAm control. The enhancement in mechanical stiffness was attributed to intimate molecular interactions and complexation between cellulose and DMAm.

Chauhan et al. [43] have reported the use of cellulose extracted from pine needles for the preparation of IPN hydrogels. They used hydroxypropyl cellulose and other cellulosics such as cyanoethyl cellulose, hydroxyethyl cellulose (HEC), hydrazino-deoxy cellulose, and cellulose phosphate.

Another cellulose derivative used is the cellulose acetate butyrate (CAB). Laskar et al. [44] used CAB to produce polycarbonate/CAB IPNs in order to obtain transparent materials with better damping properties than the single polycarbonate network. They found that the storage moduli of IPNs are hardly changed when the CAB content increases but the damping properties are improved as well as UV and chemical resistance. Fichet et al. [45] have also used CAB networks. They synthesized polydimethylsiloxane–CAB IPNs through a one-pot shot process in which all components are first mixed together and the networks are then formed independently and quasi-simultaneously.

Other IPNs based on cellulose derivatives include CAB photocrosslinkable IPNs containing 50% by weight loading level of cross-linked vinyl polymers [46], high (>70%) water content IPNs from combinations of CAB and nitrogen-containing hydrophilic vinyl monomers [47], polyaniline and poly(3,4-ethylenedioxythiophene) conductive IPNs [48,49], and hydroxyethyl cellulose and polyacrylamide IPNs [50], CAB–dimethacrylate IPNs [51], and CAB–polyethylene glycol dimethacrylate IPNs [52].

8.3.10
Polyhydroxyalkanoates

Polyhydroxyalkanoates are a group of naturally occurring biodegradable and biocompatible polymers that belong to the aliphatic polyester family [53]. They are also known as bacterial polyesters or microbial polyesters. They are stored inside some types of bacteria as intracellular granules as a result of metabolic stress upon imbalanced growth due to a limited supply of an essential nutrient and the presence of an excess of a carbon source [54]. They have been used in several biomedical applications, such as sutures, cardiovascular patches, wound dressings, guided tissue repair/regeneration devices, and tissue engineering scaffolds [53].

Ralstonia eutropha and *Pseudomonas oleovorans* are among the bacteria that produce PHAs. Typical PHAs produced by these bacteria are poly(3-hydroxybutyrate) (PHB), poly(3-hydroxyvalerate) (PHV), poly(3-hydroxybutyrate-*co*-3-hydroxyvalerate) (PHBV) copolymer, poly(3-hydroxyoctanoate) (PHO), and poly(3-hydroxynonanoate) (PHN) [55].

Some of the PHAs are too rigid and brittle, whereas others are elastomeric but have very low mechanical strength. Therefore, for packaging materials, biomedical applications, tissue engineering, and other specific applications, the physical and mechanical properties of PHAs need to be diversified and improved [56-58].

Hao and Deng [59] have prepared semi-IPNs of PEG and PHB by a UV radiation technique. The prepared semi-IPNs showed improved mechanical properties compared with pure PEG hydrogels. Gursel et al. [60] prepared PHBV–HEMA IPN membranes by photopolymerization of HEMA in the presence of PHBV. The membranes exhibited novel properties such as partial hydrophilicity and hydrophobicity on the same membrane as well as significantly altered mechanical properties. Martellini et al. [61] obtained semi-IPNs based on PHB with PEG diacrylate at different compositions by radiation-induced polymerization using γ-rays from a ^{60}Co source with a total dose of 10–100 kGy. The results showed a water uptake increasing with the hydrophilic component until 25%.

8.3.11
Lactide-Derived Polymers

Polylactides are linear aliphatic thermoplastic polyesters that are synthesized from lactic acid typically derived from the fermentation of corn using suitable microorganisms such as lactobacilli [62]. Polylactides are one of the most important renewable resource polymer systems and have been investigated for a wide range of applications including biomedical, pharmaceutical, and packaging applications.

Lactic acid has two optically active configurations, the L- and D-enantiomeric forms. The polymerization of L- or D-lactic acid (or L- or D-lactide) yields isotactic semicrystalline polylactides. In contrast, a mixture of both forms of L- or D-lactic acid or *meso*-lactide can form atactic polylactide, which is amorphous with alterable properties. Rohman et al. [63,64] have obtained semi-IPNs based on poly(DL-lactide)/poly(methyl methacrylate) (PMMA) in order to prepare porous materials.

Poly(lactic-*co*-glycolic acid) (PLGA) is a copolymer of polyglycolic acid (PGA) and PLA. Hasircia et al. [65,66] have used PLGA to obtain IPNs that improved mechanical properties of PLGA used in the fabrication of bone plates. They used PLGA strengthened by a three-dimensional network of poly(propylene fumarate) (PPF). This procedure improved the mechanical properties of the bone plates as well as provided better dimensional stability.

8.4
Manufacture of IPNs

8.4.1
Casting–Evaporation Processing

The casting–evaporation method has been used to form cross-linked networks of polymer chains. Briefly, each constituent polymer is heated until dissolved and then added to the cross-linking agent. In a sequential process, one polymer solution is added directly to a cross-linking agent solution, followed by addition of the second polymer solution. In a simultaneous process, both polymer solutions are mixed together first and then added to the cross-linking agent solution. In both cases, the resultant mixtures are heated and blended. Finally, the mixture is cast and dried. IPN gels are prepared using this technique. Kosmala et al. [6] have prepared enzymatically degradable IPNs of gelatin and dextran by using the casting–evaporation method. They used glyceraldehyde as a cross-linking agent for gelatin.

8.4.2
Emulsification Cross-Linking Technique

This method is based on the phenomenon of phase separation and is used to form IPN microspheres. A single emulsion cross-linking technique is based on a water-in-oil (w/o) emulsion. It consists in dissolving water-soluble polymers in distilled water at a specific temperature in order to form a homogeneous solution by continuous stirring. Then, the homogeneous polymer solution that acts as a water phase is mixed with an oil phase to form a water-in-oil emulsion. Some chemicals are added to improve the formation of hardened microspheres. Finally, the microspheres are separated by filtration and washed repeatedly to remove the oil layer and excess amount of unreacted surfactant from the prepared emulsion (Figure 8.4).

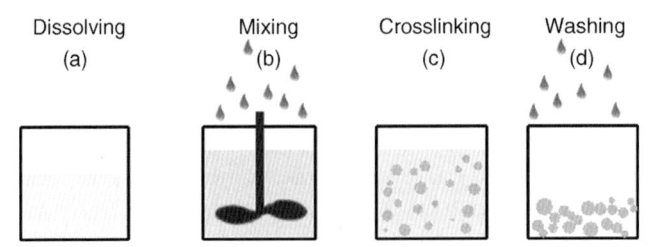

Figure 8.4 Fundamental scheme of the emulsification cross-linking technique described in four stages: (a) dissolution of water-soluble polymer solution (water phase); (b) mixing with the oil phase; (c) cross-linking of microspheres; and finally (d) washing of the isolated microspheres.

In drug delivery applications, biologically active substances such as drugs are dissolved to be completely soluble before adding to the homogeneous solution. Then, the drug-loaded homogeneous polymeric solution is mixed with an oil phase to form the water-in-oil emulsion. Babu et al. [67] synthesized semi-IPN graft copolymer made up of 2-hydroxyethyl methacrylate and chitosan in the form of microspheres using this water-in-oil emulsion method. They prepared microspheres loaded by a drug called indomethacin. Reddy et al. [68] prepared semi-IPNs of NaAlg and N-isopropylacrylamide (NIPAAm) by the water-in-oil emulsification method. These semi-IPN microspheres were encapsulated with 5-fluorouracil (5-FU). The results showed that the drug-loaded microspheres exhibited release of 5-FU up to 12 h and the microdomains have released in a controlled manner due the presence of NIPAAm in the matrix.

Recently, water-in-water (w/w) emulsion methods have been developed to form IPN structures. The advantage of this method is that there is no need of toxic organic solvents as the continuous phase of water-in-oil emulsions [29,69,70]. These organic solvents might leave toxic residues incompatible for IPN biomaterials designed for food applications. In the water-in-water emulsion method, an aqueous solution of a water-soluble polymer is emulsified as a dispersed phase in an aqueous solution of another polymer that acts as a continuous phase [29,71]. Then, the dispersed polymer phase is cross-linked to form microspheres resulting in the phase separation from continuous phase. This method has been used to form drug-loaded protein microspheres. Li et al. [29,30] prepared octenyl succinic anhydride (OSA) starch-based microspheres by the water-in-water emulsification method. The results showed that these homo-IPN microspheres show a good dispersibility and solid structure without wrinkle or concave depression on the surface and possess hydrogel character.

8.4.3
Miniemulsion/Inverse Miniemulsion Technique

Recently, these techniques have been used to synthesize multiple-component polymeric particles [8]. The miniemulsion (ME) technique allowed us to obtain monomers whose size is distributed at submicron level and polymer droplets in a continuous phase.

The process of inverse miniemulsion (IME) technique can be divided into three stages. In the first stage, stable droplets comprised of the constituent polymers are obtained by sonication using specific initiator for radical polymerization. In the second stage, one of the constituent polymers is polymerized and cross-linked employing a cross-linking agent. As a result, a semi-IPN is formed until the second stage. In the third stage, a full IPN is obtained by polymerizing and cross-linking the other constituent polymer by the addition of a second cross-linking agent.

For instance, Koul et al. [8] prepared semi-IPNs and full IPNs of gelatin/acrylic acid gels using the IME technique. The process of synthesis of these nanogels is depicted in Figure 8.5. They used ammonium persulfate (APS) as an initiator for the radical polymerization. Glutaraldehyde and N,N-methylenebisacrylamide (BIS) were used as cross-linking agents for gelatin and AA, respectively.

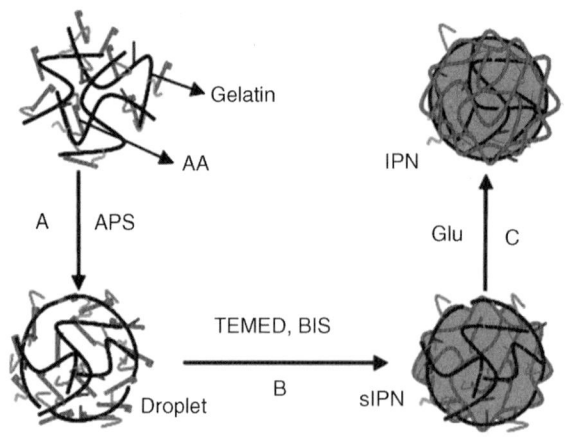

Figure 8.5 Process of synthesis of nanogels by the inverse miniemulsion technique.

8.4.4
Freeze Drying Technique

Several authors have reported the used of freeze drying as part of their procedures for preparing IPNs [12,35]. For instance, Muthyala *et al.* [72] have obtained gelatin/PVP IPNs by using a freeze drying technique. A gelatin solution and a PVP solution were mixed at a warm temperature. The mixture was cast on plastic vials and freeze dried to obtain a porous structure. Then, the gelatin/PVP porous structure was immersed in cross-linking agents such as GTA and EDC in order to cross-link the gelatin component. Finally, the cross-linked structures were freeze dried again to obtain dried IPN scaffolds.

Wen *et al.* [7] synthesized IPN hydrogels composed of konjac glucomannan (KGM) and cross-linked PAA. The KGM/PAA IPN hydrogels were prepared by immersing cylindrical tablets of prefabricated KGM gel in an aqueous solution containing the cross-linking agent, until the liquid was absorbed into the hydrogel. Then, the hydrogel was warmed to start the polymerization reaction. Finally, the IPN hydrogels obtained were washed and freeze dried.

8.5
Characterization of IPNs

8.5.1
Morphological and Structural Characterization

Scanning electron microscopy (SEM) as well as atomic force microscopy (AFM) has been used to characterize the structure and morphology of IPNs. SEM allows the analysis of the interior morphology and homogeneity of IPNs. In addition, the X-ray

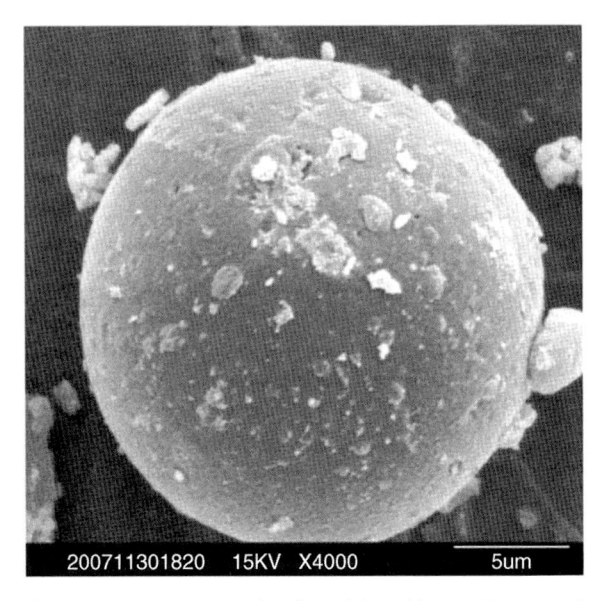

Figure 8.6 SEM micrographs of starch-based homo-IPN microspheres.

diffraction (XRD) pattern allowed identifying the type of structure and the presence of undesired constituents in the resulting IPNs.

Homo-IPNs that possess the simplest structure have been characterized morphologically by several authors [73]. Lie *et al.* [73] studied the detailed morphology of OSA starch-based homo-IPN microspheres prepared by a water-in-water emulsification–cross-linking technique. The OSA starch was cross-linked by trisodium trimetaphosphate (TSTP). SEM micrographs of these IPNs showed a spherical shape and a solid structure without wrinkle or concave depression on the surface (Figure 8.6).

The type of processing technique has an influence on the formation of the IPN structure. Wen *et al.* [7] prepared KGM/PAA IPN hydrogels in two steps. KGM hydrogel is synthesized in the first step, showing a porous honeycomb-like structure with large pores. After PAA is introduced into the previous gel, the pore size becomes smaller since PAA enters into the pores and cross-links to form an IPN. As a result, the average pore diameter decreases with the increase of the cross-linking density [7] (Figure 8.7).

8.5.2
FTIR Spectroscopy

FTIR spectroscopy is used to obtain information about the intermolecular interactions in IPNs as well as typical absorption of proteins and peptides such as amide bands. Mandal *et al.* [3] fabricated semi-IPN hydrogels of silk

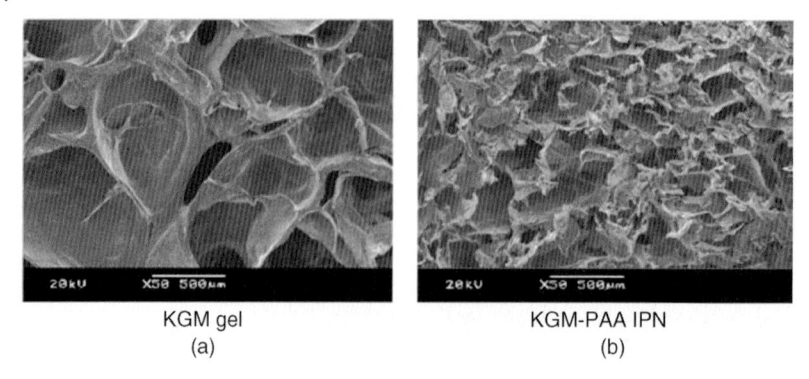

Figure 8.7 Cross-sectional morphology of the freeze-dried KGM IPN hydrogel (a) and KGM/PAA IPN hydrogel (b) [7].

fibroin/polyacrylamide. They could identify possible intermolecular interactions between silk fibroin and polyacrylamide by FTIR analysis. The typical amide bands of silk fibroin were shifted due to the presence of polyacrylamide.

Kumar and Singh [41] synthesized sodium alginate/gelatin IPN microgels by the chemical cross-linking technique using glutaraldehyde as a cross-linking agent. FTIR analysis suggested the formation of hydrogen bonds between –COO– groups of sodium alginate and –$CONH_2$ groups of gelatin as well as between –OH groups of sodium alginate and –NH_2 groups of gelatin.

8.5.3
Mechanical Characterization

IPNs can improve the mechanical properties of the biopolymers used for their production [74]. Mechanical properties of IPNs have been characterized using both tensile test and dynamic mechanical analysis (DMA). Lodha and Netravali [75] studied an IPN obtained by incorporating Phytagel®, a biopolymer gellan, in an SPI resin. The highly cross-linked structure of the IPN increased the mechanical properties of the SPI resin. The storage and loss modulus of the IPN were higher than the modulus of the SPI resin. In addition, the IPN structure showed a higher thermal stability than the unmodified SPI resin, increasing the glass transition temperature. Gil and Hudson [76] prepared protein/synthetic polymer hybrid IPNs composed of poly(N-isopropylacrylamide) (PNIPAAm) and *Bombyx mori* fibroin (SF). The results showed that those IPNs improved their storage and loss moduli due to the β-sheet crystalline structure of SF induced by methanol. Mandal *et al.* [3] prepared semi-IPN hydrogels of silk fibroin/polyacrylamide. They found an increase in both the compressive strength and modulus with the increase in polyacrylamide amount in the hydrogel. The results showed that the compressive strength of IPN hydrogels was much higher than that reported for fibroin-based hydrogel alone.

8.5.4
Rheological Characterization

Oscillatory shear rheology (OSR) is used to quantify both the viscous-like and the elastic-like properties of soft IPNs. Choudhary et al. [77] reported the preparation and rheological characterization of alginate/hydrophobically modified ethyl hydroxyethyl cellulose (HMEHEC) IPN hydrogels. The rheology of these materials can be easily tuned. The results showed that the storage modulus of IPNs strongly depends on the relative ratio of the constituent polymers. Aalaie et al. [11] synthesized semi-IPN hydrogels composed of partially hydrolyzed polyacrylamide (PHAM) and cross-linked scleroglucan (Scl). The gelation process was characterized using dynamic rheometry. OSR tests showed that the increase in scleroglucan concentration produced an increase in the limiting storage modulus of these semi-IPN gels and a decrease in the loss factor revealing that the viscous properties of the gelling system decreased more than its elastic properties.

8.5.5
Swelling Behavior Characterization

Multicomponent superabsorbent hydrogels from natural polymers are able to absorb and conserve considerable amounts of aqueous fluids under certain conditions of heat and pressure. Varguese et al. [12] prepared semi-IPN hydrogels composed of chondroitin sulfate (ChS) and PNIPAAm. They introduced ChS to increase the water absorption of PNIPAAm hydrogel. The results showed that the swelling ratio of the PNIPAAm hydrogel was greatly enhanced (2200%) by semi-IPN formation with ChS. Figure 8.8 shows that the hydrogels retained their shapes during the temperature cycling experiments. In addition, these IPNs showed a high reversibility or durability of the changes in volume.

The influence of the ChS polymer chains is shown in the swelling and deswelling process. The high swelling ratio of PNIPAAm/ChS semi-IPN hydrogels was attributed to two factors: the synergistic effect of the anionic nature of the ChS chains and the interconnected porous morphology. The fast deswelling of these

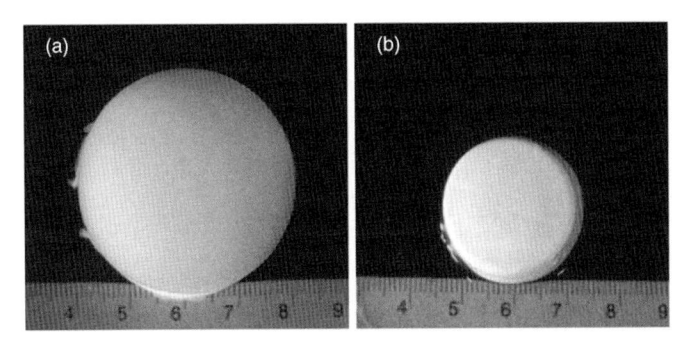

Figure 8.8 Photographs of IPN hydrogels when they were swollen at 20 °C (a) and deswelled at 45 °C (b).

hydrogels is mainly due to the presence of free or unbound water within the interconnected porosity of the hydrogels.

Wang et al. [15] synthesized a new type of semi-IPN superabsorbent hydrogels composed of NaAlg-g-PNaA network and linear PVP, where PVP was penetrated throughout the NaAlg-g-PNaA network. The introduction of PVP in the semi-IPN structure improved the water absorption and swelling rate of these hydrogels.

8.5.6
Thermal Characterization

Thermal transitions and thermal stability are studied using differential scanning calorimetry (DSC). The glass transition temperature (T_g) of IPNs is an important parameter that controls different properties of the resulting materials, such as mechanical and swelling behavior. T_g values of IPNs are evaluated by DSC and DMA. Dielectric analysis (DEA) is used to ascertain the glass transition temperature as well. Thermal decompositions can be evaluated by thermogravimetric analysis (TGA).

Kim et al. [78] synthesized IPN hydrogels composed of polyallylamine and chitosan. The DSC test revealed that melting temperatures of IPNs were observed with increasing chitosan content. Thermogravimetric analysis was carried out showing that the thermal decomposition started at around 270 °C. DEA tests showed two glass transitions indicating the phase separation in the IPN.

Lee et al. [79] prepared IPN hydrogels of poly(ethylene glycol) macromer (PEGM) with chitosan. DSC and DEA were used to evaluate glass transitions. The results showed that all the samples have two glass transition temperatures (T_g), revealing the presence of phase separation in the IPNs. In contrast, Li et al. [29,30] prepared semi-IPNs composed of acrylic acid in cationic starch and poly(methacryloyloxyethyl trimethylammonium chloride) (PDMC) solution. DSC tests revealed that the semi-IPN hydrogel has only one T_g, demonstrating a good miscibility with their components.

8.6
Applications of IPNs

8.6.1
Drug Delivery Applications

Wu and Wang [80] prepared pH-responsive IPN hydrogels comprised of silk sericin and poly(methacrylic acid) (PMAA). Swelling analysis revealed that hydrogels displayed definite pH sensitivity under physiological conditions. These hydrogels were loaded with bovine serum albumin (BSA) and tested under simulated gastric and intestinal pH conditions. The results showed that the release rate of BSA was lower in acidic media (pH 2.6) and higher in basic media (pH 7.4).

Drug delivery systems based on polysaccharides have been lately studied at different scales. Wen et al. [7] prepared IPN hydrogels composed of konjac glucomannan and cross-linked PAA by N,N-methylenebisacrylamide. The results showed that these hydrogels can be used as potential carriers for colon-specific drug delivery. Sodium alginate/gelatin IPN gels have been studied for the controlled release mucoadhesive drug delivery system of tramadol [41]. Semi- and full IPN-type hydrogels composed of acrylic acid, acrylamide, and chitosan were prepared to be used in controlled release of piperacillin–tazobactam [2]. Gupta and Ravi Kumar [81] fabricated semi-IPN microspheres of chitosan and PEG for controlled release of drugs. These microspheres have 93% drug loading capacity.

8.6.2
Scaffolds for Tissue Engineering

Natural polymers have been employed in tissue engineering in order to develop biological substitutes that restore, maintain, or improve tissue function [82]. The specific structure of IPNs can help to form suitable biomaterials for specific areas in the field of tissue engineering.

Type 1 diabetes is a chronic disease that leads to hyperglycemia. It is caused by the destruction of insulin-producing cells within the pancreas. IPN structures can be used as artificial scaffolds in the field of pancreatic tissue engineering. Muthyala et al. [72] prepared semi-IPN scaffolds composed of gelatin (G) and gelatin/PVP (GP). These scaffolds have the ability to maintain the viability and function of mouse pancreatic islet cells *in vitro*. The results showed that the islets are able to adhere and remain intact for 1 month in GP scaffolds cross-linked with 1-ethyl-3-[3-dimethylaminopropyl]carbodiimide hydrochloride (EDC). These IPN scaffolds have the potential to support and maintain islet cells for prolonged period.

Tiğli and Gümüşderelioğlu [83] prepared alginate and alginate:chitosan semi-IPN scaffolds. The attachment and proliferation properties of ATDC5 murine chondrogenic cells were assessed on these scaffolds demonstrating the potential use of chitosan semi-IPNs in alginate scaffolds for cartilage tissue engineering.

IPNs have also been used in bone tissue engineering. Barber et al. [84,85] prepared biomimetic implant surfaces based on IPN structures for studying the process of bone development *in vitro* and *in vivo*. Barber et al. [84] prepared IPNs composed of poly(acrylamide-*co*-ethylene glycol/acrylic acid) (p(AAm-co-EG/AA)) functionalized with an Arg–Gly–Asp-containing peptide extracted from rat bone sialoprotein (bsp-RGD(15)). These IPNs were grafted to titanium implants to modulate osteoblast behavior *in vitro*. The results showed that peptide-modified implants have greater mineralization compared to the base titanium and other control surface and can enhance the kinetics of osteoblastic cell differentiation. Recent studies [85] used these peptide-modified implants to modulate bone formation in the peri-implant region in the rat femoral ablation model.

8.6.3
Other Biomedical Applications

The presence of IPN biomaterials is not limited to clinical applications. IPN biomaterials can be employed to investigate the mechanisms of follicle development. Shikanov et al. [9] prepared fibrin/alginate IPN gel for *in vitro* ovarian follicle development. The IPN matrices promoted the follicle growth and the increase of the number of meiotically competent oocytes relative to either fibrin or alginate alone. These results showed a potential reproductive option for women who face cancer diagnosis.

IPNs have been used to form biodegradable polymer scaffolds for *in vitro* cancer cell culture. Shanmugasundaram et al. [10] prepared collagen/chitosan IPN scaffolds for *in vitro* culture of human epidermoid carcinoma cells (HEp-2). *In vitro* culture studies suggested the suitability of these scaffolds as substrates to culture HEp-2 cells and *in vitro* model for testing anticancer drugs.

8.6.4
Antibacterial Applications

IPNs can take advantage of the antibacterial properties of some particles. Vimala et al. [86] synthesized semi-IPN hydrogel of cross-linked polyacrylamide and carbohydrates such as acacia gum, carboxymethyl cellulose (CMC), and starch. Silver nanoparticles were distributed uniformly within the network via *in situ* reduction of silver nitrate ($AgNO_3$) using sodium borohydride ($NaBH_4$) as a reducing agent. The natural polymer-based networks were able to regulate the silver nanoparticle size. The results showed that all the IPN hydrogels exhibited similar antibacterial activity in an *E. coli* medium. Gils et al. [37] reported the preparation of semi-IPN hydrogels of GA and cross-linked copolymer of poly(2-hydroxyethyl methacrylate-*co*-acrylic acid). This hydrogel was loaded with silver nanoparticles and was tested in an *E. coli* medium. The results concluded that silver nanoparticles were immobilized throughout the hydrogel network. The silver ions underwent a strong localization within the network due to the complexation of the ions by functional groups of GA and the oxygen atoms of carboxylic groups of acrylic acid. The hydrogel-stabilized silver nanoparticles showed good antibacterial properties.

8.6.5
Sensor, Actuators, and Artificial Muscle Applications

Biological elements used as bioreceptors are the most important part of biosensors. They depend on the analytes, the storage capacity of the element, and the environmental and operational stability of the element. Biological elements include enzymes, DNA, antibodies, whole cells, dead cells, and so on [87]. These biological elements must be able to form immobilization matrices that can be functional at ambient temperatures. Stimuli-responsive biomaterials have been developed with sensor–actuator functions. These biomaterials are based on natural polymers with IPN structures that have the potential to be used as bioreceptors. Zeng et al. [88] prepared semi-IPN hydrogels of polyacrylamide and chitosan for the immobilization of redox protein hemoglobin (Hb). The results showed that the semi-IPN structure of the hydrogel film preserved the native structure of Hb without any change in its secondary structure and provided a favorable environment around Hb to retain its enzymatic bioactivity. Furthermore, the immobilized Hb displayed good bioelectrocatalytic activity toward H_2O_2 showing its potential application as an electrochemical sensing platform for redox proteins and enzymes. Varghese *et al.* [12] prepared PNIPAAm/ChS semi-IPN hydrogels with a filamentous porous network and high free water content. These thermoresponsive IPNs showed a change in volume of approximately 75% with stable thermoresponsive actuation (swelling/shrinking) showing their potential use as biocompatible thermoresponsive hydrogels for sensors, actuators, or artificial muscle applications.

8.7 Conclusions

In this chapter, we have reviewed the use of natural polymers in IPN structures. The remarkable properties of natural polymers along with typical properties of synthetic polymers can be combined in order to obtain improved properties and even new properties of the resulting hybrid materials. The constituent polymers, polymerization procedure, and the network structure of the resulting IPNs are factors that determine the final physical properties of these novel materials. These multipolymer materials have been assessed for their potential uses in different fields such as ambient, biomedical, and electronic applications.

References

1 Suthar, B., Xiao, H.X., Klempner, D., and Frisch, K.C. (1997) A review of kinetic studies on the formation of interpenetrating polymer networks, in *IPNs Around the World: Science and Engineering* (eds S.C. Kim and L.H. Sperling), John Wiley & Sons, Ltd, Chichester, UK.

2 Pulat, M, Tan, N., and Onurdağ, F.K. (2011) Swelling dynamics of IPN hydrogels including acrylamide–acrylic acid–chitosan and evaluation of their potential for controlled release of piperacillin–tazobactam. *J. Appl. Polym. Sci.*, **120**, 441–450.

3 Mandal, B.B., Kapoor, S., and Kundu, S.C. (2009) Silk fibroin/polyacrylamide semi-interpenetrating network hydrogels for controlled drug release. *Biomaterials*, **30**, 2826–2836.

4 Sperling, L.H. (1981) *Interpenetrating Polymer Networks and Related Materials*, Plenum Press, New York.

5 Gutowska, A., Bae, Y.H., Jacobs, H., Feijen, J. and Kim, S.W. (1994) Thermosensitive Interpenetrating Polymer Networks: Synthesis, Characterization, and Macromolecular Release. *Macromolecules*, **27**, 4167–4175.

6 Kosmala, J.D., Henthorn, D.B., and Brannon-Peppas, L. (2000) Preparation of interpenetrating networks of gelatin and dextran as degradable biomaterials. *Biomaterials*, **21**, 2019–2023.

7 Wen, X., Cao, X., Yin, Z., Wang, T., and Zhao, C. (2009) Preparation and characterization of konjac glucomannan–poly(acrylic acid) IPN hydrogels for controlled release. *Carbohydr. Polym.*, **78**, 193–198.

8 Koul, V, Mohamed, R., Kuckling, D., Adler, H.J.P., and Choudhary, V. (2011) Interpenetrating polymer network (IPN) nanogels based on gelatin and poly(acrylic acid) by inverse miniemulsion technique:

9 Shikanov, A., Xu, M., Woodruff, T.K., and Shea, L.D. (2009) Interpenetrating fibrin–alginate matrices for *in vitro* ovarian follicle development. *Biomaterials*, **30**, 5476–5485.

10 Shanmugasundaram, N., Ravichandran, P., Reddy, P.N., Ramamurty, N., Pal, S., and Rao, K.P. (2001) Collagen–chitosan polymeric scaffolds for the *in vitro* culture of human epidermoid carcinoma cells. *Biomaterials*, **22**, 1943–1951.

11 Aalaie, J., Rahmatpour, A., and Vasheghani-Farahani, E. (2009) Rheological and swelling behavior of semi-interpenetrating networks of polyacrylamide and scleroglucan. *Polym. Adv. Technol.*, **20**, 1102–1106.

12 Varghese, J.M., Ismail, Y.A., Lee, C.K., Shin, K.M., Shin, M.K., Kim, S.I., So, I., and Kim, S.J. (2008) Thermoresponsive hydrogels based on poly(N-isopropylacrylamide)/chondroitin sulfate. *Sens. Actuators B*, **135**, 336–341.

13 Kim, S.J., Yoon, S.G., Lee, S.M., Lee, J.H. and Kim, S.I. (2003) Characteristics of electrical responsive alginate/poly (diallyldimethylammonium chloride) IPN hydrogel in HCl solutions. *Sensors and Actuators B: Chemical*, **96**, 1–5.

14 Matricardi, P., Pontoriero, M., Coviello, T., Casadei, M.A. and Alhaique, F. (2008) In Situ Cross-Linkable Novel Alginate-Dextran Methacrylate IPN Hydrogels for Biomedical Applications: Mechanical Characterization and Drug Delivery Properties. *Biomacromolecules*, **9**, 2014–2020.

15 Wang, W. and Wang, A. (2010) Synthesis and swelling properties of pH-sensitive semi-IPN superabsorbent hydrogels based on sodium alginate-g-poly(sodium acrylate) and polyvinylpyrrolidone. *Carbohydrate Polymers*, **80**, 1028–1036.

16 Pescosolido, L., Vermonden, T., Malda, J., Censi, R., Dhert, W.A.J., Alhaique, F., Hennink, W.E. and Matricardi, P. (2011) In situ formin IPN hydrogels of calcium alginate and dextran-HEMA for biomedical applications. *Acta Biomaterialia*, **7**, 1627–1633.

17 Bao, X., Hayashi, K., Li, Y., Teramoto, A., and Abe, K. (2010) Novel agarose and agar fibers: fabrication and characterization. *Mater. Lett.*, **64**, 2435–2437.

18 DeKosky, B.J., Dormer, N.H., Ingavle, G.C., Roatch, C.H., Lomakin, J., Detamore, M.S., and Gehrke, S.H. (2010) Hierarchically designed agarose and poly (ethylene glycol) interpenetrating network hydrogels for cartilage tissue engineering. *Tissue Eng. Part C*, **16**, 1533–1542.

19 Lomakin, J. (2008) Mechanical Performance and Microstructure of Biomimetic PEG-Agarose Interpenetrating Networks as Determined by Dynamic Mechanical Analysis and AFM. AIChE Annual Meeting, Philadelphia

20 Peng, T., Yao, K.D., Yuan, C. and Goosen, M.F.A. (1994) Structural changes of pH-sensitive chitosan/polyether hydrogels in different pH solution. *J. Polym. Sci. Part A: Polym. Chem.*, **32**, 591–596.

21 Park, H. and Kim, D. (2006) Swelling and mechanical properties of glycol chitosan/ poly(vinyl alcohol) IPN-type superporous hydrogels. *J. Biomedical Mater. Res. Part A*, **78A**, 662–667.

22 Lee, S.J., Kim, S.S. and Lee, Y.M. (2000) Interpenetrating polymer network hydrogels based on poly(ethylene glycol) macromer and chitosan. *Carbohydrate Polymers*, **41**, 197–205.

23 Shin, M., Kim, S.J., Park, S.J., Lee, Y.H. and Kim, S.I. (2002) Synthesis and characteristics of the interpenetrating polymer network hydrogel composed of chitosan and polyallylamine, *J. App. Polym. Sci.*, **86**, 498–503.

24 Risbud, M.V., Hardikar, A.A., Bhat, S.V. and Bhonde, R. (2000) pH-sensitive freeze-dried chitosan–polyvinyl pyrrolidone hydrogels as controlled release system for antibiotic delivery. *Journal of Controlled Release*, **668**, 23–30.

25 Yang, H., Jiang L., Huang, A. and Deng, X. (2007) Synthesis and drug release behavior of polylactide/chitosan IPN film. *Journal of Jinan University Natural Science and Medicine Edition*, **28**, 494–497.

26 Rani, M., Agarwal, A., Maharana, T. and Negi, Y.S. (2010) A comparative study for interpenetrating polymeric network (IPN) of chitosan-amino acid beads for controlled drug release. *African Journal of Pharmacy and Pharmacology*, **4**, 35–54.

27 Murthy, P.S.K., Mohan, Y.M., Sreeramulu, J. and Raju, K.M. (2006) Semi-IPNs of starch and poly(acrylamide-co-sodium

28 Dufresne, A. and Vignon, M.R. (1998) Improvement of Starch Film Performances Using Cellulose Microfibrils. *Macromolecules*, **31**, 2693–2696.
29 Li, B.-Z., Wang, L.-J., Li, D., Bhandari, B., Li, S.-J., Lan, Y., Chen, X.D., and Mao, Z.-H. (2009) Fabrication of starch-based microparticles by an emulsification–crosslinking method. *J. Food Eng.*, **92**, 250–254.
30 Li, X., Xu, S., Wang, J., Chen, X., and Feng, S. (2009) Structure and characterization of amphoteric semi-IPN hydrogel based on cationic starch. *Carbohydr. Polym.*, **75**, 688–693.
31 Bajpai, S.K. and Saxena, S. (2004) Dynamic release of riboflavin from a starch-based semi IPN via partial enzymatic degradation: part II. *Reactive and Functional Polymers*, **61**, 115–129.
32 Lee, M., Swanson, B.G. and Baik, B. (2001) Influence of Amylose Content on Properties of Wheat Starch and Breadmaking Quality of Starch and Gluten Blends. *Cereal Chemistry*, **78**, 701–706.
33 Cao, X. and Zhang, L. (2005) Effects of Molecular Weight on the Miscibility and Properties of Polyurethane/Benzyl Starch Semi-Interpenetrating Polymer Networks. *Biomacromolecules*, **6**, 671–677.
34 Cadee, J.A, van Luyn, M.J., Brouwer, L.A., Plantinga, J.A., van Wachem, P.B., Groot, C. J., Otter, W.D., and Hennink, W.E. (2000) In vivo biocompatibility of dextran-based hydrogels. *J. Biomed. Mater. Res.*, **50**, 397–404.
35 Liu, Y. and Chan-Park, M.B. (2009) Hydrogel based on interpenetrating polymer networks of dextran and gelatin for vascular tissue engineering. *Biomaterials*, **30**, 196–207.
36 Dror, Y. and Cohen, Y. (2006) Structure of gum arabic in aqueous solution. *J. Polym. Sci. Part B*, **44**, 3265–3271.
37 Gils, P.S., Ray, D., and Sahoo, P.K. (2010) Designing of silver nanoparticles in gum arabic based semi-IPN hydrogel. *Int. J. Biol. Macromol.*, **46**, 237–244.
38 Ali, B.H., Ziada, A., and Blunden, G. (2009) Biological effects of gum arabic: a review of some recent research. *Food Chem. Toxicol.*, **47**, 1–8.
39 Abd-Allah, A.R., Al-Majed, A.A., Mostafa, A.M., Al-Shabanah, O.A., Din, A.G., and Nagi, M.N. (2002) Protective effect of arabic gum against cardiotoxicity induced by doxorubicin in mice: a possible mechanism of protection. *J. Biochem. Mol. Toxicol.*, **16**, 254–259.
40 Verbeken, D., Dierckx, S. and Dewettinck, K. (2003) Exudate gums: occurence, production, and applications. *Applied Microbiology and Biotechnology*, **63**, 10–21.
41 Kumar, P. and Singh, I. (2010) Formulation and characterization of tramadol-loaded IPN microgels of alginate and gelatin: optimization using response surface methodology. *Acta Pharm.*, **60**, 295–310.
42 Williamson, S.L., Armentrout, R.S., Porter, R.S. and McCormick, C.L. (1998) Microstructural Examination of Semi-Interpenetrating Networks of Poly(N,N-dimethylacrylamide) with Cellulose or Chitin Synthesized in Lithium Chloride/N, N-Dimethylacetamide. *Macromolecules*, **31**, 8134–8141.
43 Chauhan, G.S., Guleria, L.K. and Mahajan, S. (2001) A study in sorption of some metal ions on novel hydrogels based on modified cellulosics and 2-acrylamido-2-methyl propane sulphonic acid. *Desalination*, **141**, 325–329.
44 Laskar, J., Vidal, F., Fichet, O., Gauthier, C. and Teyssie, D. (2004) Synthesis and characterization of interpenetrating networks from polycarbonate and cellulose acetate butyrate. *Polymer*, **45**, 5047–5055.
45 Vidal, F., Fichet, O., Laskar, J. and Teyssie, D. (2006) Polysiloxane–Cellulose acetate butyrate cellulose interpenetrating polymers networks close to true IPNs on a large composition range. Part II. *Polymer*, **47**, 3747–3753.
46 Kamath, M., Kincaid, J. and Mandal, B.K. (1998) Interpenetrating polymer networks of photocrosslinkable cellulose derivatives. *J. App. Polym. Sci.*, **59**, 45–50.
47 Corkhill, P. H. and Tighe, B.J. (1990) Synthetic hydrogels: 7. High EWC semi-interpenetrating polymer networks based on cellulose esters and N-containing hydrophilic monomers. *Polymer*, **31**, 1526–1537.
48 Yin, W., Li, J., Li, Y., Wu, Y., Gu, T. and Liu, C. (1997) Conducting IPN Based on

Polyaniline and Crosslinked Cellulose. *Polymer International*, **42**, 276–280.

49 Randriamahazaka, H., Vidal, F., Dassonville, P., Chevrot, C. and Teyssié, D. (2002) Semi-interpenetrating polymer networks based on modified cellulose and poly(3,4-ethylenedioxythiophene). *Synthetic Metals*, **128**, 197–204.

50 Buyanov, A.L., Revel'Skaya, L.G., Kuznetzov, Y. P. and Khripunov, A.K. (2001) Cellulose–poly(acrylamide–acrylic acid) interpenetrating polymer network membranes for the pervaporation of water–ethanol mixtures. II. Effect of ionic group contents and cellulose matrix modification. *J. App. Polym. Sci.*, **80**, 1452–1460.

51 Nguyen, Q., Leger, C., Billard, P. and Lochon, P. (1997) Novel membranes made from a semi-interpenetrating polymer network for ethanol–ETBE separation by pervaporation. *Polym. Adv. Tech.*, **8**, 487–495.

52 Billard, P., Nguyen, Q.T., Leger, C. and Clement, R. (1998) Diffusion of organic compounds through chemically asymmetric membranes made of semi-interpenetrating polymer networks. *Separation and Purification Technology*, **14**, 221–232.

53 Misra, S.K., Valappil, S.P., Roy, I. and Boccaccini, A.R. (2006) Polyhydroxyalkanoate (PHA)/Inorganic Phase Composites for Tissue Engineering Applications. *Biomacromolecules*, **7**, 2249–2258.

54 Lenz, R.W. and Marchessault, R.H. (2005) Bacterial Polyesters:. Biosynthesis, Biodegradable Plastics and Biotechnology. *Biomacromolecules*, **6**, 1–8.

55 Gross, R.A., DeMello, C., Lenz, R.W., Brandl, H. and Fuller, R.C. (1989) The biosynthesis and characterization of poly(β-hydroxyalkanoates) produced by Pseudomonas oleovorans. *Macromolecules*, **22**, 1106–1115.

56 Hrabak, O. (1992) Industrial production of poly-β-hydroxybutyrate; *FEMS Microbiol.*, **103**, 251–256.

57 Hazer, B. (2002) Chemical modification of bacterial polyester. *Curr. Trends Polym. Sci.*, **7**, 131–138.

58 Hazer, B. (2003) Chemical modification of synthetic and biosynthetic polyesters. *Biopolymers*, **10**, 181–208.

59 Hao, J., Deng, X. (2001) Semi-interpenetrating networks of bacterial poly(3-hydroxybutyrate) with net-poly(ethylene glycol). *Polymer*, **42**, 4091–4097.

60 Gursel, I., Balcik, C., Arica, Y., Akkus, O., Akkas, N., and Hasirci, V. (1998) Synthesis and mechanical properties of interpenetrating networks of polyhydroxybutyrate-*co*-hydroxyvalerate and polyhydroxyethyl methacrylate. *Biomaterials*, **13**, 1137–1143.

61 Martellini, F., Innocentini, L.H., Lora, S., Carenza, M. (2004) Semi-interpenetrating polymer networks of poly(3-hydroxybutyrate) prepared by radiation-induced polymerization. *Radiation Physics and Chemistry*, **71**, 257–262.

62 Gupta, A.P. and Kumar, V. (2007) New emerging trends in synthetic biodegradable polymers – Polylactide: A critique. *Eur. Polym. J.*, **43**, 4053–4074.

63 Rohman, G., Grande, D., Laupêtre, F., Boileau, S. and Guérin, P. (2005) Design of Porous Polymeric Materials from Interpenetrating Polymer Networks (IPNs):. Poly(DL-lactide)/Poly(methyl methacrylate)-Based Semi-IPN Systems. *Macromolecules*, **38**, 7274–7285.

64 Rohman, G., Laupêtre, F., Boileau, S., Guérin, P. and Grande, D. (2007) Poly(d,l-lactide)/poly(methyl methacrylate) interpenetrating polymer networks: Synthesis, characterization, and use as precursors to porous polymeric materials. *Polymer*, **48**, 7017–7028.

65 Hasirci, V., Lewandrowski, K.U., Bondre, S.P., Gresser, J.D., Trantolo, D.J. and Wise, D.L. (2000) High strength bioresorbable bone plates:preparation, mechanical properties and in vitro analysis. *Bio-Medical Materials and Engineering*, **10**, 19–29.

66 Hasirci, V., Lewandrowski, K., Gresser, J. D., Wise, D.L. and Trantolo, D.J. (2001) Versatility of biodegradable biopolymers: degradability and an in vivo application. *Journal of Biotechnology*, **86**, 135–150.

67 Babu, V.R., Sairam, M., Hosamani, K.M., and Aminabhavi, T.M. (2007) Preparation and characterization of novel semi-interpenetrating 2-hydroxyethyl methacrylate-g-chitosan copolymeric microspheres for sustained release of indomethacin. *J. Appl. Polym. Sci.*, **106**, 3778–3785.

68 Reddy, K.M., Babu, V.R., Rao, K.S.V.K., Subha, M.C.S., Rao, K.C., Sairam, M., and

Aminabhavi, T.M. (2008) Temperature sensitive semi-IPN microspheres from sodium alginate and *N*-isopropylacrylamide for controlled release of 5-fluorouracil. *J. Appl. Polym. Sci.*, **107**, 2820–2829.

69 Hamdi, G., Ponchel, G., and Duchêne, D. (1998) An original method for studying *in vitro* the enzymatic degradation of cross-linked starch microspheres. *J. Control. Release*, **55**, 193–201.

70 Latha, M.S., Lal, A.V., Kumary, T.V., Sreekumar, R., and Jayakrishnan, A. (2000) Progesterone release from glutaraldehyde cross-linked casein microspheres: *in vitro* studies and *in vivo* response in rabbits. *Contraception*, **61**, 329–334.

71 Bos, G.W., Verrijk, R., Franssen, O., Bezemer, J., Hennink, W.E., and Crommelin, D.J.A. (2001) Hydrogels for the controlled release of pharmaceutical proteins. *Pharm. Dev. Technol.*, **25**, 110–120.

72 Muthyala, S., Bhonde, R.R., and Nair, P.D. (2010) Cytocompatibility studies of mouse pancreatic islets on gelatin–PVP semi IPN scaffolds *in vitro*. *Islets*, **2**, 357–366.

73 Li, B., Wang, L., Li, D., Bhandari, B., Li, S., Lan, Y., Chen, X.D. and Mao, Z. (2009) Fabrication of starch-based microparticles by an emulsification-crosslinking method. *Journal of Food Engineering*, **92**, 250–254.

74 Banerjee, S., Ray, S., Maiti, S., Sen, K.K., Bhattacharyya, U.K., Kaity, S., and Ghosh, A. (2010) Interpenetrating polymer network (IPN): a novel biomaterial. *Int. J. Appl. Pharm.*, **2**, 28–34.

75 Lodha, P. and Netravali, A.N. (2005) Characterization of Phytagel® modified soy protein isolate and unidirectional flax yarn reinforced "green" composites. *Polym. Compos.*, **26**, 647–659.

76 Gil, E.S. and Hudson, S.M. (2007) Effect of silk fibroin interpenetrating networks on swelling/deswelling kinetics and rheological properties of poly(*N*-isopropylacrylamide) hydrogels. *Biomacromolecules*, **8**, 258–264.

77 Choudhary, S., White, J.C., Stoppel, W.L., Roberts, S.C., and Bhatia, S.R. Gelation behavior of polysaccharide-based interpenetrating polymer network (IPN) hydrogels. *Rheol. Acta*. doi: 10.1007/s00397-010-0499-9.

78 Kim, S.J., Park, S.J., Shin, M.S., Lee, Y.H., Kim, N.G., and Kim, S.I. (2002) Thermal characteristics of IPNs composed of polyallylamine and chitosan. *J. Appl. Polym. Sci.*, **85**, 1956–1960.

79 Lee, S.J., Kim, S.S., and Lee, Y.M. (2000) Interpenetrating polymer network hydrogels based on poly(ethylene glycol) macromer and chitosan. *Carbohydr. Polym.*, **41**, 197–205.

80 Wu, W. and Wang, D.-S. (2010) A fast pH-responsive IPN hydrogel: synthesis and controlled drug delivery. *React. Funct. Polym.*, **70**, 684–691.

81 Gupta, K.C. and Ravi Kumar, M.N.V. (2001) pH dependent hydrolysis and drug release behavior of chitosan/poly(ethylene glycol) polymer network microspheres. *J. Mater. Sci. Mater. Med.*, **12**, 753–759.

82 Langer, R. and Vacanti, J. (1993) Tissue engineering. *Science*, **260**, 920–926.

83 Tiğli, R.S and Gümüşderelioğlu, M. (2009) Evaluation of alginate–chitosan semi IPNs as cartilage scaffolds. *J. Mater. Sci. Mater. Med.*, **20**, 699–709.

84 Barber, T.A., Gamble, L.J., Castner, D.G., and Healy, K.E. (2006) *In vitro* characterization of peptide-modified p(AAm-*co*-EG/AAc) IPN-coated titanium implants. *J. Orthop. Res.*, **24**, 1366–1376.

85 Barber, T.A., Ho, J.E., De Ranieri, A., Virdi, A.S., Sumner, D.R., and Healy, K.E. (2007) Peri-implant bone formation and implant integration strength of peptide-modified p(AAM-*co*-EG/AAC) interpenetrating polymer network-coated titanium implants. *J. Biomed. Mater. Res. A*, **80**, 306–320.

86 Vimala, K., Sivudu, K.S., Mohan, Y.M., Sreedhar, B., and Raju, K.M. (2009) Controlled silver nanoparticles synthesis in semi-hydrogel networks of poly (acrylamide) and carbohydrates: a rational methodology for antibacterial application. *Carbohydr. Polym.*, **75**, 463–471.

87 D'Souza, S.F. (2001) Microbial biosensors. *Biosens. Bioelectron.*, **16**, 337–353.

88 Zeng, X., Wei, W., Li, X., Zeng, J., and Wu, L. (2007) Direct electrochemistry and electrocatalysis of hemoglobin entrapped in semi-interpenetrating polymer network hydrogel based on polyacrylamide and chitosan. *Bioelectrochemistry*, **71**, 135–141.

9
Associating Biopolymer Systems and Hyaluronate Biomaterials
Deborah Blanchard and Rachel Auzély-Velty

9.1
Introduction

Hydrophobically modified water-soluble polymers that associate in solution via physical interactions are often efficient rheology modifiers [1–3]. These are used as thickening agents in many fields of applications such as paints, cosmetics, foods, and oil recovery. The main microstructural feature of such polymers is their ability to give rise to weak intra- and intermolecular hydrophobic interactions in aqueous solutions. In the semidilute regime, these intermolecular associations are predominant. Very viscous solutions or physical gels exhibiting a shear thinning behavior can thus be obtained under certain experimental conditions. With hydrophobically modified polyelectrolytes, the viscosity of the solution can also be enhanced by several orders of magnitude upon addition of salt, unlike nonassociating polyelectrolytes with which a decrease in the viscosity is usually observed. Compared to the synthetic polyelectrolytes, natural charged polysaccharides exhibit interesting and original properties that make them a special class among polymer materials. They are generally biocompatible and biodegradable and may also exhibit a biological activity that can be advantageously exploited for applications in the biomaterials field [4]. This is highlighted in the case of chitosan (CHI) and hyaluronic acid (HA), which are two charged polysaccharides with various biomedical and cosmetic applications (Figure 9.1).

Chitosan is produced by the deacetylation of chitin obtained from the shells of crustaceans. It is one of the most abundant natural biopolymers and the only natural polycationic polysaccharide. Chitosan has great potential as a biomaterial because of its biocompatible properties. It is hydrophilic, non-antigenic, and has a low toxicity toward mammalian cells [5,6]. Chitosan is known to facilitate drug delivery across cellular barriers partly due to its mucoadhesive properties. In addition, it increases the healing rate of open wounds by stimulating the immune response and tissue reconstruction. It is also a suitable substrate for cell culture and additionally stimulates cell growth. Hyaluronic acid or hyaluronan is a natural linear acidic polysaccharide, belonging to the glycosaminoglycan family. HA exists in human and

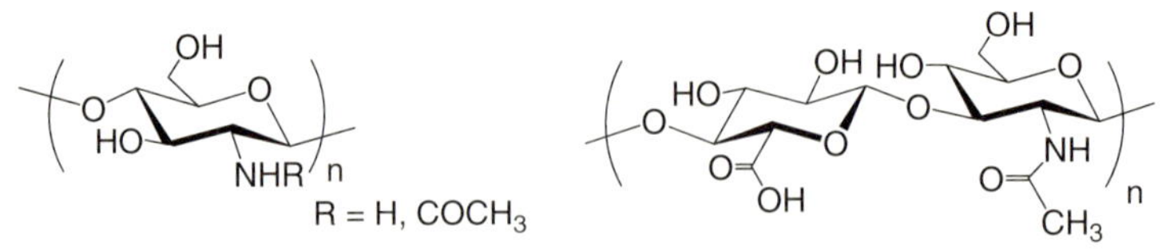

Chitosan

(1→4)-β-D-glucosamine and 1→4)-
β-N-acetyl-D-glucosamine copolymer

Hyaluronic acid

[(1→4)-β-D-glucuronic acid-(1→4)-β-
N-acetyl-D-glucosamine]$_n$

Figure 9.1 Chemical structures of chitosan and hyaluronic acid.

animal tissues as the major component of extracellular matrices (ECMs), particularly in synovial fluid, umbilical cords, and the vitreous humor of the eye [7]. Hyaluronic acid is already used in several biomedical applications such as ophthalmological surgery, treatment of the osteoarthritis of the knee, and plastic surgery [8,9]. In cataract surgery, for example, the role of HA solution is to facilitate the procedure and to protect the corneal endothelium. In the treatment of inflammatory and degenerative joint diseases, a viscoelastic HA gel is injected in the osteoarthritic knee with the purpose of restoring the lubricating and protective properties normally ensured by the presence of HA in the synovial fluid [10]. Along with fulfilling structural roles related to its lubricating and water retaining properties, HA plays an important role in many biological processes including cell proliferation, cell differentiation, morphogenesis, inflammation, and wound repair [11–14]. Therefore, over the past 15 years, much attention has been paid on the chemical modification of this polysaccharide as well as chitosan for the design of new biomaterials for drug delivery and tissue engineering. Such biomaterials include micelles, particles, thin films, or hydrogels. Recent reviews have summarized research progress in the chemical modification of hyaluronic acid [15–17] and chitosan [5,6,18–20] from the viewpoint of biomedical applications.

This chapter will focus on hydrophobically modified derivatives of CHI and HA, in view of providing a perception of the potential of such systems in the pharmaceutical and biomedical fields. A wide variety of such derivatives have been synthesized. Their self-association properties leading to either physical gels or nanoparticles (NPs) in aqueous solution can be influenced by different physical and chemical parameters including polymer concentration, salt concentration, molar mass of the polysaccharide, temperature, and size of the hydrophobic group. Thus, for a given polymer concentration radical changes in inter- and intramolecular associations may be observed only by modifying the nature of the hydrophobic group or the molar mass of the polysaccharide. Moreover, the strength of the associations can also be modulated by adding cosolute molecules such as surfactants or cyclodextrins to the polymer solution. In aqueous mixtures of a hydrophobically modified polysaccharide and a surfactant, the association strength can be increased or weakened depending on the level of surfactant addition [21]. The addition of cyclodextrin molecules to the polymer solution provides decoupling of hydrophobic interactions via inclusion complex formation with the polymer hydrophobic moieties, and this leads to a dramatic reduction of the viscoelastic response [21]. However, by using cyclodextrin polymers, it is possible to build up associative network structures [22]. In this case, cyclodextrin cavities of the cyclodextrin polymer form inclusion complexes with pendent hydrophobic groups of the polysaccharide. Acting as physical bridges between adjacent polymer chains, these inclusion complexes, in contrast to cyclodextrin monomers, strengthen the intermolecular associations as demonstrated in systems based on chitosan or hyaluronic acid [23–25]. Such polymers grafted with cyclodextrin molecules constitute original supramolecular assemblies having potential as hydrogels, gel nanoparticles, or thin films in tissue engineering and drug delivery [22]. Since this subject has been developed in recent articles [21,22], this chapter will cover self-assembling systems based solely on hydrophobically modified chitosan or hyaluronic acid. This chapter is composed of two parts. First, the synthesis and self-association of hydrophobically modified derivatives of chitosan and hyaluronic acid in aqueous solution are described and discussed. Then, recent contributions made in drug delivery applications of hydrophobically modified HA as well as in cell biology and tissue engineering of HA hydrogels are reviewed.

9.2
Synthesis and Self-Association of Hydrophobically Modified Derivatives of Chitosan and Hyaluronic Acid in Aqueous Solution

9.2.1
General Aspects of Association in Amphiphilic Polyelectrolytes

Over the past two decades, "amphiphilic polyelectrolytes" or "hydrophobically modified polyelectrolytes" have attracted considerable attention in both academic and industrial communities for their scientific and technological interest. An important reason for the interest of these polymers is the recognition that they undergo hydrophobically driven self-association in aqueous solution to form well-defined spatial structures on a nanometer scale. Depending on the polymer concentration, the structural and macromolecular characteristics of the amphiphilic polyelectrolytes, nanoparticles, or hydrogels can be obtained. A characteristic feature of hydrophobic self-associations in these systems is that hydrophobic interaction competes with electrostatic repulsion within the same polymer chain and/or between different polymer chains. As a result, the balance of attractive hydrophobic interaction and electrostatic repulsion is a critical parameter in the determination of whether the polymer undergoes hydrophobic self-association. Solution properties of amphiphilic polyelectrolytes may therefore be sensitive to changes in conditions such as ionic strength (added salt) and pH as well as shear stress and temperature. Such stimuli-responsive properties can be advantageously used to design biomaterials able to capture and deliver molecules in a controlled way.

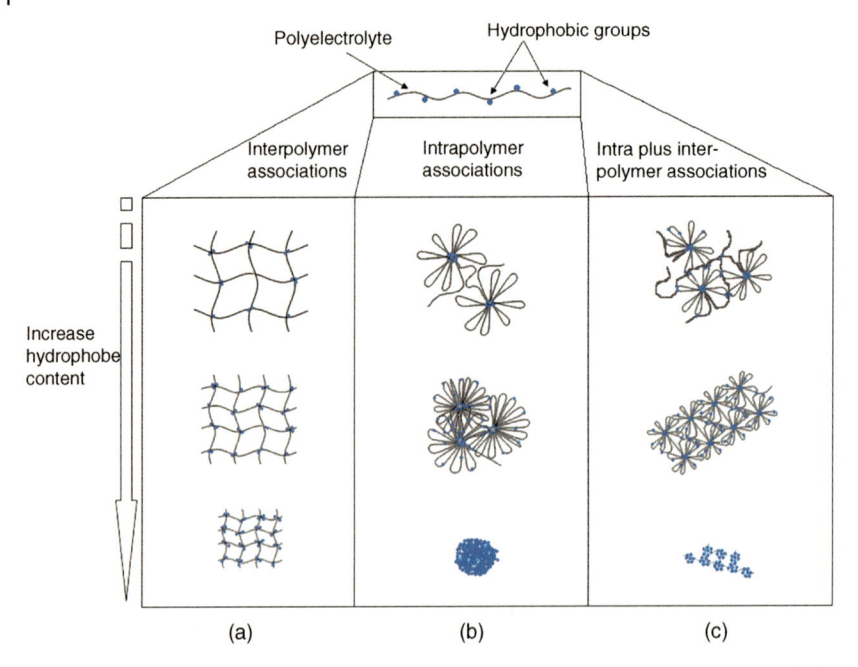

Figure 9.2 Schematic representation of hydrophobic associations of hydrophobically modified water-soluble polymers.

The type of nanostructures formed from self-association of polyelectrolytes bearing pendent hydrophobic groups in aqueous solution is strongly dependent on whether hydrophobic associations occur in an interchain or intrachain fashion. Figure 9.2a illustrates an extreme case where all polymer-bound hydrophobes undergo interchain association. Polyelectrolyte polymers with a low level of hydrophobic groups show a strong tendency for interpolymer associations even at very low polymer concentrations [26], resulting in a situation where several polymer chains are cross-linked. This hydrophobic cross-linking causes a large increase in solution viscosity, which may be followed by gelation upon further increasing polymer concentration. Figure 9.2b illustrates the other extreme case where all polymer-bound hydrophobes undergo completely intrachain associations. In this extreme case, hydrophobic associations lead to the formation of single molecular self-assemblies ("unimolecular micelles" or "unimer micelles") [27]. When the content of hydrophobes in a polymer is sufficiently low, a "flower-like" unimolecular micelle may be formed, which consists of hydrophobic core surrounded by hydrophilic loops. As the content of hydrophobes in amphiphilic polyelectrolyte is increased, the flower-like micellar structure is expected to become instable because a large portion of the surface of the hydrophobic core is exposed to the water phase. This would lead to a further collapse into a more compact micelle with a third-order structure due to secondary association of hydrophobic cores of the flower-like micelles. Between these two extreme cases, an intermediate case where intrachain associations mainly occur but a portion of hydrophobes undergoes interchain associations can also be considered (Figure 9.2c). In the latter situation, which may be more realistic, intermolecularly bridged flower-like structure may be formed. The extent of such micelle bridging may depend strongly on the content of hydrophobes in the polymer as well as the polymer concentration. As discussed below, the formation of intra- or intermolecular associations depends on different chemical (degree of substitution, nature of the hydrophobic group, and molar mass of the polysaccharide backbone) and physical parameters (polymer concentration, temperature, etc.).

The understanding of hydrophobically modified polyelectrolyte polysaccharides is an important challenge as they are encountered in many biological systems and have potential useful applications in biomedical and pharmaceutical applications as discussed below.

9.2.2
Synthesis and Behavior in Aqueous Solution of Hydrophobically Modified Water-Soluble Derivatives of Chitosan

The potential uses of chitosan derive from its unique chemistry: it is a polycation at pH <6 by protonation of the $-NH_2$ function on the C2 position of the D-glucosamine repeating unit ($pK_a = 6.0$) [28,29]. It is the only pseudonatural cationic polyelectrolyte that is largely used in different domains as aqueous solutions, hydrogels, films, and fibers. Hydrophobic chitosans were prepared by grafting alkyl chains. The procedure generally used for the synthesis of such derivatives relies on a reductive amination reaction using aldehydic alkyl chains [30,31]. It has the advantage of producing selectively modified polymers under mild and homogeneous conditions, that is, allowing a random substitution with no modification of the degree of acetylation nor of the polymerization degree.

Preliminary rheological, dynamical, and structural properties of such chitosan derivatives bearing dodecyl (C_{12}) chains with degrees of substitution (DS, number of moles of substituent per mole of repeating unit) of 0.025, 0.05, and 0.1 were reported by Nyström and coworkers [30,32]. The authors observed the formation of physical gels by increasing polymer concentration. Notably, the value of the gel concentration was found to decrease with increasing hydrophobicity of the polymer indicating that physical gelation is related to the formation of hydrophobic domains made of alkyl chains playing the role of interchain junctions. Moreover, it was shown that a temperature rise resulted in a weakening of the networks. However, contrasting results were obtained by Desbrières who observed an increase in viscosity or a constant value of the viscosity with the temperature from alkylated derivatives of chitosan modified also with C_{12} chains as well as with hexyl (C_6) and octyl (C_8) chains [31]. Considering heterogeneity between batches of chitosan resulting by the deacetylation of chitin under heterogeneous conditions and the fact that initial chitosan chains tend to associate by forming hydrophobic domains [33,34], the association process may differ according to the origin of the chitosan sample as well as the degree of substitution.

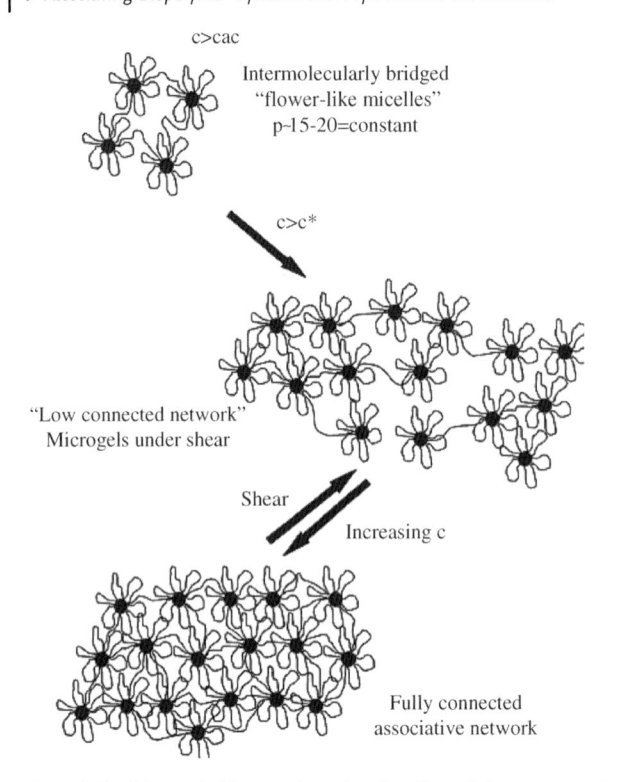

Figure 9.3 Schematic diagram depicting the effect of the concentration and of the shear on the structure of the associative network [35].

In another work focusing on the structural and rheological properties of chitosan bearing octyl (C_8) chains with a DS of 0.02, Esquenet *et al.* reported the formation of flower-like micelles of chitosan bearing octyl (C_8) chains with a DS of 0.02 [35,36]. According to the authors, as the DS is low, the average distance between two hydrophobic groups (~25 nm) allows the chain to easily form loops and then flower-like micelles. By varying the polymer concentration, the authors identified three regions on the phase diagram of the chitosan derivative in aqueous solution (Figure 9.3): (i) a supernatant phase (unimer phase) at low polymer concentration ($C_p <$ cac, cac being the polymer concentration for an onset of intermolecular association); (ii) solutions of intermolecularly bridged flower-like micelles at intermediate polymer concentration (cac $< C_p < c^*$, with c^* being the onset concentration for gel-like behavior); (iii) an associative gel phase at high polymer content ($C_p > c^*$). It was found that the unimer concentration depends on the total polymer concentration, whereas the aggregation number seems to be constant over the whole dilute micellar regime.

The formation of intermolecularly bridged flower-like micelles at intermediate polymer concentration was also suggested by Hu *et al.* who synthesized and investigated the self-aggregation behavior in aqueous solution and antibacterial activity of chitosan derivatives that were selectively N-acylated with acetic, propionic, and hexanoic anhydrides under homogeneous conditions (aqueous acetic acid) [37]. It was found that the *in vitro* inhibitory activity of chitosan and its N-acylated derivatives (DS ~0.5) against Gram-positive and Gram-negative bacteria at pH 5.4 decreased in the following order: chitosan > N-hexanoyl chitosan > N-propionyl chitosan ≥ N-acetyl chitosan. The lower inhibitory effect of the N-acylated derivatives compared to the native chitosan could be explained by the decrease in the positively charged density by N-substitution reaction, in which the amino groups were partly converted to the amide groups. The influence of the length of acyl chains on the antimicrobial activity was attributed to the self-aggregation of N-acylated chitosans by the authors.

Indeed, by comparing the viscosity of solutions of N-acylated chitosans at high polymer concentrations, they observed a much larger increase in the solution viscosity for the N-hexanoyl chitosan derivative compared to the N-acetyl and N-propionyl chitosan derivatives as well as to the native chitosan (Figure 9.4a). This revealed the formation of hydrophobic interactions. Considering that intrachain associations are predominant in the dilute regime, intermolecularly bridged flower-like micelles were assumed to form at intermediate polymer concentration (Figure 9.4b). As a result, it was hypothesized that the density of protonated amino groups increased, thereby promoting the antimicrobial activity. In this work, the stronger association was obtained from the chitosan sample bearing long alkyl chains with a relatively high DS (~0.5).

From Figure 9.4a, it can be noticed that the viscosity values at different shear rates measured for the solutions of N-acetyl and N-propionyl chitosan are lower than

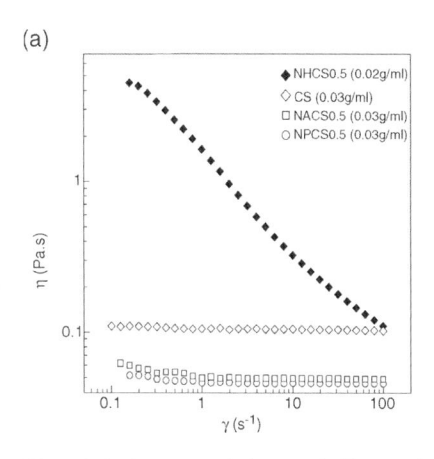

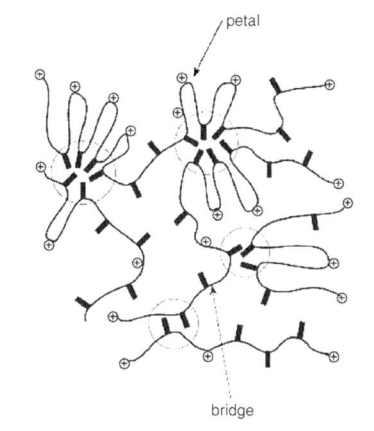

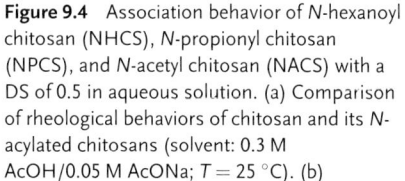

Figure 9.4 Association behavior of N-hexanoyl chitosan (NHCS), N-propionyl chitosan (NPCS), and N-acetyl chitosan (NACS) with a DS of 0.5 in aqueous solution. (a) Comparison of rheological behaviors of chitosan and its N-acylated chitosans (solvent: 0.3 M AcOH/0.05 M AcONa; $T = 25\ °C$). (b) Schematic representation of intermolecularly bridged "flower-like" micelles formed at intermediate polymer concentration for N-acylated chitosans with desired amount of hydrophobic acyl chains. (Adapted from Ref. [37].)

those of the parent polymer solution, suggesting the formation of more or less compact aggregates. Indeed, although the methyl and propionyl chains are too short to form hydrophobic domains allowing the formation of a three-dimensional network, the alkyl chains tend to associate in water to minimize interaction with water, leading to aggregates.

Much stronger hydrophobic associations, leading to the formation of stable nanoparticles, could be achieved by grafting bulky hydrophobic groups such as bile acid derivatives on glycol chitosan (GCHI), a commercially available derivative of chitosan possessing enhanced water solubility at physiological pH [38–43]. Notably, it was shown that GCHI particles obtained by coupling 5β-cholanic acid with chitosan samples having different molar masses (250, 100, and 20 kg/mol) showed different *in vivo* stability and tumor targeting characteristics as visualized by a noninvasive animal imaging system (Figure 9.5) [44]. The near-infrared (NIR) fluorescence intensities of the GCHI NPs prepared from chitosan samples having molar masses of 20 and 100 kg/mol rapidly decreased in the whole body within 6 h postinjection (Figure 9.5a and b), possibly due to clearance through urinary excretion. However, the high fluorescence intensity of GCHI NPs prepared from chitosan having a molar mass of 250 kg/mol was maintained in the whole body up to 3 days, implying prolonged circulation time and effective tumor accumulation (Figure 9.5c). Thus, the blood circulation time of the "GCHI-250 NP" was shown to be sufficiently long to produce tumor selectively [44]. This study thus demonstrated potential GCHI NPs for anticancer drug delivery by passive tumor targeting.

Importantly, the NPs had a sufficient deformability to pass through the smaller pore size, suggesting that their size may be changed in the bloodstream. This property was considered to explain the fact that the NPs are not trapped by the mononuclear phagocytic system in liver and spleen [45]. As a result, docetaxel-loaded NPs showed higher antitumor efficacy such as reduced tumor volume and increased survival.

9.2.3
Synthesis and Behavior in Aqueous Solution of Hydrophobically Modified Water-Soluble Derivatives of Hyaluronic Acid

Over the past 10 years, many efforts have been made to control the rheological properties of HA for broader biomedical applications. The rheological properties of native HA have been studied extensively [46–48]. Above the overlap concentration in buffer solution, that is, in the semidilute regime, it is a viscoelastic solution and an important dependence of the viscosity η occurs as a function of the shear rate. From the dynamic rheological measurements, HA thus exhibits an elastic behavior with the storage modulus (G') higher than the loss modulus (G'') in the high-frequency domain. Below a characteristic frequency ω_0, one finds $G'' > G'$, reflecting a viscous behavior in the low-frequency domain (Figure 9.6). Elastic responses can help in viscosupplementation (Figure 9.6) but complicate facile injection of HA to injury sites, and for certain applications, it would be useful to broaden the frequency range over which the material behaves elastically. Strategies recently developed for tuning

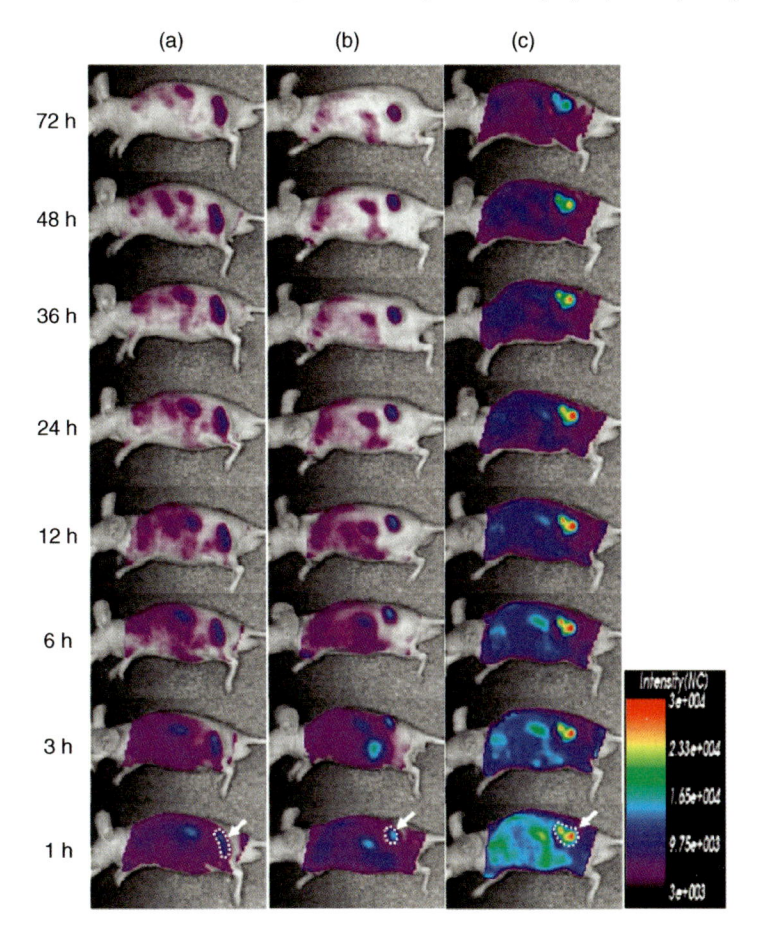

Figure 9.5 *In vivo* noninvasive NIR fluorescence images of real-time tumor targeting characteristics of glycol chitosan nanoparticles. *In vivo* fluorescence imaging of athymic nude mice bearing subcutaneous SCC7 tumors after intravenous injection of GCHI NP prepared from chitosan samples having molar masses of 20 kg/mol (a), 100 kg/mol (b), and 250 kg/mol (c). The tumor location is specified with an arrow. (Adapted from Ref. [44].)

the viscosity and conditions under which viscous or elastic responses are observed consisted in the synthesis of amphiphilic systems of HA.

As reported in a recent review on alkyl (aryl) derivatives of HA, several methodologies have been proposed for the synthesis of alkyl derivatives of HA [49]. Among these methods, the alkylation (arylation) of HA based on the reaction in polar organic solvents such as dimethyl sulfoxide (DMSO) or dimethylformamide (DMF) of an alkyl halide (i.e., ethyl, propyl, pentyl, benzyl, dodecyl, or octadecyl bromide) with the carboxylic group of HA has been extensively used [50–52] (Figure 9.7). To make HA soluble in DMSO or DMF, HA was preliminary transformed into the tetrabutylammonium (TBA) salt to shield highly anionic carboxylic acid groups and

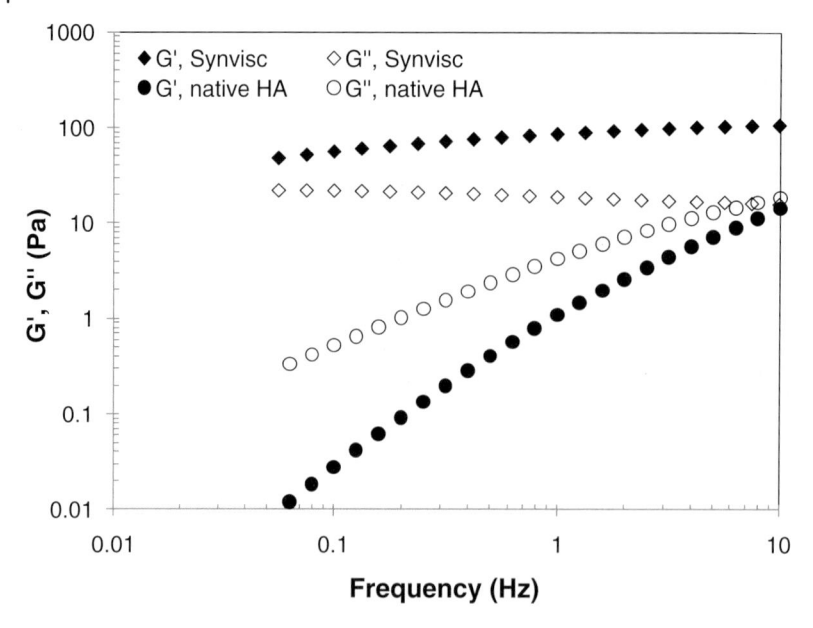

Figure 9.6 Comparison of the storage and loss moduli as a function of frequency for solutions of native HA (MW = 1×10^6 g/mol; polymer concentration $C_p = 10$ g/l) and Hylan G-F 20 (Synvisc® from Genzyme) in phosphate buffered saline (PBS, pH 7.4, 0.15 M NaCl) at 25 °C. Synvisc® was engineered to act as a synovial prosthesis for the treatment of osteoarthritis.

to disrupt inter/intramolecular hydrogen bonds in HA. Alternatively, the preparation of esters was based on the reaction of HA with palmitoyl chloride in DMF in the presence of pyridine [53]. Other methods of hydrophobization in DMSO or DMF leading to the formation of ether [54,55], carbamoyl [56], and amide [57–59] derivatives of HA were also developed. In order to avoid the problems with solvent removal, our group proposed an alkylation procedure having the advantage of producing selectively modified HA in aqueous solution under mild conditions.

This was based on the selective functionalization of HA by reactive dihydrazide groups [60,61] followed by the coupling with an aldehydic alkyl chain using reductive

Figure 9.7 Alkylation (arylation) of HA based on ester bond formation in polar organic solvents.

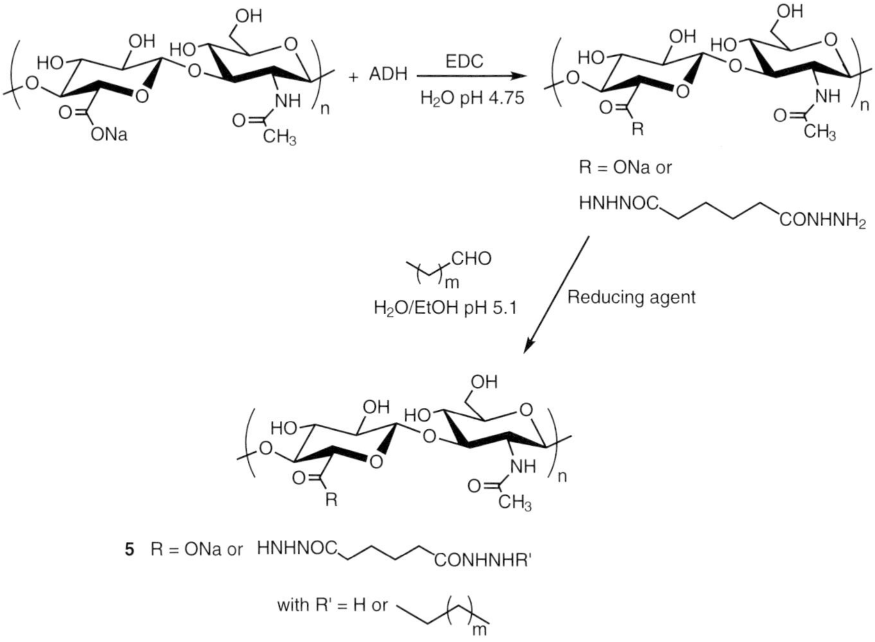

Figure 9.8 Synthesis of alkylamino hydrazide HA derivatives. HA is first modified by hydrazide groups based on its coupling reaction with adipic acid dihydrazide (ADH) using 1- ethyl-3-[3-(dimethylamino)propyl]carbodiimide (EDC). The resulting HA derivative is then reacted with aldehydic alkyl chains ($m = 4, 6, 8$) under reductive amination conditions.

amination conditions (Figure 9.8). Alkyl carbamates were also prepared in aqueous solution by the reaction of HA with alkyl amines using the cyanogen bromide activation method [62].

The alkylamino hydrazide derivatives of HA were shown to have different thickening properties in aqueous solution depending on the nature of the hydrophobic group and the degree of substitution. Moreover, different viscometric features could be observed as a function of the molar mass of HA. Alkyl derivatives prepared from a HA sample of 600 000 g/mol (HA-600) exhibited a much higher tendency to self-aggregate than their counterparts prepared from a HA sample of 200 000 g/mol (HA-200). Differences in the association behavior depending on the polymer chain lengths were also observed for other hydrophobically modified HA derivatives [63,64]. This was related to competitive forces existing between intramolecular interactions and internal stress according to intrinsic flexibility of the polymer backbone [64]. It was thus suggested that lower masses impair intramolecular interactions, because of the decrease in both the number of hydrophobic groups per macromolecule and intrinsic flexibility [64].

The best thickening properties were obtained from decylamino hydrazide derivatives, leading to the formation of elastic hydrogels in aqueous solution (Figure 9.9) [65].

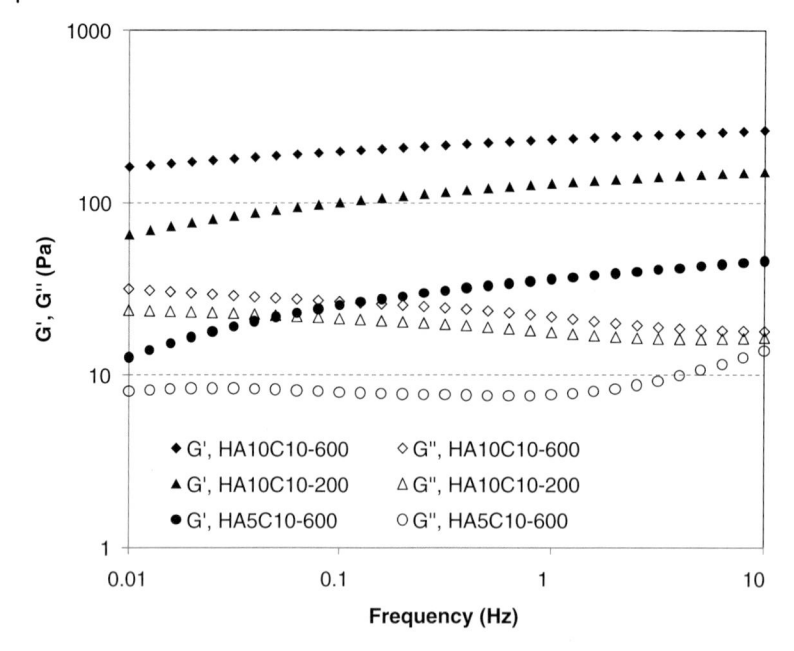

Figure 9.9 Comparison of the storage and loss moduli as a function of frequency for solutions of decylamino hydrazide derivatives of HA in PBS at 25 °C and at a polymer concentration of 10 g/l. The derivatives are designated as HAxC10, where x reflects the degree of substitution ($x = 100$ DS) of the decyl (C_{10}) chains [65].

Thickening properties were also observed by esterification of HA with dodecyl (C_{12}) or octadecyl (C_{18}) bromide [52]. The influence of the length of alkyl chain, its content on HA, and polymer concentration was well identified and it was shown that some derivatives were potential materials useful for the treatment of the osteoarthritis of the knee and cartilage repair. Dynamic rheological measurements demonstrated a gel-like behavior for aqueous solutions of HA (MW = 480 000 g/mol) modified with C_{18} chains and with a degree of substitution of 0.02 (HA-C_{18}-2) in 0.15 M NaCl (polymer concentration $C_p = 4$ and 7 g/l), contrary to the solutions of HA-C_{18}-1 (DS = 0.01) and HA-C_{12}-5 (DS = 0.05) at concentrations in the range of 6–12 g/l, which only exhibit a viscous character. Indeed, from Figure 9.10, it can be seen that the storage (elastic) modulus (G') is larger than the loss modulus (G'') within the whole range of frequencies covered for the HA-C_{18}-2 solution. Moreover, the values of the storage and loss moduli are much higher than those obtained for the solution of HA-C_{18}-1, although the concentration of the latter is higher than that of HA-C_{18}-2. These rheological data thus demonstrate the formation of physical network made of loosely cross-linked HA. Interestingly, a similar rheological behavior is observed in the case of Hylan G-F 20 (Synvisc®) (Figure 9.6). Hylan G-F 20 consists of chemically cross-linked HA that was specifically designed to have a high percentage of elasticity over the entire frequency of physiological movement [10]. Thus, when Hylan is injected into the osteoarthritic joint, it raises

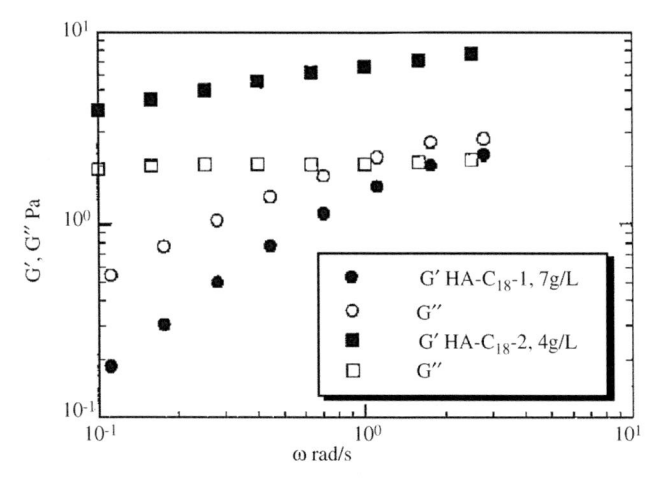

Figure 9.10 Comparison of the storage and loss moduli as a function of frequency for solutions of HA-C_{18}-1 and HA-C_{18}-2 derivatives in 0.15 M NaCl at 37 °C.

the percent of elasticity of the pathological synovial fluid (which has a viscous behavior due to a decreased molecular weight and concentration of HA) in the normal range.

Stable hydrogels at polymer concentrations higher than 3 g/l were also obtained from amide derivatives of HA possessing C_{16} side chains with a DS of 0.01–0.03 [59].

It should be noted that in the case of alkylamino hydrazide derivatives of HA, the formation of physical hydrogels was observed for derivatives possessing shorter chains (C_{10}) [65]. The ability of the latter derivatives to self-associate in aqueous solution similarly to those directly modified by longer alkyl chains (C_{16} or C_{18}) may be attributed to the presence of the adipic spacer arm and/or the different experimental conditions used for the grafting of alkyl chains. Nevertheless, it can be noticed that all these derivatives have the common feature to be modified with low degrees of substitution (DS ≤ 0.2).

Indeed, functionalization of HA with hydrophobic molecules at high degrees of substitution promotes the formation of aggregates with a compact conformation in aqueous solution as described in the case of HA esters modified with the steroidal anti-inflammatory drug 6α-methylprednisolone [64] and benzyl moieties [66]. Thus, the zero-shear viscosity of a semidilute solution of HA (MW = 150 000 g/mol) benzyl ester (DS = 0.5) was found to be lower than that of initial HA [66]. Moreover, nanosized spherical micelles were obtained from HA–paclitaxel conjugates in aqueous solution. The micelles showed an average diameter of 196 ± 9.6 nm with a narrow size distribution. Paclitaxel, which is a very hydrophobic anticancer drug, was directly conjugated to HA in DMSO by forming HA/poly(ethylene glycol) (PEG) nanocomplexes with a size of 120 ± 6.3 nm [67]. This strategy indeed allowed solubilization of HA and other biomacromolecules in DMSO by formation of inter- and intrahydrogen bonds [68]. This solubilization method was also used to graft hydrophobic and biodegradable poly(lactic acid-co-glycolic acid) (PLGA) on HA

(MW = 17 000 g/mol) [69]. The resulting HA-g-PLGA copolymers (with $0.05 \leq$ DS ≤ 0.26) self-assembled in aqueous solution to form nanosized micellar aggregates with average diameters ranging from 98.4 to 539.4 nm depending on the degree of substitution and molar mass of PLGA. Thus, the size of the micelles was found to decrease when the DS was increased. This phenomenon was also observed in the case of HA NPs that were formed by the self-assembly of HA grafted with 5β-cholanic acid [70]. As discussed earlier in this chapter, those nanoparticles may be useful as biocompatible and biodegradable carriers for the targeting of anticancer drugs.

9.3
Design of Novel Biomaterials Based on Chemically Modified Derivatives of Hyaluronic Acid

Advances in healthcare and surgery, specifically novel opportunities associated with drug delivery using micro/nanoparticles and tissue engineering using scaffold materials, have given rise to a new research direction, namely, biomaterial–tissue interaction. In this area, HA is a very attractive polymer to mimic the native cellular environment and promote communication with the host tissue. As mentioned earlier, this glycosaminoglycan is emerging more and more as a key molecule in the regulation of many cellular and biological processes, such as cell proliferation, cell differentiation, morphogenesis, inflammation, and wound repair. In particular, HA is able to stimulate cells via binding to cell surface receptors. Actually, there have been many reports on HA receptors that play important biological roles such as endocytosis, degradation, and signal transduction. Cluster determinant 44 (CD44) [71], receptor for hyaluronate-mediated motility (RHAMM) [14,72], HA receptor for endocytosis (HARE) [73,74], and lymphatic vessel endothelial hyaluronan receptor-1 (LYVE-1) [75] have been identified as HA receptors for various biological functions. The binding of HA to RHAMM triggers the signal transduction for cell trafficking [72]. Homeostasis of HA is maintained by receptor-mediated degradation at LYVE-1.

In cancer cells, binding of HA to CD44 is involved in tumor growth and spreading [76,77]. In addition, disruption of HA–CD44 binding was shown to reduce tumor progression [78–80]. Administration of exogenous HA resulted in arrest of tumor spreading [80].

Taking advantage of the unique biological properties of HA, recent studies demonstrated promising applications of HA as drug delivery vehicles and tissue engineering scaffolds. These include nanoassemblies and hydrogels depending on the clinical mechanism of action and application.

In the following, we present in the first part recent successes of nanoassemblies based on HA for anticancer therapy. We further present novel approaches for the controlled drug encapsulation and delivery based on multicompartmental systems consisting of layer-by-layer (LbL) capsules.

In the second part, we focus on the design and use of HA hydrogels as scaffolds for two- and three-dimensional cell culture and tissue engineering. Inspired by advanced understanding of how manipulations in material chemistry and structure influence cellular interactions, recent work aimed at designing HA hydrogels with tunable mechanical and chemical properties on the time- and length scales of cell development, to truly mimic the ECM. Therefore, in this part, after addressing the main cross-linking strategies of HA used to create microenvironments for cells, the engineering of biological functionality into HA hydrogels and the approaches recently developed for the patterning of HA for the presentation of multiple signals in a temporally dynamic and/or spatially patterned manner will be discussed.

9.3.1
Nanoassemblies Based on Amphiphilic Hyaluronic Acid

Controlled delivery of hydrophilic and hydrophobic drugs is currently a great challenge in the field of nanobiotechnology. Micro/nanoparticulate drug carriers can shield the drug from degradation, improve its biodistribution, and facilitate targeted delivery. The problem of delivery is especially crucial for water-insoluble organic compounds, which constitute a great part of the currently available drugs, whether anti-inflammatory or anticancer [81,82].

In particular, since HA can specifically bind to various cancer cells that overexpress CD44, many studies have focused on the pharmaceutical applications of HA for anticancer therapeutics. Most HA–drug conjugates have been developed for cancer chemotherapy as macromolecular prodrugs. HA has also been conjugated onto various drug-loaded nanoparticles for use as a targeting moiety [83,84]. These HA conjugates containing anticancer agents such as paclitaxel [85–87] and doxorubicin [88] exhibited enhanced targeting ability to the tumor and higher therapeutic efficacy compared to free anticancer agents. Based on these results, recent studies focused on the design of HA-based micro/nanoparticulate drug carriers combining the intrinsic biological properties of HA with the solubilization and transport properties of the particles. As discussed above, one promising approach is based on the covalent grafting of "bulky" hydrophobic groups on the polysaccharide backbone leading to formation of nanosized particles consisting of a hydrophobic core surrounded by a hydrophilic HA shell, able to entrap hydrophobic drug molecules (Table 9.1).

As can be seen from Table 9.1, a few self-assembled nanoparticles based on amphiphilic HA have been investigated for their potential use in cancer therapy. Since they combine the intrinsic property of HA as a targeting moiety toward cancer cells with the ability to encapsulate a large amount of hydrophobic drug, they appear as promising carriers for anticancer therapy. Paclitaxel was selected for HA–drug conjugation in organic solvents by forming nanoscale complexes using biocompatible PEG. Analyses by atomic force microscopy and transmission electron microscopy demonstrated the formation of HA–paclitaxel NPs in aqueous solution by hydrophobic interaction between paclitaxel molecules as well as slight cross-linking of HA in HA/PEG nanocomplexes by an amine–acid coupling process. *In vitro* cytotoxicity of HA NPs against three different cell lines (HCT-116, MCF-7, and NIH-3T3) monitored by a cell viability assay indicated a higher cytotoxic effect of HA

Table 9.1 Self-assembled hyaluronic acid nanoparticles for active tumor targeting.

Type of NP	Strategy of coupling with HA	MW (HA) (kg/mol)	DS of modified HA	Average diameter of NPs	Nature of the drug loaded	Biological properties
HA–paclitaxel [67]	Amide bond formation on HA/PEG complex in DMSO	64	0.05	~230 nm	Paclitaxel	Cytotoxicity of NPs toward HCT-116 cancer lines overexpressing HA receptor
HA–PLGA [69]	Amide bond formation on HA/PEG complex in DMSO	17, 64	0.05–0.26	From ~100 to ~540 nm	Doxorubicin	Cytotoxicity of doxorubicin-loaded NPs toward HCT-116 cancer lines overexpressing HA receptor
HA–cholanic acid [70]	Amide bond formation from aminoethyl 5β-cholanoamide in aqueous solution	234.4	0.02–0.10	From ~237 to ~424 nm	—	High targeting efficiency of HA NPs to SCC7 tumor tissue

NPs than paclitaxel in the Taxol formulation (Cremophor EL) with HCT-116 and MCF-7 cells overexpressing CD44. This result was attributed to HA receptor-mediated endocytosis. Confocal microscopy observations together with flow cytometry experiments suggested that the NPs significantly enhanced the extent of apoptosis-induced cell death. The formation of self-assembled HA NPs was also observed by replacing paclitaxel by PLGA. From AFM and TEM images, these HA NPs, prepared by a dialysis method, exhibited a spherical shape. It was found that the degree of substitution and molecular weight of grafted PLGA chains mainly affected the diameter of HA–PLGA nanoparticles, ranging from 98.4 to 539.4 nm from DLS measurements. Confocal microscopy observations suggested that the HA NPs were taken up by HCT-116 cells via an endocytic pathway. The role of HA receptor-mediated endocytosis in efficient intracellular delivery of HA–PLGA nanoparticles was additionally demonstrated by a competitive inhibition experiment performed by adding free HA in the incubating media to block HA receptors on the surface of HCT-116 cells. Similar to HA–paclitaxel NPs, flow cytometry experiments showed that the HA–PLGA NPs loaded with doxorubicin exhibited much higher cytotoxicity than the free drug, which was related to the enhanced cellular uptake of HA-g-PLGA NPs.

5β-Cholanic acid was also used recently to prepare amphiphilic HA conjugates, as mentioned above. The latter derivatives lead after sonication in water to spherical particles whose size decreases as the DS increases (Figure 9.11) [70]. Interestingly, the mean diameters of the nanoparticles exhibited no significant changes over the course of 7 days when stored under physiological conditions, suggesting they are highly stable. Moreover, studies by confocal microscopy of the HA NPs fluorescently

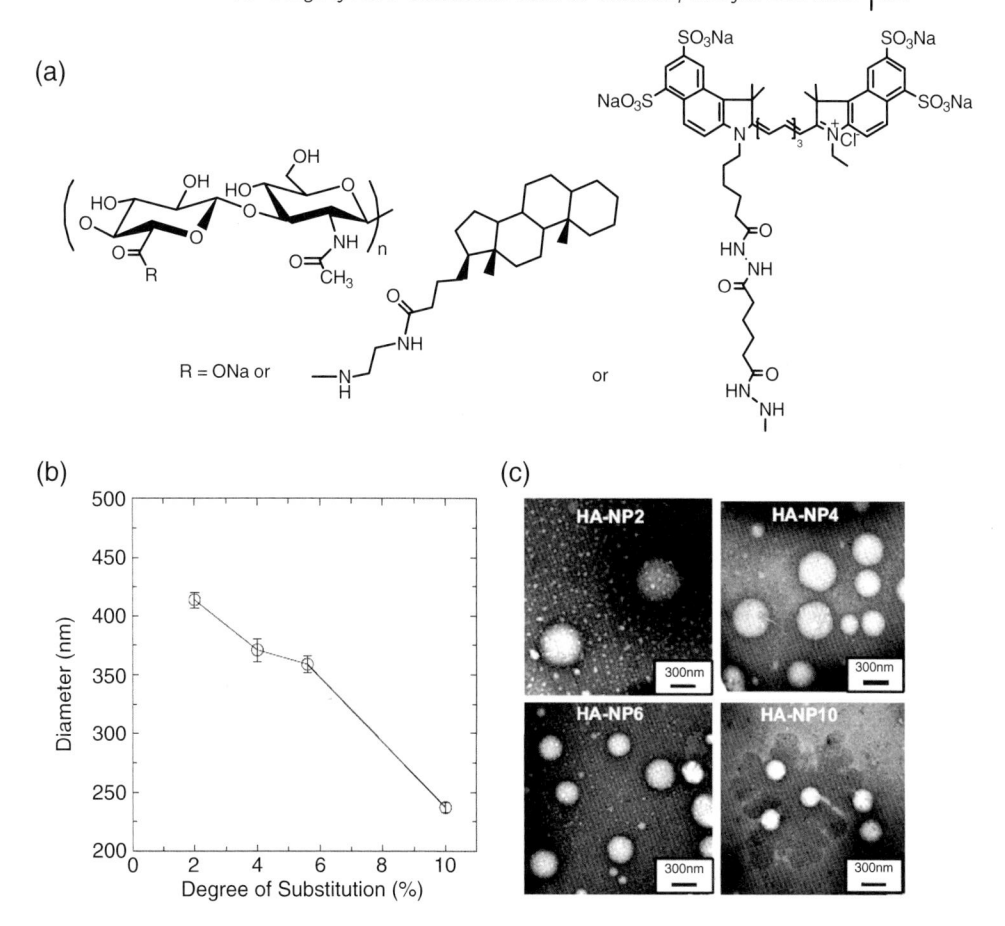

Figure 9.11 Hyaluronic acid nanoparticles obtained from HA–cholanic acid. (a) Structure of Cy5.5-labeled HA–cholanic acid. (b) Particle size of HA NPs with different amounts of 5β-cholanic acid. (c) Morphology of HA NPs, measured using transmission electron microscopy. (Adapted from Ref. [70].)

labeled with the NIR dye, cyanine 5.5 (Cy5.5), incubated with cancer cells (SCC7) overexpressing CD44 demonstrated cellular uptake.

No significant differences were found in uptake among the HA NPs prepared from HA having different DS, indicating that chemical modification of HA with 5β-cholanic acid does not affect binding affinity of HA to CD44. In contrast, when the SCC7 cells were pretreated with a high dose of free HA to block CD44 prior to Cy5.5-labeled HA NP treatment, cellular uptake of HA NPs was found to be inhibited. Following systemic administration of Cy5.5-labeled HA NPs into a tumor-bearing mouse, their biodistribution was monitored as a function of time using a non-invasive near-infrared fluorescence imaging system. Irrespective of the particle size, significant amounts of HA NPs circulated for 2 days in the bloodstream and were

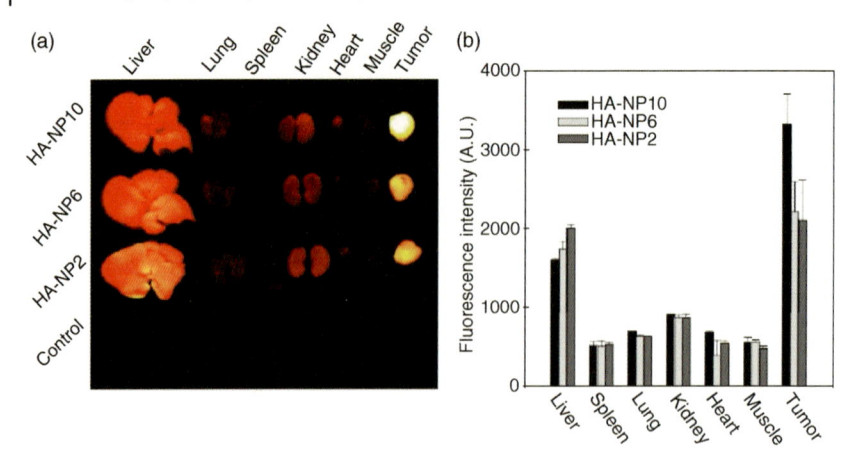

Figure 9.12 (a) *Ex vivo* fluorescence images of normal organs and tumors collected at 2 days postinjection of HA NPs. (b) Quantification of the *ex vivo* tumor targeting characteristics of HA NPs in tumor-bearing mice. Error bars represent standard deviation ($n = 3$) [70].

selectively accumulated into the tumor site (Figure 9.12). The smaller HA NPs were able to reach the tumor site more effectively than larger HA NPs. Interestingly, the concentration of HA NPs in the tumor site was dramatically reduced when mice were pretreated with an excess of free HA. This study thus showed that HA NPs were effectively accumulated into the tumor site by a combination of passive and active targeting mechanisms. However, a significant portion of HA NPs was also found in the liver site, possibly owing to their cellular uptake by phagocytic cells of the reticuloendothelial system and by liver sinusoidal endothelial cells expressing another HA receptor (HARE) [89]. Therefore, PEGylation of HA NPs was performed by varying the DS of PEG in order to control their surface property [89].

Although PEGylation of HA NPs reduced their cellular uptake *in vitro*, larger amounts of nanoparticles were taken up by cancer cells overexpressing CD44 than by normal fibroblast cells. The *ex vivo* images of the organs using an optical imaging technique after the intravenous injection of Cy5.5-labeled nanoparticles into normal mice demonstrated that PEGylation could effectively reduce the liver uptake of HA NPs and increase their circulation time in the blood. When the nanoparticles were systemically administered into tumor-bearing mice for *in vivo* real-time imaging, the strongest fluorescence signals were detected at the tumor site of the mice for the whole period of time studied, indicating their high tumor targetability. Interestingly, PEGylated HA NPs were more effectively accumulated into the tumor tissue up to 1.6-fold higher than bare HA NPs. These results suggest that PEGylated HA NPs can be useful as a means for cancer therapy and diagnosis.

Besides nanoparticles, LbL hollow microcapsules have recently emerged as a novel potential therapeutic tool. These capsules, also called multilayer capsules, are formed by deposition of an LbL film onto a sacrificial template that is dissolved after film deposition. One of the advantages of these multicompartmental systems is the possibility to introduce a high degree of multifunctionality within their nanoshell by the nature of the polyelectrolytes and assembly conditions used for their preparation. The shell can consist of various types of polymers, but these have to sustain the core removal step. Recent developments of polyelectrolyte microcapsules in life sciences include the use of polypeptides and polysaccharides as shell components, as these are biocompatible and biodegradable but these require the core to be removed under mild conditions. Nevertheless, polyelectrolyte microcapsules based on polysaccharides are only emerging, due to the inherent difficulties in preparing such capsules, which are related to the "hydrogel type" and softness of the polysaccharide multilayer films [90]. In this context, previous work in our laboratory focused on the design of biocompatible and biodegradable capsules based on hyaluronic acid [91–93]. In spite of its highly hydrated nature and tendency to form soft hydrogel-type multilayer films, capsules consisting of HA and poly(allylamine) (PAH) could be successfully obtained without chemically cross-linking the layers [93]. These capsules were prepared using calcium carbonate particles as sacrificial templates due to their biocompatibility and their ability to be readily dissolved under mild conditions. As the goal was to obtain fully biodegradable capsules, PAH was then replaced by a biocompatible and biodegradable polypeptide, namely, poly(L-lysine) (PLL), but shell cross-linking was required to improve capsule stability [92]. These capsules containing HA could be taken up by the phagocyting cells and, in contrast to other types of capsules, deformed fast inside of cells, which offers perspective of using such capsules as intracellular delivery carriers. Based on these results, we then proposed to prepare capsules made entirely of polysaccharides, combining their biodegradable and biocompatible properties together with specific biological activities. To this end, we selected chitosan as a polycationic partner of HA. To circumvent the inherent drawback of CHI (low solubility in physiological conditions due to its pK_a at 6), which hinders the film buildup at neutral pH as well as causes problems for subsequent core dissolution, we synthesized water-soluble quaternized chitosan (QCHI) derivatives. The latter cationic polymers could be successfully LbL assembled with HA to form stable capsules without cross-linking the multilayer assembly [91]. We next used these water-soluble quaternized chitosan derivatives to prepare capsules that can selectively encapsulate poorly water-soluble drugs within the nanoshell, leaving the aqueous cavity available for incorporation of water-soluble active substances. For this purpose, we developed a versatile method based on the very high affinity of hydrophobic molecules for hyaluronic acid modified with alkyl chains [65] (Figure 9.13).

The formation of microcapsule from alkylated hyaluronic acid and QCHI in aqueous conditions was successful and highly efficient for highly substituted polysaccharide samples (i.e., QCHI with $DS = 1.1$ or 1.3 and HA20C10 with $DS = 0.20$) (Figure 9.14) [94]. Furthermore, these hydrophobic nanoshells exhibited high trapping capacity for the hydrophobic model drug nile red, as probed by its fluorescence. In addition, they exhibited a very high stability in a physiological medium. Given the versatility of the LbL assembly to produce nanoshells and the durability of drug entrapment, these hydrophobic polysaccharide nanoshells open new avenues for applications in nanomedicine. Such systems offer opportunity to simultaneously deliver several types of hydrophilic and hydrophobic drugs.

254 | *9 Associating Biopolymer Systems and Hyaluronate Biomaterials*

Figure 9.13 Solubilization of a hydrophobic dye, nile red (NR), in hydrophobic nanodomains of alkyl chains grafted on HA, monitored by fluorescence intensity measurements of aqueous solutions of alkylated HA derivatives ($C_p = 2$ g/l in PBS). The different polymers obtained from the linear alkyl chains are designated as HAxCy, where x reflects the degree of substitution ($x = 100$ DS) and y is the number of carbon atoms of the grafted alkyl chain. The derivatives bearing branched citronellyl chains are designated as HAxC10Br [65].

9.3.2
Hydrogels for Cell Biology and Tissue Engineering

9.3.2.1 Strategies for the Cross-Linking of HA to Obtain Scaffolds for Cells

HA has been chemically cross-linked to create mechanically robust materials as scaffolds for cell culture and tissue engineering by several strategies, which can be

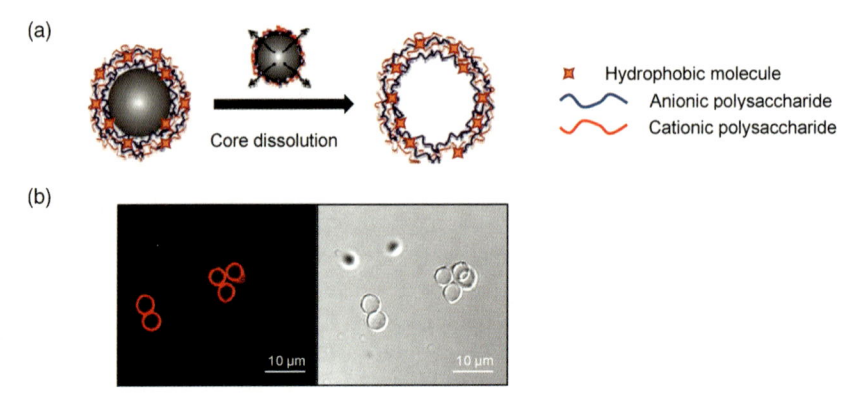

Figure 9.14 Hydrophobic shell loading of polysaccharide capsules. (a) Polysaccharide capsules containing hydrophobic molecules in the shell are prepared by layer-by-layer deposition of oppositely charged polysaccharides (i.e., quaternized chitosan and alkylated HA) on calcium carbonate particles followed by core dissolution under mild conditions. The hydrophobic molecules are embedded in the shell by precomplexation with alkylated HA. (b) Confocal microscopy images of (HA20C10/QCHI)$_5$ microcapsules exhibiting a highly localized red fluorescence in their nanoshell due to the specific entrapment of nile red molecules in hydrophobic nanodomains.

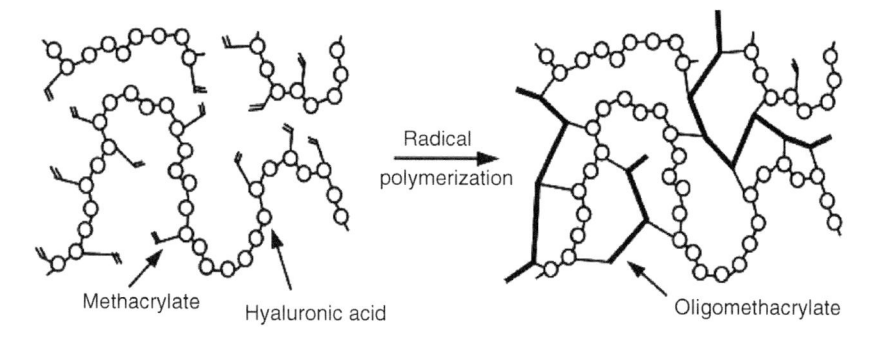

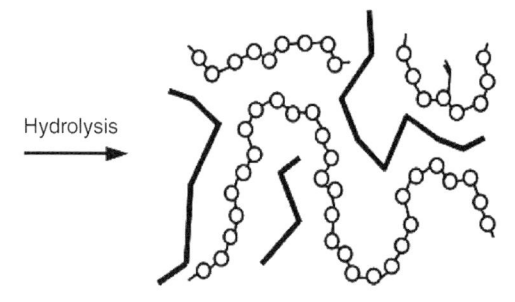

Figure 9.15 Schematic representation of the formation of HA-based hydrogels by radical polymerization of methacrylate groups grafted on HA. (Adapted from Ref. [97].)

broadly grouped into photopolymerization reactions and "click"-type reactions including thiol-ene and Diels–Alder reactions.

Photopolymerization is an attractive technique to create hydrogels as the conversion of liquid polymer solution to a gel occurs rapidly, under physiological temperature, with minimal heat production, and can be controlled in time and space [95]. In addition, this methodology served as an easy means to encapsulate cells uniformly in a 3D hydrogel matrix. As illustrated by Figure 9.15, hydrogel formation results from the radical polymerization of methacrylate groups grafted on HA. Radicals are generated after exposure of an aqueous solution of HA macromer (HA–MA) containing a water-soluble photoinitiator (typically 2-hydroxy-1-[4-(2-hydroxyethoxy)phenyl]-2-methyl-1-propanone that has shown biocompatibility with cells [96]). The photopolymerization of HA modified with methacrylate groups provided hydrogels with tunable mechanical strength and degradation properties.

HA–MA derivatives can be prepared by reaction of HA with methacrylic anhydride [98–101] and glycidyl methacrylate (GMA) [12,102–105] leading to ester derivatives. In the case of reactions performed with GMA, different esters resulting from a reversible transesterification through the primary hydroxyl group and an irreversible ring-opening conjugation through the carboxylic acid group toward the highest substituted carbon of epoxide were assumed to be formed simultaneously (Figure 9.16). Their respective amounts were reported to be time dependent, the concentration of ring-opening products thus increasing with time.

Figure 9.16 Methacrylation of HA with GMA leading to HA–GMA macromers by a competition reaction between ring opening and transesterification.

It was reported that the difference between the ester linkages may be responsible for the difference in hydrolytic degradation behavior observed from cross-linked HA–GMA hydrogels (Figure 9.17) [106]. From ^{13}C NMR analysis, it was suggested that the "slow degrading hydrogels" may have the methacrylated product formed via transesterification while the "fast degrading hydrogels" may have an additional glyceryl spacer.

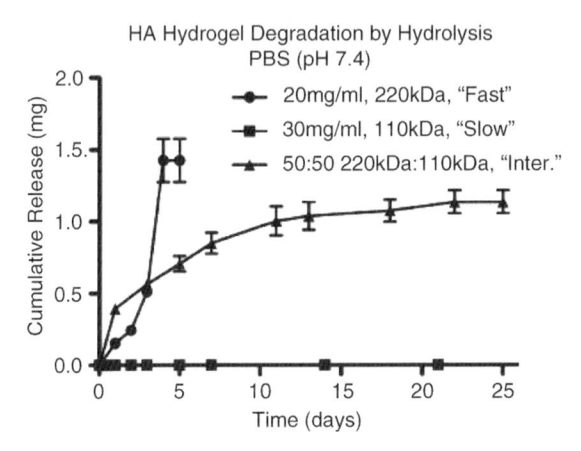

Figure 9.17 Hydrolytic degradation for different HA–GMA hydrogel formulations targeting fast, intermediate, and slow degradation rates. Results are shown as mean and standard error from triplicate samples. (Adapted from Ref. [106].)

The high viability of various cells (>75%), including mesenchymal stem cells (MSCs), 3T3 fibroblasts, and chondrocytes, after photoencapsulation in methacrylate-modified HA hydrogels demonstrated that these hydrogels could be successfully used as a biomimetic extracellular matrix [98,107,108]. Interestingly, it was shown by comparison with synthetic poly(ethylene glycol) hydrogels that photo-cross-linked HA hydrogels could provide a favorable niche for human MSC (hMSC) chondrogenesis. This was attributed to favorable cell–scaffold interactions, likely involving the CD44 receptor expressed by hMSCs [107]. Moreover, methacrylate-modified HA hydrogels were found to be a better matrix for osteogenic differentiation of encapsulated goat MSCs than methacrylated hyperbranched polyglycerol hydrogels [108]. The MSCs indeed exhibited better stretching in HA gels, accompanied by a significantly higher expression of alkaline phosphatase, after both 2 and 3 weeks. Regarding degradation of HA hydrogels, *in vivo* studies of photo-cross-linked HA–GMA hydrogels delivering bone morphogenetic protein-2 (BMP-2) as an osteoinductive molecule indicated that the scaffold degradation rate affected the organization of the collagen matrix but had no effect on the amounts of mineral formation [106]. It should be noted that degradation of scaffolds *in vivo* is an important issue in tissue remodeling (i.e., in the reorganization of existing tissues) and regeneration. Stimuli-responsive degradation is ideal because the degradation rates of the scaffolds can be modulated by the tissue regeneration rate [109]. In addition to hydrolysis due to ester linkage, HA–methacrylate hydrogels can be biodegraded by hyaluronidase. Minimal new bone formation was observed for the HA-based hydrogels alone indicating that they do not provide enough activity to regenerate bones *in vivo*. On the other hand, the amount of mineralized tissue formed was increased by the codelivery of vascular endothelial growth factor (VEGF), one of the most potent angiogenic molecules, in conjunction with BMP-2. This may be related to the ability of VEGF to increase blood vessel formation and thus nutrient delivery to the regenerating tissue.

Hydrogels based on HA have been alternatively synthesized using "click"-type reactions. Most of them consisted in thiol-ene reactions [110] between thiol groups and vinyl/(meth)acrylate/maleimide groups. These have been applied to prepare under physiological conditions two-component hybrid hydrogels, composed of HA and homobifunctional or multifunctional poly(ethylene glycol) as cross-linkers (Figure 9.18 and Table 9.2). As can be seen from Table 9.2, fast gelation can occur from some two-component mixtures (entries 2, 3, and 5) offering the opportunity to use them as cell-seeded injectable tissue materials. It should be noted that a different cross-linker concentration may be used to change the gelation time or degree of cross-linking to tailor the material to a particular application [111,112]. In particular, although the thiols grafted on HA backbone (entries 3–5) can act as latent cross-linking agents by forming disulfide bonds upon exposure to air, it was demonstrated that the formation of stable networks depended strictly on PEG diacrylate concentration, indicating that the Michael addition-mediated cross-linking dominated the network formation [111].

It was shown that the acrylated HA/PEG-$(SH)_4$ system can be advantageously used as a scaffold for BMP-2 and human MSCs for *in vivo* rat calvarial defect regeneration

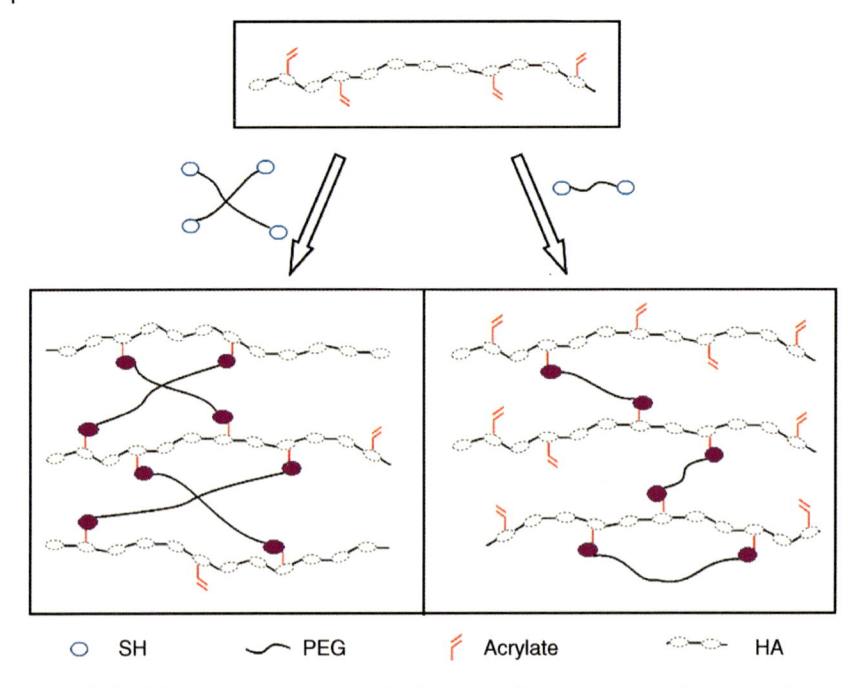

Figure 9.18 Schematic representation of the formation of two-component hybrid hydrogels composed of HA and homobifunctional or multifunctional poly(ethylene glycol) as cross-linkers based on thiol-ene reactions.

Table 9.2 Chemical cross-linking strategies for hyaluronic acid based on thiol-ene reactions.

Entry	HA derivative	Cross-linker	Gelation time	Application	References
1	Methacrylated HA	Four-arm PEG-$(SH)_4$; six-arm PEG-$(SH)_6$	5.5 h[a]	—	[113]
2	Acrylated HA	Four-arm PEG-$(SH)_4$	10 min[a]	3D cell culture with human MSCs; bone regeneration	[114,115]
3	Thiol-modified HA	PEG bismaleimide	<0.5 min[b]	3D cell culture with fibroblasts and hepatocytes	[116]
4	Thiol-modified HA	PEG diacrylate	36[b]		[111,116]
5	Thiol-modified HA	Four-arm PEG-(vinyl sulfone)$_4$	~1–10 min[b]	3D cell culture with chondrocytes; cartilage repair	[112]

a) Derived from rheometry measurements.
b) Measured using the test tube inversion method [116].

[114]. *In vitro* cell viability of the HA hydrogel was found to be 72% without BMP-2 and 81% with BMP-2. In contrast to goat MSCs embedded in photocrosslinked HA–methacrylate hydrogels as mentioned above [108], the cell morphology in the *in situ* cross-linkable hydrogels was shown to be round. Nevertheless, bone regeneration was observed *in vivo* for the hydrogel embedded with MSCs as well as the hydrogel combined with MSCs and BMP-2. In the latter case, bone formation was clearly enhanced indicating that this mixture exerted a synergistic effect on new bone formation. Interestingly, the VEGF that directs the proliferation of endothelial cells was found to be expressed in hMSCs mixed with hydrogels [114]. This result means that stem cells in hydrogel implanted in bone defects can differentiate into vascular cells or secrete the recruiting signal of endothelial progenitor cells from the circulating blood. Moreover, culturing chondrocyte–hydrogel constructs prepared from HA-SH cross-linked with four-arm PEG–(vinyl sulfone)$_4$ for 3 weeks *in vitro* revealed that the cells were viable and that cell division took place [112]. Gel–cell matrices degraded in approximately 3 weeks, as shown by a significant decrease in dry gel mass. This was attributed to the presence of chondrocytes that are capable of expressing hyaluronidase [117,118]. At day 21, glycosaminoglycans and collagen type II were found to have accumulated in hydrogels, indicating that these injectable hydrogels have a high potential for cartilage tissue engineering.

Injectable HA hydrogels capable of phase transition in response to a change in temperature have also been reported as promising scaffolds for tissue regeneration [119]. In these systems, the formation of physical junctions required for gelation is due to the autoassociation of thermosensitive polymer chains possessing a lower critical solution temperature (LCST) in aqueous solution, grafted along the HA backbone. Above this temperature, which is designed to be below body temperature, a phase separation occurs as a result of an increase in the hydrophobic character of the polymer. Poly(*N*-isopropylacrylamide) (PNIPAAm) is a typical paradigm of thermosensitive polymers that undergo a coil-to-globule phase transition at 32 °C [120,121]. This polymer has been grafted on HA modified with reactive hydrazide groups under peptide-like coupling conditions [122]. The resulting HA–PNIPAAm derivatives having weight ratios of PNIPAAM of 28 and 53% exhibited a sol–gel transition at 30 °C from rheological experiments (Figure 9.19).

Encapsulation of human adipose-derived stem cells (ASCs) within hydrogels showed that the HA–PNIPAAm copolymers were noncytotoxic and preserved the viability of the entrapped cells. A preliminary *in vivo* study demonstrated the usefulness of the HA–PNIPAAm copolymer with 53% PNIPAAm as an injectable hydrogel for adipose tissue engineering.

This work thus indicated that the thermosensitive HA–PNIPAAm copolymers may have potential uses in adipose regeneration and other soft tissues.

Censi *et al.* recently combined thermal gelling and Michael addition cross-linking to obtain *in situ* forming hydrogels of HA [123]. (Meth)acrylate bearing ABA triblock copolymers consisting of a PEG middle block, flanked by thermosensitive blocks of random *N*-isopropylacrylamide/*N*-(2-hydroxypropyl)methacrylamide dilactate, was thus reacted with thiol-modified HA at thiol/(meth)acrylate groups of 1/1, leading to *in situ* gelling systems that were progressively stabilized as the Michael addition

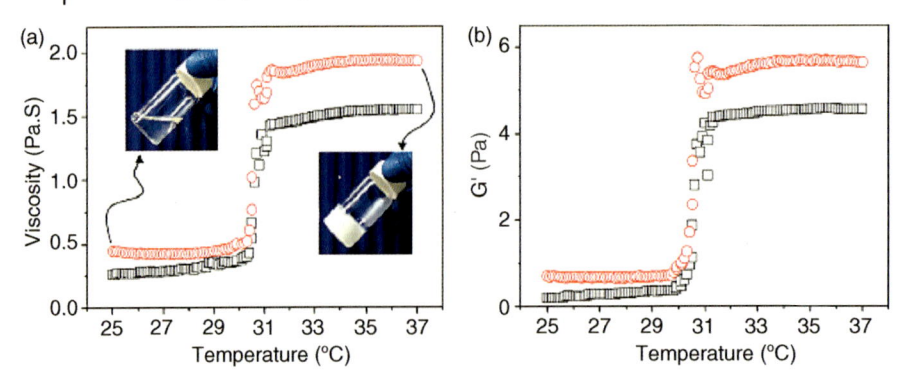

Figure 9.19 Viscosity (a) and storage modulus G' (b) of HA–PNIPAAm copolymer (DS = 0.28 (□) and 0.53 (○) with 5 wt% concentration in PBS as a function of temperature. Insets are *in situ* gelation of HA–PNIPAAm hydrogel in PBS [122].

between the (meth)acrylate and thiol groups proceeded. This study pointed out the potential of this tandem system for biomedical application, such as delivery of biologically active molecules.

9.3.2.2 Engineering Biological Functionality in Hyaluronic Acid-Based Scaffolds

In order to regulate cell function and subsequent tissue formation, some groups exploited the chemistry approaches described above to functionalize HA scaffolds with short peptide sequences that are found in ECM-derived biomolecular signals, such as cell adhesion sites or protease substrate sites.

Shu *et al.* prepared cell-adhesive HA-based hydrogels using thiol-ene coupling reactions involving thiol-modified HA (3,3′-dithiobis(propanoic dihydrazide) [HA–DTPH]), PEG diacrylate, and thiol-modified peptides containing the Arg–Gly–Asp (RGD) sequence (CRGDS and CCRGDS) [124]. By reaction of the RGD peptides onto the PEG diacrylate, only a fraction of the double bonds was functionalized, while at the same time a sufficient number of RGD groups were immobilized to influence cell adhesion. Indeed, attachment and spreading of human and murine fibroblasts were significantly enhanced on the surface of hydrogels that had been modified by covalent attachment of an RGD-containing peptide. The concentration and structure of RGD peptides and the length of PEG spacer significantly influenced cell attachment and spreading on the hydrogel surface, demonstrating that surface bioavailable RGD groups controlled adhesion and spreading. Murine fibroblasts were successfully encapsulated *in situ* into a gelling mixture of HA–DTPH and PEG diacrylate, with or without covalently attached RGD peptides. Cells remained viable and proliferated in *in vitro* culture for up to 15 days. However, although RGD peptides significantly promoted cell proliferation on the hydrogel surfaces, increase in cell proliferation within the hydrogels *in vitro* was minimal. In contrast, *in vivo* studies using nude mice indicated that inclusion of covalently attached RGD peptides accelerated the formation of fibrous tissue at 4 weeks postinjection.

Park *et al.* also demonstrated the ability to modulate the HA properties from cell nonadhesive to adhesive by incorporation of RGD peptides [125]. In this study, HA-based hydrogels were prepared via copolymerization of pendent methacrylate groups of HA with PEG diacrylate. HA-based hydrogels were then made cell adhesive, by reaction of a cysteine-containing RGD sequence (GCGYGRGDSPG) onto the PEG diacrylate, similar to the approach used by Shu *et al.* This study also led to the conclusion that HA alone cannot provide a sufficient signal for cell adhesion and spreading.

Using the thiol-ene chemistry approach, Kim *et al.* prepared HA hydrogels by conjugation with two different peptides: cell adhesion peptides (RGD) and a crosslinker with matrix metalloproteinase (MMP) degradable peptides to mimic the remodeling characteristics of natural ECMs by cell-derived MMPs [126]. The authors showed that cells in MMP-sensitive hydrogels spread inside the gel by degrading surrounding hydrogels with MMPs whereas cells in the control, insensitive hydrogel remained round (Figure 9.20). When cells were cultured in MMP-insensitive hydrogels functionalized with the integrin adhesive peptide Arg–Gly–Asp (RGD), cell spreading was not dramatic as shown in MMP-sensitive hydrogel samples. Cells in RGD-immobilized, MMP-sensitive hydrogels showed more spindle-like shapes sprouting filopodia into the hydrogels (Figure 9.20d).

9.3.2.3 Patterning of Hyaluronic Acid

Microscale approaches have been used to control topographical features and spatial presentation of surface molecules for the development of cell and protein arrays. These arrays have been used for drug discovery, diagnostic assays, biosensors, and fundamental cell biology. By soft lithographic application of HA using microcontact printing (μCP) and molding (also known as capillary force lithography [127]), Suh *et al.* constructed well-defined patterns of proteins and cells on various substrates including glass, silicon dioxides, poly(HEMA), polystyrene culture dishes, and biodegradable PLGA [128]. As shown in Figure 9.21, these soft lithographic techniques commonly utilize a microstructured surface made with an elastomeric material, polydimethylsiloxane (PDMS) [129,130], to generate patterns on surfaces. To pattern HA using μCP, oxygen plasma-treated PDMS stamps were coated with HA by spin coating. Then, the HA layer was transferred to the substrate surface. In the molding process, HA was spin coated on the substrate surface and a PDMS stamp was brought into conformal contact with the substrate.

This study suggested that HA could be used as a general platform on hydrophilic substrates for cell and protein patterning or even on hydrophobic substrates provided that the surface is made hydrophilic through simple modification of NaOH or oxygen plasma treatment. In addition, it revealed that HA is highly resistant to protein adhesion including bovine serum albumin (BSA), fibronectin (FN), and goat anti-rabbit immunoglobulin G (IgG) and to cell adhesion for fibroblasts (Figure 9.22).

Based on this result, this method was then exploited to create patterned cocultures as shown in Figure 9.23 [131–133]. HA was micropatterned on a glass substrate by using capillary force lithography. The exposed region of a glass substrate was coated

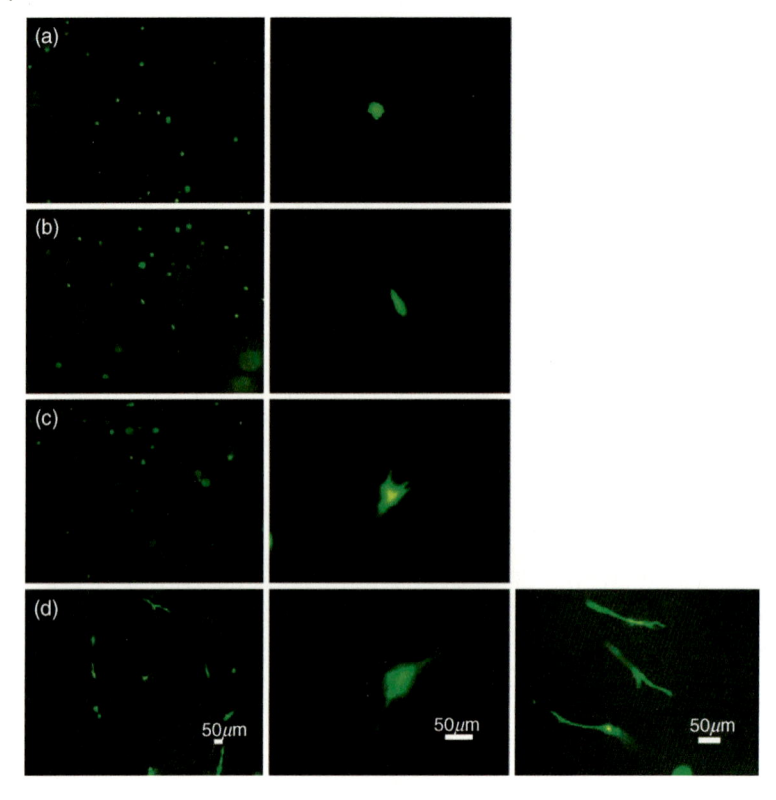

Figure 9.20 Cell viability test of human MSCs in the peptide–hyaluronic acid-based hydrogels by live and dead assay: photographs of human MSCs in the hyaluronic acid-based hydrogel. Human MSCs were cultured for 3 days in hyaluronic acid-based hydrogels: 1×10^5 human MSCs per construct. The membranes of live cells were stained with green fluorescence and the nuclei of dead cells with red: (a) MMP-insensitive peptide hyaluronic acid-based hydrogel; (b) MMP-insensitive peptide hyaluronic acid-based hydrogel + RGD peptides; (c) MMP-sensitive peptide hyaluronic acid-based hydrogel; (d) MMP-sensitive peptide hyaluronic acid-based hydrogel + RGD peptides. All samples are peptide–50 kg/mol HA-based hydrogels. Scale bar = 50 μm [106].

with fibronectin. Cells were then selectively adhered to the FN-coated regions. The HA-coated surface was complexed with poly(L-lysine) or collagen, allowing for the subsequent adhesion of secondary cells.

This technique was shown to be independent of the differences between primary and secondary seeded cells. It was used to localize cells efficiently within patterned cocultures as illustrated in Figure 9.24 [132]. Figure 9.24 shows images of cocultures of embryonic stem (ES) cells with NIH-3T3 fibroblast in patterned cocultures of circular FN islands. In this study, collagen was deposited on HA by ionic complexation to switch the HA surface from cell repulsive to cell adherent. Both light and fluorescent images indicate that ES cells were clearly patterned in dense spheroid aggregates against a background monolayer of NIH-3T3 cells. Fluorescent images

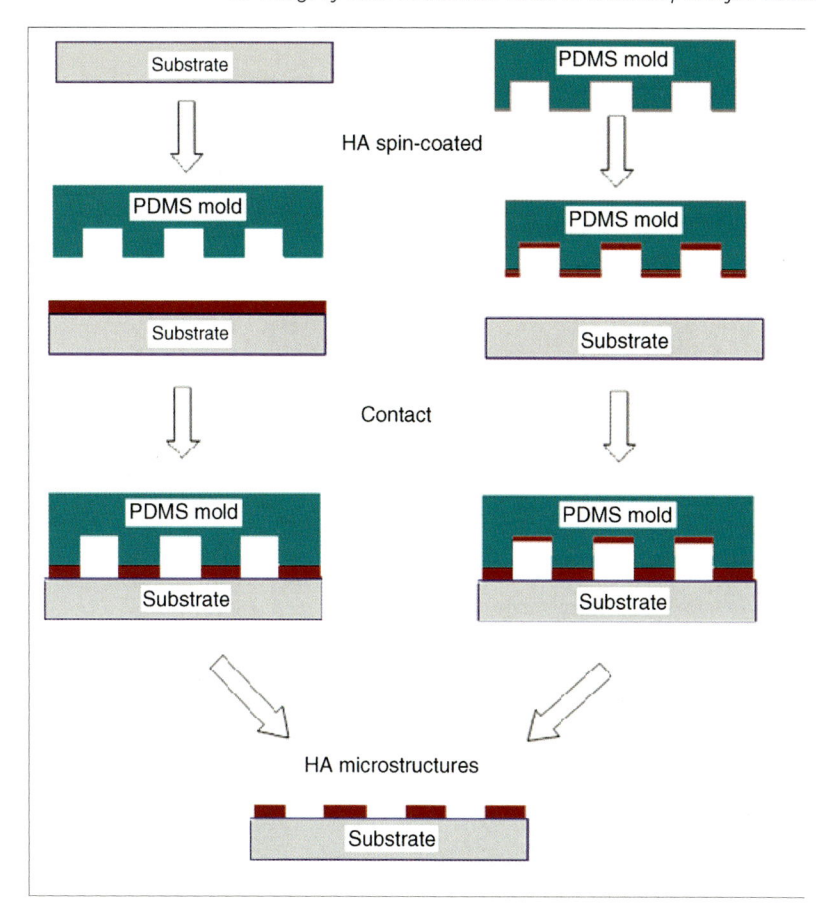

Figure 9.21 Schematic illustration of soft lithographic methods used for HA patterning: µCP (right) and molding (left). µCP utilizes direct transfer from a stamp to a substrate whereas molding deals with pattern formation from a uniform polymer film into the features of the stamp.

confirmed that NIH-3T3 cells were restricted to the HA-coated regions and were not seeded on top of ES cells as indicated by the lack of yellow regions on the aggregates (indicating the presence of both ES and NIH-3T3 cells).

Micropatterned HA surfaces were also prepared by photolithography, based on the photoimmobilization of the polysaccharide or sulfated derivatives functionalized with photoreactive moieties (azidophenyl groups) using photomasks of suitable dimensions (Figure 9.25) [134–136]. This involved the modification of glass substrates through aminosilanization. By UV irradiation, the grafted azidophenyl groups (Ph–N$_3$) are photolyzed to generate highly reactive nitrene (Ph–N) groups forming covalent bonds with the amine groups on the glass surface. Different stripe patterns of 10, 25, 50, and 100 µm width and ranging from 300 nm to 1 µm could be obtained. Interestingly, glass-sulfated HA microstructures were shown to influence

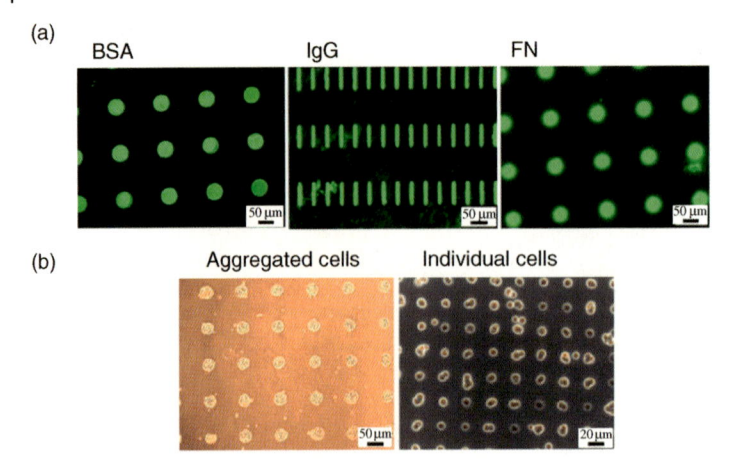

Figure 9.22 Patterns of proteins and cells obtained from a microstructured platform based on HA. (a) Fluorescent images of patterns of BSA, IgG, and FN on glass substrate. (b) Optical images of cell arrays on glass substrate; using FN as an adhesion layer, NIH-3T3 cells were seeded on glass with 150 μm holes (left) and 15 μm holes (right), leading to aggregated and individual cell arrays depending on the feature size. (Adapted from Ref. [128].)

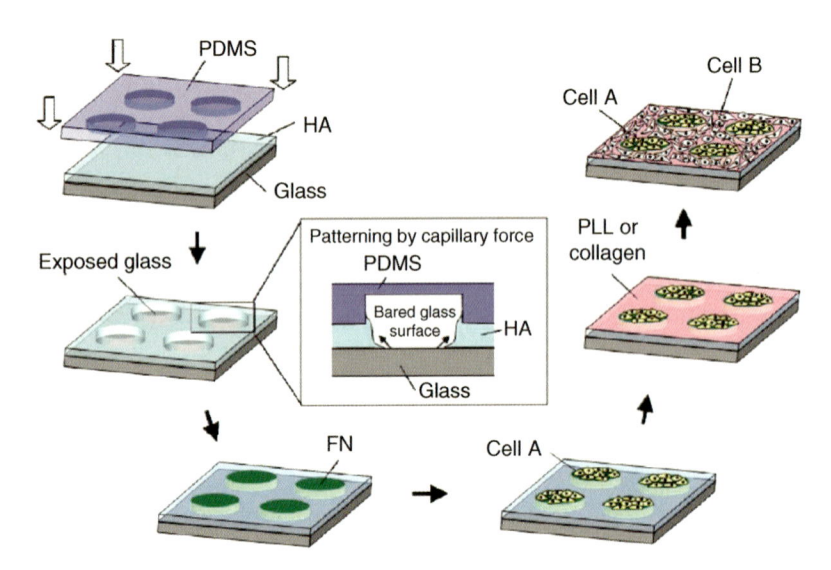

Figure 9.23 Schematic illustration of the fabrication of the coculture system using capillary force lithography and layer-by-layer deposition. A few drops of HA solution were spun coated onto a glass slide, and a PDMS mold was immediately placed on the thin layer of HA. HA under the void space of the PDMS mold receded until the glass surface became exposed. The exposed region of a glass substrate was coated with FN, where primary cells could be selectively adhered. Subsequently, the HA surface was complexed with collagen or poly(L-lysine), allowing for the subsequent adhesion of secondary cells. (Adapted from Ref. [131].)

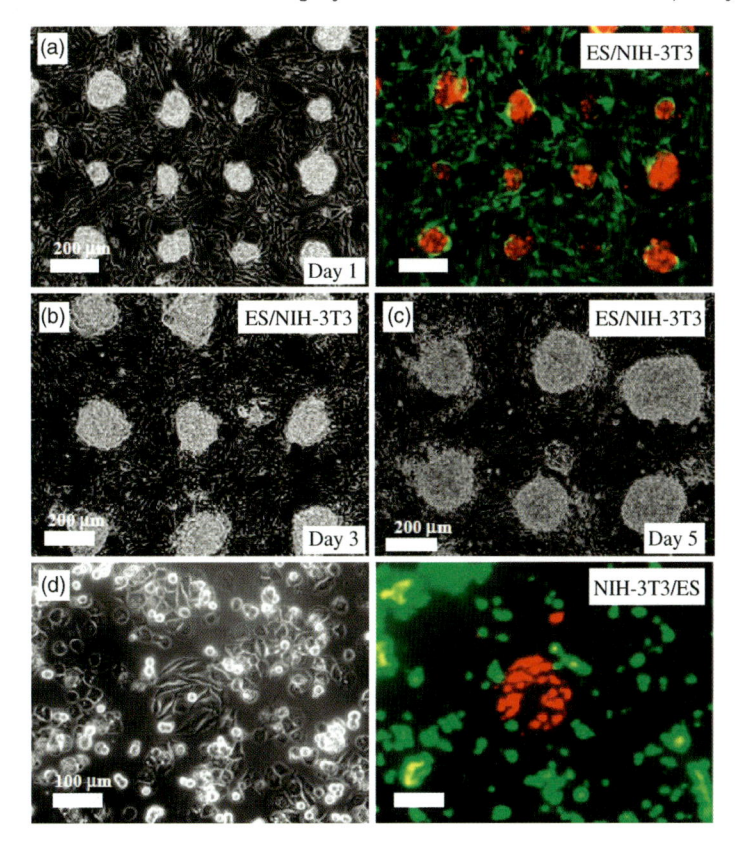

Figure 9.24 Patterned cocultures of ES cells with fibroblasts. Part (a) represents light (left) and fluorescent (right) images of patterned cocultures of ES cells (red) with NIH-3T3 fibroblasts (green) after 1 day. Parts (b) and (c) illustrate that cocultures remained stable for 3 and 5 days, respectively. Part (d) represents the reversal in the order of cell seeding in which NIH-3T3 fibroblasts (red) were initially seeded followed by ES cells as the secondary cells. Therefore, the technique is independent of the type of cell initially seeded [132].

endothelial cell behavior [134]. Decreasing the stripe dimensions leads to a more fusiform shape of the adhered endothelial cells. At the same time, the cell locomotion and orientation were increased. In contrast, investigations performed on sulfated HA microstructures on sulfated HA continuous substrate showed that cells behave as they were on a homogeneous substrate, revealing that the chemical control is much more important than the topographic one [134].

In contrast to the above studies that exploited the bioresistant behavior of HA to control cell adhesion on surfaces, Dickinson *et al.* showed the ability to direct cancer cells to adhere preferentially on HA-presenting regions [137,138]. For this purpose, the authors combined soft lithography and carbodiimide chemistry glass substrates in patterned monolayers presenting HA surrounded by PEG–silane. PEG was used to prevent cell adhesion. Combining carbodiimide linking chemistry and µCP, they generated well-defined, discrete patterned regions of HA chemically bound to the

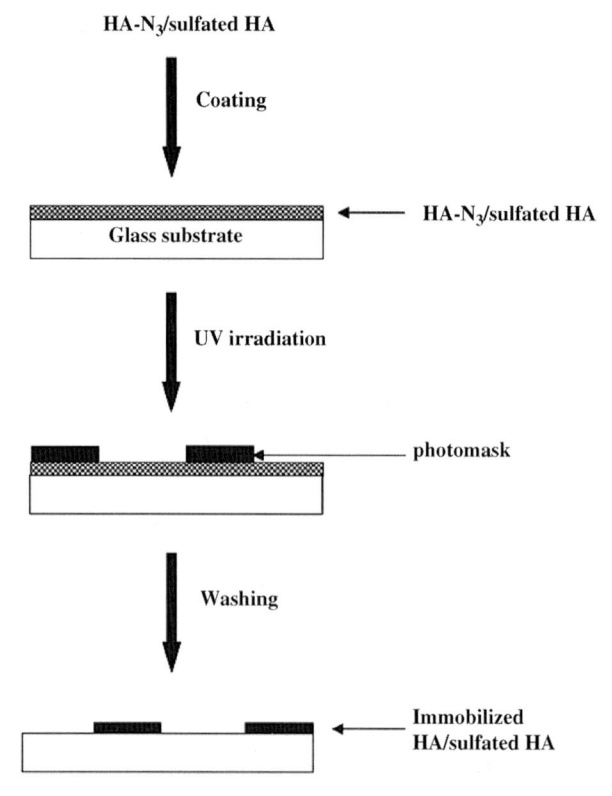

Figure 9.25 Schematic representation of photoimmobilization process of azidophenylamino-derivatized HA or sulfated HA. (Adapted from Ref. [135].)

glass substrate. Colon and breast cancer cells were seeded on sterile fluorescein-labeled HA (FL-HA) patterned surfaces. Within 24 h, colonies of colon cancer cells were observed within the HA-presenting squares. After 48 h, the colon cancer colonies grew in size, spreading beyond the perimeter of the HA patterns and extending to connect neighboring colonies (Figure 9.26a(i) and (ii)). Immunofluorescence analysis confirmed the localization of the cells within and surrounding the FL-HA patterns (Figure 9.26a(iii) and (iv)). Breast cancer cells attached to the HA patterns within 24 h, while proliferation and minimal spreading outside the patterns was observed within 48 h (Figure 9.26b(i) and (ii)). Breast cancer cells remained mostly restricted to HA regions, as compared to the colon cancer cells, which grow in colonies and exceeded the 80 μm × 80 μm HA patterns.

As both colon and breast cancer cells were characterized by membrane expression of CD44, it was examined whether blocking CD44 affected adhesion onto HA surfaces. Colon and breast cancer cells seeded onto functional HA surfaces in the presence of anti-CD44 were unable to attach onto the HA surfaces within 24 h, confirming that adhesion occurs through surface receptor CD44. Overall, functional

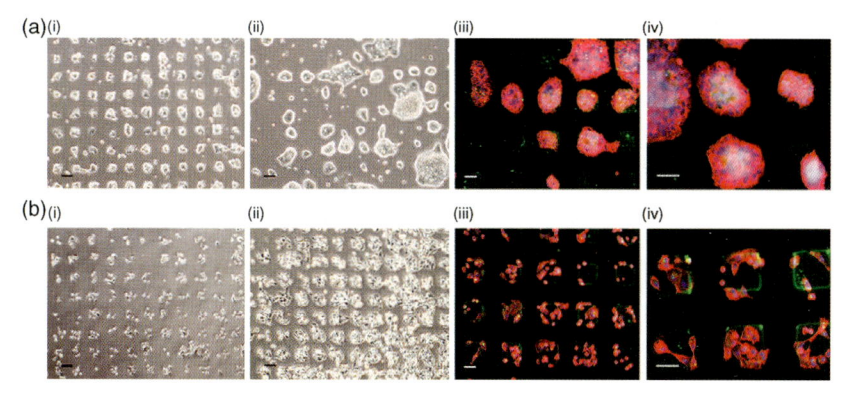

Figure 9.26 HA surfaces for the culture of cancer cells. (a) Colon cancer cells seeded on HA-patterned surfaces (i) formed colonies within the 80 μm × 80 μm squares after 24 h and (ii) grew in size, with some spreading outside the squares after 48 h of culture. Immunofluorescence analysis verified the CD44$^+$ colon cancer cells (red; nuclei in blue) within FL-HA squares (green) at (iii) low and (iv) high magnifications. (b) Breast cancer cells seeded on HA-patterned surfaces (i) attached within the 80 μm × 80 μm squares after 24 h and (ii) grew outside the squares after 48 h of culture. Immunofluorescence analysis verified the CD44$^+$ breast cancer cells (red; nuclei in blue) within FL-HA squares (green) at (iii) low and (iv) high magnifications. Scale bars are 100 μm in (a)(i) and (ii) and (b)(i) and (ii) and 50 μm in (a)(iii) and (iv) and (b)(iii) and (iv) [138].

HA surfaces were shown to be a useful tool to perform high-resolution visualization to analyze cancer cell attachment and migration at a single-cell level.

In all these approaches, cells were patterned on 2D surfaces, which lack a 3D microenvironment and the appropriate cell–ECM interactions. By combining photo-cross-linkable HA with micromolding, Khademhosseini et al. fabricated hydrogel microstructures (Figure 9.27) [139].

This approach was shown to be useful for capturing cells inside HA microwells or for encapsulating cells directly into the HA gels.

By the sequential cross-linking of HA modified with acrylate moieties with the use of photomasks, Khetan et al. showed the ability to spatially control the behavior of cells encapsulated into 3D hydrogels [140,141]. The sequential cross-linking strategy is illustrated in Figure 9.28. In the primary cross-linking step, a "−UV" hydrogel is formed using Michael-type reactivity between HA macromers and bifunctional, proteolytically degradable peptides. Monofunctional, pendant RGDS-containing peptides are also added (prior to cross-linking) to incorporate cell adhesion sites. With these components (adhesion and proteolytic degradability), this hydrogel was expected to support cellular remodeling as discussed above [126]. The primary addition step was performed in the presence of a photoinitiator (at this point, nonreactive) and designed so that only a portion of total available acrylate groups are consumed, making secondary free radical cross-linking possible. In the secondary step, "−UV" hydrogels were exposed to light to initiate free radical photopolymerization of the remaining acrylate groups. The resulting "+UV" hydrogels were

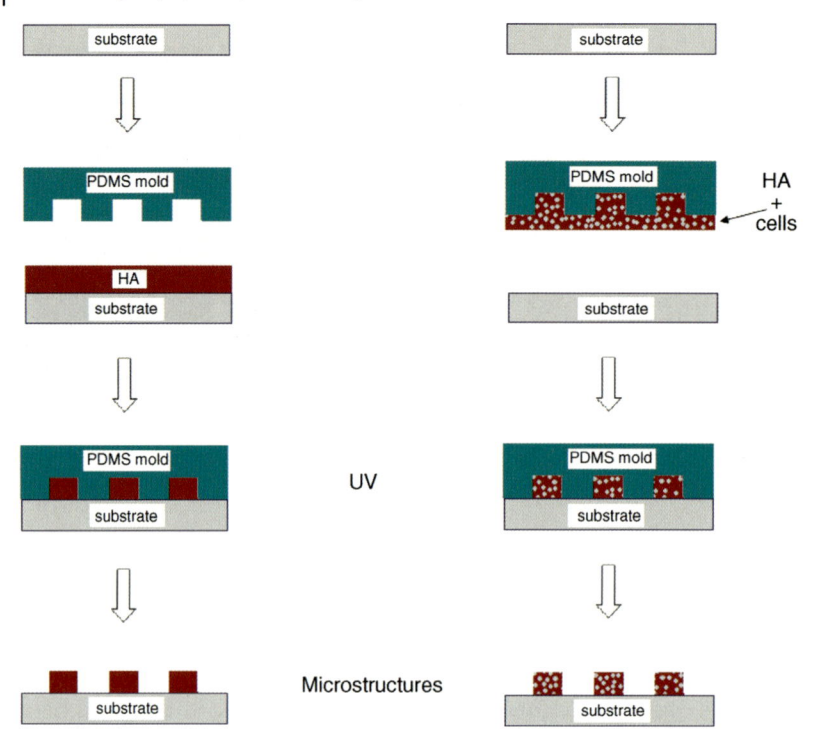

Figure 9.27 Schematic diagram of the HA micromolding process. PDMS molds were used to mold a layer of HA into the void regions of the stamp. The polymer was then cured with exposure to UV light to fabricate HA microstructures. To fabricate HA microstructures without cells, a thin polymer film on the substrate was molded (left). To fabricate HA microstructures that encapsulated cells (right), the HA solution was transferred from the PDMS mold onto the substrate and subsequently cross-linked.

expected to prevent remodeling due to the incorporation of nondegradable covalent cross-links from kinetic chain formation. Since mesh sizes in the +UV hydrogels are orders of magnitude smaller than typical cell diameters [126,142], secondary cross-linking was also predicted to prevent cellular outgrowth from tissue and to confine encapsulated cells to a rounded morphology.

The switch of such HA hydrogel from a "permissive" to "inhibitory" state was demonstrated by investigating the behavior (i.e., spreading) of encapsulated hMSCs [140,141]. Variations in the ratio of the two cross-link types in individual constructs controlled the degradation and mechanical properties of the hydrogels, as well as the degree of spreading of encapsulated cells (Figure 9.29). Cell spreading was further controlled spatially with the use of photomasks (Figure 9.29). The ability to spatially control cell spreading was also shown using cells from tissue (aortic arches) [140].

A similar dual cross-linking method that supports major changes in mechanics (from several to ~100 kPa) was also utilized to explore the effects of mechanics of HA hydrogels on 2D hMSC behavior (i.e., spreading and proliferation) [143].

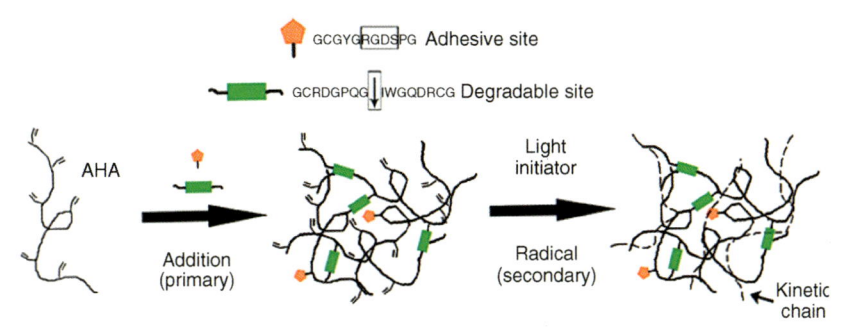

9.3 Design of Novel Biomaterials Based on Chemically Modified Derivatives

Figure 9.28 Schematic representation of sequential cross-linking of acrylated HA using a primary addition reaction and a secondary radical polymerization [141]. (Reproduced with permission of the Royal Society of Chemistry.)

Michael addition cross-linking (via dithiothreitol, DTT) was performed initially (−UV group) to consume all or a fraction of methacrylate groups grafted on HA and then followed by radical cross-linking (+UV group) to further consume reactive groups. A highly functionalized HA (∼100% modified) was used to allow for large changes in mechanics at a uniform concentration (3 wt%) (see Figure 9.30a). A clear dependence of hMSC spreading on the mechanics of the substrate as spreading increases was observed (Figure 9.30b).

In addition, the spatial patterning of mechanics was realized by regionally restricting light exposure during the second cross-linking. After 24 h, cells seeded on gels with photopatterned mechanics acquired morphologies reminiscent of the

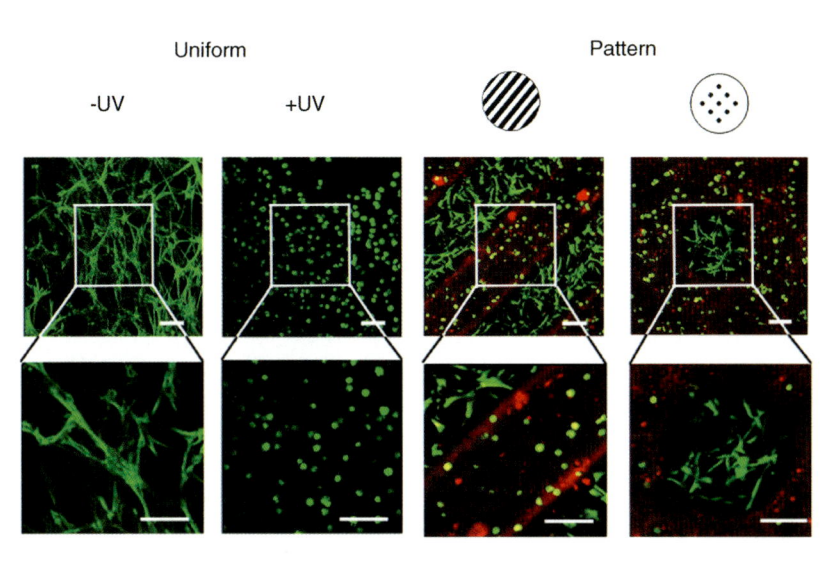

Figure 9.29 Images of encapsulated hMSCs (stained with calcein) in uniform and photopatterned acrylated HA hydrogels. Spreading only occurs in the permissive −UV regions. Scale bars = 100 μm. All cultures were for 14 days [140].

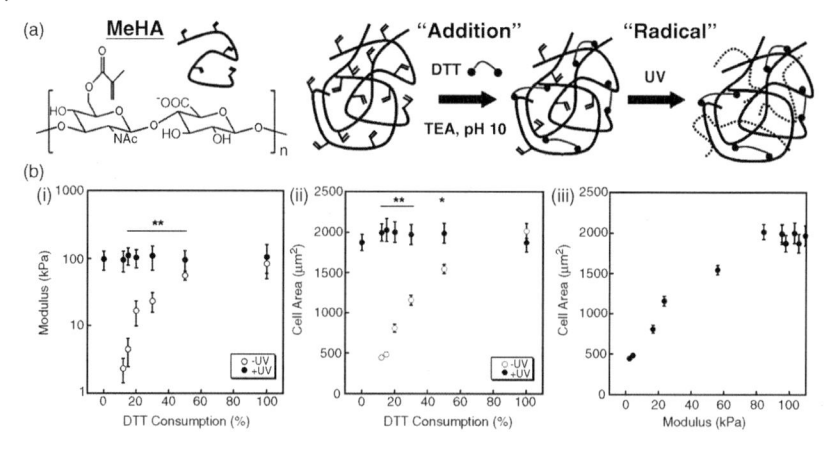

Figure 9.30 (a) Chemical structure of methacrylated hyaluronic acid (HA–MA) and dual cross-linking mechanism. Michael addition of methacrylates with DTT (dithiol cross-linker) induces partial cross-linking of a solution of HA–MA and triethanolamine at pH 10. Remaining methacrylates undergo radical polymerization when exposed to UV light in the presence of a photoinitiator (dotted lines are kinetic chains) to increase cross-linking density (i.e., mechanics). (b) Characterization of mechanics (assessed with AFM) and hMSC response to hydrogels with uniform properties. (i) Hydrogel modulus for variable DTT consumption (theoretical values shown, based on molar ratio of thiols on DTT to methacrylates on HA–MA) before (−UV, white) and after (+UV, black) light exposure. The mechanics can be tailored over two orders of magnitude with this system and result in a peak modulus of ∼100 kPa. (b) hMSC spread area 24 h after seeding for the same hydrogel systems. Spreading increased with increasing DTT consumption and UV-exposed gels show similar spread area responses. (c) hMSC spread area 24 h after seeding versus mechanics shows increased cell area with increasing mechanics until a plateau is reached at ∼80 kPa (Reproduced from Ref. 143 with permission of the Royal Society of Chemistry.)

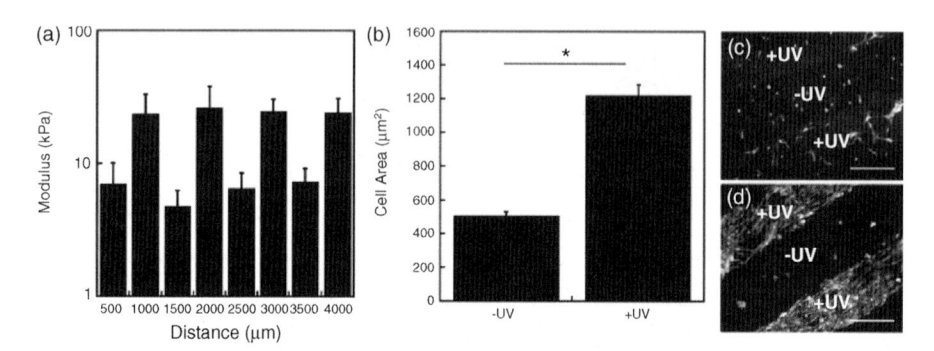

Figure 9.31 Spatially controlled mechanics (a) and hMSC spreading (b) on photopatterned stripes (500 μm width) on 12% DTT hydrogels. The mechanics vary with space across the hydrogel depending on whether it was exposed to light and is correlated to hMSC response. The cellular morphology on patterns after 1 (c) and 7 (d) days illustrates the importance of mechanics on cellular behavior, including spreading and proliferation. Scale bar = 400 μm [143]. (reproduced with permission of the Royal Society of Chemistry.)

uniform gels on the corresponding mechanical environments (i.e., rounded on "soft" regions and highly spread on "stiff" regions, Figure 9.31). Cells maintained a rounded morphology on the softer −UV regions and were highly spread on the stiffer +UV regions (Figure 9.31).

Although this is only a preliminary step toward the utility of these systems for actual tissue engineering constructs, this novel system allows for spatial control of matrix mechanics for the purposes of driving stem cell behavior.

9.4
Conclusions

In this chapter, we have reported recent advances in the design of biomaterials from chemically modified chitosan and hyaluronic acid. Hydrophobically modified derivatives of these polysaccharides demonstrate interesting self-associating properties in aqueous solution that may be advantageously used for developing new applications in the biomedical field. Hydrogels as well as nanoparticles may be obtained. Their formation seems to be closely dependent on some chemical parameters including the degree of hydrophobic content, the nature/size of the hydrophobic moiety, and the molar mass of the polysaccharide backbone. However, further research is required to more fully understand the impact of these parameters on the associating behavior of these polymers, allowing the development of specific applications, especially in tissue engineering, cell culture, and drug delivery. As discussed in this chapter, hydrogels of hyaluronic acid appear as suitable materials for cell culture and tissue engineering. Further trends go toward the development of smart hydrogels and gel microstructures that can trigger specific cellular responses. Therefore, future developments of HA biomaterials for biological studies and *in vivo* applications will require specific chemical modifications to spatially and/or temporally control their mechanical and swelling properties, degradability, and cell presentation of biomolecular signals.

References

1 Glass, J.E. (ed.) (1989) *Polymers in Aqueous Media: Performance Through Association*, Advances in Chemistry Series, vol. **223**, American Chemical Society, 194th National Meeting of the American Chemical Society, New Orleans, LA, August 30–September 4, 1987.

2 Glass, J.E., Schulz, D.N., and Zukoski, C. F. (1991) Polymers as rheology modifiers. An overview. *ACS Symp. Ser.*, **462**, 2–17.

3 Winnik, M.A. and Yekta, A. (1997) Associative polymers in aqueous solution. *Curr. Opin. Colloid Interface Sci.*, **2**, 424–436.

4 Baldwin, A.D. and Kiick, K.L. (2010) Polysaccharide-modified synthetic polymeric biomaterials. *Biopolymers*, **94**, 128–140.

5 Muzzarelli, R.A.A. and Muzzarelli, C. (2005) Chitosan chemistry: relevance to the biomedical sciences. *Adv. Polym. Sci.*, **186**, 151–209.

6 Kim, I.-Y., Seo, S.-J., Moon, H.-S., Yoo, M.-K., Park, I.-Y., Kim, B.-C., and Cho, C.-S.

(2008) Chitosan and its derivatives for tissue engineering applications. *Biotechnol. Adv.*, **26**, 1–21.

7 Laurent, T.C. (ed.) (1998) *The Chemistry, Biology and Medical Applications of Hyaluronan and Its Derivatives*, Wenner-Gren International Series, vol. **72**, Portland Press.

8 Kogan, G., Soltes, L., Stern, R., and Gemeiner, P. (2007) Hyaluronic acid: a natural biopolymer with a broad range of biomedical and industrial applications. *Biotechnol. Lett.*, **29**, 17–25.

9 Balazs, E.A. (2009) Therapeutic use of hyaluronan. *Struct. Chem.*, **20**, 341–349.

10 Abatangelo, G. and Weigel, P.H. (eds) (2000) *New Frontiers in Medical Sciences: Redefining Hyaluronan*, International Congress Series 1196, Elsevier, Proceedings of the Symposium held in Padua, Italy, June 17–19, 1999.

11 Chen, W.Y.J. (2002) Functions of hyaluronan in wound repair, in *Hyaluronan*, vol. **2**, Proceedings of the 12th International Cellucon Conference, Wrexham, UK, 2000, pp. 147–156.

12 Jia, X., Burdick, J.A., Kobler, J., Clifton, R.J., Rosowski, J.J., Zeitels, S.M., and Langer, R. (2004) Synthesis and characterization of *in situ* crosslinkable hyaluronic acid-based hydrogels with potential application for vocal fold regeneration. *Macromolecules*, **37**, 3239–3248.

13 Takahashi, Y., Li, L., Kamiryo, M., Asteriou, T., Moustakas, A., Yamashita, H., and Heldin, P. (2005) Hyaluronan fragments induce endothelial cell differentiation in a CD44- and CXCL1/GRO1-dependent manner. *J. Biol. Chem.*, **280**, 24195–24204.

14 Toole, B.P. (2001) Hyaluronan in morphogenesis. *Semin. Cell Dev. Biol.*, **12**, 79–87.

15 Oh, E.J., Park, K., Kim, K.S., Kim, J., Yang, J.-A., Kong, J.-H., Lee, M.Y., Hoffman, A.S., and Hahn, S.K. (2010) Target specific and long-acting delivery of protein, peptide, and nucleotide therapeutics using hyaluronic acid derivatives. *J. Control. Release*, **141**, 2–12.

16 Prestwich, G.D. and Kuo, J.-W. (2008) Chemically-modified HA for therapy and regenerative medicine. *Curr. Pharm. Biotechnol.*, **9**, 242–245.

17 Yadav, A.K., Mishra, P., and Agrawal, G.P. (2008) An insight on hyaluronic acid in drug targeting and drug delivery. *J. Drug Target.*, **16**, 91–107.

18 Kumar, M.N.V.R., Muzzarelli, R.A.A., Muzzarelli, C., Sashiwa, H., and Domb, A.J. (2004) Chitosan chemistry and pharmaceutical perspectives. *Chem. Rev.*, **104**, 6017–6084.

19 Peniche, C., Arguelles-Monal, W., Peniche, H., and Acosta, N. (2003) Chitosan: an attractive biocompatible polymer for microencapsulation. *Macromol. Biosci.*, **3**, 511–520.

20 Sashiwa, H. and Aiba, S. (2004) Chemically modified chitin and chitosan as biomaterials. *Prog. Polym. Sci.*, **29**, 887–908.

21 Nyström, B., Kjoniksen, A.-L., Beheshti, N., Zhu, K., and Knudsen, K.D. (2009) Rheological and structural aspects on association of hydrophobically modified polysaccharides. *Soft Matter*, **5**, 1328–1339.

22 Auzély-Velty, R. (2011) Self-assembling polysaccharide systems based on cyclodextrin complexation: synthesis, properties and potential applications in the biomaterials field. *C. R. Chim.*, **14**, 167–177.

23 Charlot, A. and Auzély-Velty, R. (2007) Synthesis of novel supramolecular assemblies based on hyaluronic acid derivatives bearing bivalent β-cyclodextrin and adamantane moieties. *Macromolecules*, **40**, 1147–1158.

24 Charlot, A. and Auzély-Velty, R. (2007) Novel hyaluronic acid based supramolecular assemblies stabilized by multivalent specific interactions: rheological behavior in aqueous solution. *Macromolecules*, **40**, 9555–9563.

25 Charlot, A., Auzély-Velty, R., and Rinaudo, M. (2003) Specific interactions in model charged polysaccharide systems. *J. Phys. Chem. B*, **107**, 8248–8254.

26 Yusa, S.-i., Hashidzume, A., and Morishima, Y. (1999) Interpolymer association of cholesterol pendants linked to a polyelectrolyte as studied by quasielastic light scattering and fluorescence techniques. *Langmuir*, **15**, 8826–8831.

27 Yamamoto, H. and Morishima, Y. (1999) Effect of hydrophobe content on intra-

and interpolymer self-associations of hydrophobically modified poly(sodium 2-(acrylamido)-2-methylpropanesulfonate) in water. *Macromolecules*, **32**, 7469–7475.

28 Rinaudo, M., Pavlov, G., and Desbrières, J. (1999) Influence of acetic acid concentration on the solubilization of chitosan. *Polymer*, **40**, 7029–7032.

29 Rinaudo, M., Pavlov, G., and Desbrières, J. (1999) Solubilization of chitosan in strong acid medium. *Int. J. Polym. Anal. Charact.*, **5**, 267–276.

30 Kjoniksen, A.-L., Nyström, B., Iversen, C., Nakken, T., Palmgren, O., and Tande, T. (1997) Viscosity of dilute aqueous solutions of hydrophobically modified chitosan and its unmodified analog at different conditions of salt and surfactant concentrations. *Langmuir*, **13**, 4948–4952.

31 Desbrières, J., Martinez, C., and Rinaudo, M. (1996) Hydrophobic derivatives of chitosan: characterization and rheological behavior. *Int. J. Biol. Macromol.*, **19**, 21–28.

32 Nyström, B., Kjoniksen, A.-L., and Iversen, C. (1999) Characterization of association phenomena in aqueous systems of chitosan of different hydrophobicity. *Adv. Colloid Interface Sci.*, **79**, 81–103.

33 Philippova, O.E., Volkov, E.V., Sitnikova, N.L., Khokhlov, A.R., Desbrières, J., and Rinaudo, M. (2001) Two types of hydrophobic aggregates in aqueous solutions of chitosan and its hydrophobic derivative. *Biomacromolecules*, **2**, 483–490.

34 Amiji, M.M. (1995) Pyrene fluorescence study of chitosan self-association in aqueous solution. *Carbohydr. Polym.*, **26**, 211–213.

35 Esquenet, C., Terech, P., Boue, F., and Buhler, E. (2004) Structural and rheological properties of hydrophobically modified polysaccharide associative networks. *Langmuir*, **20**, 3583–3592.

36 Esquenet, C. and Buhler, E. (2001) Phase behavior of associating polyelectrolyte polysaccharides. 1. Aggregation process in dilute solution. *Macromolecules*, **34**, 5287–5294.

37 Hu, Y., Du, Y., Yang, J., Tang, Y., Li, J., and Wang, X. (2007) Self-aggregation and antibacterial activity of N-acylated chitosan. *Polymer*, **48**, 3098–3106.

38 Kim, K., Kim, J.-H., Kim, S., Chung, H., Choi, K., Kwon, I.C., Park, J.H., Kim, Y.-S., Park, R.-W., Kim, I.-S., and Jeong, S.Y. (2005) Self-assembled nanoparticles of bile acid-modified glycol chitosans and their applications for cancer therapy. *Macromol. Res.*, **13**, 167–175.

39 Kim, K., Kwon, S., Park, J.H., Chung, H., Jeong, S.Y., Kwon, I.C., and Kim, I.-S. (2005) Physicochemical characterizations of self-assembled nanoparticles of glycol chitosan–deoxycholic acid conjugates. *Biomacromolecules*, **6**, 1154–1158.

40 Kwon, S., Park, J.H., Chung, H., Kwon, I.C., Jeong, S.Y., and Kim, I.-S. (2003) Physicochemical characteristics of self-assembled nanoparticles based on glycol chitosan bearing 5β-cholanic acid. *Langmuir*, **19**, 10188–10193.

41 Park, J.H., Cho, Y.W., Chung, H., Kwon, I.C., and Jeong, S.Y. (2003) Synthesis and characterization of sugar-bearing chitosan derivatives: aqueous solubility and biodegradability. *Biomacromolecules*, **4**, 1087–1091.

42 Son, Y.J., Jang, J.-S., Cho, Y.W., Chung, H., Park, R.-W., Kwon, I.C., Kim, I.-S., Park, J.Y., Seo, S.B., Park, C.R., and Jeong, S.Y. (2003) Biodistribution and anti-tumor efficacy of doxorubicin loaded glycol–chitosan nanoaggregates by EPR effect. *J. Control. Release*, **91**, 135–145.

43 Yoo, H.S., Lee, J.E., Chung, H., Kwon, I.C., and Jeong, S.Y. (2005) Self-assembled nanoparticles containing hydrophobically modified glycol chitosan for gene delivery. *J. Control. Release*, **103**, 235–243.

44 Park, K., Kim, J.-H., Nam, Y.S., Lee, S., Nam, H.Y., Kim, K., Park, J.H., Kim, I.-S., Choi, K., Kim, S.Y., and Kwon, I.C. (2007) Effect of polymer molecular weight on the tumor targeting characteristics of self-assembled glycol chitosan nanoparticles. *J. Control. Release*, **122**, 305–314.

45 Hwang, H.-Y., Kim, I.-S., Kwon, I.C., and Kim, Y.-H. (2008) Tumor targetability and antitumor effect of docetaxel-loaded hydrophobically modified glycol chitosan nanoparticles. *J. Control. Release*, **128**, 23–31.

46 Krause, W.E., Bellomo, E.G., and Colby, R.H. (2001) Rheology of sodium hyaluronate under physiological conditions. *Biomacromolecules*, **2**, 65–69.

47 Milas, M., Rinaudo, M., Roure, I., Al-Assaf, S., Phillips, G.O., and Williams, P.

A. (2001) Comparative rheological behavior of hyaluronan from bacterial and animal sources with cross-linked hyaluronan (hylan) in aqueous solution. *Biopolymers*, **59**, 191–204.

48 Milas, M., Roure, I., and Berry, G.C. (1996) Crossover behavior in the viscosity of semiflexible polymers: solutions of sodium hyaluronate as a function of concentration, molecular weight, and temperature. *J. Rheol.*, **40**, 1155–1166.

49 Sedova, P., Knotkova, K., Dvorakova, J., and Velebny, V. (2007) Review: water soluble and insoluble alkylderivatives of hyaluronic acid. *Prog. Biopolym. Res.*, 77–105.

50 Callegaro, L. and Bellini, D. (1998) Hyaluronic acid esters, threads and biomaterials containing them, and their use in surgery. WO (PCT) Application, p. 30 (to Fidia Advanced Biopolymers s.r.l., Italy).

51 Pelletier, S., Hubert, P., Lapicque, F., Payan, E., and Dellacherie, E. (2000) Amphiphilic derivatives of sodium alginate and hyaluronate: synthesis and physicochemical properties of aqueous dilute solutions. *Carbohydr. Polym.*, **43**, 343–349.

52 Pelletier, S., Hubert, P., Payan, E., Marchal, P., Choplin, L., and Dellacherie, E. (2001) Amphiphilic derivatives of sodium alginate and hyaluronate for cartilage repair: rheological properties. *J. Biomed. Mater. Res.*, **54**, 102–108.

53 Kawaguchi, Y., Matsukawa, K., Gama, Y., and Ishigami, Y. (1992) The effects of polysaccharide chain length in coating liposomes with partial palmitoyl hyaluronates. *Carbohydr. Polym.*, **18**, 139–142.

54 Mlcochova, P., Hajkova, V., Steiner, B., Bystricky, S., Koos, M., Medova, M., and Velebny, V. (2007) Preparation and characterization of biodegradable alkylether derivatives of hyaluronan. *Carbohydr. Polym.*, **69**, 344–352.

55 Benesova, K., Pekar, M., Lapcik, L., and Kucerik, J. (2006) Stability evaluation of n-alkyl hyaluronic acid derivatives by DSC and TG measurement. *J. Therm. Anal. Calorim.*, **83**, 341–348.

56 Mariotti, P., Navarini, L., Stucchi, L., Vinkovic, V., and Sunjic, V. (2002) New derivatives of hyaluronan. WO 098923.

57 Bellini, D. and Topai, A. (2011) Preparation and use of hyaluronic acid amides or other derivatives. US Patent 7,884,087.

58 Hamilton, R.G., Fox, E.M., Acharya, R.A., and Watts, A.E. (1990) Water-insoluble biocompatible derivatives of hyaluronic acid, their manufacture and use. US Patent 4,937,270.

59 Finelli, I., Chiessi, E., Galesso, D., Renier, D., and Paradossi, G. (2009) Gel-like structure of a hexadecyl derivative of hyaluronic acid for the treatment of osteoarthritis. *Macromol. Biosci.*, **9**, 646–653.

60 Kirker, K.R. and Prestwich, G.D. (2004) Physical properties of glycosaminoglycan hydrogels. *J. Polym. Sci. Polym. Phys.*, **42**, 4344–4356.

61 Prestwich, G.D., Marecak, D.M., Marecek, J.F., Vercruysse, K.P., and Ziebell, M.R. (1998) Controlled chemical modification of hyaluronic acid: synthesis, applications, and biodegradation of hydrazide derivatives. *J. Control. Release*, **53**, 93–103.

62 Mlcochova, P., Bystricky, S., Steiner, B., Machova, E., Koos, M., Velebny, V., and Krcmar, M. (2006) Synthesis and characterization of new biodegradable hyaluronan alkyl derivatives. *Biopolymers*, **82**, 74–79.

63 Mravec, F., Pekar, M., and Velebny, V. (2008) Aggregation behavior of novel hyaluronan derivatives – a fluorescence probe study. *Colloid Polym. Sci.*, **286**, 1681–1685.

64 Taglienti, A., Valentini, M., Sequi, P., and Crescenzi, V. (2005) Characterization of methylprednisolone esters of hyaluronan in aqueous solution: conformation and aggregation behavior. *Biomacromolecules*, **6**, 1648–1653.

65 Kadi, S., Cui, D., Bayma, E., Boudou, T., Nicolas, C., Glinel, K., Picart, C., and Auzély-Velty, R. (2009) Alkylamino hydrazide derivatives of hyaluronic acid: synthesis, characterization in semidilute aqueous solutions, and assembly into thin multilayer films. *Biomacromolecules*, **10**, 2875–2884.

66 Ambrosio, L., Borzacchiello, A., Netti, P.A., and Nicolais, L. (1999) Properties of new materials: rheological study on hyaluronic acid and its derivative solutions. *J. Macromol. Sci. Pure Appl. Chem.*, **A36**, 991–1000.

67 Lee, H., Lee, K., and Park, T.G. (2008) Hyaluronic acid–paclitaxel conjugate micelles: synthesis, characterization, and antitumor activity. *Bioconjug. Chem.*, **19**, 1319–1325.

68 Mok, H., Kim, H.J., and Park, T.G. (2008) Dissolution of biomacromolecules in organic solvents by nano-complexing with poly(ethylene glycol). *Int. J. Pharm.*, **356**, 306–313.

69 Lee, H., Ahn, C.-H., and Park, T.G. (2009) Poly[lactic-*co*-(glycolic acid)]-grafted hyaluronic acid copolymer micelle nanoparticles for target-specific delivery of doxorubicin. *Macromol. Biosci.*, **9**, 336–342.

70 Choi, K.Y., Chung, H., Min, K.H., Yoon, H.Y., Kim, K., Park, J.H., Kwon, I.C., and Jeong, S.Y. (2010) Self-assembled hyaluronic acid nanoparticles for active tumor targeting. *Biomaterials*, **31**, 106–114.

71 Aruffo, A., Stamenkovic, I., Melnick, M., Underhill, C.B., and Seed, B. (1990) CD44 is the principal cell surface receptor for hyaluronate. *Cell*, **61**, 1303–1313.

72 Entwistle, J., Hall, C.L., and Turley, E.A. (1996) HA receptors: regulators of signalling to the cytoskeleton. *J. Cell. Biochem.*, **61**, 569–577.

73 Asayama, S., Nogawa, M., Takei, Y., Akaike, T., and Maruyama, A. (1998) Synthesis of novel polyampholyte comb-type copolymers consisting of a poly(L-lysine) backbone and hyaluronic acid side chains for a DNA carrier. *Bioconjug. Chem.*, **9**, 476–481.

74 Takei, Y., Maruyama, A., Ferdous, A., Nishimura, Y., Kawano, S., Ikejima, K., Okumura, S., Asayama, S., Nogawa, M., Hashimoto, M., Makino, Y., Kinoshita, M., Watanabe, S., Akaike, T., Lemasters, J.J., and Sato, N. (2004) Targeted gene delivery to sinusoidal endothelial cells: DNA nanoassociate bearing hyaluronan-glycocalyx. *FASEB J.*, **18**, 699–701.

75 Schledzewski, K., Falkowski, M., Moldenhauer, G., Metharom, P., Kzhyshkowska, J., Ganss, R., Demory, A., Falkowska-Hansen, B., Kurzen, H., Ugurel, S., Geginat, G., Arnold, B., and Goerdt, S. (2006) Lymphatic endothelium-specific hyaluronan receptor LYVE-1 is expressed by stabilin-1$^+$, F4/80$^+$, CD11b$^+$ macrophages in malignant tumours and wound healing tissue *in vivo* and in bone marrow cultures *in vitro*: implications for the assessment of lymphangiogenesis. *J. Pathol.*, **209**, 67–77.

76 Birch, M., Mitchell, S., and Hart, I.R. (1991) Isolation and characterization of human melanoma cell variants expressing high and low levels of CD44. *Cancer Res.*, **51**, 6660–6667.

77 Penno, M.B., August, J.T., Baylin, S.B., Mabry, M., Linnoila, R.I., Lee, V.S., Croteau, D., Yang, X.L., and Rosada, C. (1994) Expression of CD44 in human lung tumors. *Cancer Res.*, **54**, 1381–1387.

78 Bartolazzi, A., Peach, R., Aruffo, A., and Stamenkovic, I. (1994) Interaction between CD44 and hyaluronate is directly implicated in the regulation of tumor development. *J. Exp. Med.*, **180**, 53–66.

79 Guo, Y., Ma, J., Wang, J., Che, X., Narula, J., Bigby, M., Wu, M., and Sy, M.S. (1994) Inhibition of human melanoma growth and metastasis *in vivo* by anti-CD44 monoclonal antibody. *Cancer Res.*, **54**, 1561–1565.

80 Zeng, C., Toole, B.P., Kinney, S.D., Kuo, J.-W., and Stamenkovic, I. (1998) Inhibition of tumor growth *in vivo* by hyaluronan oligomers. *Int. J. Cancer*, **77**, 396–401.

81 Agarwal, A., Lvov, Y., Sawant, R., and Torchilin, V. (2008) Stable nanocolloids of poorly soluble drugs with high drug content prepared using the combination of sonication and layer-by-layer technology. *J. Control. Release*, **128**, 255–260.

82 Smith, R.C., Riollano, M., Leung, A., and Hammond, P.T. (2009) Layer-by-layer platform technology for small-molecule delivery. *Angew. Chem., Int. Ed.*, **48**, 8974–8977.

83 Platt, V.M. and Szoka, F.C. (2008) Anticancer therapeutics: targeting macromolecules and nanocarriers to hyaluronan or CD44, a hyaluronan receptor. *Mol. Pharm.*, **5**, 474–486.

84 Lapcik, L., De Smedt, S., Demeester, J., and Chabrecek, P. (1998) Hyaluronan: preparation, structure, properties, and applications. *Chem. Rev.*, **98**, 2663–2684.

85 Auzenne, E., Ghosh, S.C., Khodadadian, M., Rivera, B., Farquhar, D., Price, R.E., Ravoori, M., Kundra, V., Freedman, R.S., and Klostergaard, J. (2007) Hyaluronic

acid-paclitaxel: antitumor efficacy against CD44$^+$ human ovarian carcinoma xenografts. *Neoplasia*, **9**, 479–486.

86 Luo, Y. and Prestwich, G.D. (1999) Synthesis and selective cytotoxicity of a hyaluronic acid–antitumor bioconjugate. *Bioconjug. Chem.*, **10**, 755–763.

87 Luo, Y., Ziebell, M.R., and Prestwich, G.D. (2000) A hyaluronic acid–taxol antitumor bioconjugate targeted to cancer cells. *Biomacromolecules*, **1**, 208–218.

88 Eliaz, R.E. and Szoka, F.C. (2001) Liposome-encapsulated doxorubicin targeted to CD44: a strategy to kill CD44-overexpressing tumor cells. *Cancer Res.*, **61**, 2592–2601.

89 Choi, K.Y., Min, K.H., Yoon, H.Y., Kim, K., Park, J.H., Kwon, I.C., Choi, K., and Jeong, S.Y. (2011) PEGylation of hyaluronic acid nanoparticles improves tumor targetability *in vivo*. *Biomaterials*, **32**, 1880–1889.

90 Picart, C. (2008) Polyelectrolyte multilayer films: from physico-chemical properties to the control of cellular processes. *Curr. Med. Chem.*, **15**, 685–697.

91 Cui, D., Szarpak, A., Pignot-Paintrand, I., Varrot, A., Boudou, T., Detrembleur, C., Jerome, C., Picart, C., and Auzely-Velty, R. (2010) Contact-killing polyelectrolyte microcapsules based on chitosan derivatives. *Adv. Funct. Mater.*, **20**, 3303–3312.

92 Szarpak, A., Cui, D., Dubreuil, F., De Geest, B.G., De Cock, L.J., Picart, C., and Auzély-Velty, R. (2010) Designing hyaluronic acid-based layer-by-layer capsules as a carrier for intracellular drug delivery. *Biomacromolecules*, **11**, 713–720.

93 Szarpak, A., Pignot-Paintrand, I., Nicolas, C., Picart, C., and Auzély-Velty, R. (2008) Multilayer assembly of hyaluronic acid/poly(allylamine): control of the buildup for the production of hollow capsules. *Langmuir*, **24**, 9767–9774.

94 Cui, D., Jing, J., Boudou, T., Pignot-Paintrand, I., De Koker, S., De Geest, B. G., Picart, C., and Auzély-Velty, R. (2011) *Adv. Mater.*, in press.

95 Nguyen, K.T. and West, J.L. (2002) Photopolymerizable hydrogels for tissue engineering applications. *Biomaterials*, **23**, 4307–4314.

96 Bryant, S.J., Nuttelman, C.R., and Anseth, K.S. (1999) The effects of crosslinking density on cartilage formation in photocrosslinkable hydrogels. *Biomed. Sci. Instrum.*, **35**, 309–314.

97 Hennink, W.E. and van Nostrum, C.F. (2002) Novel crosslinking methods to design hydrogels. *Adv. Drug Deliv. Rev.*, **54**, 13–36.

98 Burdick, J.A., Chung, C., Jia, X., Randolph, M.A., and Langer, R. (2005) Controlled degradation and mechanical behavior of photopolymerized hyaluronic acid networks. *Biomacromolecules*, **6**, 386–391.

99 Pitarresi, G., Pierro, P., Palumbo, F.S., Tripodo, G., and Giammona, G. (2006) Photo-cross-linked hydrogels with polysaccharide–poly(amino acid) structure: new biomaterials for pharmaceutical applications. *Biomacromolecules*, **7**, 1302–1310.

100 Shah, D.N., Recktenwall-Work, S.M., and Anseth, K.S. (2008) The effect of bioactive hydrogels on the secretion of extracellular matrix molecules by valvular interstitial cells. *Biomaterials*, **29**, 2060–2072.

101 Smeds, K.A., Pfister-Serres, A., Miki, D., Dastgheib, K., Inoue, M., Hatchell, D.L., and Grinstaff, M.W. (2001) Photocrosslinkable polysaccharides for *in situ* hydrogel formation. *J. Biomed. Mater. Res.*, **54**, 115–121.

102 Leach, J.B., Bivens, K.A., Collins, C.N., and Schmidt, C.E. (2004) Development of photocrosslinkable hyaluronic acid–polyethylene glycol–peptide composite hydrogels for soft tissue engineering. *J. Biomed. Mater. Res. A*, **70**, 74–82.

103 Leach, J.B., Bivens, K.A., Patrick, Jr., C.W., and Schmidt, C.E. (2003) Photocrosslinked hyaluronic acid hydrogels: natural, biodegradable tissue engineering scaffolds. *Biotechnol. Bioeng.*, **82**, 578–589.

104 Oudshoorn, M.H.M., Rissmann, R., Bouwstra, J.A., and Hennink, W.E. (2007) Synthesis of methacrylated hyaluronic acid with tailored degree of substitution. *Polymer*, **48**, 1915–1920.

105 Trudel, J. and Massia, S.P. (2002) Assessment of the cytotoxicity of photocrosslinked dextran and hyaluronan-based hydrogels to vascular smooth muscle cells. *Biomaterials*, **23**, 3299–3307.

106 Patterson, J., Siew, R., Herring, S.W., Lin, A.S.P., Guldberg, R., and Stayton, P.S.

(2010) Hyaluronic acid hydrogels with controlled degradation properties for oriented bone regeneration. *Biomaterials*, **31**, 6772–6781.

107 Chung, C. and Burdick, J.A. (2009) Influence of three-dimensional hyaluronic acid microenvironments on mesenchymal stem cell chondrogenesis. *Tissue Eng. Part A*, **15**, 243–254.

108 Fedorovich, N.E., Oudshoorn, M.H., van Geemen, D., Hennink, W.E., Alblas, J., and Dhert, W.J.A. (2008) The effect of photopolymerization on stem cells embedded in hydrogels. *Biomaterials*, **30**, 344–353.

109 Patterson, J., Martino, M.M., and Hubbell, J.A. (2010) Biomimetic materials in tissue engineering. *Mater. Today (Oxford, UK)*, **13**, 14–22.

110 van Dijk, M., Rijkers, D.T.S., Liskamp, R. M.J., van Nostrum, C.F., and Hennink, W. E. (2009) Synthesis and applications of biomedical and pharmaceutical polymers via click chemistry methodologies. *Bioconjug. Chem.*, **20**, 2001–2016.

111 Ghosh, K., Shu, X.Z., Mou, R., Lombardi, J., Prestwich, G.D., Rafailovich, M.H., and Clark, R.A.F. (2005) Rheological characterization of *in situ* cross-linkable hyaluronan hydrogels. *Biomacromolecules*, **6**, 2857–2865.

112 Jin, R., Teixeira, L.S.M., Krouwels, A., Dijkstra, P.J., van Blitterswijk, C.A., Karperien, M., and Feijen, J. (2010) Synthesis and characterization of hyaluronic acid–poly(ethylene glycol) hydrogels via Michael addition: an injectible biomaterial for cartilage repair. *Acta Biomater.*, **6**, 1968–1977.

113 Kim, G.-W., Choi, Y.-J., Kim, M.-S., Park, Y., Lee, K.-B., Kim, I.-S., Hwang, S.-Y., and Noh, I. (2007) Synthesis and evaluation of hyaluronic acid–poly (ethylene oxide) hydrogel via Michael-type addition reaction. *Appl. Phys.*, **7S1**, e28–e32

114 Kim, J., Kim, I.S., Cho, T.H., Lee, K.B., Hwang, S.J., Tae, G., Noh, I., Lee, S.H., Park, Y., and Sun, K. (2007) Bone regeneration using hyaluronic acid-based hydrogel with bone morphogenic protein-2 and human mesenchymal stem cells. *Biomaterials*, **28**, 1830–1837.

115 Kim, J., Park, Y., Tae, G., Lee, K.B., Hwang, C.M., Hwang, S.J., Kim, I.S., Noh, I., and Sun, K. (2008) Characterization of low-molecular-weight hyaluronic acid-based hydrogel and differential stem cell responses in the hydrogel microenvironments. *J. Biomed. Mater. Res. A*, **88**, 967–975.

116 Vanderhooft, J.L., Mann, B.K., and Prestwich, G.D. (2007) Synthesis and characterization of novel thiol-reactive poly(ethylene glycol) cross-linkers for extracellular-matrix-mimetic biomaterials. *Biomacromolecules*, **8**, 2883–2889.

117 Flannery, C.R., Little, C.B., Hughes, C.E., and Caterson, B. (1998) Expression and activity of articular cartilage hyaluronidases. *Biochem. Biophys. Res. Commun.*, **251**, 824–829.

118 Tanimoto, K., Suzuki, A., Ohno, S., Honda, K., Tanaka, N., Doi, T., Nakahara-Ohno, M., Yoneno, K., Nakatani, Y., Ueki, M., Yanagida, T., Kitamura, R., and Tanne, K. (2004) Hyaluronidase expression in cultured growth plate chondrocytes during differentiation. *Cell Tissue Res.*, **318**, 335–342.

119 Klouda, L. and Mikos Antonios, G. (2008) Thermoresponsive hydrogels in biomedical applications. *Eur. J. Pharm. Biopharm.*, **68**, 34–45.

120 Kim, J.H., Lee, S.B., Kim, S.J., and Lee, Y. M. (2002) Rapid temperature/pH response of porous alginate-g-poly(*N*-isopropylacrylamide) hydrogels. *Polymer*, **43**, 7549–7558.

121 Wang, L.-Q., Tu, K., Li, Y., Zhang, J., Jiang, L., and Zhang, Z. (2002) Synthesis and characterization of temperature responsive graft copolymers of dextran with poly(*N*-isopropylacrylamide). *React. Funct. Polym.*, **53**, 19–27.

122 Tan, H., Ramirez, C.M., Miljkovic, N., Li, H., Rubin, J.P., and Marra, K.G. (2009) Thermosensitive injectable hyaluronic acid hydrogel for adipose tissue engineering. *Biomaterials*, **30**, 6844–6853.

123 Censi, R., Fieten, P.J., di Martino, P., Hennink, W.E., and Vermonden, T. (2010) *In situ* forming hydrogels by tandem thermal gelling and Michael addition reaction between thermosensitive triblock copolymers and thiolated hyaluronan. *Macromolecules (Washington, DC)*, **43**, 5771–5778.

124 Shu, X.Z., Ghosh, K., Liu, Y., Palumbo, F. S., Luo, Y., Clark, R.A., and Prestwich, G.

D. (2004) Attachment and spreading of fibroblasts on an RGD peptide-modified injectable hyaluronan hydrogel. *J. Biomed. Mater. Res. A*, **68**, 365–375.

125 Park, Y.D., Tirelli, N., and Hubbell, J.A. (2003) Photopolymerized hyaluronic acid-based hydrogels and interpenetrating networks. *Biomaterials*, **24**, 893–900.

126 Kim, J., Park, Y., Tae, G., Lee, K.B., Hwang, S.J., Kim, I.S., Noh, I., and Sun, K. (2008) Synthesis and characterization of matrix metalloprotease sensitive-low molecular weight hyaluronic acid based hydrogels. *J. Mater. Sci. Mater. Med.*, **19**, 3311–3318.

127 Suh, K.Y., Kim, Y.S., and Lee, H.H. (2001) Capillary force lithography. *Adv. Mater. (Weinheim)*, **13**, 1386–1389.

128 Suh, K.Y., Khademhosseini, A., Yang, J.M., Eng, G., and Langer, R. (2004) Soft lithographic patterning of hyaluronic acid on hydrophilic substrates using molding and printing. *Adv. Mater. (Weinheim)*, **16**, 584–588.

129 Xia, Y. and Whitesides, G.M. (1998) Soft lithography. *Angew. Chem., Int. Ed.*, **37**, 550–575.

130 Kane, R.S., Takayama, S., Ostuni, E., Ingber, D.E., and Whitesides, G.M. (1999) Patterning proteins and cells using soft lithography. *Biomaterials*, **20**, 2363–2376.

131 Fukuda, J., Khademhosseini, A., Yeh, J., Eng, G., Cheng, J., Farokhzad, O.C., and Langer, R. (2006) Micropatterned cell co-cultures using layer-by-layer deposition of extracellular matrix components. *Biomaterials*, **27**, 1479–1486.

132 Khademhosseini, A., Suh, K.Y., Yang, J.M., Eng, G., Yeh, J., Levenberg, S., and Langer, R. (2004) Layer-by-layer deposition of hyaluronic acid and poly-L-lysine for patterned cell co-cultures. *Biomaterials*, **25**, 3583–3592.

133 Takahashi, S., Yamazoe, H., Sassa, F., Suzuki, H., and Fukuda, J. (2009) Preparation of coculture system with three extracellular matrices using capillary force lithography and layer-by-layer deposition. *J. Biosci. Bioeng.*, **108**, 544–550.

134 Barbucci, R., Lamponi, S., Magnani, A., and Pasqui, D. (2002) Micropatterned surfaces for the control of endothelial cell behaviour. *Biomol. Eng.*, **19**, 161–170.

135 Barbucci, R., Magnani, A., Lamponi, S., Pasqui, D., and Bryan, S. (2003) The use of hyaluronan and its sulfated derivative patterned with micrometric scale on glass substrate in melanocyte cell behavior. *Biomaterials*, **24**, 915–926.

136 Barbucci, R., Torricelli, P., Fini, M., Pasqui, D., Favia, P., Sardella, E., d'Agostino, R., and Giardino, R. (2005) Proliferative and re-differentiative effects of photo-immobilized micro-patterned hyaluronan surfaces on chondrocyte cells. *Biomaterials*, **26**, 7596–7605.

137 Dickinson, L.E. and Gerecht, S. (2010) Micropatterned surfaces to study hyaluronic acid interactions with cancer cells. *J. Vis. Exp.*, (46), e246.

138 Dickinson, L.E., Ho, C.-C., Wang, G.M., Stebe, K.J., and Gerecht, S. (2010) Functional surfaces for high-resolution analysis of cancer cell interactions on exogenous hyaluronic acid. *Biomaterials*, **31**, 5472–5478.

139 Khademhosseini, A., Eng, G., Yeh, J., Fukuda, J., BlumlingJ III, J., Langer, R., and Burdick, J.A. (2006) Micromolding of photocrosslinkable hyaluronic acid for cell encapsulation and entrapment. *J. Biomed. Mater. Res. A*, **79**, 522–532.

140 Khetan, S. and Burdick, J.A. (2010) Patterning network structure to spatially control cellular remodeling and stem cell fate within 3-dimensional hydrogels. *Biomaterials*, **31**, 8228–8234.

141 Khetan, S., Katz, J.S., and Burdick, J.A. (2009) Sequential crosslinking to control cellular spreading in 3-dimensional hydrogels. *Soft Matter*, **5**, 1601–1606.

142 Chung, C., Mesa, J., Miller, G.J., Randolph, M.A., Gill, T.J., and Burdick, J.A. (2006) Effects of auricular chondrocyte expansion on neocartilage formation in photocrosslinked hyaluronic acid networks. *Tissue Eng.*, **12**, 2665–2673.

143 Marklein, R.A. and Burdick, J.A. (2010) Spatially controlled hydrogel mechanics to modulate stem cell interactions. *Soft Matter*, **6**, 136–143.

10
Polymer Gels from Biopolymers

Esra Alveroglu, Ali Gelir, and Yasar Yilmaz

10.1
Introduction

In this chapter, an experimental technique based on the doping of gels with some molecules having counterions is improved and used mainly for studying polymerization, gelation, sol–gel transition, gel imprinting, internal morphology, and swelling of the gels. When the hydrogels are doped with aromatic molecules having subgroups, they are used for probing the polymerization and gelation reactions. Counterions of the subgroups of the dye molecules or some other salt molecules behave as free ions in the gel. Upon swelling of the gels, these free ions become mobile and serve as charge carriers. The gels doped with negative and positive counterions show interesting behavior like rectifying the current similarly to traditional p–n junctions. At the same time, the dye molecules shift their emission spectra when they are bonded chemically to the gel over one of the subgroups during the polymerization. This allows us to follow the monomer conversion accurately as a function of polymerization time. By this way, many physical parameters are examined with high accuracy. In this chapter, some studies about theory and application of the hydrogels are presented in different sections.

10.2
Experimental Methods

The monomer acrylamide (AAm), the initiator ammonium persulfate (APS), the activator tetramethylethylenediamine (TEMED), and the cross-linker methylenebisacrylamide (BIS) were supplied by Merck (Darmstadt, Germany), and the monomer N-isopropylacrylamide (NIPA) and methacrylamidopropyl trimethylammonium chloride (MAPTAC) were supplied by Kohjin Co. Ltd (Japan). Pyranine was supplied by Fluka. All chemicals were used as received. Bidistilled water was used in the experiments.

Gels and polymers used in Section 10.3 were synthesized at 25 °C by using various amounts of AAm, ranging between 0.45 and 5 M, and BIS, ranging between 3.84

and 32 mM, by dissolving them in 25 cm³ of water. The concentrations of TEMED and APS were kept fixed for all samples at 10 μl per 25 cm³ of water and 7 mM, respectively. The gels used in Section 10.4 were synthesized in a similar way but the AAm and BIS concentrations were kept fixed at 2 M and 65 mM, respectively. The concentration of the fluorescence molecule, pyranine, used for probing the polymerization/gelation processes and the internal morphology was 10^{-4} M for all samples synthesized in Sections 10.3 and 10.4.

AAm concentrations of the gels used in Sections 10.6 and 10.7 were changed from 0.5 to 4 M and the BIS and APS concentrations were chosen so as to synthesize the gels at normal stoichiometry [1,2], where the ratios of AAm/BIS and AAm/APS were 31 and 43, respectively. Pyranine concentrations were changed from 0 M (plain gel) to 10^{-2} M. The gelation experiments were performed at 60 °C and no activator molecule, TEMED, was used.

Fluorescence measurements were carried out by using LS50 Perkin & Elmer fluorescence spectrometer for the experiments in Section 10.3 and USB2000 CCD array spectrometer of Ocean Optics for the experiments in Sections 10.4 and 10.5. Electrical measurements given in Sections 10.6 and 10.7 were performed by using Keithley 6487 model picoammeter.

After bubbling nitrogen for 10 min, each prepolymer and pregel solution was poured into a $1 \times 1 \times 4$ cm³ glass tube; thereafter, the glass tube was sealed and put into the sample holder of the spectrometer. Fluorescence measurements during the polymerization and gelation were made at 90° position and slit widths were kept at 5 nm. Pyranine was excited at 340 nm wavelength of light and variations in the fluorescence spectra were monitored as a function of polymerization or gelation times. Swelling experiments were also performed in bidistilled water, and the fluorescence measurements were carried out during the swelling processes of the gel samples.

The NIPA gels used in Section 10.7 (called n-type gels) were synthesized at 60 °C. NIPA concentrations of these gels were 2 and 4 M. BIS was used as a cross-linker and AIBN was used as an initiator. NIPA/BIS and NIPA/AIBN ratios were fixed at 69 and 103, respectively, for all gels. MAPTAC was added while preparing the pregel solution. MAPTAC concentrations were changed from 0 M (plain gel) to 3×10^{-2} M.

The gels used in Section 10.5 for imprinting studies were prepared by free radical polymerization of NIPA (6 M) with BIS (100 mM), MAPTAC (30 mM), and pyranine (10 mM) in DMSO. Here, NIPA, MAPTAC, and pyranine were chosen as main monomer unit, the receptor monomer with functional group, and the target molecule, respectively, which were studied earlier by Tanaka and coworkers [3]. After the addition of AIBN (10 mM), the solutions were immediately transferred into a tube containing micropipettes with inner diameter of 0.5 mm and into narrowly spaced glass plates in order to obtain cylindrical and slab gels, respectively. The solutions were then degassed under vacuum. The polymerization was carried out at 60 °C for 24 h. For the imprinted gel, 30 mM MAPTAC and 10 mM pyranine were mixed in DMSO to form a complex structure before adding the pregel solution. After the synthesis, all the gels were washed with frequently refreshed aqueous solution of 0.1 M HCl and 0.1 M NaOH for about 2 weeks.

Pieces of dried cylindrical gels were weighed (4–10 mg) and placed in aqueous solutions of pyranine (3 ml) of 10^{-5} M concentration. Adsorption studies were performed at different temperatures (30, 40, and 60 °C) and with different replacement molecule (NaCl) concentrations, 50 and 200 mM. All the adsorption experiments were performed in the temperature-controlled sample holder of the fluorescence spectrophotometer and the fluorescence spectra of the target molecule, pyranine, were monitored during the adsorption process. The excitation wavelength was chosen as 400 nm and the emission wavelength of 515 nm was monitored during the adsorption process.

10.3
Polymerization and Gelation Kinetics

A polymer is a chain of monomers – two functional molecules – connected by chemical or physical interactions. The number of monomers in a polymer, so-called the degree of polymerization, may reach up to 10^{10} monomers, one of the largest DNA molecules [4]. Depending on the chemical identity of the monomer and the degree of polymerization, polymers have different properties and applications, from simple gas (number of C atoms: 1–4) to tough plastic solids (number of C atoms: >1000) [4].

The whole structure of a polymer is generated during the polymerization. The microstructure, isomerism, and tacticity may become less important for long polymer chains. Thus, they can be considered as long flexible threads from the viewpoint of physics. Average size defined as the radius of gyration of a polymer chain varies with the monomer–monomer and monomer–solvent interactions. Attractive interactions make the polymer chain more open, while repulsive interactions can cause the chain to be in its collapsed state.

Branched polymers can be formed in the presence of a cross-linker, a monomer with three or more functional groups. Increasing cross-linker concentration will generally result in a three-dimensional network, the so-called gel. The gel is a solid-like material that can have properties ranging from soft and weak to hard and tough. Although the gels are mostly liquid, they behave like solids due to a three-dimensional cross-linked network within the liquid.

A hydrogel is a network of highly absorbent natural or synthetic polymers in which water is the dispersion medium. They can contain over 99% water that may have degree of flexibility very similar to natural tissue. Therefore, they can be used in tissue engineering. Environment-sensitive hydrogels that are also known as "smart gels" can have the ability to sense changes in pH, temperature, or reversible adsorption and release target molecules as a result of such a change. They may be designed so as to be responsive to some specific molecules. Diversity of monomers and the molecules that can be used as target put forward almost infinite number of possibilities in the application of these gels.

Physical properties of the polymeric gels have prompted researchers to investigate gels as sensors, membranes for separations, drug delivery agents, artificial muscles, and actuators [5,6]. Moreover, the polymer gels can simulate biological tissues; they have an enormous importance in biomedical applications [7–9]. Polyacrylamide (PAAm) hydrogels are widely used because of this biocompatibility and unique chemical and physical properties. Therefore, they attract the attention of researchers working on PAAm hydrogels as superabsorbents [10], electrophoresis matrices, encapsulating agents for drugs, enzymes and proteins in drug delivery systems, biosensors, and so on [11,12].

Monomer conversion is the key variable for polymerization studies. To understand the physical nature of polymerization processes, to regulate the compositions, and to test the existing theories, one must follow the reaction kinetics via monomer conversion.

Different methods such as gravimetry [13], densitometry [14], calorimetry [15], and ultrasound velocity [16] have been used for the measurement of monomer conversion. These techniques are carried out offline, resulting in a measurement delay, which is undesirable for real-time control. Some analytical techniques such as Raman [17], dielectric [18], and Fourier transform infrared [19] spectroscopy have been developed in the past two decades for inline or online monitoring of the compositions in polymerization processes. However, for the samples with very high conversion (especially for gels), all these methods provide a relatively poor accuracy in monomer conversion.

Recently, we developed a new approach, based on the steady-state fluorescence spectroscopy, for *in situ* monitoring of the monomer conversion based on the chemical interaction of a fluorescence probe (aromatic hydrocarbons that show fluorescence properties, abbreviated as *fluoroprobe* or *dye*) with polymer chains during the free radical polymerization or gelation of PAAm [20,21]. The probe molecule binds chemically to the polymer strands *during the polymerization* and thus changes its emission spectra [20]. It allows us to monitor the polymerization process in real time with great sensitivity. Thus, we are able to measure some of the physical parameters. These are measuring the monomer conversion [21], monitoring the network formation in real time, and testing the network theories discussing the universality of the gel at the sol–gel transition point [22,23].

Almost all the universality discussions near the sol–gel threshold have been tested by computer simulations. Using this new approach, the results of the percolation theory have been tested for the first time in the laboratory by measuring the universal exponents [22,23] β and γ, defined for the average cluster size and the strength of the infinite network, and the fractal dimension [23] d_f of the gel during the gelation.

Fluoroprobes are widely used for monitoring various processes and functions on a microscopic level [24,25]. This technique is based on the interpretation of the change in anisotropy, emission spectra, or intensity, and viewing the lifetimes of injected fluoroprobes to monitor the change in their microenvironment [26–29].

Fluoroprobes can be used in two ways for the studies on polymerization and gelation. In the first approach, one can add a fluoroprobe to the system (extrinsic fluoroprobe) and by this way it is possible to measure some physical parameters of the polymerizing system, such as polarity [30,31], viscosity [32–35], and hydrophobicity [36]. In the second approach, the fluoroprobe is covalently attached to the polymer and serves as a polymer bond label (intrinsic fluoroprobe) [37], where the polymer–fluoroprobe association depends on some factors including chemical bonding, Coulombic interactions, and hydrophobicity of the polymer–fluoroprobe pair.

These techniques have been successfully used to perform the experiments on polymerization [38,39], chemical gel formation [40–42], swelling of the gels [43], slow release of the probe molecules from a gel [44,45], metal ion detection via metal ion-templated polymeric gel [46], affinity of the gels to the target molecules [47], and examination of the collapsed state phases and volume phase transition of the polymeric gels [48,49].

We show that the fluoroprobe pyranine (8-hydroxypyrene-1,3,6-trisulfonic acid, trisodium salt) can be used for *in situ* monitoring of the free radical polymerization of the acrylamide system [20]. It binds chemically to the growing polyacrylamide chains over –OH group and electrostatically to the protonated amide groups on the polymer chains over SO_3^- groups [21].

10.3.1
Fluoroprobe–Polymer Interactions

The molecular structure and typical fluorescence spectra of pyranine at different stages during the free radical polymerization of AAm are shown in Figure 10.1.

As seen from Figure 10.1, before adding APS to the pregel solution, the fluorescence emission of the pyranine is similar to the fluorescence emission of pyranine in pure water. But after adding APS for initiating the polymerization, the intensity of 515 nm peak starts decreasing and a new peak appears at around 406 nm. While the intensity of 515 nm peak decreases, the intensity of the new peak increases and the maximum of this new peak gradually shifts to 428 nm from 406 nm as the polymerization progresses. This shift saturates at 428 nm and no

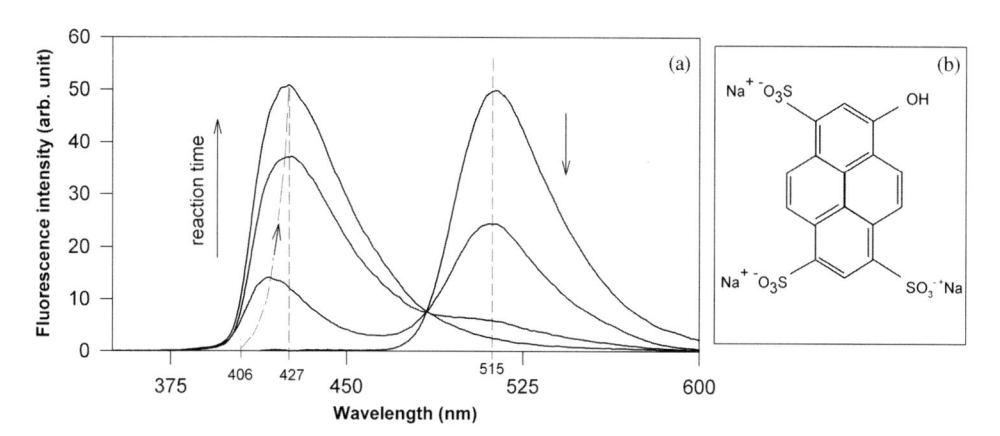

Figure 10.1 (a) Typical fluorescence spectra of pyranine at different stages of free radical cross-linking copolymerization of AAm with BIS. (b) Molecular structure of pyranine. (Adapted from Ref. [20].)

further change is observed as the polymerization continues. Here, it should be noted that the blueshift from 515 to 406 nm is *not* due to the pH effect since the maximum of the fluorescence spectra of pyranine in pure water below pH 1 is clearly different [20] from short-wavelength peak (around 406 nm, appears only if the polymerization is initiated). It was also shown in the literature that [44,45] the time for gradual shift from 406 to 428 nm is a function of AAm concentration and it decreases when the AAm concentration is increased. For AAm concentrations below 0.5 M, the gradual shift is a linear function of the polymerization time, while it is an exponential function for AAm concentrations above 0.5 M. It was observed that the samples below 0.5 M AAm did not turn into the gel; they become viscous fluids with varying viscosity depending on the polymer concentration.

This blueshift from 515 to 406 nm is due to C—O ether bond formation between the hydroxylic oxygen, which is the oxygen atom included in the univalent radical or group OH, of pyranine and a terminal C atom of the growing AAm chain, as depicted in Scheme 10.1. Also, the protonation effect produced by water, as depicted

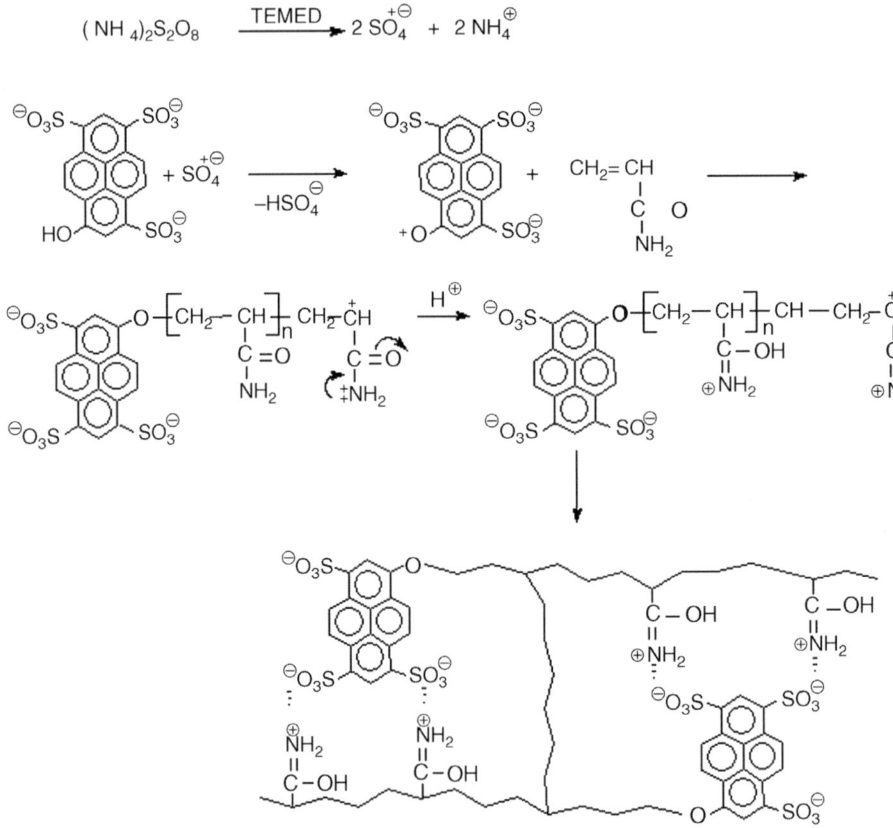

Scheme 10.1 Representation of the multiple-point interaction of pyranine with PAAm chains during the polymerization.

in Scheme 10.1, leads to formation of macroradical carrying positive charge on the polymer chain as a result of the localization of an unpaired electron on the amide groups of the acrylamide chains [50]. Therefore, the complex structure that consists of mutual attraction between sulfate groups of the probe and amide groups with positive charges on polyacrylamide chain is formed.

It was shown in Refs [44,45] that pyranines can only bind to the polymer strands during the polymerization process in a way represented in Scheme 10.1. In these references, some diffusion and washing experiments were performed to prove this binding. In the diffusion experiments, no change was observed in the emission spectra of pyranine diffused into a plain gel at different pH values. By the plain gel we mean the gels that do not include pyranine when they are synthesized. In the washing experiments, although the gels, synthesized with pyranine, were washed many times by water, acid, and base, still fluorescence emission spectra continue to present in the washed gels with a maximum at 410 nm. These two experiments clearly showed that pyranines added to the pregel solution can bind chemically over –OH group to the polymer strands only during the polymerization.

The second point in this binding is the blueshift from 515 to 406 nm. It is a well-known property of the pyranine in the literature that when the deprotonation rate of the excited pyranine is small enough, that is, it is longer than the fluorescence lifetime, the fluorescence emission is observed at around 406 nm [48,49]. In Refs [48,49], pyranines chemically bind to the polymer strands over –OH group and this structure is the analogue of the deprotonated pyranine, so a blueshift from 515 to 406 nm is observed upon this chemical binding.

To find out possible reasons of the gradual shift from 406 to 428 nm, some swelling experiments on the gels and dilution experiments on the linear polymers of AAm synthesized with pyranine were performed in Ref. 9. When the washed gel, which includes bonded pyranines emitting at 428 nm peak in the collapsed state of the gel, swells in water, the emission maximum at 428 nm shifts to the 410 nm gradually (reverse of the shift observed during the synthesis) depending on the swelling degree of the gel. During the dilution of the polymers synthesized with pyranine, similar results were observed and thus a reverse gradual shift from 428 nm to 410 nm, as observed in the swelling of the gels. When the polymers are precipitated in methanol, the emission maximum of the pyranines shifts to 406 nm, as observed at the beginning of the polymerization. These experiments clearly show that the number of SO_3^-–amide contacts decreases – in contrast to the gelation/polymerization process – because the gel/polymer becomes more open and the probability of SO_3^- groups interacting electrostatically with the amide groups decreases as the gel swells or the polymer is diluted. Also when the sample is precipitated, the electrostatic interaction between SO_3^- and amide groups was destroyed since the charged sites on the polymer chains that are induced by water disappear when methanol is replaced with water. Therefore, it can be concluded that the gradual shift from 406 to 428 nm upon polymerization and the reverse gradual shift upon swelling and dilution are due to the number of the electrostatic interactions between the SO_3^- groups of pyranine and the amide groups of the polymer strands as depicted in Scheme 10.1 and the binding probability of a SO_3^- group to an amide group increases with monomer conversion or polymerization time.

10.3.2
Real-Time Monitoring of Monomer Conversion

As mentioned earlier, chemical binding of pyranines to the polymer chains causes a blueshift, from 515 to 406 nm, and a gradual redshift, from 406 to 428 nm, is observed as the polymerization progresses. Here we show that, both theoretically and by comparing with gravimetric measurements, the fluorescence intensity of pyranine (added prepolymer solution in trace amount) monitored during the polymerization process can be used for *in situ* monitoring of the monomer conversion – the percentage of the monomers converted to the polymer or to gel at any instant of the reaction time – with great sensitivity.

It has been proved that the monomer conversion, $\chi(t)$, is a linear function of the fluorescence intensity of the bonded probes, $I(t)_{bonded}$, measured at the same instant of time during the polymerization:

$$\chi(t) = \frac{n(t)_{bonded}}{M_0 P_p} \propto \frac{I(t)_{bonded}}{M_0 P_p} = kI(t)_{bonded}, \tag{10.1}$$

where $n(t)_{bonded}$ is the number of bonded probes, M_0 is the total number of the monomers, P_p is simply the fraction of the probe molecules in the sample, and k is the proportionality constant including M_0 and P_p that are fixed for each experiment [48,49]. Here, it should be noted that this relation is true only if nonradiative processes of the bonded species remain unaltered during the polymerization. This condition could be fulfilled in low concentrated samples. Otherwise, another contribution related with the microviscosity should be added.

In Refs [48,49], the reliability of Equation 10.1 was confirmed by the gravimetric measurements. In these references, identical samples were prepared for fluorescence and gravimetric measurements for gelation and polymerization and the polymerization was initiated simultaneously under the same environmental conditions, as mentioned in Section 10.3.1. When the polymerizations were progressing and the fluorescence of the pyranine was being observed *in situ*, identical samples of gravimetric measurement were quenched one by one at varying reaction times by pouring the samples quickly into a vessel including 10 ml methanol and then putting in liquid nitrogen to stop the polymerization reaction. The polymer parts in each sample were precipitated in methanol for 1 week and filtered and dried in the furnace at 40 °C until the equilibrium is reached (about 2 weeks). Then the mass of the precipitated (and dried) parts, $m(t)_{precipitated}$, was measured and recorded as a function of the reaction time.

The monomer conversion was then calculated as $\chi(t) = m(t)_{precipitated}/m_{total}$, where m_{total} is the total initial mass of the polymer constituents AAm, BIS, and APS. The percentage of measured monomer conversions and the normalized fluorescence intensities, $I(t)_{bonded}$, measured in real time against the reaction

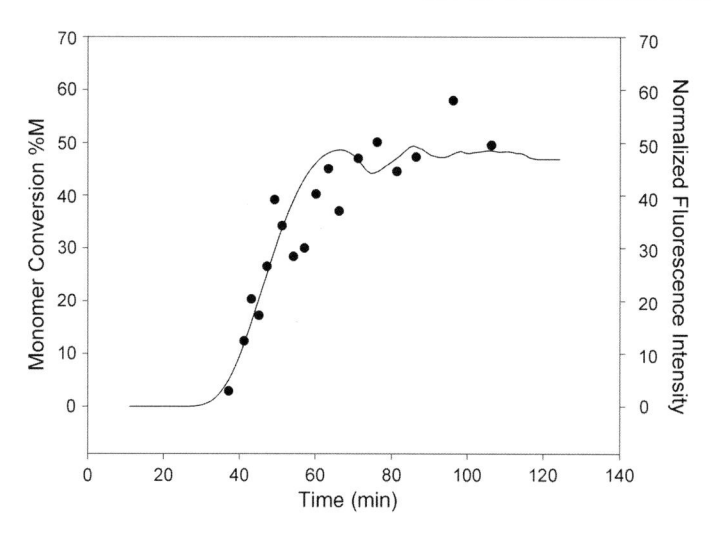

Figure 10.2 The fluorescence intensity of the bonded pyranines (solid line) and percent monomer conversion (•) obtained by the gravimetric measurements for 0.5 mol/l AAm linear polymer. (Taken from Ref. [21].)

time are plotted in Figure 10.2 for 0.5 M AAm polymerization. Similar results were observed for the gel samples as discussed in Refs [48,49].

As seen from Figure 10.2, the normalized fluorescence intensity fits, in experimentally acceptable errors, well to the monomer conversion data measured by means of the macroscopic experiments.

10.4
Sol–Gel Transition and Universality Discussion

In percolation theory, a chemical bond between two monomers is represented by an edge between two neighboring lattice sites. Each bond is formed randomly with probability p. The predictions of Flory–Stockmayer (mean-field) [51,52] and percolation theories [53–58] about the critical exponents for the sol–gel transition are different from the viewpoint of the universality.

Consider, for example, the exponents γ and β for the weight average degree of polymerization, DP_w, and the gel fraction G (average cluster size, S, and the strength of the infinite network, P_∞, in percolation language near the gel point p_c). They are defined as

$$DP_w \propto (p_c - p)^{-\gamma}, \quad p \to p_c^-, \quad (10.2)$$

$$G \propto (p - p_c)^{\beta}, \quad p \to p_c^+, \quad (10.3)$$

where mean-field theory [51,52] gives $\beta = \gamma = 1$, independent of the dimension (D), while percolation theory gives γ and β values of around 1.8 and 0.42, respectively, for $D = 3$ [54,55,57].

The comparison of the percolation results with real gelation has long remained inconclusive because of insensitivity and inaccuracy of the experimental methods. We surmount the experimental difficulties using pyranine as a probe of polymerization as discussed above. First, we measured directly the critical exponents γ and β together in laboratory with great sensitivity and accuracy. By this way, the result of the classical and percolation theories could have been tested adequately with real experiments.

The experiments were repeated for different AAm and BIS contents, and for varying temperatures. The gel points were determined by the dilatometric technique, where a steel sphere was moved in the sample up and down slowly by means of a piece of magnet applied from the outer face of the sample cell. The time at which the motion of the sphere stopped was considered as the onset of the gel point, t_c. The gel point was also tested by measuring the fractal dimension of the gel at the sol–gel threshold. This work will be shown in a later section.

No considerable deviation is observed when the areas under the spectra (see Figure 10.1) are calculated instead of the maximum intensities. Therefore, we had the chance to get relatively more data in time, just monitoring the small part (including the maxima) of the spectra. We then used these data to evaluate the critical point behavior of the sol–gel transition [22].

We proved that the total fluorescence intensity from the bonded pyranines monitors the weight average degree of polymerization and the growing gel fraction for below and above the gel point, respectively:

$$I \propto \mathrm{DP_w} = C^+(t_c - t)^{-\gamma}, \quad t \to t_c^-, \tag{10.4}$$

$$I - I_{ct} \propto G = B(t - t_c)^\beta, \quad t \to t_c^+, \tag{10.5}$$

where C^+ and B are the critical amplitudes.

These relations were proved in Ref. 22 under the assumption that the monomers occupy the sites of an imaginary periodic lattice.

If the temperature and concentration are kept fixed, then p will be directly proportional to the reaction time, t. This proportionality is not linear over the whole range of reaction time, but it can be assumed that in the critical region, that is, around the critical point, $(p - p_c($ is linearly proportional to $(t - t_c($ [59–61].

Therefore, below the gel point, that is, for $t < t_c$, the fluorescence intensity I measures the weight average degree of polymers (or average cluster size). Above t_c, if the intensity I_{ct} from finite clusters distributed through the infinite network is subtracted from the total intensity, then the corrected intensity $I - I_{ct}$ measures solely the gel fraction G, the fraction of the monomers that belong to the macroscopic network as discussed in detail in Ref. 22.

Using Equations 10.4 and 10.5, and the measured values for t_c, exponents γ and β were calculated as a function of AAm concentration, BIS concentration, and temperature. A typical example of the procedure for measuring the exponents will be given in a later section – to avoid the repetition – together with the measurements of the fractal dimension of the gel at the sol–gel transition point. Here we would like to argue the results.

The exponents γ and β agree best with the Flory–Stockmayer theory in the concentration range of 0.65–1 M, and agree with percolation theory for higher AAm concentrations above 1 M. It seems that 0.65 and 2 M concentrations are crossover concentrations for exponents γ and β: 0.65 M from nonuniversal to mean-field values, and 2 M from mean-field to percolation values for the acrylamide system. The nonuniversal behavior for extremely small concentrations may be due to the fact that we are extremely far away from the critical region for these experimental conditions.

Using percolation theory, it was argued that [55–57] the largest clusters at p_c have a fractal behavior,

$$M \propto L^{d_f}, \tag{10.6}$$

defined for the first time by Mandelbrot [58]. Here M is the mass of the cluster and L is the lattice size. The fractal dimension, d_f, of the infinite cluster formed at $p = p_c$ is about 2.5 [56,57]. The finite clusters formed at p_c also have the same fractal dimension for length scales smaller than or equal to the size of the clusters. Above p_c, both the infinite clusters and the finite clusters are no longer fractal; they form three-dimensional objects. It is argued [56] on the basis of percolation theory that below p_c the fractal dimension of the finite clusters is around 2, $d_f \sim 2$, in three dimensions.

Recently, a new technique based on the steady-state fluorescence spectroscopy was proposed to measure, for first time, the fractal dimension of PAAm hydrogel during the gelation process at any instant of the polymerization, and tested the percolation picture against real experiments [23].

In situ monitoring of the intensity of the fluorescence emission of the bonded pyranines, which corresponds to 428 nm, during the polymerization at different slit widths of the excitation source of the spectrophotometer is the main point of this technique. The results of such an experiment are given in Figure 10.3 [22].

As previously discussed in Ref. 22, the fluorescence intensity from the gelling sample is directly proportional to the average cluster size S (or the weight average degree of polymerization, DP_w) at the same instant t: $I(t) \propto S(t)$. Therefore, the intensities given in Figure 10.3 will be proportional to the mass M of the average cluster formed at time t: $I(t) \propto M(t)$.

In the experiments in Ref. 22, the illuminated region of the samples is a rectangular prism in form. The height and the depth of this illuminated region were kept fixed for each sample and the width was changed by changing the slit width of the slit regulator. Therefore, the relation defining the fractal dimension $M \propto L^{d_f}$ takes the following form:

$$I = A w^{d_f/3}, \tag{10.7}$$

where w is the width of the slit and A is a proportionality constant. Note that the exponent in Equation 10.7 changes to $d_f/3$ since the illuminated region is scaled only in one dimension.

Using the data in Figure 10.3, the intensities corresponding to certain times were recorded for each sample, and the log–log plots of Equation 10.7,

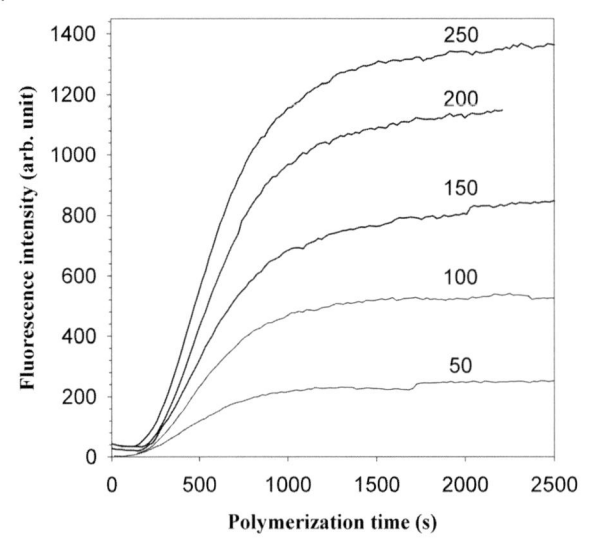

Figure 10.3 The fluorescence emission intensities at 427 nm from five identical samples during the gelation processes. The numbers on the curves represent the corresponding slit width, in μm unit, of the exciting light. (Taken from Ref. [23].)

$\log I = \log A + (d_f/3)\log w$, were formed for different time steps, as given in Figure 10.4. The slope of each curve in Figure 10.4, thus, gives $d_f/3$ for the corresponding gelation time.

The d_f is plotted in Figure 10.5 as a function of the gelation time. At the initial stage of the polymerization, the monomer conversion is small and thus the number

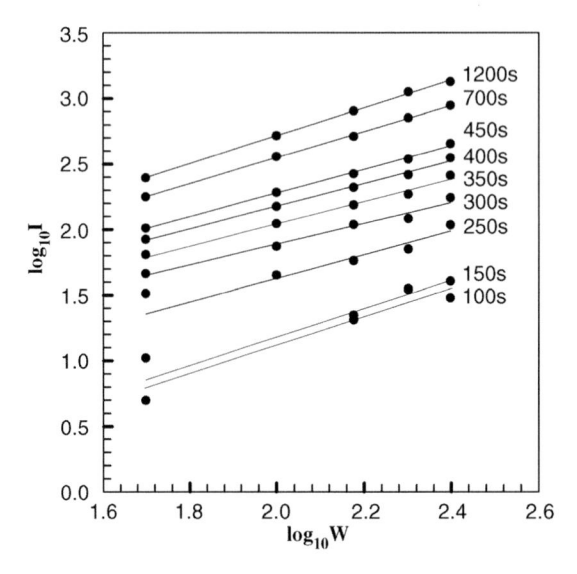

Figure 10.4 The log–log plots of Equation 10.7 formed, using the data in Figure 10.3, for different time steps between 100 and 1200 s. (Taken from Ref. [23].)

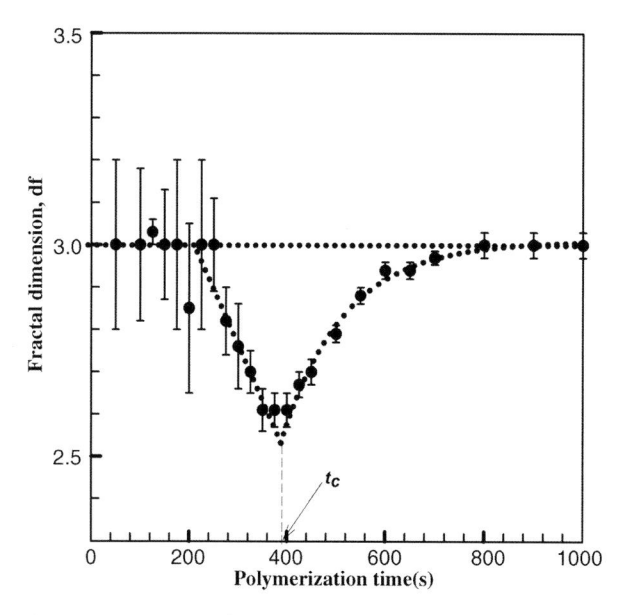

Figure 10.5 Measured fractal dimension d_f during the polymerization. (Taken from Ref. [23].)

of pyranines attached to the polymer strands will be small as well, since the number of bonded pyranines increases with increasing monomer conversion [22,23]. Therefore, the intensities at the beginning of polymerization include big random errors. As the monomer conversion increases, the measured fluorescence intensities increase, and thus the experimental errors on d_f decrease as seen from Figure 10.5. In spite of big random errors in the initial stage of the gelation, it is obvious from Figure 10.5 that the fractal dimension passes through a minimum value of ~2.5 at a critical time t_c.

Here, the time t_c corresponding to the minimum of the d_f–t curve is, at the same time, close to the bond percolation threshold, $p_c = 0.2488$ [56], for a 3D cubic lattice. The critical conversion p_c is defined as the critical ratio of *the number of monomers that have been polymerized to the total number of monomers*. Since the number of monomers converted to polymers will be proportional to the fluorescence intensity I, p_c can be defined as $p_c = I(t_c)/I(t_{max})$ in terms of the fluorescence intensities taken from the gel at $t = t_c$ and $t = t_{max}$. Using the data in Figure 10.4 and taking t_c as the time corresponding to the minimum of d_f (Figure 10.5), the average value of p_c was found to be 0.254 ± 0.003, which is very close to the percolation threshold [56].

The exponents γ and β defined in Equations 10.2 and 10.3 near the critical point t_c were calculated in Ref. 22. Figure 10.6 represents the log–log plots of Equations 10.4 and 10.5, where the slopes of the straight lines, near the gel point, give the exponents γ and β. We repeated this procedure for different times thought as the critical points and observed that the average values of γ and β coincide with percolation results, $\gamma = 1.8$ and $\beta = 0.45$, only if t_c is chosen as the time corresponding to the minimum of d_f–t curve presented in Figure 10.5.

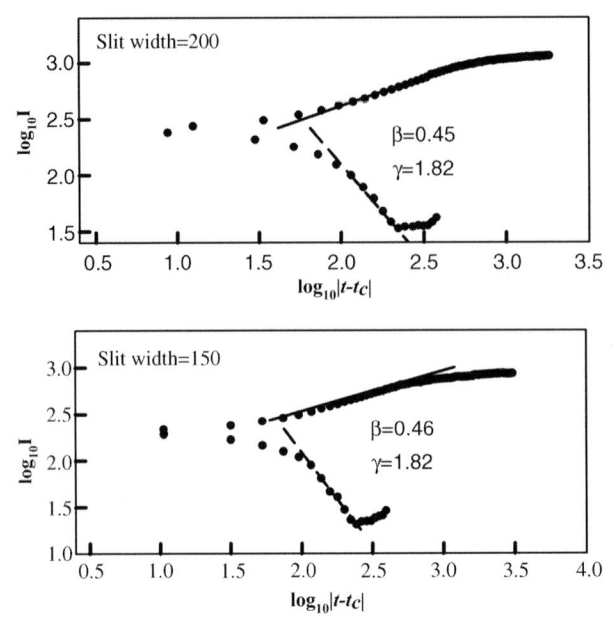

Figure 10.6 Typical log–log plots of Equations 10.4 and 10.5 for the the slit widths of 150 and 200 µm. (Taken from Ref. [23].)

It should also be noted that we are *not* able to measure the fractal dimension of the polymer clusters formed *before the gelation threshold*, where d_f is expected to be close to 2 [57], but it *seems* that we measure it as 3 as can be seen from the initial part of Figure 10.5. This unexpected result can be evaluated as follows. The radius of polymer clusters is roughly 0.05 µm [4] on an average, which is smaller than our minimum length of scale (50 µm). Therefore, we are not able to observe the inside of small clusters. The picture thus seems to be a three-dimensional object at the beginning of the polymerization and we measure d_f as 3 instead of 2. As the clusters grow up and become sensible with our length scales, the crossover in d_f occurs from 3 to ~2.5 when the infinite cluster starts to appear. Therefore, the minimum of d_f denoted as t_c is the sol–gel transition point or the so-called percolation threshold.

10.5
Imprinting the Gels

Molecular recognition is defined as reversible adsorption – by means of different physical and chemical interactions – of a specific molecule in a medium in which different kinds of molecules exist together. The interactions between target and receptor molecules are generally noncovalent interactions such as van der Waals interactions, hydrogen bonds, and hydrophobic interactions [62].

Recognition is a vital process in biological systems. The specific interaction of antigen and antibody was the first molecular recognition process observed by Landsteiner in 1936. This interaction was first named as "molecular imprinting" by Pauling in 1940.

Molecular recognition was first defined in the literature by Cram, Lehn, and Pedersen in 1987 in their work on selective interactions between molecules based on the structure of the molecules, which was rewarded by Nobel Prize.

Studies on molecular recognition/imprinting were started by Polyakov in 1930s by the experiments on the special structures in silica matrix that can select different doped molecules [63]. The basis of the modern molecular imprinting studies was provided by Wulff, Mosbach, and Shea in 1980s and 1990s [64–67]. After the pioneering papers of these researchers, the experimental and theoretical studies in this field started to increase rapidly.

Polymeric systems are the most suitable systems by which molecular recognition processes can be studied experimentally. The specially designed polymeric systems to study the molecular recognition processes are called "imprinted polymers." Polymeric gels are one of the most important classes of the polymeric systems because of their important chemical and physical properties. Special geometrical structures and interactions can be constructed inside a gel to recognize a molecule. They can respond to external environmental changes such as temperature [53,54,68], solvent composition [55–57,69], pH [55–57,68,70,71], electric field [68,72], or photon energy [68,73]. First imprinting studies in gels (especially in hydrogels) were started theoretically by Dusek and Patterson in 1968 [74] and experimentally by Tanaka in 1978 [75].

To realize molecular recognition in imprinted polymeric systems, there must be one or combinations of the three interactions, covalent [76], noncovalent [77], and semicovalent (hybrid) [78], between receptor and target molecules. Imprinted systems with covalent interactions are the most sensitive but irreversible systems. After binding of targets to receptors, it is not so easy to take out the targets again [76]. In imprinted systems with noncovalent interactions, reversible adsorption of the target molecules is possible. Before the synthesis of this type of imprinted gels, a complex between target and receptor molecules must be formed and the gel synthesis performed in the presence of these complexes. During the synthesis, some of the complexes are decomposed, so the imprinting efficiency of this type of system gets lower than the efficiency of the systems in which the covalent interactions occur [77]. In imprinted systems with semicovalent interactions, covalent interactions occur between target and receptor molecules during the synthesis. After synthesis, the target molecules may be dissociated by some chemical processes and noncovalent interactions between target and the receptor molecules take place of the covalent interactions [78].

In the literature, different imprinted polymers and gels were synthesized for different kinds of targets such as metals [46,79], organic molecules [80], and proteins [81].

Imprinted polymers and gels have important technological applications such as controlled drug delivery and transportation [82,83], biotechnological applications

[84,85], food industry [86], chromatography [67], environmental applications [87], and sensor/biosensor designs [46,88].

Characterization of imprinted polymers and gels is the key point in these studies. For this purpose, different methods are used such as optical methods [89], surface plasmon resonance [90], and electrochemical methods [91]. Some of these methods are very expensive to apply and some of them are used only on a limited scale [92]. Because of these limitations, a low-cost, easy-to-apply, and wide-range method is needed. Fluorescence spectroscopy is a suitable candidate for characterization of the imprinted systems because this method is sensitive, not expensive, and gives comprehensive information about the interactions [93,94]. Nowadays, there are many papers in the literature about the usage of this system in imprinting [92,95].

In the fluorescence method, a fluoroprobe is required and it may be present (or bonded) in target molecules [3,47] or in receptor molecules [46,79]. After the interaction of target molecules with receptor molecules, the fluorescence properties of the fluoroprobe change and by evaluating these changes in the fluorescence, it is possible to carry out the recognition process. These changes can be observed in the intensity [46,79,96] or in the emission/excitation wavelength of the fluoroprobe, or in the anisotropy of the molecules due to the rotations of the excited molecules [97].

The geometrical shapes of the gels used for the diffusion experiments were cylindrical in form. The diffusion of a molecule is given by the following equation [98]:

$$\frac{M_t}{M_\infty} = 1 - \sum_{n=1}^{\infty} \frac{4\alpha(1+\alpha)}{4 + 4\alpha + \alpha^2 q_n^2} e^{-Dq_n^2 t/r^2}, \tag{10.8}$$

where M_t and M_∞ are the amounts of molecules diffused at time t and $t = \infty$, respectively, α is the ratio of the cross-sectional areas of the gel and the tube in which the experiment is performed ($\alpha = A_t/A_j$), r is the radius of the gel, D is the diffusion coefficient, and q_n are the roots of the following equation:

$$\alpha q_n J_0(q_n) + 2 J_1(q_n) = 0, \tag{10.9}$$

where J_0 and J_1 are the zeroth- and first-order Bessel functions.

Since the diffusion process was monitored via fluorescence spectra, a relation between the fluorescence intensity and the amount of molecules diffused into the gel can be derived. When the concentration range of the fluoroprobe is selected in the region where the fluorescence intensity against the concentration of the fluoroprobe changes linearly, as done in this work, then the relation $I \propto M$ can be written as

$$M_t = I_0 - I(t) \quad \text{and} \quad M_\infty = I_0 - I_\infty, \tag{10.10}$$

where I_0, I_∞, and $I(t)$ are the fluorescence intensities taken from the solution of fluoroprobe in which the diffusion experiment is performed at times $t = 0$, $t = \infty$, and t, respectively.

By putting the relations given in Equation 10.10 in Equation 10.8, the following equation is obtained:

$$\frac{I(t) - I_\infty}{I_0 - I_\infty} = \sum_{n=1}^{\infty} \frac{4\alpha(1+\alpha)}{4 + 4\alpha + \alpha^2 q_n^2} e^{-Dq_n^2 t/r^2}. \tag{10.11}$$

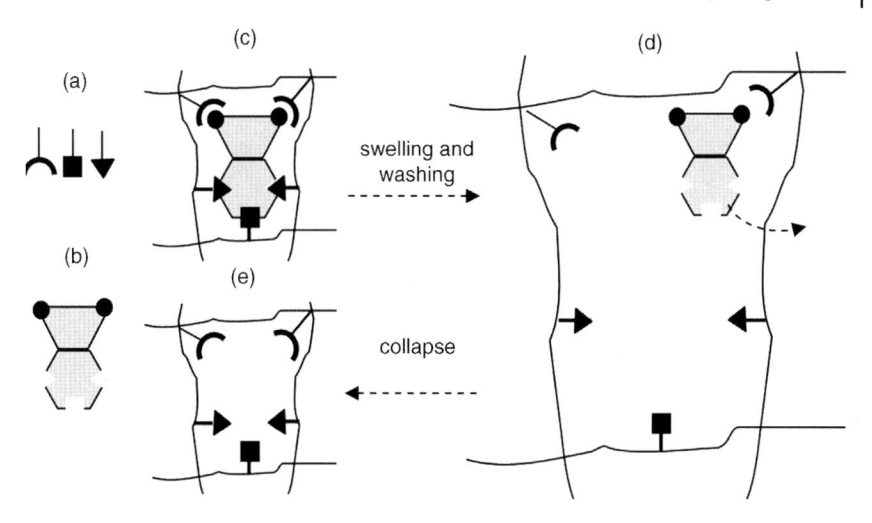

Scheme 10.2 Schematic representation of construction of an imprinted site in a gel: (a) functional groups with different interactions, (b) target molecule, (c) imprinting a cavity, (d) swelling and washing, and (e) collapsing.

If Equation 10.11 is written for $n = 1$ [99] and natural logarithm is taken, the relation between the fluorescence intensity and the diffusion coefficient can be written as

$$\ln\left(\frac{I(t) - I_\infty}{I_0 - I_\infty}\right) = \ln\left(\frac{4\alpha(1+\alpha)}{4 + 4\alpha + \alpha^2 q_1^2}\right) - \frac{D q_1^2}{r^2} t. \tag{10.12}$$

When the fluorescence intensities taken from the solution are plotted against time, the diffusion coefficients can be calculated from the slopes of these plots.

In Scheme 10.2, a general picture of the imprinting process in polymeric gels is represented. A few examples of functional groups with different interactions such as electrostatic, van der Waals, covalent, and so on are given in Scheme 10.2a. To construct a complex structure between the receptor and the target molecules (Scheme 10.2b), both of them must have similar type of functional groups. When the gel is synthesized in the presence of the complex between the target molecule and the receptor monomer, this complex binds to the polymer chains over the receptor monomers, and as a result a special cavity that has a geometrical structure of target molecule is constructed (Scheme 10.2c). By swelling and washing, the target molecules can be washed out of the gel (Scheme 10.2d), and after collapsing the gel, the vacant cavities for the target molecules preserve their geometrical structures (Scheme 10.2e). The affinity of these vacant imprinted sites to the target molecules is very high and they can select and adsorb the target molecules between different kinds of molecules in a solution [3,47]. In this chapter, the receptor monomer is MAPTAC that has a positively charged N^+ group and the target molecule is pyranine that has three negatively charged SO_3^- groups (Scheme 10.3). The complex structure between the three receptor monomers and

Scheme 10.3 Molecular structures of (a) receptor monomer (MAPTAC) and (b) target molecule (pyranine).

the target molecule is constructed via the electrostatic interactions between N^+ and SO_3^- groups.

In Figure 10.7, the emission spectra of pyranine in DMSO and water are given and the emission peaks are observed at 415 and 510 nm in DMSO and water, respectively. The spectroscopic properties of this molecule are well known from many works [45,78,82–86,90,96,98].

In Figure 10.8, the time evolution of the ratios of the excimer emission [94] at 480 nm to monomer emission at 430 nm of pyranine during the gelation processes of the imprinted and random gels is presented. As seen from this figure, the ratio for the random gel increases more than that for the imprinted gel as the pregel solution turns to the gel state. This clearly shows that the pyranines in the random gel have much more tendency to come close to each other compared with the imprinted gel. The reason for this difference is the complex structure between the MAPTAC and

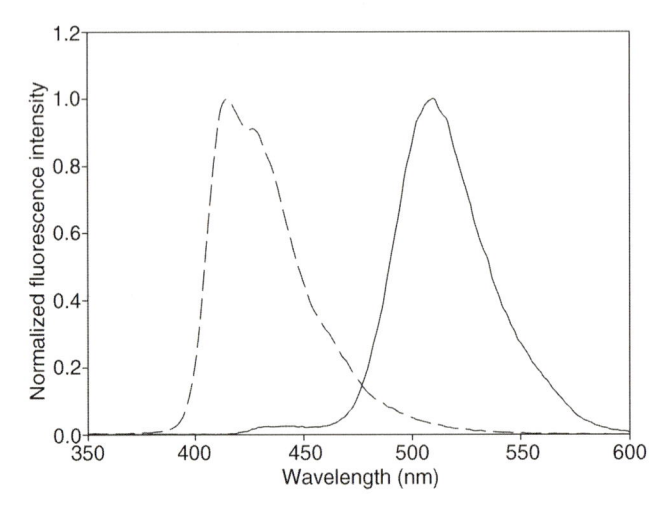

Figure 10.7 Fluorescence emission spectra of pyranine in DMSO (dashed line) and in water (solid line). The excitation wavelength is 400 nm.

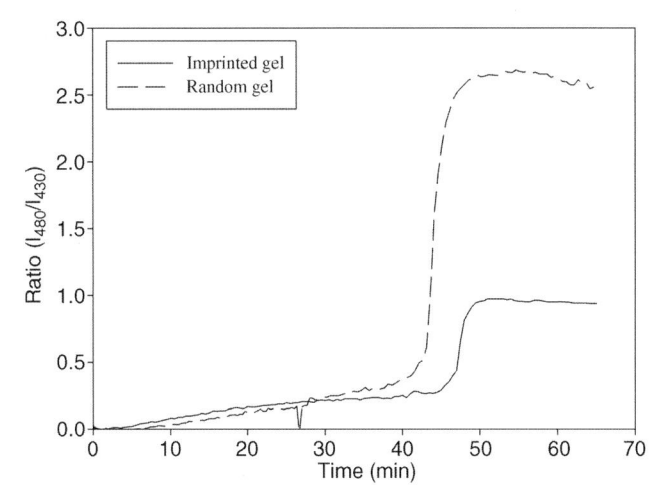

Figure 10.8 Time evolution of the ratio of excimer emission intensity to monomer emission intensity during the gelation for random (dashed line) and imprinted (solid line) gels. The excitation wavelength is $\lambda_{exc} = 400$ nm.

the pyranine in the imprinted gel as discussed earlier. This complex structure prevents most of the pyranines to come close to each other in the imprinted gel. Nevertheless, presence of an excimer emission in the imprinted gel shows that a portion of the complexes was decomposed during the gelation. However, this figure clearly proves that most of the complexes in the imprinted gel preserve their structures.

In Figure 10.9, the results of the adsorption studies at different temperatures are given with respect to time. In the swollen state of the gels (30 °C), no difference was observed between the random and imprinted gels. Swelling exterminates the difference between the imprinted and random gels by widening the network of the gels. At the beginning of the collapsed state (40 °C) and at the collapsed state at 60 °C, which is the synthesis temperature of the gels, a difference between the random and imprinted gels was observed. The network structures of the gels at 40 °C are the initial stages of their collapsed states and the diffusion of pyranines to the random gels is larger than the diffusion to the imprinted gel. This is due to the difference between the network structures of the random and imprinted gels, meaning that the sizes of the cavities in the random gel are larger than those in the imprinted gel, or the imprinted gel is in a more collapsed state than the random gel. This result may indicate that the imprinted gel tends to go to its initial collapsed state, which it had at the end of the synthesis, faster than the random gel. At 60 °C, a considerable difference between the random and imprinted gels was observed. At this temperature, the imprinted gel reaches its more regular state in which special cavities having appropriate size, shape, and functional groups for binding the target molecules are present. Therefore, most of the pyranines were absorbed by the imprinted gel.

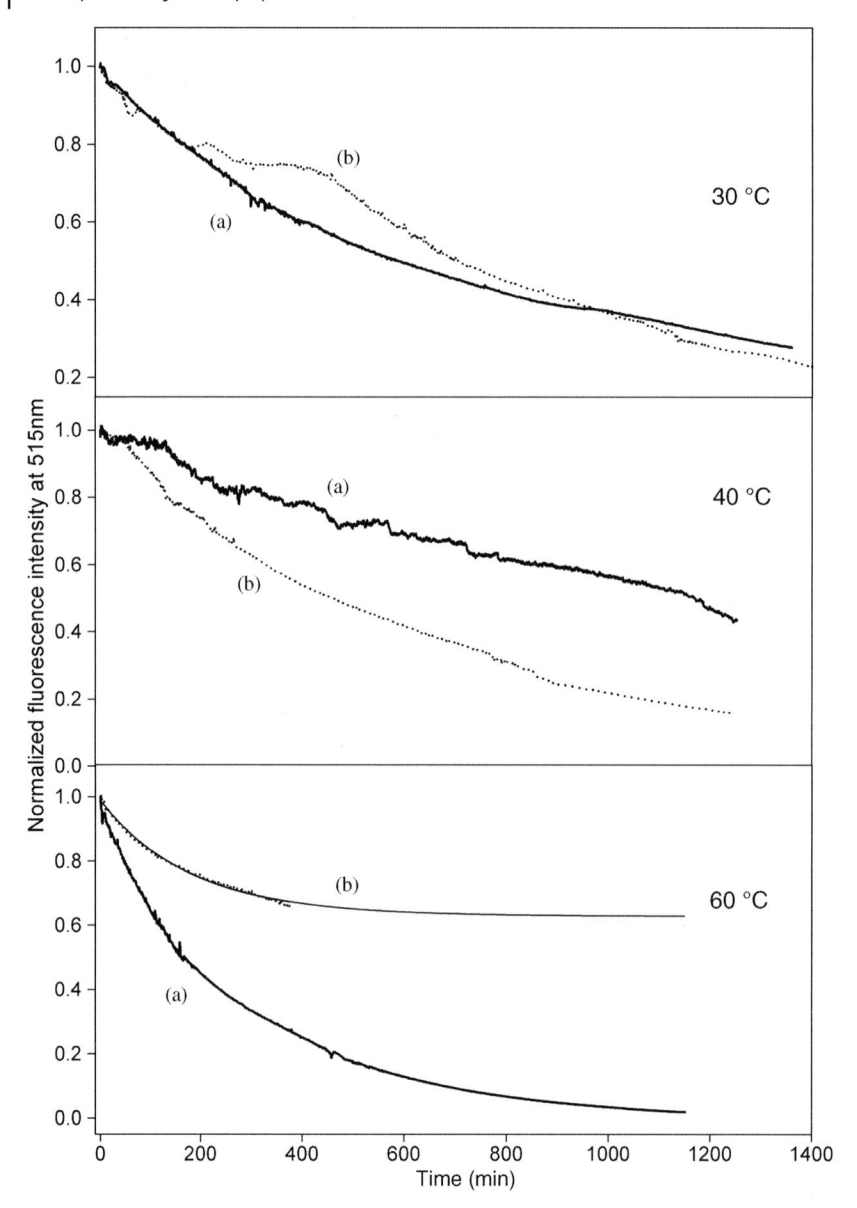

Figure 10.9 Time evolution of the fluorescence emission intensity of pyranine at 515 nm taken from the solution during the diffusion for imprinted (a) and random (b) gels at different temperatures. The excitation wavelength is $\lambda_{exc} = 400$ nm.

The diffusion coefficients were calculated by using Equation 10.12 as 3.58 and 5.56 for the imprinted and random gels, respectively. Higher diffusion coefficient for the random gel should be due to the presence of the large vacant blobs in this gel. At the beginning of the diffusion, a large amount of pyranines is expected to diffuse into these vacant blobs in the random gel in a short time. But this process results in reducing the difference of the charge density between inside and outside of the gel, which is an important factor for the diffusion expressed by the Donnan potential [20]. Since this difference is reduced, the total amount of diffused pyranines gets lower in the random gel compared to the imprinted gel after a long time period. As a result, it can be concluded that the large vacant blobs in the random gel or disordered structure compared to the imprinted gel increase the diffusion coefficient but decrease the total amount of diffused pyranines.

Salt (NaCl, replacement molecule) plays a key role in the molecular recognition process and the presence of an appropriate amount of salt results in an efficient recognition between receptor and target molecules [3]. Salt pulls out the unwanted molecules from the imprinted sites and binds itself to the receptor molecule due to its higher affinity. But the affinity of the target molecules to the imprinted sites is larger than that of the salt, and once a target is bound to the receptor molecule on an imprinted site, salt cannot pull out the target molecule easily [3]. Therefore, the sensitivity of the target to receptor should be tested in the presence of the salt to elucidate the sole effect of imprinting.

In Figure 10.10a and b, the fluorescence spectra of the pyranine diffused into the gel for different salt concentrations in the solution at which the diffusion experiments were performed are represented for random and imprinted gels, respectively. When the solid lines of Figure 10.10a and b are compared for low salt concentration, it is clearly seen that the emission peaks are at the same wavelength (525 nm) for both random and imprinted gels. It is a very interesting result because in the random gel since no imprinted cavity for pyranines is present, it is expected that the pyranines would emit at around 510 nm. But it seems that in the random gel the pyranines are interacting over three SO_3^- groups, as in imprinted gels. When the salt concentration is increased, a considerable change was observed in the emission spectrum taken from the random gel while the change in the emission spectrum taken from the imprinted gel was not so much (dashed lines in Figure 10.10a and b). In the emission spectrum taken from the random gel, a new peak around 430 nm was observed. This new peak may be due to the emission of the pyranines in which the hydrogen atom of –OH group was replaced by Na. As shown earlier in Ref. 96, the pyranine with –ONa group instead of –OH group emits at around 430 nm in water. In addition to this, the electrostatic interactions between the salt clusters and the SO_3^- groups of the pyranines also shift the emission spectrum to the long wavelengths as shown in Ref. 20 and this may be the reason for the appearance of emission at 525 nm in the random gel. In the imprinted gel, the fact that there is a small change in the emission spectrum after increasing the salt concentration indicates that the pyranines are bound to the imprinted sites in the imprinted gel and salt cannot pull out most of these pyranines, so only small amount of the pyranines may be pulled out. It can be concluded from these results that in the random gel since the pyranines are free, salt molecules can affect their fluorescence properties easily, but in imprinted gel it is not possible because the pyranines are bonded to the imprinted sites.

All the above observations clearly show that the fluorescence technique can be a better tool for searching the imprinted gels where target molecules are aromatic

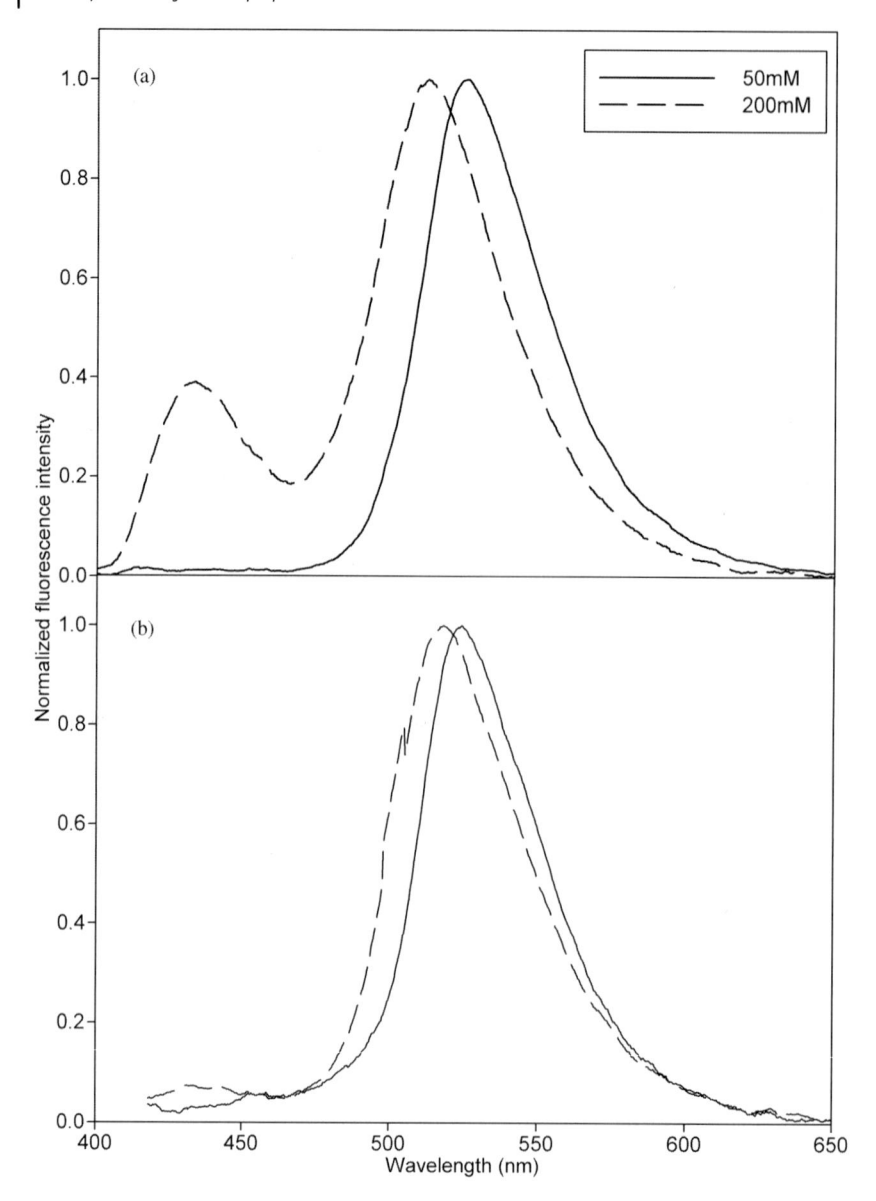

Figure 10.10 The fluorescence emission spectra of pyranine diffused into the random (a) and imprinted (b) gels at different salt concentrations. Fluorescence spectra were taken from the gels. The excitation wavelength is 400 nm.

hydrocarbons. When the fluorescence technique is compared with the UV/Vis spectroscopic technique used in Ref. 93, it can be easily seen that analysis of the fluorescence measurements is easier and gives much more information about the recognition.

10.6
Heterogeneity of Hydrogels

Experimental studies on PAAm hydrogels have shown that they are naturally heterogeneous [100–108], because of the inhomogeneous cross-linking distribution [109–113]. Highly intermolecular cross-linked structures are formed in the pregel period due to the high extent of cyclization [114]. Thus, the structure of PAAm gel represents itself as a set of blobs and long polymer chains connecting these blobs to each other [115]. This is due to different affinity to the interactions between AAm–AAm and AAm–BIS molecules [115] during the reaction. The reactivity of AAm–BIS is higher than that of AAm–AAm. The difference in the reactivities results in cross-link-rich regions, which causes the local heterogeneity in PAAm gels. This heterogeneous structure contributes to the phenomenon of nonergodicity of a gel as discussed in theoretical and experimental works in the literature [116–123]. It was also shown that partially hydrolyzed PAAm hydrogels undergo a phase transition as a result of changes in solvent composition, temperature, and pH of the solution [124]. The increase in the cross-linker concentration increases the compactness of these junctions without essentially changing distances between the blobs [125].

Dielectric spectroscopy measurements of conductivity [126] are performed for understanding internal morphology of neutral and permanently charged PAAm hydrogels during the swelling process. In that study, some dominant peaks in conductivity were observed during the swelling of these gels with pure water. Each of the peaks corresponds to a different swelling stage of the neutral gels. These peaks are related to the number of generations of the "blobs" appearing in a microstructure of the given PAAm gel; each peak corresponds to a generation (the blobs that have more or less the same size) of the blobs. This behavior (specific peaks in the conductivity) becomes almost negligible when the gels are charged with ionic groups. The heterogeneity decreases due to the internal electric field for the charged gels [126].

10.6.1
Effect of Ion Doping on Swelling Properties and Network Structure of Hydrogels

The morphology and heterogeneity affect swelling properties of the PAAm hydrogels. Tanaka and Filmore initiated the swelling experiments on PAAm gels in 1970s [127]. They observed that the final swelling ratio of the PAAm gels changes as a function of solvent composition. In other studies, the effects of free ionic groups on the swelling behavior of the hydrogels have been studied [128–130]. It has been shown that the gels containing free ions undergo discontinuous volume transition during swelling, depending on the concentration of the ionic group. The swelling behavior of PAAm hydrogels has been studied extensively for different solvent compositions and the ionic groups that are free in solution [107,108,131–137]. Blobs in PAAm gels can more or less swell depending on the cross-linker concentration and the nature of the solvent [138,139]. Some "correlation length" can be attributed to the average size between the clusters, which are not entangled with the smaller ones. This length approximately coincides with the typical size of "voids" distributed in the blob clusters. The average size of a single blob strongly depends on the temperature, presence of an additional charge, and the cross-linker concentration.

In most studies, the osmotic pressure produced by the difference between the ion concentrations inside and outside of the gels was taken into account but the swelling kinetics have not been evaluated from the viewpoint of the internal morphology (the effect of heterogeneity) of the gels. We studied the swelling kinetics of PAAm gels synthesized, as described in Section 10.2, with/without pyranines that were chemically bonded or entrapped (free) in the gel. The effect of the free and bonded ions on the swelling of PAAm gels was examined in detail [1]. In swelling curves, a stepwise behavior was observed for a certain monomer and bonded ion concentration. Moreover, this gel shows an abrupt swelling up to 1300-fold of its dried volume, 2 M AAm and 10^{-2} M pyranine [1]. The stepwise behavior is not observed for the neutral and other charged gels, because a continuous size distribution of heterogeneous regions was probably seen during the gelation process. The swelling kinetics of this special gel does not obey the classical Li–Tanaka model. It seems that the Li–Tanaka equation can be applied to only those gels where the gel is completely homogenous or the gel is so heterogeneous that the generations of the blobs are not distinguishable, that is, blobs of all sizes exist at the same time.

10.6.2
Current Measurements for Searching the Internal Morphology of the Gels

Estimating the size or density distribution of "blobs" in the heterogeneous gel, we have developed a model [140]. In this work, a theoretical model based on the measured current–time plots of the neutral and charged gels was developed for estimating the weight fraction and the number of generations of the blobs. According to this model, the current density per unit mass, J/m_{gel}, decreases with time as a series of exponentially decreasing terms, where the exponent B and the multiplier A measure the density of the blob and the fraction of the corresponding blob generation, respectively:

$$J/m_{gel} = \sum_{i,j} A_{ij} e^{-B_{ij}t}. \tag{10.13}$$

This theoretical model was confirmed experimentally via DC current measurements. The gels prepared, as described in Section 10.2, were in cylindrical form and placed between two platinum electrodes that were fixed gently in a cylindrical glass tube. The open ends of the tube were sealed by a Teflon stopper. The voltage was applied between these electrodes and current through the gel was measured by using Keithley 6487 multimeter. The mass of the gels was also measured just before and after the current measurements to be sure that there is no considerable change in the water content of the gel due to possible drying effect. No considerable changes in the initial mass of the gels were observed after the measurements. We examined our model for PAAm systems prepared with/without charged groups and found that each blob generation adds a new time-dependent exponentially decreasing term to

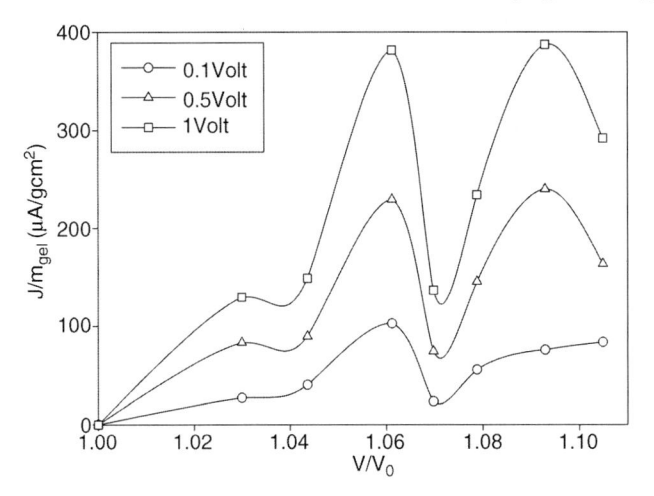

Figure 10.11 Variation in J/m_{gel} against V/V_0 of neutral gel under constant applied voltages of 0.1, 0.5, and 1 V.

the current. The results agree with the recent literature results [126], where the different peaks in the conductivity are related to the distribution of dense polymer regions defined as the "blobs" appearing in a microstructure of the given PAAm gels. The number of the terms in Equation 10.13 that fits best with the experimental data corresponds to the number of peaks in the conductivity curves [126], or, equivalently, to the number of peaks in $J/m_{gel}-V/V_0$ curves given in Figure 10.11.

10.7
Ionic p-Type and n-Type Semiconducting Gels

The silicon p–n junction is a landmark in the development of modern microelectronics. Since the discovery of the silicon p–n junction, it has become the cornerstone of modern microelectronics, because it is capable of being used in a diversity of electronic devices such as transistors, light-emitting diodes, and photovoltaic cells. Silicon and other inorganic semiconductors are the predominant materials in electronic manufacturing; however, the development of organic-based electronic components has been a focus of intensive research and development. The fact that the conjugated polymers become electrically conducting initiated many fundamental investigations and created the possibility of a new kind of materials combining the typical properties of plastics with the electrical conductivity of metals [141–144]. Today, the development of organic-based electronic equipment is an attractive research area for the production of electronic devices. So, organic materials have vast potential for integration in low-cost microelectronic devices. Circuits and displays based on organic electronics may also be flexible, low-weight, and environment-friendly. Finally, some water-tolerant organic devices may be biocompatible; that is, they may be used in circuits that directly reside on or are implanted inside animal and human tissue and could perform various sensing, interfacing, and controlling functions for drug delivery, prosthetics, and neural–electronic integration.

The effect of electrically conductive polymers and their advantages continue to enhance applications in a wide range of arising technologies from polymer-based electronics to nanotechnologies. This is a very active research subject spanning many areas, including materials development, device design, deposition processes, and modeling with conductive polymers. Electronic devices, which were produced by organic semiconductors, can be scalable, less expensive, and less complex. Therefore, polymeric materials have the potential for integration in inexpensive microelectronic devices.

Moreover, the great advantage of conjugated polymer semiconductors is their ability to quickly change the carrier density electrochemically in the presence of mobile counterions. These features of organic semiconductors provide new mechanism for time-dependent and field-dependent properties [145].

10.7.1
Electrical Properties of Ionic p-Type and n-Type Semiconducting Gels

Insulating polymeric gels have been made conductive by using various methods such as ion implantation [146–149], press contacting [150,151], and photochemical doping [152]. These materials have the ability to manipulate electronic properties by changing their molecular structure [153].

Recently, we synthesized [154] new kinds of semiconducting polymeric hydrogels having negative (n-type) and positive (p-type) counterions as charge carriers. The PAAm hydrogel was doped with Na^+ ions as charge carriers, to make the so-called ionic p-type gel.

For doping the gel with Na^+ ions, we synthesized the gel in the presence of pyranine. The pyranine binds chemically to the polymer chains, over its –OH group, during the polymerization process, as discussed in Section 10.3.1, and they form stable charged sites doped with positive counterions. Thus, the polyacrylamide gel is doped with pyranine having SO_3^- ions as side groups and Na^+ as counterions. These counterions can move under the electric field and act as positive charged ions. So, this type of gel is called a p-type semiconducting gel.

N-Isopropylacrylamide hydrogel was doped with Cl^- ions as charge carriers, to obtain an ionic n-type semiconducting gel in a similar way to the p-type gel. The methods of synthesis are described in Section 10.2.

After the gelation processes, the gels were washed and dried [47,49]. Thus, only charged molecules, which are bound chemically to the polymer strands, remained in the gel. After the gels dried, they were cut into thin slices of $\sim$1 mm thickness and swollen with pure water to certain degree for electrical measurements. All measurements were made at room temperature (23 °C). No current was observed in the collapsed state. The gels become conductive only upon water uptake, because counterions cannot move in the collapsed state.

The current density per unit mass of the gel, J/m_{gel}, decreases exponentially with time under constant voltage and mass ratio. The ionic conductivity is the effective mechanism for conduction in these gels. Conductivity decreases as the ionic charge carriers are accumulated on the electrode surfaces. The decrease in the conductivity is not due to the drying of the gel since the experimental system is sealed against drying. The initial amplitude of a current–time plot increases with increasing doping concentration, showing that the net charge in the gel increases with increasing doping concentration [154].

When the electrodes are short circuited after $\sim$300 s while the constant voltage is applied to the gel, first the current density suddenly becomes negative and then it goes to zero exponentially with time. This is also the proof that ionic conductivity plays a role in these gels.

As discussed earlier in this chapter and in Refs [126,140], some peaks in J/m_{gel}–m/m_0 plots have been observed that are related to internal heterogeneity of neutral and charged gels. When gels start to swell, ions in the less dense region can move and contribute to the current. As time goes on, charge carriers are depleted from dense blobs. Thus, current densities decrease until more dense blobs are swollen and newly freed ions contribute to current.

10.7.2
Polymeric Gel Diodes with Ionic Charge Carriers

There are a lot of studies that show that conjugated semiconducting polymer composites that contain immobile anions as counterions for the oxidized form of the polymer and mobile cations can be used for electronic junctions and devices [145,155–168]. These devices are made in temporary [164,169] or fixed state via frozen junction [167,170], electrochemical disproportionation and trapping methods [159], or radical-induced polymerization of ion pair monomers as counterions [163]. The current rectifying properties of these junctions arise from redox reactions occurring either across a membrane [171] or at the interface with the electrodes [172,173]. Within some polymeric diodes known as electrolyte diodes [174–176], a uniform hydrogel or film, freely transporting the ion, separates two depots containing acidic and basic media. The disadvantages of bipolar membranes or these electrolyte diodes are that they include water-filled chambers. Moreover, gradients of electrolyte and/or pH affect their operation; for example, over time these devices will stop rectifying. But, still these devices can be used for special applications because they have the potential of biocompatibility. Differently current rectifying in semiconductor diodes via diffusion of ionic charges across a p–n junction sets up a built-in potential; thus, a unidirectional current is observed across a membrane where a built-in potential arises from diffusion of anions and cations in organic junctions [159].

We designed the junction between two hydrogels, containing oppositely charged counterions [154]. These ionic p–n junction prototypes operate on the (counter) ionic conductance; that is, charge carriers do not move through the polymer strands, but move over the whole medium under the electric field. This constitutes a different approach to construct a junction, in contrast to earlier studies [159,163,166,177–180].

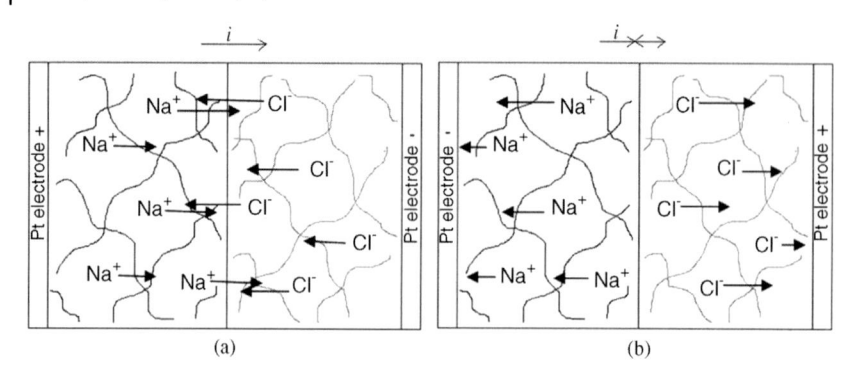

Scheme 10.4 Schematic representation of the working mechanism of a polymeric p–n junction under forward bias (a) and reverse bias (b).

For constructing the ionic p–n junction, ionic p- and ionic n-type slice gels, synthesized as described in Section 10.2, are put one over the other and brought in close contact with small pressure. Normalized current density versus voltage (J/m_{gel}–U curves) characteristics of these junctions are studied with Keithley 6487 picoammeter/voltage source. The working mechanism of the diode formed by bringing two opposite charged gels in contact is shown in Scheme 10.4. When direct bias is applied, current passes through the interface between the gels. Na^+ counterions, weakly bound to pyranine, can move toward the negative electrode, and similarly, Cl^- counterions, bound to MAPTAC, can move toward the positive electrode. Pyranine and MAPTAC molecules cannot move because they are chemically bound to the polymer strands; only counterions can move under the electric field.

In the backward direction, there are no carriers passing through the interface between the gels, so no current should be observed. Na^+ and Cl^- counterions are attracted to the nearest electrode and do not cross the interface between the gels. In addition to the counterions, hydroxide and hydrogen ions in water cause a residual current. They contribute to the current under both direct and reverse biases. For higher mass ratios, the current rectification must disappear because the large number of hydrogen and hydroxide ions will screen the effect of the counterions.

Figure 10.12 shows J/m_{gel}–U behavior for the junctions constructed with neutral gels for varying monomer concentration and for the junctions constructed with oppositely charged gel with varying doping concentration. All experiments given in Figure 10.12 were performed for the gels swollen in pure water, $m/m_0 = 1.3$. The current values were measured 5 s after the voltage was applied.

Figure 10.12a presents the J/m_{gel}–U behavior for junctions formed with two neutral gels (PAAm and NIPA) having different monomer molarities. In this case, there is no rectification of the current. However, we observed considerable rectification for the junctions formed with ionic p- and ionic n-type gels, where the rectification ratio increases with increasing doping concentration.

J/m_{gel}–U characteristics of the gel p–n junctions were also tested for increasing mass ratios of the gels. It can be seen from this experiment that the rectification ratio

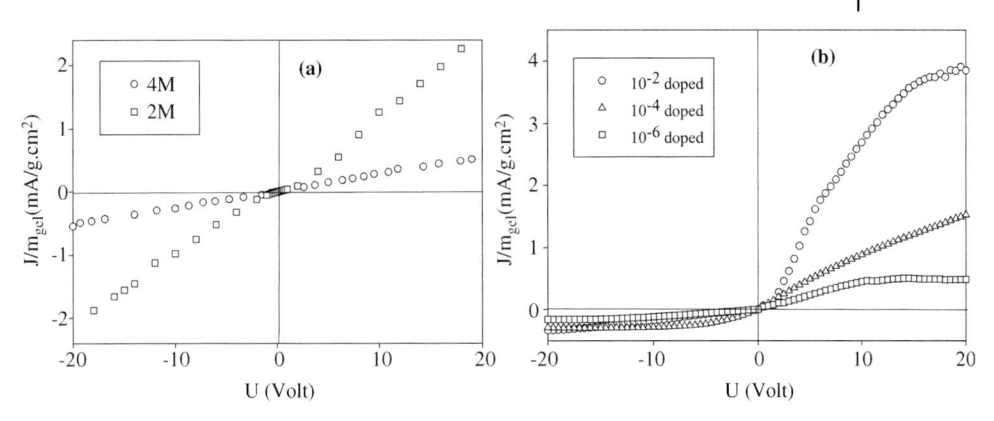

Figure 10.12 J/m_{gel}–U characteristics for (a) junctions constructed with neutral gels for varying monomer concentration and (b) junctions constructed with oppositely charged gel for varying doping concentration. All gels were kept at a fixed mass ratio, $m/m_0 = 1.3$.

first increases with increasing water uptake and then decreases with further water uptake. This shows that above certain water uptake, H and OH ions from water screen the effect of doped ions; that is, the sole effects of the doped ions are concealed by foreign ions, H and OH. It is clearly seen that the J/m_{gel}–U characteristics of these p–n junctions are adjustable by changing doping concentration and mass ratio of the gels on either side of the junction.

10.8 Conclusions

In this chapter, PAAm and NIPA gels were studied from the viewpoints of gelation kinetics, polymer–fluoroprobe interaction, heterogeneity, doping of gels as p-type and n-type, and imprinting. In the beginning, experimental results are discussed and confirmed by some theoretical models. Then, some applications of the gels such as polymeric gel diode and imprinting are presented. The conclusion of this chapter can be summarized as follows:

1) The fluoroprobe pyranine (8-hydroxypyrene-1,3,6-trisulfonic acid, trisodium salt) can be used for *in situ* monitoring of the free radical polymerization of the acrylamide system. It binds chemically to the growing PAAm chains over –OH group and electrostatically to the protonated amide groups on the polymer chains over SO_3^- groups.
2) The fluorescence intensity of pyranine (added prepolymer solution in trace amount) monitored during the polymerization process can be used for *in situ* monitoring of the monomer conversion with great sensitivity.
3) The steady-state fluorescence measurements can be used for real-time measurements of the fractal dimension and the critical exponents γ and β simultaneously during the gelation process.

4) We characterized the recognition process in the imprinted gel via steady-state fluorescence spectroscopy. Also, the effects of the temperature and the salt concentration on the recognition were determined.
5) We developed a theoretical model based on the measured current–time plots of the neutral and charged gels for estimating the weight fraction and the number of generations of heterogeneous regions of the gels.
6) We synthesized new kinds of semiconducting polymeric hydrogels having negative (ionic n-type) and positive (ionic p-type) counterions as charge carriers and we designed the junction between these two types of hydrogels. The current–voltage characteristics of the polymeric p–n junction can be varied by changing doping concentration and mass ratio.

References

1 Alveroglu, E. et al. (2008) Swelling behavior of chemically ion-doped hydrogels. *48th Microsymposium on Polymer Colloids – From Design to Biomedical and Industrial Applications*, vol. 281, pp. 174–180.

2 Alveroglu, E. and Yilmaz, Y. (2011) Estimation of the generation and the weight fraction of dense polymer regions in heterogeneous hydrogels. *Macromol. Chem. Phys.*, 212 (14), 1451–1459.

3 Ito, K. et al. (2003) Multiple point adsorption in a heteropolymer gel and the Tanaka approach to imprinting: experiment and theory. *Prog. Polym. Sci.*, 28 (10), 1489–1515.

4 Rubinstein, M. and Colby, R.H. (2005) *Polymer Physics*, Oxford University Press, New York.

5 Hassan, C.M. and Peppas, N.A. (2000) Structure and applications of poly(vinyl alcohol) hydrogels produced by conventional crosslinking or by freezing/thawing methods, in *Biopolymers/PVA Hydrogels/Anionic Polymerisation Nanocomposites* (ed. K. Dusek), Springer, pp. 37–65.

6 Lopez, D. et al. (2001) Magnetic applications of polymer gels. *Macromol. Symp.*, 166, 173–178.

7 Peppas, N.A. and Langer, R. (1994) New challenges in biomaterials. *Science*, 263 (5154), 1715–1720.

8 Verhoeven, J. et al. (1986) Necessary steps in the design of hydrogels for biomedical application. *Pharm. Weekblad Sci. Ed.*, 8 (1), 98.

9 Berger, J. et al. (2005) Pseudo-thermosetting chitosan hydrogels for biomedical application. *Int. J. Pharm.*, 288 (2), 197–206.

10 Philippova, O.E. et al. (2003) Reinforced superabsorbent polyacrylamide hydrogels. *Macromol. Symp.*, 200, 45–53.

11 Bera, P. and Saha, S.K. (1998) Redox polymerisation of acrylamide on aqueous montmorillonite surface: kinetics and mechanism of enhanced chain growth. *Polymer*, 39 (6–7), 1461–1469.

12 Lin, H.R. (2001) Solution polymerization of acrylamide using potassium persulfate as an initiator: kinetic studies, temperature and pH dependence. *Eur. Polym. J.*, 37 (7), 1507–1510.

13 Allen, N.S. et al. (1996) A gravimetric and spectroscopic study on the photoinduced radical co-polymerisation of methyl and allyl methacrylates: kinetic and structural properties. *J. Photopolym. Sci. Technol.*, 9 (1), 143.

14 Schork, F.J. and Ray, W.H. (1981) On-line monitoring of emulsion polymerization reactor dynamics, in *Emulsion Polymers and Emulsion Polymerization*, Proceedings of the ACS Symposium Series (eds D.R. Bassett and A.E. Hamielec), American Chemical Society, Washington, DC.

15 Moritz, H.U.I. and Geiseler, R.W. (eds) (1989) *Polymer Reaction Engineering*, Wiley-VCH Verlag GmbH, Weinheim.

16 Siani, A. et al. (1995) in *DECHENA Monographs*, vol. **131**, p. 149, (eds K.H. Reichert and H.U. Moritz), Wiley-VCH Verlag GmbH, Weinheim.

17 Wang, C. et al. (1993) Use of water as an internal standard in the direct monitoring of emulsion polymerization by fiberoptic raman-spectroscopy. *Appl. Spectrosc.*, **47**, 928.

18 Crowley, T.J. and Choi, K.Y. (1995) In-line dielectric monitoring of monomer conversion in a batch polymerization reactor. *J. Appl. Polym. Sci.*, **55** (9), 1361–1365.

19 Chatzi, E.G. et al. (1997) Use of a midrange infrared optical-fiber probe for the on-line monitoring of 2-ethylhexyl acrylate/styrene emulsion copolymerization. *J. Appl. Polym. Sci.*, **63** (6), 799–809.

20 Yilmaz, Y. et al. (2009) Elucidation of multiple-point interactions of pyranine fluoroprobe during the gelation. *Spectrochim. Acta A*, **72** (2), 332–338.

21 Kizildereli, N. et al. (2010) Theoretical confirmation of *in situ* monitoring of monomer conversion during acrylamide polymerization via pyranine fluoroprobe. *J. Appl. Polym. Sci.*, **115** (4), 2455–2459.

22 Kaya, D. et al. (2004) Direct test of the critical exponents at the sol–gel transition. *Phys. Rev. E*, **69** (1), 016117.1–016117.10.

23 Yilmaz, Y. et al. (2008) Testing percolation theory in the laboratory: measuring the critical exponents and fractal dimension during gelation. *Phys. Rev. E*, **77** (5), 051121.1–051121.4.

24 Mayer, A. and Neuenhofer, S. (1994) Luminescent labels – more than just an alternative to radioisotopes. *Angew. Chem., Int. Ed.*, **33** (10), 1044–1072.

25 Rettig, W.Lapouyarde, R. and Lakowicz, J.R. (eds) (1994) *Probe Design and Chemical Sensing, Topics in Fluorescence Spectroscopy*, Plenum Press, New York.

26 Barrow, G.M. (1962) *Introduction to Molecular Spectroscopy*, McGraw-Hill, New York.

27 Birks, J.B. (1965) *Photophysics of Aromatic Molecules*, Wiley–Interscience, London.

28 Weber, G. and Herculus, D.M. (eds) (1965) *Fluorescence and Phosphorescence Analysis*, John Wiley & Sons, Inc., New York.

29 Galanin, M.D. (1995) *Luminescence of Molecules and Crystals*, Cambridge International Science Publishing Ltd, Cambridge, UK.

30 Jager, W.F. et al. (1995) Solvatochromic fluorescent probes for monitoring the photopolymerization of dimethacrylates. *Macromolecules*, **28**, 8153.

31 Schaeken, T.C. and Warman, J.M. (1995) Radiation-induced polymerization of a monoacrylate and diacrylate studied using a fluorescent molecular Probe. *J. Phys. Chem.*, **99**, 6145.

32 Royal, J.S. and Torkelson, J.M. (1993) Physical aging effects on molecular-scale polymer relaxations monitored with mobility-sensitive fluorescent molecules. *Macromolecules*, **26**, 5331.

33 Miller, K.E. et al. (1995) Mobility-sensitive fluorescence probes for quantitative monitoring of water sorption and diffusion in polymer coatings. *J. Polym. Sci. B*, **33**, 2343.

34 Vatanparast, R. et al. (2000) monitoring of curing of polyurethane polymers with fluorescence method. *Macromolecules*, **33**, 438.

35 Warman, J.M. et al. (1997) Maleimido-fluoroprobe: A dual-purpose fluorogenic probe of polymerization dynamics. *J. Phys. Chem. B*, **101**, 4913.

36 Ercelen, S. et al. (2002) Ultrasensitive fluorescent probe for the hydrophobic range of solvent polarities. *Anal. Chim. Acta*, **464**, 273.

37 Jager, W.F. et al. (1999) Functionalized 4-(dialkylamino)-4'-nitrostilbenes as reactive fluorescent probes for monitoring the photoinitiated polymerization of mma. *Macromolecules*, **32**, 8791.

38 Pekcan, O. et al. (1997) Real time monitoring of polymerization rate of methyl methacrylate using fluorescence probe. *Polymer*, **38** (7), 1693–1698.

39 Yilmaz, Y. et al. (2001) Fluorescence technique to study free-radical polymerization of 2-vinylnaphthalene. *J. Macromol. Sci. Pure Appl. Chem.*, **38** (7), 741–749.

40 Pekcan, O. et al. (1994) Fluorescence technique for studying the sol–gel transition in the free-radical cross-linking copolymerization of methyl methacrylate and ethylene glycol dimethacrylate. *Chem. Phys. Lett.*, **229** (4–5), 537–540.

41 Serrano, B. et al. (1996) Studies of polymerization of acrylic monomers using luminescence probes and differential scanning calorimetry. *Polym. Eng. Sci.*, **36** (2), 175–181.

42 Okay, O. et al. (1999) Heterogeneities during the formation of poly(sodium acrylate) hydrogels. *Polym. Bull.*, **43** (4–5), 425–431.

43 Pekcan, O. and Yilmaz, Y. (1997) Modeling of swelling by fluorescence technique in poly(methyl methacrylate) gels. *J. Appl. Polym. Sci.*, **63** (13), 1777–1784.

44 Pekcan, O. et al. (1997) In situ fluorescence study of slow release and swelling processes in gels formed by solution free radical copolymerization. *Polym. Int.*, **44** (4), 474–480.

45 Yilmaz, Y. and Pekcan, O. (1998) In situ fluorescence experiments to study swelling and slow release kinetics of disc-shaped poly(methyl methacrylate) gels made at various crosslinker densities. *Polymer*, **39** (22), 5351–5357.

46 Guney, O. et al. (2002) Metal ion templated chemosensor for metal ions based on fluorescence quenching. *Sens. Actuators B*, **85** (1–2), 86–89.

47 Oya, T. et al. (1999) Reversible molecular adsorption based on multiple-point interaction by shrinkable gels. *Science*, **286** (5444), 1543–1545.

48 Yilmaz, Y. (2002) Fluorescence study on the phase transition of hydrogen-bonding gels. *Phys. Rev. E*, **66** (5), 052801.1–052801.4.

49 Yilmaz, Y. (2007) Transition between collapsed state phases and the critical swelling of a hydrogen bonding gel: poly(methacrylic acid-*co*-dimethyl acrylamide). *J. Chem. Phys.*, **126** (22), 224501.1–224501.5.

50 Kurenkov, V.F. and Antonovich, O.A. (2003) Radical polymerization of acrylamide in aqueous-dimethyl sulfoxide solutions in the presence of sodium acetate. *Russ. J. Appl. Chem.*, **76** (2), 280–283.

51 Flory, P.J. (1941) Molecular size distribution in three dimensional polymers. *J. Am. Chem. Soc.*, **63**, 3083–3100.

52 Stockmayer, W. (1943) Theory of molecular size distribution and gel formation in branched-chain polymers. *J. Chem. Phys.*, **11**, 45.

53 de Gennes, P.G. (1976) On a relation between percolation theory and the elasticity of gels. *J. Phys. Lett.*, **37**, 1–2.

54 Stauffer, D. (1976) Gelation in concentrated critically branched polymer solutions. Percolation scaling theory of intramolecular bond cycles. *J. Chem. Soc.*, **72**, 1354–1364.

55 Stauffer, D. et al. (1982) Gelation and critical phenomena. *Adv. Polym. Sci.*, **44**, 193.

56 Bunde, A. and Havlin, S. (1991) *Fractals and Disordered Systems*, Springer, Berlin.

57 Stauffer, D. and Aharony, A. (1994) *Introduction to Percolation Theory*, Taylor & Francis, London.

58 Mandelbrot, B.B. (1982) *The Fractal Geometry of Nature*, Freeman, San Francisco, CA.

59 Yilmaz, Y. et al. (2002) Slow regions percolate near glass transition. *Eur. Phys. J. E*, **9**, 135–141.

60 Yilmaz, Y. et al. (1998) Critical exponents and fractal dimension at the sol–gel phase transition via *in situ* fluorescence experiments. *Phys. Rev. E*, **58** (6), 7487–7491.

61 Tuzel, E. et al. (2000) A new critical point and time dependence of bond formation probability in sol–gel transition: a Monte Carlo study in two dimension. *Eur. Polym. J.*, **36** (4), 727–733.

62 Lehn, J.-M. et al. (1996) *Comprehensive Supramolecular Chemistry*, Pergamon Press, Oxford.

63 Polyakov, M.V. (1940) Adsorption properties and structure of silica gel. *Zh. Fiz. Khim. Akad. SSSR*, **2**, 799–805.

64 Wulff, G. (2002) Enzyme-like catalysis by molecularly imprinted polymers. *Chem. Rev.*, **102** (1), 1–27.

65 Haupt, K. and Mosbach, K. (2000) Molecularly imprinted polymers and their use in biomimetic sensors. *Chem. Rev.*, **100** (7), 2495–2504.

66 Whitcombe, M.J. and Vulfson, E.N. (2001) Imprinted polymers. *Adv. Mater.*, **13** (7), 467–478.
67 Shea, K.J. (1994) Molecular imprinting of synthetic network polymers – the *de-novo* synthesis of macromolecular binding and catalytic sites. *Abstr. Pap. Am. Chem. Soc.*, **208**, 467–471.
68 Qiu, Y. and Park, K. (2001) Triggering in drug delivery systems: environment-sensitive hydrogels for drug delivery. *Adv. Drug Deliv. Rev.*, **53**, 321–339.
69 Ohmine, I. and Tanaka, T. (1982) Salt effects on the phase transitions of polymer gels. *J. Chem. Phys.*, **77**, 5725–5729.
70 Sahimi, M. (1994) *Application of Percolation Theory*, Taylor & Francis, London.
71 Siegel, R.A. and Firestone, B.A. (1988) pH-dependent equilibrium swelling properties of hydrophobic polyelectrolyte copolymer gels. *Macromolecules*, **21**, 3254–3259.
72 Aharony, A. (1980) Universal critical amplitude ratios for percolation. *Phys. Rev. B*, **22**, 400–414.
73 Colby, R.H. and Rubinstein, M. (1993) Dynamics of near-critical polymer gels. *Phys. Rev. E*, **48**, 3712.
74 Dusek, K. and Patterson, D. (1968) Transition in swollen polymer networks induced by intramolecular condensation. *J. Polym. Sci.*, **6**, 1209–1216.
75 Tanaka, T. (1978) Collapse of gels and critical endpoint. *Phys. Rev. Lett.*, **40**, 820–823.
76 Wulff, G. (1982) Selective binding to polymers via covalent bonds – the construction of chiral cavities as specific receptor sites. *Pure Appl. Chem.*, **54**, 2093–2102.
77 Sellergren, B. (1989) Molecular imprinting by noncovalent interactions – tailor-made chiral stationary phases of high selectivity and sample load capacity. *Chirality*, **1**, 63–68.
78 Umpleby, R.J. *et al.* (2000) Measurement of the continuous distribution of binding sites in molecularly imprinted polymers. *Analyst*, **125**, 1261–1265.
79 Guney, O. (2003) Multiple-point adsorption of terbium ions by lead ion templated thermosensitive gel: elucidating recognition of conformation in gel by terbium probe. *J. Mol. Recognit.*, **16** (2), 67–71.
80 Rachkov, A. and Minoura, N. (2001) Towards molecularly imprinted polymers selective to peptides and proteins. The epitope approach. *BBA Protein Struct. Mol. Enzymol.*, **1544** (1–2), 255–266.
81 Shi, H.Q. *et al.* (1999) Template-imprinted nanostructured surfaces for protein recognition. *Nature*, **398** (6728), 593–597.
82 Peppas, N.A. *et al.* (1999) Poly(ethylene glycol)-containing hydrogels in drug delivery. *J. Control. Release*, **62** (1–2), 81–87.
83 Bures, P. *et al.* (2001) Surface modifications and molecular imprinting of polymers in medical and pharmaceutical applications. *J. Control. Release*, **72** (1–3), 25–33.
84 Lye, G.J. and Woodley, J.M. (1999) Application of *in situ* product-removal techniques to biocatalytic processes. *Trends Biotechnol.*, **17** (10), 395–402.
85 Ramstrom, O. *et al.* (1998) Applications of molecularly imprinted materials as selective adsorbents: emphasis on enzymatic equilibrium shifting and library screening. *Chromatographia*, **47** (7–8), 465–469.
86 Dela Cruz, E.O. *et al.* (1999) Molecular imprinting of methyl pyrazines. *Anal. Lett.*, **32** (5), 841–854.
87 Janotta, M. *et al.* (2001) Molecularly imprinted polymers for nitrophenols – an advanced separation material for environmental analysis. *Int. J. Environ. Anal. Chem.*, **80** (2), 75–86.
88 Kriz, D. *et al.* (1995) Introducing biomimetic sensors based on molecularly imprinted polymers as recognition elements. *Anal. Chem.*, **67** (13), 2142–2144.
89 Baird, C.L. and Myszka, D.G. (2001) Current and emerging commercial optical biosensors. *J. Mol. Recognit.*, **14** (5), 261–268.
90 Homola, J. (2003) Present and future of surface plasmon resonance biosensors. *Anal. Bioanal. Chem.*, **377** (3), 528–539.
91 Warsinke, A. *et al.* (2000) Electrochemical immunoassays. *Fresenius J. Anal. Chem.*, **366** (6–7), 622–634.
92 Altschuh, D. *et al.* (2006) Fluorescence sensing of intermolecular interactions

and development of direct molecular biosensors. *J. Mol. Recognit.*, **19** (6), 459–477.
93 Lakowicz, J.R. (1983) *Principles of Fluorescence Spectroscopy*, Plenum Press, New York.
94 Valeur, B. (2002) *Molecular Fluorescence: Principles and Applications*, Wiley-VCH Verlag GmbH, Weinheim.
95 Flint, N.J. *et al.* (1998) Fluorescence investigations of "smart" microgel systems. *J. Fluoresc.*, **8** (4), 343–353.
96 Marchesi, J.R. (2003) A microplate fluorimetric assay for measuring dehalogenase activity. *J. Microbiol. Methods*, **55** (1), 325–329.
97 Nielsen, K. *et al.* (2000) Fluorescence polarization immunoassay: detection of antibody to *Brucella abortus*. *Methods*, **22** (1), 71–76.
98 Crank, J. (1979) *The Mathematics of Diffusion*, Oxford University Press.
99 Gelir, A. *et al.* (2007) *In situ* monitoring of the synthesis of a pyranine-substituted phthalonitrile derivative via the steady-state fluorescence technique. *J. Phys. Chem. B*, **111** (2), 478–484.
100 Geissler, E. and Hecht, A.M. (1981) The Poisson ratio in polymer gels macromolecules. *Macromolecules*, **14** (1), 185–188.
101 Baselga, J. *et al.* (1987) Elastic properties of highly cross-linked polyacrylamide gels. *Macromolecules*, **20** (12), 3060–3065.
102 Patel, S.K. *et al.* (1989) Mechanical and swelling properties of polyacrylamide gel spheres. *Polymer*, **30** (12), 2198–2203.
103 Hu, Z. *et al.* (1993) The scaling exponents of polyacrylamide and acrylamide sodium acrylate copolymer gels. *J. Chem. Phys.*, **99**, 7108.
104 Baker, J.P. *et al.* (1994) Swelling equilibria for positively ionized polyacrylamide hydrogels. *Macromolecules*, **27**, 1446.
105 Richards, E.G. and Temple, C.J. (1971) Some properties of polyacrylamide gels. *Nature (Phys. Sci.)*, **22**, 92.
106 Hsu, T.P. and Cohen, C. (1983) Effects of inhomogeneities in polyacrylamide gels on thermodynamic and transport properties. *Polymer*, **25**, 1419.
107 Baselga, J. *et al.* (1989) Effect of crosslinker on swelling and thermodynamic properties of polyacrylamide gels. *Polym. J.*, **21**, 467.
108 Hooper, H.H. *et al.* (1990) Swelling equilibria for positively ionized polyacrylamide hydrogels. *Macromolecules*, **23**, 1096.
109 Weiss, N. *et al.* (1974) Permeability of heterogeneous gels. *J. Polym. Sci. Polym. Phys. Ed.*, **17**, 2229.
110 Weiss, N. and Silberberg, A. (1975) Permeability of gels. *Polym. Prepr. Am. Chem. Soc. Div. Polym. Chem.*, **16**, 289.
111 Janas, V.F. *et al.* (1980) Aging and thermodynamics of polyacrylamide gels. *Macromolecules*, **13** (4), 977–983.
112 Matsuo, E.S. *et al.* (1994) Origin of structural inhomogeneities in polymer gels. *Macromolecules*, **27** (23), 6791–6796.
113 Tobita, H. and Hamielec, A.E. (1990) Cross-linking kinetics in polyacrylamide networks. *Polymer*, **31** (8), 1546–1552.
114 Pekcan, O. *et al.* (1998) In situ photon transmission technique for studying ageing in acrylamide gels due to multiple swelling. *Polymer*, **39**, 4453.
115 Naghash, H.J. and Okay, O. (1996) Formation and structure of polyacrylamide gels. *J. Appl. Polym. Sci.*, **60** (7), 971–979.
116 Geissler, E. and Brown, W. (1993) in *Dynamic Light Scattering* (ed. W. Brown), pp. 471–511., Oxford University Press, Oxford, UK.
117 Dusek, K. (1971) in *Polymer Networks* (eds A.J. Chompff and S. Newman), p. 245, Plenum Press, New York.
118 Goldbart, P. and Goldfeld, N. (1987) Rigidity and ergodicity of randomly cross-linked macromolecules. *Phys. Rev. Lett.*, **58**, 2676.
119 Pusey, P.N. and Megen, W.V. (1989) Dynamic light scattering by non-ergodic media. *Physica A*, **157**, 705.
120 Chu, B. (1991) *Laser Light Scattering*, Academic Press, London.
121 Joosten, J.G. *et al.* (1991) Dynamic and static light scattering by aqueous polyacrylamide gels. *Macromolecules*, **24**, 6690.
122 Fang, L. and Brown, W. (1992) Static and dynamic properties of polyacrylamide gels and solutions in mixtures of water and

122. glycerol – a comparison of the application of mean-field and scaling theories. *Macromolecules*, **215**, 6897.
123. Skouri, R. *et al.* (1993) Frozen-in intensity fluctuations of light scattered by partially ionized gels. *Europhys. Lett.*, **23**, 635.
124. Li, Y. and Tanaka, T. (1989) Study of the universality class of the gel network system. *J. Chem. Phys.*, **90**, 5161.
125. Pekcan, O. *et al.* (1998) *In situ* photon transmission technique for studying ageing in acrylamide gels due to multiple swelling. *Polymer*, **39** (18), 4453–4456.
126. Yilmaz, Y. *et al.* (2006) Dielectric study of neutral and charged hydrogels during the swelling process. *J. Chem. Phys.*, **125** (23), 234705.1–234705.12.
127. Tanaka, T. and Filmore, D. (1979) Kinetics of swelling of gels. *Chem. Phys.*, **70**, 1214.
128. Zhao, Y. *et al.* (2007) Swelling behavior of ionically cross-linked polyampholytic hydrogels in varied salt solutions. *Colloid Polym. Sci.*, **285**, 1395–1400.
129. Caykara, T. *et al.* (2003) Network structure and swelling behavior of poly (acrylamide/crotonic acid) hydrogels in aqueous salt solutions. *J. Polym. Sci. Part B: Polym. Phys.*, **41**, 1656–1664.
130. Okay, O. *et al.* (1998) Swelling behavior of anionic acrylamide-based hydrogels in aqueous salt solutions: comparison of experiment with theory. *J. Appl. Polym. Sci.*, **70**, 567–575.
131. Peters, A. and Candau, S. (1986) Kinetics of swelling of polyacrylamide gels. *Macromolecules*, **19**, 1952.
132. Benjar, P.Y. and Wu, Y.S. (1997) Effect of counter-ions on swelling and shrinkage of polyacrylamide-based ionic gels. *Polymer*, **38** (10), 2557.
133. Kayaman, N. *et al.* (1997) Swelling of polyacrylamide gels in aqueous solutions of ethylene glycol oligomers. *Polym. Gels*, **5**, 339.
134. Kayaman, N. *et al.* (1998) Swelling of polyacrylamide gels in polyacrylamide solutions. *J. Polym. Sci. Part B: Polym. Phys.*, **36**, 1313.
135. Fernandez, E. *et al.* (2005) Viscoelastic and swelling properties of glucose oxidase loaded polyacrylamide hydrogels and the evaluation of their properties as glucose sensors. *Polymer*, **46**, 2211.
136. Suzuki, A. *et al.* (2005) Effects of guest microparticles on the swelling behavior of polyacrylamide gels. *J. Polym. Sci. Part B: Polym. Phys.*, **43**, 1696.
137. Roger, P. *et al.* (2007) Effect of the incorporation of a low amount of carbohydrate-containing monomer on the swelling properties of polyacrylamide hydrogels. *Polymer*, **48**, 7539.
138. Pekcan, O. and Kara, S. (2003) Swelling of acrylamide gels made at various onset temperatures: an optical transmission study. *Polym. Int.*, **52** (5), 676–684.
139. Mohomeda, K. *et al.* (2005) A broad spectrum analysis of the dielectric properties of poly(2-hydroxyethyl methacrylate). *Polymer*, **46** (11), 3847–3855.
140. Alveroglu, E. and Yilmaz, Y. (2011) Estimation of the generation and the weight fraction of dense polymer regions in heterogeneous hydrogels. *Macromol. Chem. Phys.*, **212**, 1451–1459.
141. McNeill, R. *et al.* (1963) Electronic conduction in polymers. I. The chemical structure of polypyrrole. *Aust. J. Chem.*, **16** (6), 1056–1075.
142. McGinnes., J. *et al.* (1974) Amorphous–semiconductor switching in melanins. *Science*, **183** (4127), 853–855.
143. Shirakawa, H. *et al.* (1977) Synthesis of electrically conducting organic polymers – halogen derivatives of polyacetylene, $(CH)_x$. *J. Chem. Soc., Chem. Commun.*, (16), 578–580.
144. MacDiarmid, A.C. and Heeger, A.J. (1979) in *Molecular Metals* (ed. W.E. Hatfield), p. 161, Plenum Press, New York.
145. Pillai, R.G. *et al.* (2008) Field-induced carrier generation in conjugated polymer semiconductors for dynamic, asymmetric junctions. *Adv. Mater.*, **20** (1), 49.
146. Lin, S.H. *et al.* (1989) Ion-beam modification of polyacetylene films. *Nucl. Instrum. Methods B*, **39** (1–4), 778–782.
147. Wada, T. *et al.* (1985) Fabrication of a stable p–n junction in a polyacetylene film by ion implantation. *J. Chem. Soc., Chem. Commun.*, (17), 1194–1195.
148. Koshida, N. and Wachi, Y. (1984) Application of ion implantation for doping of polyacetylene films. *Appl. Phys. Lett.*, **45** (4), 436–437.

149 Wang, W.M. et al. (1991) Diode characteristics and degradation mechanism of ion-implanted polyacetylene films. *Nucl. Instrum. Methods B*, **61** (4), 466–471.

150 Chiang, C.K. et al. (1978) Polyacetylene, $(CH)_x$ – n-type and p-type doping and compensation. *Appl. Phys. Lett.*, **33** (1), 18–20.

151 Usuki, A. et al. (1987) Photovoltaic effect of normal-type polyacetylene junctions. *Synth. Met.*, **18** (1–3), 705–710.

152 Yamashita, K. et al. (1995) Fabrication of an organic p–n homojunction diode using electrochemically cation-doped and photochemically anion-doped polymer. *Jpn. J. Appl. Phys. Part 1*, **34**, 3794–3797.

153 Eckhardt, H. et al. (1989) The electronic and electrochemical properties of poly(phenylene vinylenes) and poly(thienylene vinylenes) – an experimental and theoretical study. *J. Chem. Phys.*, **91** (2), 1303–1315.

154 Alveroglu, E. and Yilmaz, Y. (2010) Synthesis of p- and n-type gels doped with ionic charge carriers. *Nanoscale Res. Lett.*, **5** (3), 559–565.

155 Faid, K. et al. (1995) Localization effects in asymmetrically substituted polythiophenes – controlled generation of polarons, dimerized polarons, and bipolarons. *Macromolecules*, **28** (1), 284–287.

156 Macinnes, D. et al. (1981) Organic batteries – reversible n-type and p-type electrochemical doping of polyacetylene, $(CH)_x$. *J. Chem. Soc., Chem. Commun.*, (7), 317–319.

157 Angelopoulos, M. (2001) Conducting polymers in microelectronics. *IBM J. Res. Dev.*, **45** (1), 57–75.

158 Burroughes, J.H. et al. (1988) New semiconductor device physics in polymer diodes and transistors. *Nature*, **335** (6186), 137–141.

159 Cheng, C.H.W. and Lonergan, M.C. (2004) A conjugated polymer pn junction. *J. Am. Chem. Soc.*, **126** (34), 10536–10537.

160 Gurunathan, K. et al. (1999) Electrochemically synthesised conducting polymeric materials for applications towards technology in electronics, optoelectronics and energy storage devices. *Mater. Chem. Phys.*, **61** (3), 173–191.

161 Patil, A.O. et al. (1987) Water-soluble conducting polymers. *J. Am. Chem. Soc.*, **109** (6), 1858–1859.

162 Bidan, G. et al. (1988) Conductive polymers with immobilized dopants: ionomer composites and auto-doped polymers – a review and recent advances. *J. Phys. D: Appl. Phys.*, **21** (7), 1043–1054.

163 Leger, J.M. et al. (2006) Self-assembled, chemically fixed homojunctions in semiconducting polymers. *Adv. Mater.*, **18** (23), 3130.

164 Pei, Q. et al. (1995) Light-emitting electrochemical cells – reply. *Science*, **270** (5237), 719.

165 Lovrecek, B. et al. (1958) Electrolytic junctions with rectifying properties. *J. Phys. Chem.*, **63**, 750–751.

166 Cayre, O.J. et al. (2007) Polyelectrolyte diode: nonlinear current response of a junction between aqueous ionic gels. *J. Am. Chem. Soc.*, **129** (35), 10801–10806.

167 Yu, G. et al. (1998) Polymer light-emitting electrochemical cells with frozen p-i-n junction at room temperature. *Adv. Mater.*, **10** (5), 385–388.

168 Shao, Y. et al. (2007) Long-lifetime polymer light-emitting electrochemical cells. *Adv. Mater.*, **19** (3), 365.

169 Pei, Q. et al. (1996) Polymer light-emitting electrochemical cells: *in situ* formation of a light-emitting p–n junction. *J. Am. Chem. Soc.*, **118** (16), 3922–3929.

170 Yang, C.H. et al. (2003) Ionic liquid doped polymer light-emitting electrochemical cells. *J. Phys. Chem. B*, **107** (47), 12981–12988.

171 Buck, R.P. et al. (1992) Liquid/solid polyelectrolyte diodes and semiconductor analogs. *J. Electrochem. Soc.*, **139** (1), 136–144.

172 Pickup, P.G. et al. (1984) Redox conduction in single and bilayer films of redox polymer. *J. Am. Chem. Soc.*, **106** (7), 1991–1998.

173 Leventis, N. et al. (1990) Characterization of a solid-state microelectrochemical diode employing a poly(vinyl alcohol) phosphoric acid solid-state electrolyte – rectification at junctions between WO_3

and polyaniline. *Chem. Mater.*, **2** (5), 568–576.

174 Hegedus, L. *et al.* (1999) Nonlinear effects of electrolyte diodes and transistors in a polymer gel medium. *Chaos*, **9** (2), 283–297.

175 Lindner, J. *et al.* (2002) Modelling of ionic systems with a narrow acid base boundary. *Phys. Chem. Chem. Phys.*, **4** (8), 1348–1354.

176 Ivan, K. *et al.* (2004) Electrolyte diodes and hydrogels: determination of concentration and pK value of fixed acidic groups in a weakly charged hydrogel. *Phys. Rev. E*, **70** (6), 061402-1–061402-11.

177 Bernards, D.A. *et al.* (2006) Observation of electroluminescence and photovoltaic response in ionic junctions. *Science*, **313** (5792), 1416–1419.

178 Cheng, C.H.W. *et al.* (2004) Unidirectional current in a polyacetylene hetero-ionic junction. *J. Am. Chem. Soc.*, **126** (28), 8666–8667.

179 Lovrecek, B. *et al.* (1959) Electrolytic junctions with rectifying properties. *J. Phys. Chem.*, **63**, 750–751.

180 Leger, J.M. *et al.* (2008) Polymer photovoltaic devices employing a chemically fixed p-i-n junction. *Adv. Funct. Mater.*, **18** (8), 1212–1219.

11
Conformation and Rheology of Microbial Exopolysaccharides
Jacques Desbrieres

11.1
Introduction

Polysaccharides are of great interest in industry because of their unusual physico-chemical properties. They are used mainly as thickening or gelling agents but also to stabilize emulsions or suspend solid particles. These functionalities promote their use in various domains, such as food, biomedical, and oilfield industries, as well as cosmetics.

Among polysaccharides, bacterial ones, also defined as exopolysaccharides, present specificities: the stereoregularity with which the sequence of saccharidic units repeats allows them, in many cases, to adopt under certain conditions an ordered conformation. The conformation that a polysaccharide can adopt depends directly on the chemical structure, principally the position and anomeric nature of the osidic bond, and on the nature of the side chains. Thus, using molecular modeling, Rees and Scott [1] predicted for different homopolymers (glucans, mannans, xylans, etc.) constituted of one type of bond (1→2), (1→3), (1→4), or (1→6) four types of conformations:

- Type A: extended and ribbon-like chain.
- Type B: flexible helical chain.
- Type C: rigid and crumpled chain.
- Type D: very flexible chain.

Solution properties of biopolymers strongly depend on the conformation of the macromolecular chains in solution. The latter are related first to the solid-state conformation and then to the experimental conditions (nature of the solvent, temperature, ionic strength, etc.). A conformational transition is observed for major exopolysaccharides as a function of these experimental conditions. As a consequence, the applications strongly depend on these solution properties and conditions. Moreover, the nature of the stable conformations may have biological implications [2]. All these aspects will be covered in this chapter. The solid-state conformations of exopolysaccharides will be presented before their progress in solution and related properties and applications.

11.2
Conformation of Polysaccharides

In the case of homopolymers formed by (1→3), (1→4), or (1→6) bonds (the most frequently observed in the bacterial polysaccharides), only A, B, and D conformations were observed, C being obtained for structures based on (1→2) bonds. As examples, the expected conformation is of type A for (1→4)-β-D-glucan (cellulose), whereas it is of type B for (1→4)-α-D-glucan (amylose). Through the position and the anomery, it is the relative orientation of the two intersugar bonds within the macromolecular chain that can provoke the change in orientation of the chain and lead to conformation B when amylose is considered. In the case of (1→6) bonds, the presence of an additional covalent bond between monomeric units introduces a supplemental degree of freedom and leads to a larger number of possible orientations of the bond, and hence a larger flexibility of the chain and the absence of an ordered conformation. Predictions by Rees and Scott [1] are in good agreement with experimental results obtained with polysaccharides. Since the polysaccharides are strongly hydroxylated, these structures are particularly stabilized by hydrogen bonds.

The conformation depends on the interactions present in the system. In solutions, these may be polymer–polymer or polymer–solvent interactions and competition occurs between them. As a consequence, the quality of the solvent is an important parameter and its nature is of great interest. Moreover, the nature of the side chains grafted on the macromolecular backbone has to be taken into account [3].

From the ordered conformation of one macromolecular chain, the ability of association and organization of several chains may be considered. As an example, curdlan may be found in triple helix (intertwined chains) stabilized by hydrogen bonds (OH2 . . . O2) sitting in the core of the triple helix [4].

These cases are simple. The alternation of different sugar units, different types of bonds and ionic groups, more or less long lateral chains, and so on directly influence the aptitude of the macromolecular chain not only to adopt such an ordered conformation but also to create arrangements between themselves. The external conditions, such as the hydration degree or the nature of the counterion in the case of polyelectrolytes, also have an important role in the secondary structure.

11.3
Secondary Solid-State Structures for Microbial Polysaccharides

Considering that the solid-state conformations have to be kept in solution, their knowledge is of great importance. X-ray diffraction is the most suitable and the most used technique to determine the molecular structure and intermolecular interactions in the solid state. Moreover, molecular modeling based on information from the X-ray analysis, the chemical structure (geometry of monomers, steric constraints), and the existence of interactions (hydrogen bonding, van der Waals interactions, and ionic interactions) allows confirming an assumption or choosing between different ones pointing out the most likely conformation (by minimizing

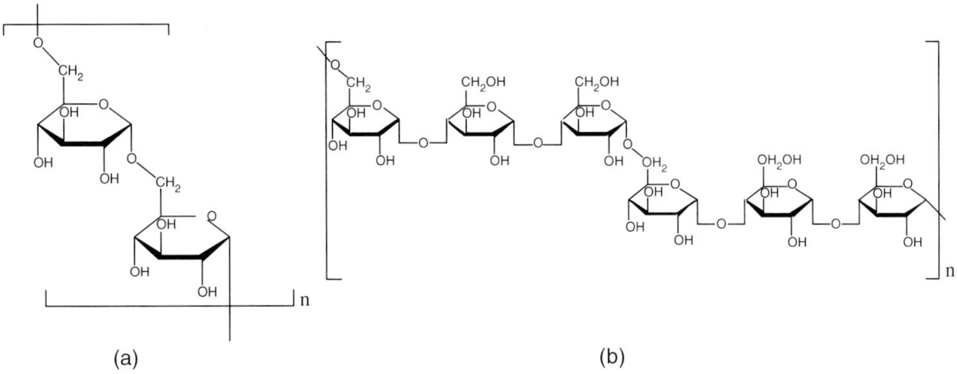

Figure 11.1 Chemical structures of dextran (a) and pullulan (b).

the potential energy of the macromolecule). The stable conformations adopted by polysaccharides are often of helical type.

The different solid-state conformations observed for major exopolysaccharides are now discussed.

11.3.1
No Secondary Solid-State Structure

Some of exopolysaccharides do not present any ordered solid-state conformation. As examples, dextran and pullulan are amorphous polymers. The chemical structure of dextran is mainly linear [5] based on the sequence of glucose units bound by $\alpha(1\rightarrow6)$ links with some $(1\rightarrow3)$-α-D-glucose branching (Figure 11.1a). Pullulan has the same structure as amylose $((1\rightarrow4)$-α-D-glucan) but on average one $\alpha(1\rightarrow4)$ bond in three is replaced by one $\alpha(1\rightarrow6)$ link (Figure 11.1b). Several $(1\rightarrow6)$-α-D-glucose connections are also present [6]. The amorphous state of these polysaccharides is due to the great presence of $\alpha(1\rightarrow6)$ bonds that possess a great conformational freedom. In pullulan, the presence of one $\alpha(1\rightarrow6)$ bond in three within the repeating unit avoids the formation of the helical conformation as observed in amylose [6].

11.3.2
Single-Chain Conformation

The chemical structure of $(1\rightarrow4)$-β-D-glucuronan corresponds to cellulose but D-glucose is replaced by D-glucuronic acid (Figure 11.2).

The first study of the solid-state structure performed by Heyraud *et al.* [7] has shown, for the reduced polysaccharide, a diffraction pattern matching that of cellulose II (twofold helical symmetry due to the ribbon conformation associating into antiparallel sheets). A molecular modeling study gave a good agreement with

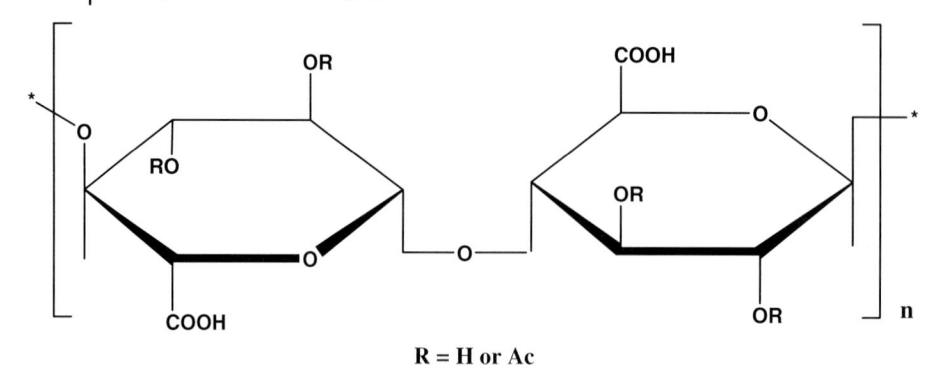

Figure 11.2 Chemical structure of glucuronan.

experimental data [8,9] (Figure 11.3). In the salt form, a threefold symmetry was obtained. Poly(α-L-guluronic acid), present within the alginate structure, also has a highly buckled (zigzag) ribbon conformation but with a lower periodicity (the helix pitch p is equal to 8.7 Å, preserved under salt form) due to diaxial bonds. As a consequence, in the case of poly(α-L-guluronic acid), regular cavities are created in which the counterions stay, in interaction with carboxylate groups. As will be seen, this difference is inconsequential on the gelling properties and gelation mechanisms of these two polymers.

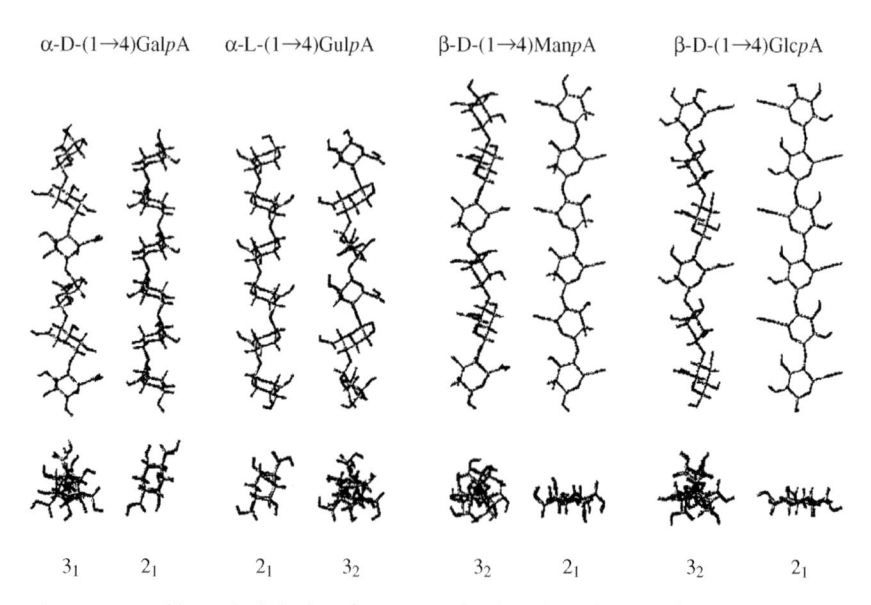

α-D-(1→4)Gal$p$A   α-L-(1→4)Gul$p$A   β-D-(1→4)Man$p$A   β-D-(1→4)Glc$p$A

3_1 2_1 2_1 3_2 3_2 2_1 3_2 2_1

Figure 11.3 Stable regular helical conformations of acidic polysaccharides. The helices are represented using projection parallel and orthogonal to their axes. (Reproduced from Ref. [9] with permission from Pergamon Press.)

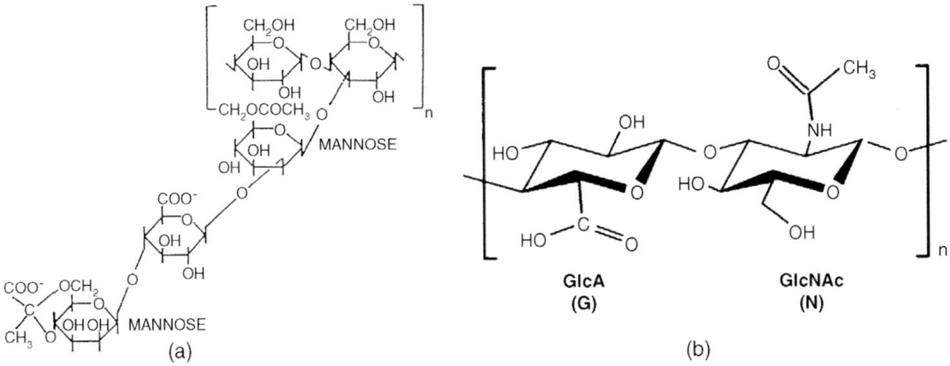

Figure 11.4 Chemical structures of xanthan (a) and hyaluronic acid (b).

11.3.3
Simple or/and Double Helices

The primary structure of xanthan has a cellulose backbone on which a three-sugar side chain is grafted through a α(1→3) bond (Figure 11.4a). Analysis of X-ray diffraction images shows a right-handed fivefold helical symmetry with a pitch p equal to 47 Å [10]. In this conformation, the trisaccharide side chain is aligned with the backbone and stabilizes the overall conformation by noncovalent interactions, principally hydrogen bonding (four intramolecular bonds and one intermolecular H-bond). Nevertheless, although all the authors agree on the observations, proposed molecular models differ, for example, simple helix stabilized by hydrogen bonds [10] or double helix formed by two antiparallel chains [11]. Grafting a lateral chain in position 3 on the backbone is the critical parameter leading to the loss of the twofold structure observed with cellulose. It avoids the formation of intrachain hydrogen bonds stabilizing this conformation. Moreover, all sugar units of the lateral chain, but the mannose one attached to the cellulosic backbone, do not occur in stabilizing the fivefold structure. Xanthan deacetylation does not modify the solid-state conformational structure.

Hyaluronic acid is a non-sulfated glycosaminoglycan (Figure 11.4b). It is constituted by two sugars per repeating unit: N-acetyl-β-D-glucosamine (GlcNAc) and β-D-glucuronic acid (GlcA). Many X-ray analyses have shown many helical conformations according to the nature of the counterions, pH, and moisture content. Chandrasekaran [4] has demonstrated at least seven different secondary structures brought together in three major categories:

- fourfold left helix either stretched or tightened;
- threefold left helix always stretched;
- fourfold left antiparallel double helix.

Left-handed fourfold helices were found in sodium and potassium environments, and left-handed threefold helices with calcium [12]. An extended twofold helix has been proposed for the structure of HA under acidic conditions in the presence of

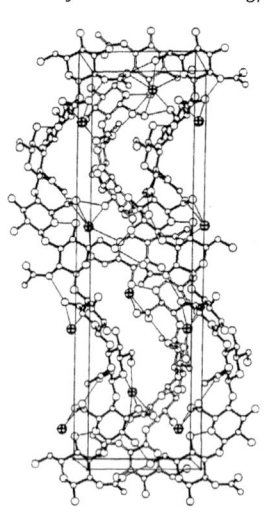

Figure 11.5 Antiparallel arrangement of sodium hyaluronate helix with a low relative humidity. (Reproduced from Ref. [4] with permission from Elsevier Science.)

most cations [13], except potassium and ammonium salts, which form left-handed fourfold antiparallel double helices [14]. Considering the first two groups, the different forms differ in the size of the crystalline mesh or how the polymeric chains arrange laterally within this mesh. These differences can concern the hydrogen bonds and the number of water molecules or counterions. In Figure 11.5, an example of the first group is given, the nonhydrated sodium hyaluronate [4]. Molecular modeling calculations demonstrated that the formation of a duplex by chain folding is possible. However, the antiparallel arrangement of the chains in the double helix is more favorable than the parallel one for HA [15]. The evaluation of the energy of the X-ray models, in comparison with the established theoretical model, indicated that the first model originally proposed by Sheehan *et al.* [16] is not viable, and that the second model proposed by Arnott *et al.* is much more reasonable [14], with maybe an overevaluated hydrogen bond network.

The helix is stabilized by periodic intramolecular H-bonds. The first one involves the NH function of the acetamido group and the carboxylate group across the $\beta(1\rightarrow 4)$ bond and the second one the O5 of the ring and alcohol C4–O4H across the $\beta(1\rightarrow 3)$ bond [17] (Figure 11.6). Sodium ion is octahedrally coordinated near carboxylate groups; it is bound to six ligands coming from three neighboring chains. If antiparallel helices are linked by carboxylate . . . sodium . . . hydroxymethyl interactions and O3 . . . O6 hydrogen bonds, two other O2 . . . O6 and O6 . . . O . . . carbonyl) hydrogen bonds connect nearby helices. Such direct interactions in the absence of water lead to a very compact arrangement.

When HA fibers are hydrated, the dimensions of the mesh increase uniaxially along the *a*-axis. Such helix can be characterized by the fact that it is stabilized by two intramolecular H-bonds, O3 . . . O5 across the $\beta(1\rightarrow 4)$ bond and O4 . . . O5 across

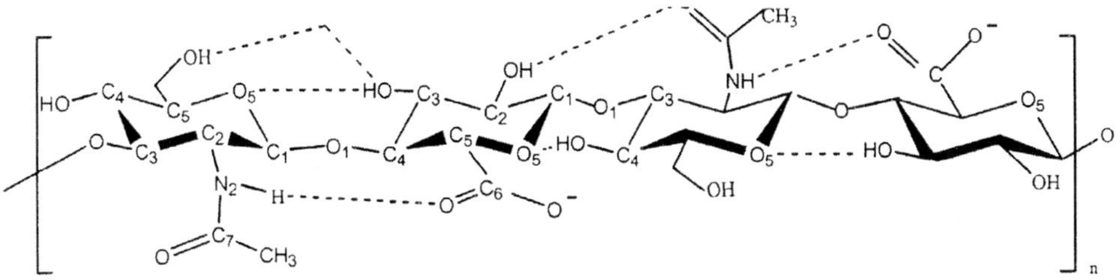

Figure 11.6 Repeating unit of hyaluronic acid with a representation of H-bonds generally cited in the literature.

$\beta(1\rightarrow3)$ bond. The acetamido group does not participate in the intramolecular H-bond network.

The left-handed fourfold antiparallel double-helix conformation was observed either when K^+ cation is present only at low pH (pH 3 or 4) or in the presence of NH_4^+, Cs^+, or Rb^+ cations [16]. Theoretically, four hydrogen bonds may form per repeating unit but all four were never observed simultaneously in the different conformations [4].

11.3.4
Double Helix

Gellan analogues are formed from the same backbone and differ by the side chain (Figure 11.7). X-ray analysis of deacetylated gellan was carried out with different monovalent cations (Li^+, Na^+, K^+, Rb^+, TMA^+, etc.) and has shown similar secondary structure and crystalline mesh. It is a double helix made from two left-handed threefold parallel helices that are staggered by $p/2$ [18] (Figure 11.8). The double helix is stabilized by interchain hydrogen bonds O6C . . . O62B. This structure allows the formation of a cage for the counterion that is bound by six ligands with the two chains via a water molecule. For native gellan, the change in orientation of the glucuronate carboxyl group imposed by the glyceryl substituent abolishes the binding site ("cage") for cations. As a consequence, the lateral organization of double helices, as previously observed for deacetylated gellan, is no longer possible. Lateral interactions are weaker.

11.3.5
Triple Helix

Curdlan is a linear chain consisting of $(1\rightarrow3)$-linked-β-D-glucopyranosyl units (Figure 11.9). It belongs to the $(1\rightarrow3)$-β-D-glucan family. Schizophyllan and scleroglucan have similar structures with $\beta(1\rightarrow6)$ branched side chains in around three glucose units. Many X-ray diffraction studies have generally shown triple-helix ordered structures, these polysaccharides being the only ones to form such conformations

Gellan →3)--β-D-Glc-(1→4)-β-D-GlcA-(1→4)-β-D-Glc-(1→4)-α-L-Rha-(1→
                    A   B   C   D

                              L-glycerate
                                  2
                                  ↑
Native Gellan →3)--β-D-Glc-(1→4)-β-D-GlcA-(1→4)-β-D-Glc-(1→4)-α-L-Rha-(1→
                                  ↓
                                  6
                              acetate

Welan →3)--β-D-Glc-(1→4)-β-D-GlcA-(1→4)-β-D-Glc-(1→4)-α-L-Rha-(1→
                                                      3
                                                      ↑
                                                      1
                                            α-L-Rha (or Man)

Rhamsan →3)--β-D-Glc-(1→4)-β-D-GlcA-(1→4)-β-D-Glc-(1→4)-α-L-Rha-(1→
                           6
                           ↑
                           1
                 α-D-Glc-(6←1)-β-D-Glc

Figure 11.7 Structures of gellan and its analogues.

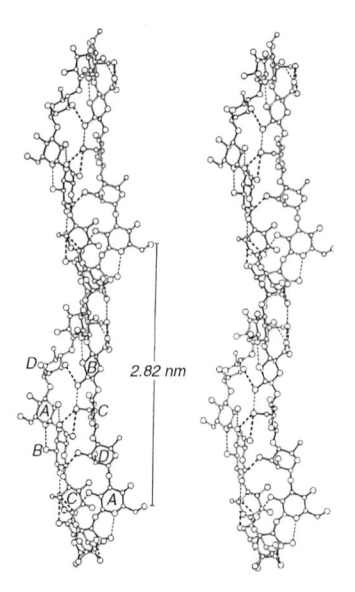

Figure 11.8 Side view of the double helix of gellan showing the OH(O hydrogen bonds within the molecule. Intrachain H-bonds are indicated by thin dashed lines and interchain H-bonds by thick dashed lines. (Reproduced from Ref. [15] with permission from Pergamon Press.)

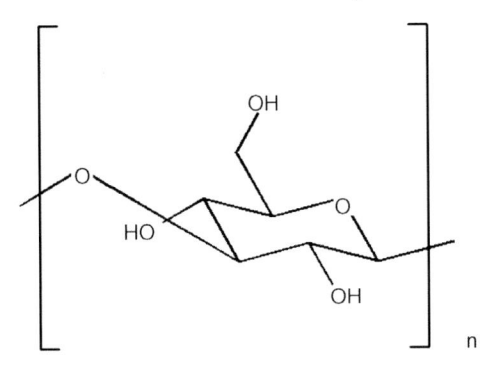

Figure 11.9 Chemical structure of curdlan.

[19]. The three strands of the triplex are linked together through triads of strong interstrand hydrogen bonds between the O2 hydroxyls stabilizing the triplex structure [20]. All the O6 hydroxymethyl groups of the glucose residues are outside of the cylinder of the triple-helical structure. The side groups of the branched (1→3)-β-D-glucan do not disturb the triple-helix formation of the backbone glucan.

11.4
Conformation in Solution: Solution Properties and Applications

When the polymer dissolves, an ordered conformation may be preserved, in particular according to the experimental conditions. So, the stability of an ordered conformation depends not only on the temperature and the solvent but also on the ionic strength, the pH, and the nature of the counterions (for polyelectrolytes). Conversion between an ordered and a disordered conformation strongly modifies the solution properties [21,22].

For different microbial polysaccharides, the existence of an ordered conformation in solution and its influence on either the rheological properties of solutions or their gelation properties will be described. Major applications of these polysaccharides will be discussed. The most widely used techniques to study the conformation of polysaccharides in solution are optical rotation and circular dichroism, but NMR spectroscopy and microscopic techniques (AFM, etc.) may also be applied.

11.4.1
Dextran and Pullulan

As previously discussed, dextran and pullulan are amorphous polymers. In aqueous solution, they do not have any ordered conformation due to the presence of the α(1→6) bonds and behave as flexible chains [23,24]. Dextran may be produced by *Leuconostoc mesenteroides* strain leading to slightly branched polymers (around 5% α(1→3)-D-glucose) and molar masses between 40×10^3 and 10^6 g/mol. Nevertheless, the macromolecular chains being very flexible, the solutions are relatively quite viscous. Pullulan is a linear polymer of fungal origin with molar masses between 10^3 and 3×10^6 g/mol [25].

Contrary to other polysaccharides, these two polysaccharides are not used for their thickening or gelling properties. Dextran of smaller molar masses was used in pharmaceutical industries as blood plasma substitute and its cross-linked derivative as molecular sieve (Sephadex). Pullulan can form water-resistant films with low oxygen permeability and therefore is used in food packaging. Indeed, its low dioxygen permeability avoids oxidation and preserves the aroma, flavor, and freshness of food. Film-forming properties allow its use as a covering agent on surfaces such as the glass bottle coating. It is of interest for recycling because it may replace aluminum layers within the packaging of beverage cartons. Moreover, pullulan may be used as starch replacement for texturing properties. It extends the conservation period of food by inhibiting mold growth [26].

11.4.2
Hyaluronan

Hyaluronan (HA) is found in the intercellular matrix of mammalian connective tissues, and it was mainly extracted from bovine vitreous humor, rooster combs, and umbilical cords. It is, nowadays, also produced by some bacteria; for example, it is a significant component of the extracellular materials of strains such as *Streptococcus zooepidemicus*, which is used to produce, on a large scale and with a good yield, pure HA having a molar mass around of $2-3 \times 10^6$ g/mol [15].

The structure in the solid state, for example, can be well demonstrated by citing the polymorphic character of hyaluronan that, depending on the conditions, presents various ordered conformations such as simple or double helices. The existence of an ordered molecular conformation in aqueous solution was demonstrated by a change in the specific optical rotation by increasing the urea concentration [27,28], which is a destabilizing agent for H-bonds, or in the presence of NaOH [29].

Using circular dichroism and analysis of the Cotton effect, Hirano and Kondo [28] have shown that HA has a left-handed conformation in aqueous solution. Another work using circular dichroism with hydroorganic solvent (20/80 ethanol/water mixture) showed a reversible conformational transition depending on pH, temperature, molar mass, and composition of the solvent [30]. It was suggested that the transition from a fourfold helix to a coil conformation was due to the breakup of an H-bond between the protonated carboxylic group and the oxygen atom of the acetamido group. NMR spectroscopy allows determining functional groups involved in the H-bonds. Use of deuterated water is not convenient but many studies are performed using DMSO-d_6. Considering sodium salt of HA oligomers obtained by enzymatic hydrolysis of animal HA, Scott *et al.* [31] have counted four H-bonds per tetrasaccharidic unit (Figure 11.10a). These results were compatible with conformational simulations [32].

However, it is of major importance to point out that no conformational transition was observed with bacterial HA. This may be due to the absence of proteins, always

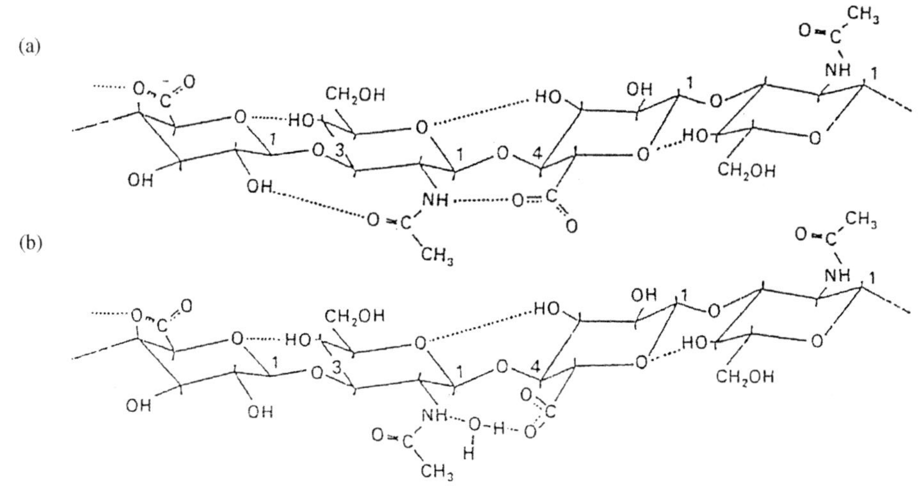

Figure 11.10 Secondary structure of hyaluronan in (a) DMSO and (b) DMSO containing water. (Reproduced from Ref. [31] with permission from Portland Press.)

present in HA from animal origin. Considering bacterial HA, when dissolution occurs, such a degree of order is not preserved. Indeed, in aqueous solution, HA presents the behavior of a disordered chain but with a stiff character [33,34]. Many groups have interpreted this behavior as the coexistence of flexible segments and many ordered segments along the macromolecular chain [33,35]. Intrachain hydrogen bonds may beat the origin of these more rigid ordered zones. Darke et al. [35] have estimated these zones by NMR at a proportion varying from 55 to 70% inside the polymer, depending on the media conditions. Heatley and Scott [2] proposed that in aqueous solutions there are four possible hydrogen bonds per repeating unit, with one of them being established via a water molecule that is between the acetamido group and the carboxylate of a neighboring unit (Figure 11.10b). According to other works [36,37], the stiffness of the isolated simple chain could be due to the restricted conformational freedom of glycosidic bonds rather than the presence of stabilizing H-bonds.

The major interest of HA resides principally in its viscosifying and hydrating properties. Initially from animal origin, it is now produced with a better purity by fermentation. As it is biocompatible, it is used not only for biomedical applications for arthrosis treatment or ophthalmic surgery but also for cosmetics where hydration power is sought. Its partially cross-linked derivative (Hylan) finds applications in surgery as postoperative implants to avoid adhesion phenomena.

For usual applications, it is important to determine the viscosity of the solution and especially how a polymer increases the viscosity of the solvent. Two types of experiments can be carried out:

- Flow experiments in which the viscosity is measured as a function of the shear rate.

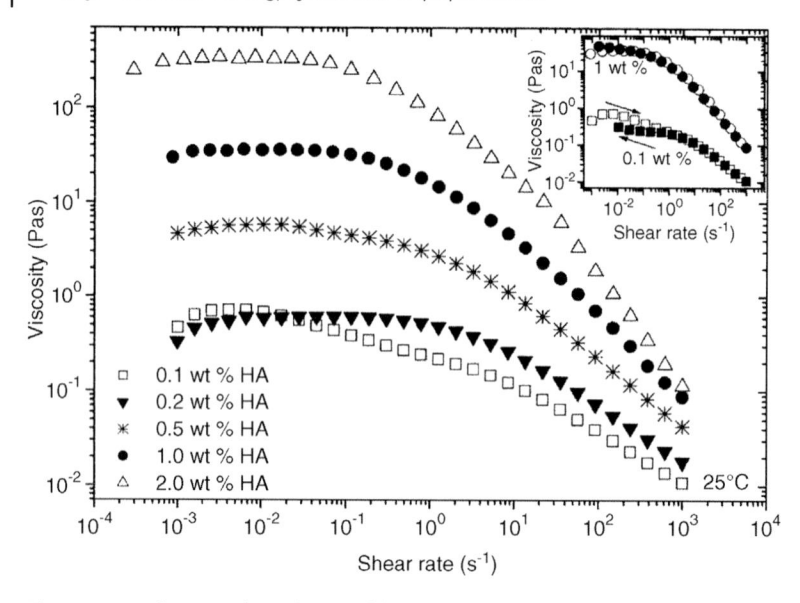

Figure 11.11 Shear rate dependences of the viscosity at 25 °C for HA solutions of various concentrations. The inset shows data obtained under increasing (open symbols) and decreasing (filled symbols) shear rates for HA solutions with concentrations of 0.1 and 1.0 wt %. (Reproduced from Ref. [38] with permission from Springer.)

- Dynamic experiments performed generally at higher polymer concentrations. A stress of defined amplitude and frequency is applied and the sinusoidal strain is measured. Storage (G') and loss (G'') moduli are considered as a function of the frequency and determine the rheological properties of the fluid, as well as its viscoelastic character.

Maleki et al. have observed anomalous viscosity behavior in dilute and semidilute aqueous solutions of hyaluronic acid in which 0.1% protein may be present (Figure 11.11) [38].

Pronounced shear thinning behavior is found for semidilute solutions of HA at high shear rates and no hysteresis effects are observed upon the subsequent return to low shear rates. In the case of dilute solutions, the shear-induced decrease in the viscosity may be related to the alignment of polymer chains. No mechanical degradation was observed for HA even if the sample was exposed to a high shear rate (1000 s^{-1}) for 10 min. When a low fixed shear rate (0.001 s^{-1}) was applied to a dilute HA solution, a significant increase in viscosity occurred over time. The growth of stronger association structures at lower HA concentrations is probably due to an easier reorganization of the chains. The assumption is that the shear-induced alignment and stretching of polymer chains favor the formation of hydrogen-bonded structures, where cooperative zipping of stretched chains yields an interconnected network.

The viscosity of hyaluronate solutions is extremely sensitive to protein contamination [39]. The rheology of NaHA solutions in phosphate-buffered saline is typical of flexible polyelectrolytes in the high-salt limit. However, there is no evidence that it

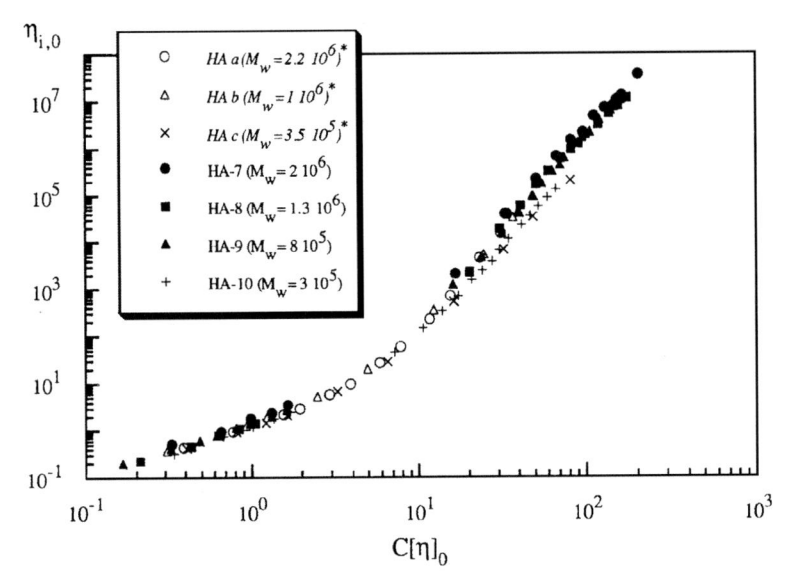

Figure 11.12 Dependence of $\eta_{sp,0}$ as a function of the overlap parameter $C[\eta]$ for hyaluronan solutions in 0.1 M NaCl at 25 °C. (Reproduced from Ref. [42] with permission from American Chemical Society.)

forms a reversible gel solution even though this is in contrast to Scott *et al.*'s suggestion [40]. Indeed, Newtonian viscosities are observed over a wide range of shear rates and all rheology results are independent of shear history. These results, combined with the weak temperature dependence of viscosity reported for NaHA solutions [41], prove that there are no strong associations between NaHA chains under physiological conditions. The gel nature of the synovial fluid is thus believed to be caused by interactions between NaHA and proteins.

Flow experiments allow the determination of a master curve between specific viscosity and the overlap parameter ($C[\eta]$), [η] being the intrinsic viscosity [42] (Figure 11.12). This permits further predictions of the rheological behavior of the solutions, determinations of the overlap concentration – an important parameter for the applications and gelation of polymeric solutions – and finally to confirm that there is no change in conformation under the experimental conditions used.

HA solutions show viscoelastic behavior for concentrations larger than the overlap concentration [43]. For semidilute solutions, in the low-frequency domain the storage modulus G' is lower than the loss modulus G''. Then above a characteristic frequency one finds $G' > G''$ (Figure 11.13). When the polymer chain is loosely cross-linked, gel-like behavior is observed with G' larger than G'' and a different variation of these moduli with frequency, in relation with their use for viscosupplementation applications.

Considering the role of pH in the rheology of HA solutions, different behaviors were observed according to the considered parameter. From molecular weight distribution analysis, a slight degradation is observed in acidic (pH 1.6) and basic

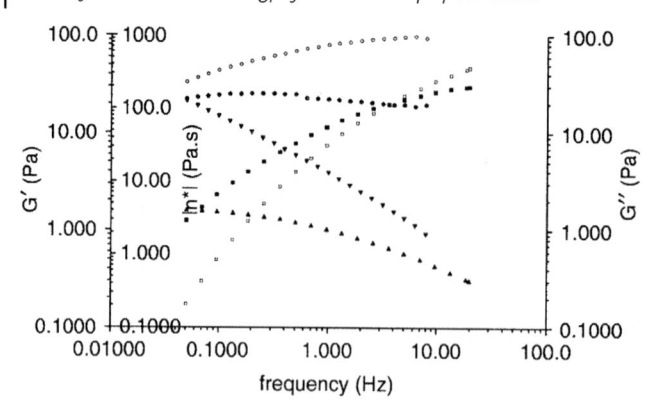

Figure 11.13 Dynamic behavior of linear (G', ■; G'', □; $|\eta|$, ▲) and loosely cross-linked (G', ○; G'', ●; $|\eta|$, ▼) HA. Polymer concentration was 10 g/l in physiological medium. (Reproduced from Ref. [43] with permission from Society of Chemical Industry.)

media (pH 12.6), but the rheological behavior is only strongly influenced by the pH within two domains (see Figure 11.14) [44]. Around pH 2.5, a thermoreversible gel-like behavior is observed and is attributed to cooperative interchain hydrogen interactions due to the reduction of the polymer net charge and maybe the protonation of the acetamido groups. At lower pH, the polymer is resolubilized and this sol–gel transition is pH-reversible.

For pH larger than 12, the decrease in viscosity is mainly attributed to a reduction in the stiffness of the polymeric backbone under alkaline conditions due to the partial breakage of the H-bond network.

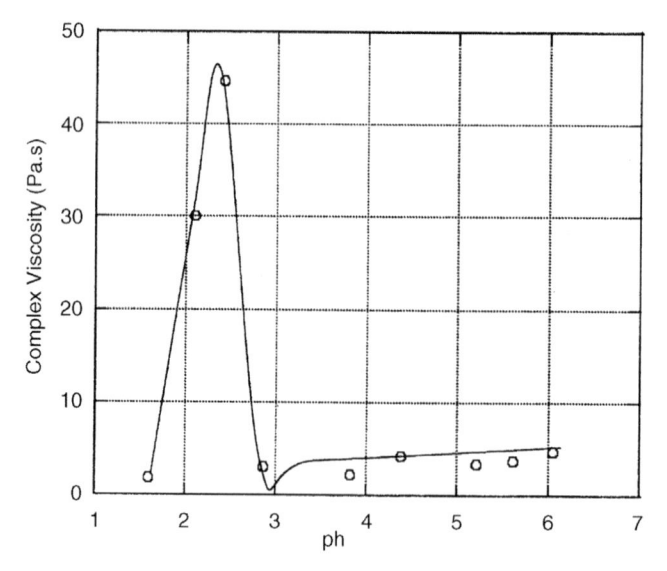

Figure 11.14 Variation of the complex viscosity of HA solutions as a function of pH. (Reproduced from Ref. [44] with permission from American Chemical Society.)

11.4.3
Xanthan

In solid state, the ordered structure of xanthan was not completely elucidated; in solution, there is the same problem. The existence of an order–disorder conformation transition was clearly established [45,46]. It is induced by a temperature increase or a decrease in the ionic strength. The ordered conformation has a stability that depends on the pyruvate and acetyl group content, the first one leading to a stabilizing effect and the second one to a destabilizing effect [47]. Many groups [46,48,49] highlight the existence of a native ordered conformation (N) obtained during biosynthesis and leading irreversibly after denaturation (D) to a renatured conformation (R):

$$N \to D \rightleftharpoons R.$$

The renatured conformation presents a stiffness [50] and viscosity much greater than the native conformation [51]. Milas *et al.* proposed a simple helix ordered conformation for the native form and a double helix formed by a single chain folded up itself for renatured form [50,51].

Rheological experiments were carried out on xanthan solutions under the order–disorder conformational transition conditions. Extensional flow can distinguish between flexible and rigid chains [52]. A xanthan solution at a polymer concentration of 1% (w/w) containing 0.008 M NaCl exhibited a conformational transition between 45 and 65 °C. When G' and G'' variations with frequency were considered, at temperatures below the conformational transition temperature (T_m), the viscoelastic characteristics of the solution are essentially similar across the studied temperature range (Figure 11.15). G' and G'' are only weakly dependent on frequency ($G' \propto \omega^{0.3}$

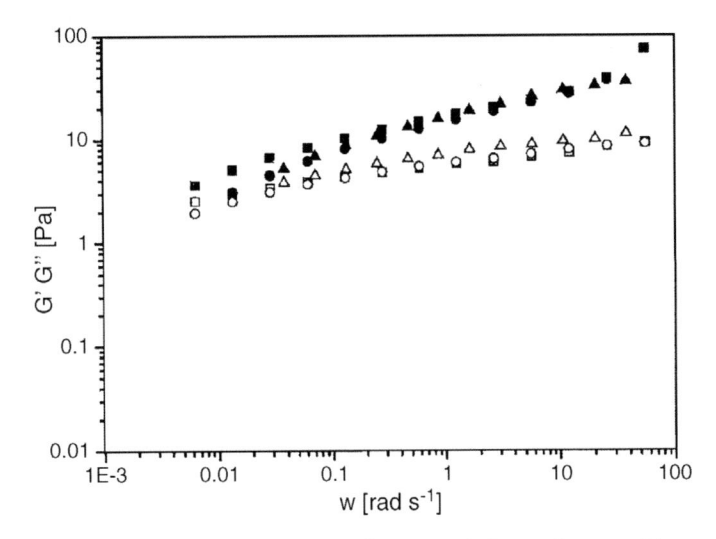

Figure 11.15 G' (solid symbols) and G'' (open symbols) as a function of the angular frequency for xanthan solution at 25 °C (■), 40 °C (●), and 50 °C (▲). (Reproduced from Ref. [45] with permission from John Wiley & Sons, Inc.)

and $G'' \propto \omega^{0.1}$), G' being greater than G''. At temperature larger than T_m, G' and G'' show a more pronounced dependence upon frequency (at 80 °C, $G' \propto \omega^{1.33}$ and $G'' \propto \omega^{0.75}$) with the values of both moduli decreasing significantly with increasing temperature.

A master curve can be obtained using the time–temperature superimposition principle. The same observations were made considering the shear and complex viscosities. The rheological description is qualitatively in accordance with the behavior of rigid polymers in concentrated solutions [53]. Superposition of the different curves indicates that while the global structure of the "network" formed by the xanthan molecules is essentially unchanged by heating well into the transition area, some structural changes do occur on a molecular scale as discussed previously.

According to the structure of the xanthan chain (degrees of pyruvate or acetate), the transition temperature and thus rheological behavior were not the same. The transition temperature varies from 60 °C for native xanthan down to 51 °C without acetate or up to 72 °C in the absence of pyruvate, in 0.1 M NaCl solution and considering a polymer concentration of 1 g/l. Under the same conditions, xanthan without acetate and pyruvate has a transition temperature equal to 61 °C [54].

Conformation has considerable impact on the material properties and applicability. We have already discussed the role of temperature and the nature of the conformation on the rheological properties but it also has a great influence on the stability against temperature, pH, or other parameters. As an example (Figure 11.16), after 30 days at 80 °C, the solution viscosity of xanthan was unchanged when the polymer chain was in the ordered helical conformation compared with the great decrease observed for the disordered coil form. In solution, in the ordered conformation, the side chains wrap around the backbone, thereby protecting the labile β(1→4) linkages from attack. It is thought that this protection is responsible for the stability of xanthan solutions [55].

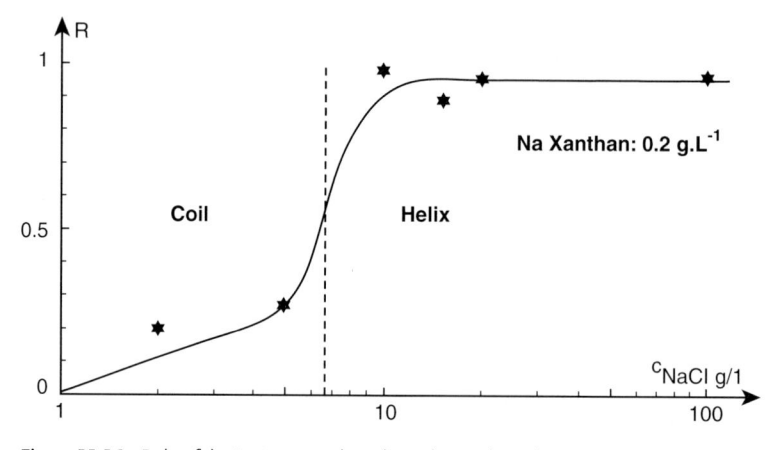

Figure 11.16 Role of the ionic strength in the polymer degradation: relation between the residual viscosity after 30 days at 80 °C and NaCl concentration (shear rate = 700 s^{-1}). (Reproduced from Ref. [56].)

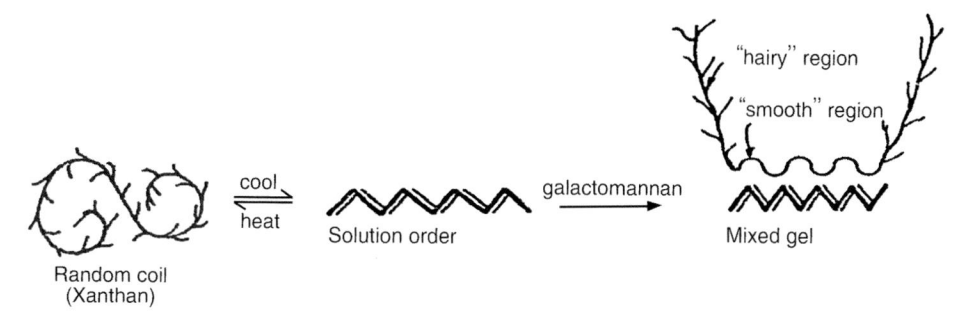

Figure 11.17 Molecular origin of xanthan interaction with galactomannans. (Reproduced from Ref. [60] with permission from Pergamon Press.)

Xanthan is the most used bacterial polysaccharide either for its high thickening power or for its marked pseudoplastic character [57]. Major applications are in food, cosmetic, paint, and petroleum industries. Xanthan may also be used for its gelling properties in the presence of trivalent ions (Cr^{3+}, Al^{3+}, Fe^{3+}) or in interaction with galactomannans or glucomannans [58,59]. In the first case, gelling is the consequence of a ligand exchange reaction between carboxylic groups of glucuronic acid and the hydration layer of chromium ions. Such a thermoreversible gel was obtained in the xanthan–galactomannan mixture, as an example. A strong synergy exists between these two nongelling polysaccharides. Galactomannans are composed of a $\beta(1\rightarrow4)$-D-mannan backbone on which $\alpha(1\rightarrow6)$-D-galactopyranosyl units are grafted at position 6. Gelling properties are based on interchain cooperative interactions and are favored by high mannose–galactose ratios and/or longer nonsubstituted sequences [60]. One of the proposed mechanisms was based on the specific interaction between the surface of the xanthan fivefold helix and nonsubstituted galactomannan sequences [60] (Figure 11.17).

Millane and Wang [61] and Chandrasekeran and Radha [62] have demonstrated by molecular modeling and X-ray diffraction analysis of mixtures that such interactions are possible, the chains being able to adapt each other and to take either a conformation close to xanthan one (5_1) [62] or a similar conformation to galactomannan (2_1) [61,62]. Many workers have studied such interactions and different mechanisms have been proposed. Among them, Mannion et al. [63] described two possible mechanisms depending on the temperature conditions: at ambient temperature, a weak gel is formed whose properties remain nearly independent of the mannose (M)/galactose (G) ratio of the galactomannan; and at higher temperature ($T>60\,°C$), a true gel forms highly dependent on the M/G ratio. Bresolin et al. [64] showed that the lower M/G ratio of the galactomannan makes interactions only involving disordered xanthan chains in contrast to M/G ratios higher than 3 that allow interactions to occur between galactomannan and both ordered and disordered xanthan chains. Moreover, Goycoolea et al. [65] suggested the occurrence of interactions between ordered xanthan and mannan blocks. Xanthan–galactomannan interactions were explained by using the release of flavor molecules (limonene) [66]. But as two distinct mechanisms involving the native helix of xanthan and the mannan chain of the galactomannan are proposed to have consequences upon limonene release, all the previous mechanisms are consistent with the conclusions drawn. Not only one mechanism is responsible for the xanthan to galactomannan synergistic interactions and it largely depends on the nature and the characteristics of both polysaccharides. Gelling properties are reinforced when xanthan is deacetylated [67]. Such mixtures are used in food applications.

11.4.4
Succinoglycan

Succinoglycan is an ionically charged polysaccharide with a relatively long lateral chain composed of four sugar units: it has acetyl, pyruvate, and succinate substituents (Figure 11.18).

In dilute solution, in the absence of external salt at temperatures lower than $10\,°C$ or in the presence of external salt, the polysaccharide has an ordered conformation that is generally admitted as being a stretched simple helix [68] with folding of the lateral chain along the macromolecular backbone. The transition is intramolecular and strongly cooperative, in particular by comparison with xanthan, and is due to the reversible partial association of the side chain. Transition temperature depends on the ionic strength, the nature of the counterion, and the concentration [22]; it is also affected by the substituent ratio [69]. As discussed for xanthan, three conformations may be considered: native, disordered, and renaturated, the transition from native to disordered being irreversible:

$$N \rightarrow D \rightleftharpoons R.$$

The stable conformation of succinoglycan either in the solid state or in solution is a simple helix stabilized by H-bonds between the backbone and lateral grafts [70]. In the ordered conformation, the lateral chains are folded up along the backbone and rigidify the conformation. During the conformational transition, the lateral grafts free themselves from the backbone and this is well observed from NMR experiments.

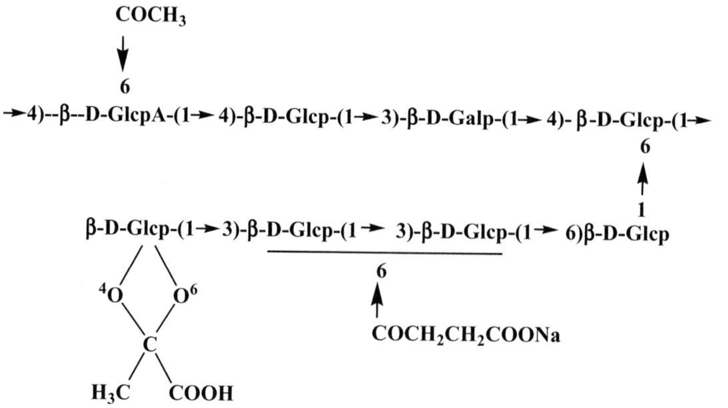

Figure 11.18 Structure of succinoglycan.

11.4 Conformation in Solution: Solution Properties and Applications | 335

This transition is reversible with temperature and the ordered conformation is obtained by cooling or increasing ionic strength. As in xanthan, the substituents modify the transition temperature. The transition temperature is 60 °C for native polymer, 63 °C without succinate, 62 °C without pyruvate, and 62 °C in the absence of both substituents, in 0.01 M NaCl solutions and a polymer concentration of 0.7 mmol repeating unit per liter [71]. This slight decrease in the transition temperature shows the small influence of the substituents on the stability of the conformation.

Contrary to xanthan, the renatured conformation may present smaller molar mass and viscosity [68,72]. The N and R forms would present the same ordered conformations. Differences may be attributed, when passing the transition in dilute solution, to an irreversible chain disaggregation or to the rupture of weak points (preexisting break points on the native chain, the chain fragments being bound together through hydrogen bonds) [73,74]. As a consequence, heating above the transition temperature decreases irreversibly the viscosity of the solution (Figure 11.19). Nevertheless, the molar masses of the renatured conformation stay high (3 to 4×10^6 g/mol) and, associated with the stiffness of the ordered conformation, lead to high viscosities. The succinate groups do not have a great influence on the viscosity of native succinoglycan solutions [71].

Succinoglycan solutions tend to form aggregates under specific conditions [69,74]. In more concentrated solutions, researchers described gelling properties [75,76], in contrast to xanthan for which, alone, gelation was never observed. Succinoglycan is used for its thickening properties and its pseudoplastic behavior, specifically under acidic conditions due to its good stability in low pH solutions.

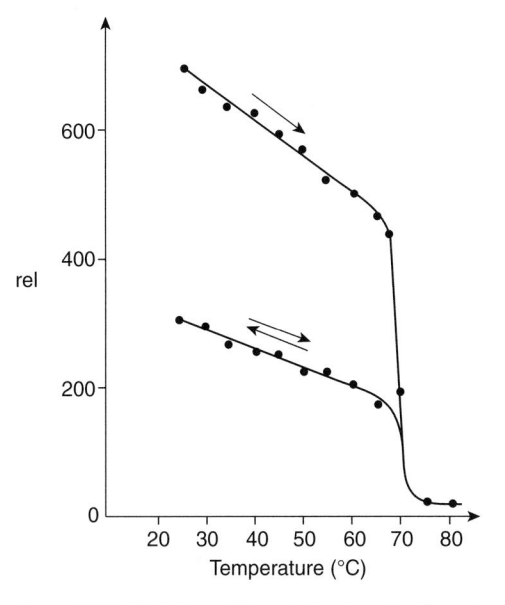

Figure 11.19 Variation of the relative viscosity of succinoglycan solutions at 1 g/l in 0.1 M NaCl as a function of temperature. (Reproduced from Ref. [72] with permission from Elsevier B.V.)

11.5
Gelling Properties in the Presence of Salts

11.5.1
β(1→4)-D-Glucuronan

The β(1→4)-D-glucuronan is a highly charged polysaccharide. Its behavior is comparable to that of other known polyuronic polymers. It is insoluble in acidic form and soluble in its saline form. Average molar masses are not very high (6×10^4 g/mol $<$ MW $< 4 \times 10^5$ g/mol).

It forms thermoreversible gels when monovalent cations are present. When divalent ions or high concentrations of monovalent ions are present, these gels are no longer reversible [77], as observed for pectins or alginates in the presence of divalent ions. Pectins are more or less esterified polygalacturonans and alginates are composed of mannuronic acid and guluronic acid blocks. Their gelation mechanism is based on the "egg-box" model [78]. Junction zones are formed by nonesterified galacturonic acid blocks (pectin) or guluronic acid blocks (alginate) that associate laterally through the complexation of divalent ions by carboxylate groups and oxygen atoms of hydroxyl groups (Figure 11.20).

In the case of polyglucuronan, the gelation mechanism is certainly different because these cavities are not created by the chain conformation. This is confirmed by the fact that, in the presence of an excess of monovalent ions, divalent ions may be exchanged. So, the presence of monovalent or divalent salt by screening the electrostatic interactions allows the lateral association of the chains, without specific interactions in the case of divalent ions. Gels formed with monovalent ions are less strong and less stable than those obtained with divalent ions.

Gel strength varies with the content of acetyl groups: the modulus increases when the acetyl group content decreases. The acetyl groups limit the aggregation [77] and

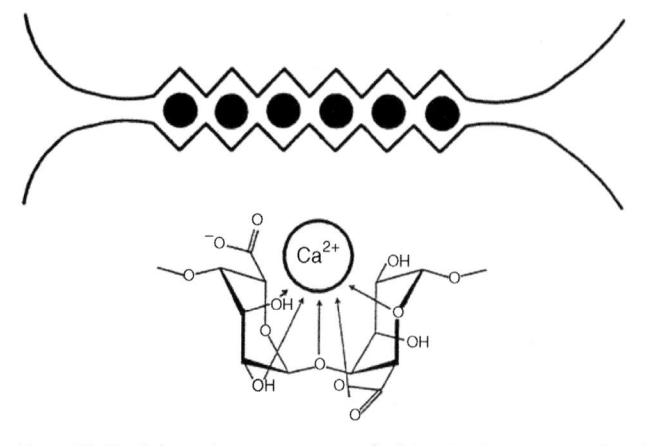

Figure 11.20 Schematic representation of calcium ion interaction with poly-L-guluronate sequences ("Egg-box" model).

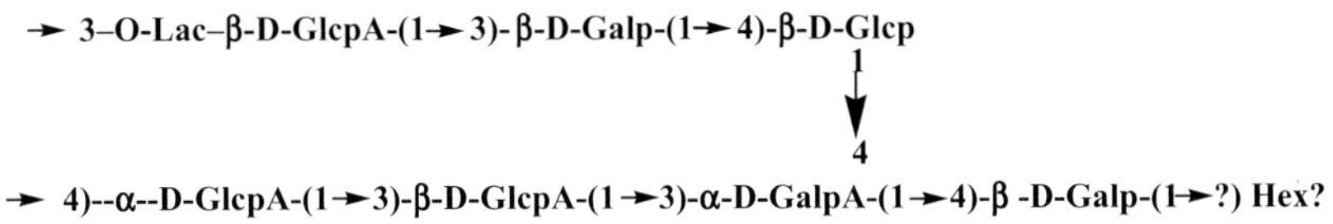

→ 3–O-Lac–β-D-GlcpA-(1→3)-β-D-Galp-(1→4)-β-D-Glcp
                                              |
                                              ↓
                                              4
→ 4)--α--D-GlcpA-(1→3)-β-D-GlcpA-(1→3)-α-D-GalpA-(1→4)-β -D-Galp-(1→?) Hex?

Figure 11.21 Structure of exopolysaccharide 1644.

disturb the ordered arrangement of the chains. As in the solid state, the deacetylated sample is better crystallized than the acetylated one [79]. Gels formed from the deacetylated polymer have comparable strengths to those of alginate.

11.5.2
Polysaccharide 1644

This bacterial polysaccharide is strongly charged: its structure is composed of an eight-sugar repeating unit that includes three uronic acids and a diacid sugar (Figure 11.21). This polymer was never studied at solid state.

Study of its polyelectrolyte properties in solution demonstrates a high selectivity between monovalent [80] and divalent ones (except Mg^{2+}) not predicted by polyelectrolyte theory, and a high affinity for divalent ions due to their selective fixation onto carboxylic sites. A conformational transition was demonstrated for the polymer in solution when NaCl and $CaCl_2$ were simultaneously present. The transition is local and not very cooperative: it was interpreted as being due to the interaction of lateral chain with macromolecular backbone.

At higher concentrations, in the presence of sufficient concentrations of divalent ions and 0.1 M NaCl, a particularly elastic and resistant clear gel was obtained. The mechanical properties of the gel are related particularly to the conformational transition; they depend on the quantity of divalent ions fixed on the polymeric chain. The gel strength also varies with the nature of the divalent cation according to the following sequence: $Ca^{2+} > Sr^{2+} > Ba^{2+} > Mg^{2+}$ [81]. The peculiar elasticity of this gel was related with the flexible nature of the intercatenary network: it would be formed by localized junctions bringing into play a limited number of very stable and small divalent ion/carboxylic site complexes (Figure 11.22).

Therefore, this proposed gelation mechanism differs markedly from those proposed for other polysaccharides that generally have long junction zones leading to more brittle gels. The gels are weaker than those of alginate, carrageenan, agarose, or gellan at similar concentrations [80]. Nevertheless, the gelling properties of this polymer remain interesting due to its strong elasticity compared with other polysaccharidic gels.

11.5.3
Gellan and Similar Polysaccharides

The study in solid state has shown a double helix similar for all polysaccharides of the gellan family. In solution, an ordered double-helix conformation was clearly established [83]. For welan, this conformation was strongly stabilized (up to 100 °C at

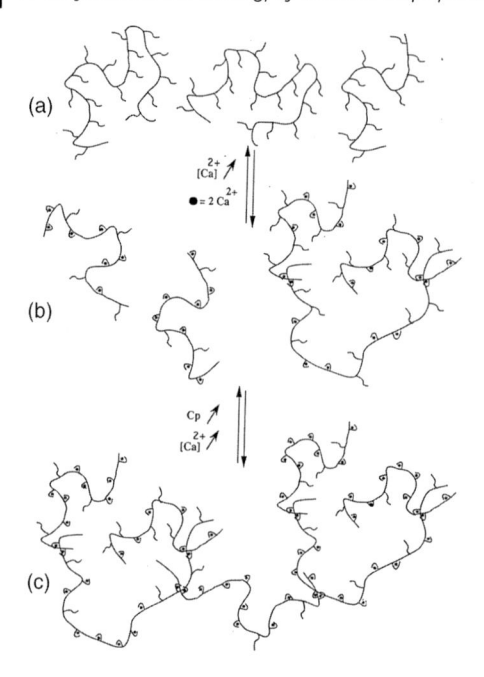

Figure 11.22 Interactions of divalent cations and gelation mechanisms proposed for polymer 1644. (Reproduced from Ref. [82].)

least) by the lateral chain screening the carboxylate group: no conformational transition was detected [83]. The ordered conformation is also stabilized by the presence of glyceryl groups in gellan [84].

Gellan shows the disorder–order transition on cooling or on adding salts, which is analogous to that of xanthan, and a gellan aqueous solution can change its behavior from a dilute polymer solution to a weak gel depending on the polymer concentration and the temperature. The shear rate dependence of the steady shear viscosity is illustrated in Figure 11.23 [85].

The 1% gellan solution showed a step-like change (T_{ch}) at 29 °C, in good agreement with the characteristic temperature where circular dichroism showed a steep change in molecular ellipticity [86]. It is suggested that gellan chains changed from coil conformations to helical ones at T_{ch}, where the viscosity showed a steep increase during cooling. Steady shear viscosity measurements indicate that gellan solutions tend to show more shear thinning behavior with the conformational change from coiled to helical because the helix may be more easily oriented along the shear flow than the coil. The range of the Newtonian plateau at low shear rates gradually becomes narrower with development of an ordered structure of the gellan solution. Stress was plotted as a function of shear rate for different concentrations and different temperatures in the absence of salt. At concentrations larger than 2%, a sol–gel transition was observed. It was demonstrated [87] that the coil–helix transition occurs concurrently with the sol–gel transition since the number of helices formed on

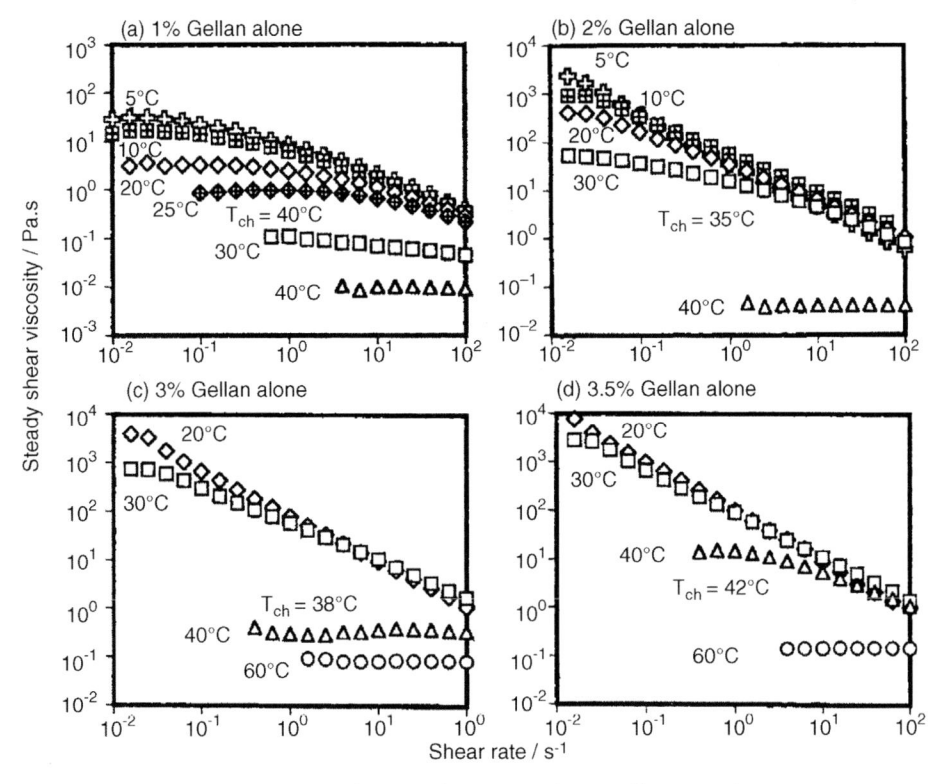

Figure 11.23 Shear rate dependence of viscosity for 1, 2, 3, and 3.5% gellan gum solutions without salt at various temperatures: (○) 60 °C; (Δ) 40 °C; (□) 30 °C; (◆) 25 °C; (◇) 20 °C; (⊡) 5 °C. (Reproduced from Ref. [85] with permission from Springer.)

cooling is enough to form a three-dimensional network throughout the whole space. Under these concentration conditions, the thermal behavior in the presence of sufficient divalent cations was quite different from that in the presence of sufficient monovalent cations. With divalent cations, the ordered structures of gellan formed on cooling became markedly thermally stable, and these ordered structures were essentially different from those in the presence of monovalent cations. The carboxyl groups in gellan molecules repulse each other by electrostatic interaction, and this hinders the tight binding of helices and also the tight aggregation of helices [88]. The introduction of cations can shield the electrostatic repulsion, thereby permitting tight binding and aggregation of helices at lower temperatures. The domain structure formed by the nested helices acts as the junction zone for the infinite networks [88].

Gellan is able to form thermoreversible as well as irreversible gels depending on the conditions. The gelation mechanism is related, as for carrageenans, to the conformational transition. In concentrated solution, when cooling, the disordered chains form double helices that aggregate and lead to a three-dimensional network. The presence of mono- or divalent ions, by screening the electrostatic repulsions, favors the aggregation of double helices and the gelation. Proposed models differ in the nature of aggregates and the role of cations. So, based on observed properties and modeling results, in the presence of divalent cations (Ca^{2+}) direct interchain interactions (like bridging) through carboxylate groups are proposed [84].

Gellan gels are rigid and brittle when they are formed from deacylated polymer and soft and elastic with the acylated chain. Acetyl and L-glyceryl substituents put at a disadvantage the helix aggregation (gelation) such as the crystallization [88]. In the absence of external salt, weak gels are obtained. The formation and properties of the gels strongly depend on the concentration and the nature of cations [87]. Gelation is favored according to the following sequence: $TMA^+ < Li^+ < Na^+ < K^+ \ll Mg^{2+} \sim Ca^{2+}$.

Deacetylated gellan (called Gelrite®) has found many applications [89]. Its gelling properties compare with carrageenans and agar ones, and they need approximately two times less polymer [90] at similar quality. Moreover, their gels are clear. It is used for its gelling and stabilizing properties, as well as its suspending power. It may replace agar as gelling agent in microbiology to prepare culture media.

Within the gellan family (Figure 11.7), the ability to form gels depends on the chemical structure (presence of substituents and side chains) that influences the formation of lateral packing in double helices and the stability of the native ordered conformation. In welan, the side chains shield the carboxylate groups and provide extra stability to the ordered conformation. In water, no condition was found to destabilize them [91] and no gel formation was observed [92]. However, welan can be converted to a disordered conformation by addition of DMSO and a gel forms when the concentration of DMSO is decreased by addition of water [93]. Rhamsan demonstrates an opposite behavior, due to the presence of two (1→6) linkages (compared to (1→3) linkage in welan); the side chain is more flexible, providing less stability to the ordered helical conformation. Then, a conformational transition occurs on deacetylated rhamsan that forms a gel in the presence of Ca^{2+}. For the acetylated rhamsan, no conformational transition or formation of gel was observed [3]. When rhamsan is compared with RMDP17 produced by the bacterium *Sphingomonas paucimobilis*, the only difference in structure is due to the uronic acid that is 2-deoxyglucuronic acid in RMDP17 and glucuronic acid in rhamsan (Figure 11.24).

They exhibit the same polyelectrolyte behavior but their main difference concerns the stability of their ordered conformation and their ability to form gels.

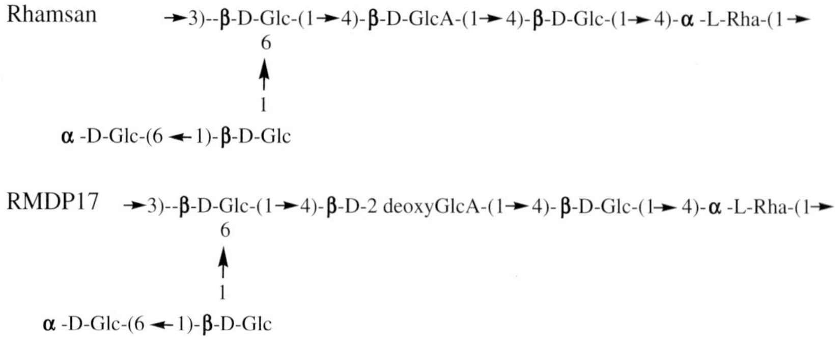

Rhamsan →3)--β-D-Glc-(1→4)-β-D-GlcA-(1→4)-β-D-Glc-(1→4)-α-L-Rha-(1→
6
↑
1
α-D-Glc-(6←1)-β-D-Glc

RMDP17 →3)--β-D-Glc-(1→4)-β-D-2 deoxyGlcA-(1→4)-β-D-Glc-(1→4)-α-L-Rha-(1→
6
↑
1
α-D-Glc-(6←1)-β-D-Glc

Figure 11.24 Structures of rhamsan and RMDP17.

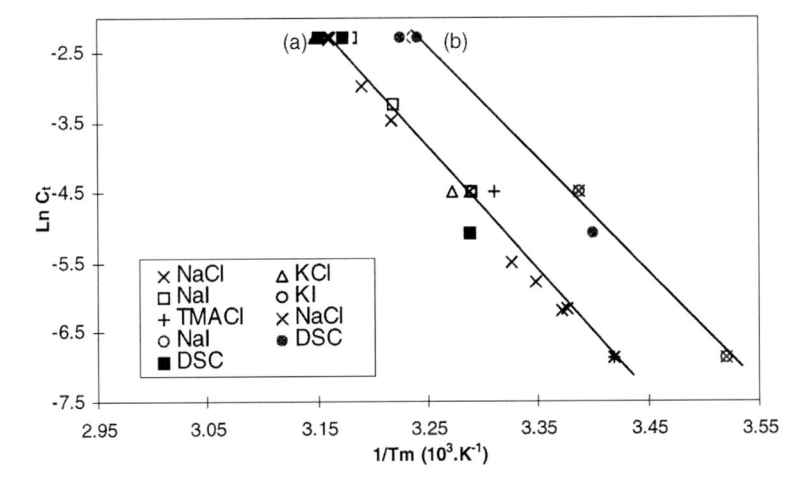

11.5 Gelling Properties in the Presence of Salts

Figure 11.25 Dependence of the inverse of the temperature of conformational transition ($1/T_m$) as a function of the logarithm of the total ionic strength (ln C_t): (a) deacetylated Na-RMDP17; (b) deacetylated rhamsan. Conformational transition was determined using optical rotation experiments (many of points) and DSC technique (when indicated DSC). (Reproduced from Ref. [94] with permission from Elsevier B.V.)

All experiments by optical rotation and DSC have shown higher stability of the double helices of the deacetylated RMDP17. The difference is about 8 °C, which is rather large in comparison with the small difference between the two chemical structures [94] (Figure 11.25).

However, from rheological studies, the deacetylated rhamsan was shown to be more efficient to form gels compared to deacetylated RMDP17. Due to the lower stability of the double helices of deacetylated rhamsan, the latter have a greater tendency to associate themselves, making the gelation easier.

11.5.4
β(1→3)-D-Glucans

These polysaccharides (curdlan, scleroglucan, and schizophyllan) are uncharged homopolymers of glucose. Curdlan is insoluble in aqueous medium, whereas the presence of ramification in scleroglucan and schizophyllan allows their solubilization. When these two polymers are solubilized at room temperature, a stiff, ordered triple-helix conformation is preserved (very similar to that observed in the solid state) [95]. Two order–disorder transitions induced by the pH or temperature increase were demonstrated [96]. The first, reversible, corresponds to the destabilization of the local structure of lateral chains bringing into play water molecules [97]. The second, irreversible, corresponds to the destruction of the triple-helix structure (triple helix → three coils) [98].

Scleroglucan and schizophyllan form gels at low temperature without prior heating, by aggregation of triple helices. However, the gels are weaker than those obtained with curdlan, the presence of lateral chains disturbing the association of ordered zones [99].

The solutions of these polysaccharides show rheological behavior similar to that of xanthan and succinoglycan [96]. They are used for their thickening and pseudoplastic properties and their suspending power. Their properties are relatively stable to temperature (from 20 to 90 °C), ionic strength, and pH (up to pH 12 due to the absence of ionic charges and the stability of their stiff ordered structure that collapses when pH >12). They may compete with xanthan in oilfield applications.

Curdlan is soluble in basic media and has a single- or multistrand helicoidal ordered structure that is destabilized by high pH (pH ~12) [100]. It is biosynthesized with low molar masses (around 5×10^4 g/mol) and shows peculiar gelling behavior. Two types of gel may be obtained [99]:

- The first is formed from a weak alkaline polymer solution, either by neutralization or by addition of divalent cations (Ca^{2+} or Mg^{2+}), or by heating an aqueous solution at a temperature around 60 °C and then cooling. These gels are thermoreversible and do not show much syneresis.
- The other type is obtained by heating an aqueous polymer suspension at a temperature higher than 80 °C. The gel strength is much greater than that for the previous type. Moreover, they are thermoreversible (they melt at temperatures of around 140 °C) and present a strong syneresis.

The gelation mechanisms are not clearly demonstrated. Kanzawa *et al.* [101] have shown that gels prepared by heating at high temperature exhibit strong associations and water release was observed. Moreover, in contrast to the previous gels, a modification of the molecular arrangement was demonstrated with a progressive change from a single-strand helical structure to a triple-strand one [102]. Saito *et al.* have shown the existence of triple-helix segments binding simple chain zones with a triple helix content increasing with the heating temperature. During gelation, association of single and triple helices in microfibrils might occur via hydrogen bonds in the first case and via hydrophobic interactions associated with dehydration of the structure for the gels prepared at high temperature [101].

Gels formed at high temperature are firm and elastic, and not greatly affected by variations in temperature, ionic strength, or pH (stable from pH 2 to 10) [103]. Furthermore, the freeze–thaw cycles do not affect their properties in comparison to other polysaccharides used in food industries. Due to their triple-helix secondary structure and high molar masses in the case of fungal β-glucans, they present interesting pharmaceutical properties such as antitumoral activity as well as activity against bacterial or viral infections.

11.5.5
YAS 34

The chemical structure of this exopolysaccharide has six neutral sugars (two of them in a side chain), one uronic acid in the main chain, and one pyruvyl substituent per repeat unit as shown in Figure 11.26.

Using different techniques such as calorimetry, optical rotation, and conductimetry, whatever the ionic form was (sodium salt or acid), these experiments gave

11.5 Gelling Properties in the Presence of Salts

→4)--β-GlcpA-(1→4)-β-Glcp-(1→4)-β-Glcp-(1→6)-α-Glcp-(1→4)-α-Galp-(1→
```
                                6
                                ↑
                                1
                    β-Galp-(1→3)-β-Galp
                           /       \
                          4         6
                            \     /
                              C
                            /   \
                         H₃C    COOH
```

Figure 11.26 Structure of YAS 34.

evidence of a conformational transition with temperature [104]. The conformational transition is reversible and cooperative for the native polymer in its acid form: heating over T_m leads to a disordered conformation that converts to a renatured form on cooling. On the contrary, in the native sodium salt form, the transition is reversible upon heating over T_m only for short heating times. Then, on cooling, the renatured conformation was recovered. SEC multidetection measurements allow us to present hypothesis on the nature of each conformation [104] (Figure 11.27). The native sodium and acid conformations are double helices, or helical dimers, formed from two chains. The conformational transition leads to the dissociation of these double helices. The renatured conformation restored below T_m is a double helix formed by a single chain folded back on itself ("hairpin-like turn") and the deacetylated chains are always in a coil conformation.

The T_m dependence with salt is, as expected from the theoretical model [105], a single linear relationship between the logarithm of the total ionic concentration (C_t)

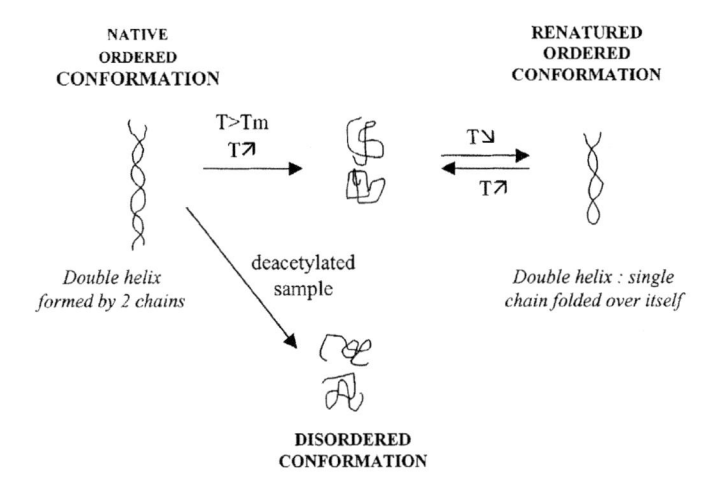

Figure 11.27 Schematic summary on the different conformations. (Reproduced from Ref. [104] with permission from Elsevier B.V.)

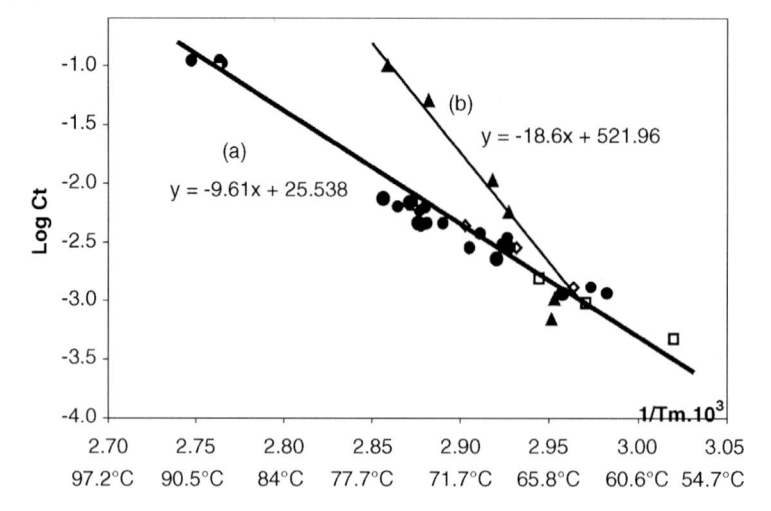

Figure 11.28 $\log(C_t) = f(1/T_m)$ for the native polymer in water or in NaCl: (a) sodium salt form, measurements by DSC (•), conductivity (◇), and optical rotation (□); (b) acid form by DSC measurements (▲). (Reproduced from Ref. [104] with permission from Elsevier B.V.)

and the reciprocal of the temperature of conformational transition (T_m^{-1}) [104] (Figure 11.28).

In the presence of divalent ions, the ordered conformation was found to be more stable as usually obtained for ionic polysaccharides. The native acid and sodium forms present different slopes. Considering that the acid form was less dissociated, T_m values of the conformational transition were less influenced by the ionic strength.

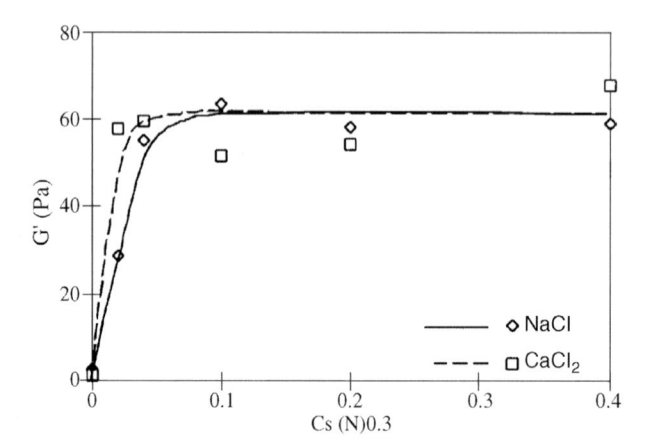

Figure 11.29 "Gel strength" (expressed by G') of YAS 34 as a function of the nature and concentration of external salt (gel prepared from sodium salt form polymer after 60 min of thermal treatment at 92 °C, $T = 25$ °C, strain 10%). (Reproduced from Ref. [106] with permission from Elsevier B.V.)

In dilute solution, under the native conformation (whatever the ionic form is), viscosity measurement shows a tendency for interchain interactions that increases the viscosity especially in the semidilute regime. Moreover, the thermally treated polysaccharide in the sodium salt form has a lower degree of helicity and forms physical gels. The mechanism of gelation was attributed to the formation of interchain interactions involving short residual helical segments of chains. This mechanism of gelation looks like that observed for gelatin gels. When the very elastic gel is formed, one observes a slow evolution of the mechanical properties in relation with a slow stabilization of interactions between residual short helical segments. An unusual behavior is observed in the presence of external salt (Figure 11.29). Above 0.04–0.05 M NaCl or $CaCl_2$, the elastic modulus becomes constant without any ionic selectivity [106]. This is very original compared with many other polysaccharides.

11.6
Conclusions

Bacterial polysaccharides present a wide variety of structures and due to their stereoregularity many adopt a solid-state secondary structure, imposed by the primary structure. Every time an ordered conformation was stabilized in solution, very interesting rheological properties were observed, coming from not only the high stiffness of the chains but also the ability to create a three-dimensional network from interchain connections of these ordered zones (gelling properties). As a consequence, these exopolysaccharide solutions can be used for specific applications covering a wide domain.

References

1 Rees, D.A. and Scott, W.E. (1971) Polysaccharide conformation. Part VI. Computer model-building for linear and branched pyranoglycans. Correlations with biological function. Preliminary assessment of inter-residue forces in aqueous solution. Further interpretation of optical rotation in terms of chain conformation. *J. Chem. Soc. B: Phys. Org.*, 469–478.

2 Heatley, F. and Scott, J.E. (1988) A water molecule participates in the secondary structure of hyaluronan. *Biochem. J.*, **254**, 489–493.

3 Campana, S., Ganter, J., Milas, M., and Rinaudo, M. (1992) On the solution properties of bacterial polysaccharides of the gellan family. *Carbohydr. Res.*, **231**, 31–38.

4 Chandrasekaran, R. (1997) Molecular architecture of polysaccharide helices in oriented fibers. *Adv. Carbohydr. Chem. Biochem.*, **52**, 311–404.

5 Jeanes, A. (1977) in *Extracellular Microbial Polysaccharides*, ACS Symposium Series 45 (eds P.A. Sandford and A. Laskins), American Chemical Society, p. 265.

6 Brant, D.A. and Burton, B.A. (1981) *Solution Properties of Polysaccharides*, American Chemical Society, pp. 81–89.

7 Heyraud, A., Dantas, L., Courtois, J., Courtois, B., Helbert, W., and Chanzy, H. (1994) Crystallographic data on bacterial (1→4)-β-D-glucuronan. *Carbohydr. Res.*, **258**, 275–279.

8 Braccini, I., Heyraud, A., and Perez, S. (1998) Three-dimensional features of the bacterial polysaccharide (1→4)-β-D-

glucuronan: a molecular modeling study. *Biopolymers*, **45**, 165–175.

9 Braccini, I., Grasso, R.P., and Perez, S. (1999) Conformational and configurational features of acidic polysaccharides and their interactions with calcium ions: a molecular modeling investigation. *Carbohydr. Res.*, **317**, 119–130.

10 Okuyama, K., Arnott, S., Moorhouse, R., Walkinshaw, M.D., Atkins, E.D., and Wolf-Ullish, C. (1980) Fiber diffraction studies of bacterial polysaccharides. *ACS Symposium* n° 141, 411–427.

11 Millane, R.P. and Wang, B. (1992) in *Gums and Stabilisers for the Food Industry*, vol. 6 (eds G.O. Phillips, P.A. Williams, and D.J. Wedlock), IRL Press, p. 541.

12 Sheehan, J.K. and Atkins, E.D.T. (1983) X-ray fiber diffraction study on the conformational change in hyaluronate induced in the presence of sodium, potassium and calcium cations. *Int. J. Biol. Macromol.*, **5**, 215–221.

13 Atkins, E.D.T., Phelps, C.F., and Sheehan, J.K. (1972) The conformation of the mucopolysaccharides, hyaluronates. *Biochem. J.*, **128**, 1255–1263.

14 Arnott, S., Mitra, A.K., and Raghunathan, S. (1983) Hyaluronic acid double helix. *J. Mol. Biol.*, **169**, 861–872.

15 Haxaire, K., Braccini, I., Milas, M., Rinaudo, M., and Milas, M. (2000) Conformational behavior of hyaluronan in relation to its physical properties as probed by molecular modeling. *Glyocobiology*, **10**, 587–594.

16 Sheehan, J.K., Gardner, K.H., and Atkins, E.D.T. (1977) Hyaluronic acid: a double helical structure in the presence of potassium at low pH and found also with the cations ammonium, rubidium and caesium. *J. Mol. Biol.*, **117**, 113–135.

17 Guss, J.M., Hukins, D.W.L., Smith, P.J.C., Winter, W.T., Arnott, S., Moorhouse, R., and Rees, D.A. (1975) Hyaluronic acid: molecular conformations and interactions in two sodium salts. *J. Mol. Biol.*, **95**, 359–364.

18 Chandrassekaran, R., Millane, R.P., Arnott, S., and Atkins, E.D.T. (1988) The crystal structure of gellan. *Carbohydr. Res.*, **175**, 1–15.

19 Chuah, C.T., Sarko, A., Deslandes, Y., and Marchessault, R.H. (1983) Triple-helical crystalline structure of curdlan and paramylon. *Macromolecules*, **16**, 1375–1382.

20 Deslandes, Y., Marchessault, R.H., and Sarko, A. (1980) Triple-helical structure of $(1\rightarrow 3)$-β-D-glucan. *Macromolecules*, **13**, 1466–1471.

21 Rinaudo, M. and Milas, M. (1987) in *Industrial Polysaccharides: Genetic Engineering, Structure/Property Relations and Applications* (ed. M. Yalpani), Elsevier Science Publishers, p. 217.

22 Gravanis, G., Milas, M., Rinaudo, M., and Clarke-Sturman, A.J. (1990) Conformational transition and polyelectrolyte behavior of a succinoglycan polysaccharide. *Int. J. Biol. Macromol.*, **12**, 195–200.

23 Kato, T., Katsuki, T., and Takahashi, A. (1984) Static and dynamic solution properties of pullulan in a dilute solution. *Macromolecules*, **17**, 1726–1730.

24 Kato, T., Okamoto, T., Tokuya, T., and Takahashi, A. (1982) Solution properties and chain flexibility of pullulan in aqueous solution. *Biopolymers*, **21**, 1623–1633.

25 Israilides, C., Smith, A., Scanlon, B., and Barnett, C. (1999) Pullulan from agro-industrial wastes. *Biotechnol. Genet. Eng. Rev.*, **116**, 309–324.

26 Yuen, S. (1974) Pullulan and its applications. *Process Biochem.*, **9**, 7–22.

27 Hirano, S. and Kondo-Ikeda, S. (1974) Molecular conformation of polysaccharides in solution. Changes in the optical rotation and in the elution pattern of gel filtration. *Biopolymers*, **13**, 1357–1366.

28 Hirano, S. and Kondo, S. (1973) Molecular conformational transition of hyaluronic acid in solution. *J. Biochem.*, **74**, 861–862.

29 Gosh, S., Kobal, I., Zanette, D., and Reed, W.F. (1993) Conformational contraction and hydrolysis of hyaluronate in sodium hydroxide solution. *Macromolecules*, **26**, 4685–4693.

30 Park, J.W. and Chakrabarti, B. (1978) Conformational transition of hyaluronic acid carboxylic group participation and

thermal effect. *Biochem. Biophys. Acta*, **541**, 263–269.

31 Scott, J.E., Heatley, F., and Hull, W.E. (1984) Secondary structure of hyaluronate in solution. A ^{1}H n.m.r. investigation at 300 and 500MHz in [^{2}H$_6$] dimethyl sulphoxide solution. *Biochem. J.*, **220**, 197–205.

32 Atkins, E.D.T., Meader, D., and Scott, J.E. (1980) Model for hyaluronic acid incorporating four intramolecular hydrogen bonds. *Int. J. Biol. Macromol.*, **2**, 318–319.

33 Morris, E.R., Reed, D.A., and Welsh, J. (1980) Conformation and dynamic interactions in hyaluronate solutions. *J. Mol. Biol.*, **138**, 383–400.

34 Reed, C.E., Li, X., and Reed, W.F. (1989) The effects of pH on hyaluronate as observed by light scattering. *Biopolymers*, **28**, 1981–2000.

35 Darke, A., Finer, E.G., Moorhouse, R., and Reed, D.A. (1975) Studies of hyaluronate solutions by nuclear magnetic relaxation measurements. Detection of covalently defined, stiff segments within the flexible chains. *J. Mol. Biol.*, **99**, 477–486.

36 Cowman, M.K., Hittner, D.M., and Feder-Davis, J. (1996) ^{13}C-NMR studies of hyaluronan: conformational sensitivity to varied environments. *Macromolecules*, **29**, 2894–2902.

37 Moulabbi, M., Broch, H., Robert, L., and Vasilescu, D. (1997) Quantum molecular modeling of hyaluronan. *J. Mol. Struct. (Theochem)*, **395–396**, 477–508.

38 Maleki, A., Kjoniksen, A.-L., and Nyström, B. (2007) Anomalous viscosity behavior in aqueous solutions of hyaluronic acid. *Polym. Bull.*, **59**, 217–226.

39 Krause, W.E., Bellomo, E.G., and Colby, R.H. (2001) Rheology of sodium hyaluronate under physiological conditions. *Biomacromolecules*, **2**, 65–69.

40 Scott, J.E., Cummings, C., Brass, A., and Chen, Y. (1991) Secondary and tertiary structures of hyaluronan in aqueous solution, investigated by rotary shadowing-electron microscopy and computer simulation. Hyaluronan is a very efficient network-forming polymer. *Biochem. J.*, **274**, 699–705.

41 Milas, M., Roure, I., and Berry, G.C. (1996) Crossover behavior in the viscosity of flexible polymers: solutions of hyaluronan as a function of concentration, molecular weight, and temperature. *J. Rheol.*, **40**, 1155–1166.

42 Fouissac, E., Milas, M., and Rinaudo, M. (1993) Shear-rate, concentration, molecular weight and temperature viscosity dependences of hyaluronate, a wormlike polyelectrolyte. *Macromolecules*, **26**, 6945–6951.

43 Rinaudo, M. (2008) Main properties and current applications of some polysaccharides as materials. *Polym. Int.*, **57**, 397–430.

44 Gatej, I., Popa, M., and Rinaudo, M. (2005) Role of the pH on hyaluronan behavior in aqueous solution. *Biomacromolecules*, **6**, 61–67.

45 Milas, M. and Rinaudo, M. (1979) Conformational investigation on the bacterial polysaccharide xanthan. *Carbohydr. Res.*, **76**, 189–196.

46 Capron, I., Brigand, G., and Muller, G. (1997) About the native and renatured conformation of xanthan exopolysaccharide. *Polymer*, **38**, 5289–5295.

47 Callet, F., Milas, M., and Rinaudo, M. (1987) Influence of acetyl and pyruvate contents on rheological properties of xanthan in dilute solution. *Int. J. Biol. Macromol.*, **9**, 291–293.

48 Milas, M. and Rinaudo, M. (1984) On the existence of two secondary structures for the xanthan in aqueous solutions. *Polym. Bull.*, **12**, 507–514.

49 Lecourtier, J., Chauveteau, G., and Muller, G. (1986) Salt-induced extension and dissociation of a native double-stranded xanthan. *Int. J. Biol. Macromol.*, **8**, 306–310.

50 Chazeau, L., Milas, M., and Rinaudo, M. (1995) Conformations of xanthan in solution: analysis by steric exclusion chromatography. *Int. J. Polym. Anal. Charact.*, **2**, 21–29.

51 Milas, M., Reed, W., and Printz, S. (1996) Conformations and flexibility of native and re-natured xanthan in aqueous solution. *Int. J. Biol. Macromol.*, **18**, 211–221.

52 Pelletier, E., Viebke, C., Meadows, J., and Williams, P.A. (2001) A rheological study of the order–disorder conformational transition of xanthan gum. *Biopolymers*, **59**, 339–346.

53 Morse, D.L. (1998) Viscoelasticity of concentrated isotropic solutions of semiflexible polymers. 2. Linear response. *Macromolecules*, **31**, 7044–7067.

54 Callet, F. (1987) Influence of structure conformation and post-fermentation treatments on properties of xanthan solution. Ph.D. thesis, University of Grenoble, France.

55 Rinaudo, M. and Milas, M. (1980) Enzymic hydrolysis of the bacterial polysaccharide xanthan by cellulose. *Int. J. Biol. Macromol.*, **2**, 45–48.

56 Lambert, F. (1983) Characterization of xanthan in solution. Application to study its thermal stability. Ph.D. thesis, University of Grenoble, France.

57 Milas, M., Rinaudo, M., Knipper, M., and Schuppiser, J.L. (1990) Flow and viscoelastic properties of xanthan gum solutions. *Macromolecules*, **23**, 2506–2511.

58 Tako, M., Asato, A., and Nakamura, S. (1984) Rheological aspects of the intermolecular interaction between xanthan and locust bean gum in aqueous media. *Agric. Biol. Chem.*, **48**, 2995–3000.

59 Higiro, J., Herald, T.J., and Alavi, S. (2006) Rheological study of xanthan and locust bean gum interaction in dilute solution. *Food Res. Int.*, **39**, 165–175.

60 Dea, I.C.M., Morris, E.R., Rees, D.A., Welsh, E.J., Barnes, H., and Price, J. (1977) Associations of like and unlike polysaccharides: mechanism and specificity in galactomannans, interacting bacterial polysaccharides, and related systems. *Carbohydr. Res.*, **57**, 249–272.

61 Millane, R.P. and Wang, B. (1990) A cellulose-like conformation accessible to the xanthan backbone and implications for xanthan synergism. *Carbohydr. Polym.*, **13**, 57–68.

62 Chandrasekaran, R. and Radha, A. (1997) Molecular modeling of xanthan: galactomannan interactions. *Carbohydr. Polym.*, **32**, 201–208.

63 Mannion, R.O., Melia, C.D., Launay, B., Cuvelier, G., Hill, S.E., Harding, S.E. *et al.* (1992) Xanthan/locust bean gum interactions at room temperature. *Carbohydr. Polym.*, **19**, 91–97.

64 Bresolin, T.M.B., Sander, P.C., Reicher, F., Sierakowski, M.R., Rinaudo, M., and Ganter, J.L.M.S. (1997) Viscometric studies on xanthan and galactomannan. *Carbohydr. Polym.*, **33**, 131–138.

65 Goycoolea, M.F., Milas, M., and Rinaudo, M. (2001) Associative phenomena in galactomannan–deacetylated xanthan systems. *Int. J. Biol. Macromol.*, **29**, 181–192.

66 Secouard, S., Grisel, M., and Malhiac, C. (2007) Flavour release study as a way to explain xanthan–galactomannan interactions. *Food Hydrocolloids*, **21**, 1237–1244.

67 Lopes, L., Andrade, C.T., Milas, M., and Rinaudo, M. (1992) Role of the conformation and acetylation of xanthan and xanthan–guar interaction. *Carbohydr. Polym.*, **17**, 121–126.

68 Dentini, M., Crescenzi, V., Fidanza, M., and Coriello, T. (1989) On the aggregation and conformational states in aqueous solution of a succinoglycan polysaccharide. *Macromolecules*, **22**, 954–959.

69 Ridout, M.J., Brownsey, G.J., York, G.M., Walker, G.C., and Morris, V.J. (1997) Effect of o-acyl substituents on the functional behaviour of *Rhizobium meliloti* succinoglycan. *Int. J. Biol. Macromol.*, **20**, 1–7.

70 Hisamatu, M., Abe, J., Amenura, A., and Harada, T. (1980) Structural elucidation on succinoglycan and related polysaccharides from *Agrobacterium* and *Rhizobium* by fragmentation with two special β-D-glycanases and methylation analysis. *Agric. Biol. Chem.*, **44**, 1049–1055.

71 Fidanza, M., Dentini, M., Crescenzi, V., and Del Vecchio, P. (1989) Influence of charged groups on the conformational stability succinoglycan in dilute aqueous solution. *Int. J. Biol. Macromol.*, **11**, 372–376.

72 Gravanis, G., Milas, M., Rinaudo, M., and Clarke Sturman, A.J. (1990) Rheological

behaviour of a succinoglycan polysaccharide in dilute and semi-dilute solutions. *Int. J. Biol. Macromol.*, **12**, 201–206.
73 Heyraud, A., Rinaudo, M., and Courtois, B. (1986) Comparative studies of extracellular polysaccharide elaborated by *Rhizobium meliloti* strain M5N1 in defined medium and in non-growing cell suspensions. *Int. J. Biol. Macromol.*, **8**, 85–88.
74 Borsali, R., Rinaudo, M., and Noirez, L. (1995) Light scattering and small-angle neutron scattering from polyelectrolyte solutions: the succinoglycan. *Macromolecules*, **28**, 1085–1088.
75 Morris, V.J., Brownsey, G.J., Gunning, A.P., and Harris, J.E. (1990) Gelation of the extracellular polysaccharide produced by *Agrobacterium rhizogenes*. *Carbohydr. Polym.*, **12**, 221–225.
76 Boutebba, A. (1998) Properties of succinoglycan: conformational transition and gelation in aqueous medium. Ph.D. thesis, University of Grenoble, France.
77 Dantas, L., Heyraud, A., Courtois, B., Courtois, J., and Milas, M. (1994) Physicochemical properties of "Exogel" exocellular β-(1–4)-D-glucuronan from *Rhizobium meliloti* strain M5N1C.S. (NCIMB 40472). *Carbohydr. Polym.*, **24**, 185–191.
78 Morris, E.R., Rees, D.A., Thom, D., and Boyd, J. (1978) Chiroptical and stoichiometric evidence of a specific, primary dimerisation process in alginate gelation. *Carbohydr. Res.*, **66**, 145–154.
79 Heyraud, A., Courtois, J., Dantas, L., Colin-Morel, P., and Courtois, B. (1993) Structural characterization and rheological properties of an extracellular glucuronan produced by a *Rhizobium meliloti* M5N1 mutant strain. *Carbohydr. Res.*, **240**, 71–78.
80 Bozzi, L., Milas, M., and Rinaudo, M. (1996) Characterization and solution properties of a new exopolysaccharide excreted by the bacterium *Alteromonas* sp. strain 1644. *Int. J. Biol. Macromol.*, **18**, 9–17.
81 Bozzi, L., Milas, M., and Rinaudo, M. (1996) Solution and gel rheology of a new polysaccharide excreted by the bacterium *Alteromonas* sp. strain 1644. *Int. J. Biol. Macromol.*, **18**, 83–91.
82 Bozzi, L. (1994)) Production and physicchemical study of new polysaccharides synthesized by marine bacteria. Ph.D. thesis, University of Grenoble, France.
83 Campana, S., Ganter, J., Milas, M., and Rinaudo, M. (1992) On the solution properties of bacterial polysaccharides of the gellan family. *Carbohydr. Res.*, **231**, 31–38.
84 Morris, E.R., Gothard, M.G.E., Hember, M.W.N., Manning, C.E., and Robinson, G. (1996) Conformational and rheological transitions of welan, rhamsan and acylated gellan. *Carbohydr. Polym.*, **30**, 165–175.
85 Miyoshi, E. and Nishinari, K. (1999) Non-Newtonian flow behavior of gellan gum aqueous solutions. *Colloid Polym. Sci.*, **277**, 727–737.
86 Tanaka, Y., Sakurai, M., and Nakamura, K. (1996) Ultrasonic velocity and circular dichroism in aqueous gellan solutions. *Food Hydrocolloids*, **10**, 133–136.
87 Miyoshi, E., Takaya, T., and Nishinari, K. (1996) Rheological and thermal studies of gel–sol transition in gellan gum aqueous solutions. *Carbohydr. Polym.*, **30**, 109–119.
88 Grasdalen, H. and Smidsrod, O. (1987) Gelation of gellan gum. *Carbohydr. Polym.*, **7**, 371–393.
89 Nussinovitch, A. (ed.) (1997) *Hydrocolloid Applications. Gum Technology in the Food and Other Industries*, Chapman & Hall, London.
90 Morris, V.J. (1995) *Food Polysaccharides and Their Applications* (ed. A.M. Stephen), Marcel Dekker, p. 341.
91 Campana, S., Andrade, C., Milas, M., and Rinaudo, M. (1990) Polyelectrolyte and rheological studies on the polysaccharide welan. *Int. J. Biol. Macromol.*, **12**, 379–384.
92 Chandrasekaran, R., Lee, E.J., Radha, A., and Thailambal, V.G. (1992) *Frontiers in Carbohydrate Research*, vol. **2** (ed. R. Chandrasekaran), Elsevier, pp. 65–84.
93 Hember, M.W.N., Richardson, R.K., and Morris, E.R. (1994) Native ordered structure of welan polysaccharide: conformational transitions and gel formation in aqueous

dimethylsulphoxide. *Carbohydr. Res.*, **252**, 209–221.
94 Villain-Simonnet, A., Milas, M., and Rinaudo, M. (1999) Comparison between the physicochemical behavior of two microbial polysaccharides: RMDP17 and rhamsan. *Int. J. Biol. Macromol.*, **26**, 55–62.
95 Yanaki, T. and Norisuye, T. (1983) Triple helix and random coil of scleroglucan in dilute solution. *Polym. J.*, **15**, 389–396.
96 Bo, S., Milas, M., and Rinaudo, M. (1987) Behaviour of scleroglucan in aqueous solution containing sodium hydroxide. *Int. J. Biol. Macromol.*, **9**, 153–157.
97 Hirao, T., Sato, T., Teramoto, A., Matsuo, T., and Suga, H. (1990) Solvent effects on the cooperative order–disorder transition of aqueous solutions of schizophyllan, a triple-helical polysaccharide. *Biopolymers*, **29**, 1867–1876.
98 Noik, C. and Lecourtier, J. (1993) Studies on scleroglucan conformation by rheological measurements versus temperature up to 150 °C. *Polymer*, **34**, 150–157.
99 Bluhm, T., Deslandes, Y., Marchessault, R. H., Perez, S., and Rinaudo, M. (1982) Solid-state and solution conformation of scleroglucan. *Carbohydr. Res.*, **100**, 117–130.
100 Ogawa, K., Watanabe, T., Tsurugi, J., and Ono, S. (1972) Conformational behavior of a gel-forming (1→3)-β-D-glucan in alkaline solution. *Carbohydr. Res.*, **23**, 399–405.
101 Kanzawa, Y., Harada, T., Koreeda, A., Harada, A., and Okuyama, K. (1989) Difference of molecular association in two types of curdlan gel. *Carbohydr. Polym.*, **10**, 299–313.
102 Saito, H., Yokoi, M., and Yoshioka, Y. (1989) Effect of hydration on conformational change or stabilization of (1→3)-β-D-glucans of various chain lengths in the solid state as studied by high-resolution solid-sate ^{13}C NMR spectroscopy. *Macromolecules*, **22**, 3892–3898.
103 Konno, A., Azechi, Y., and Kimura, H. (1979) Properties of curdlan gel. *Agric. Biol. Chem.*, **43**, 101–104.
104 Villain-Simonnet, A., Milas, M., and Rinaudo, M. (2000) A new bacterial polysaccharide (YAS34). I. Characterization of the conformations and conformational transition. *Int. J. Biol. Macromol.*, **27**, 65–75.
105 Record Jr, M.T. (1975) Effects of Na^{2+} and Mg^{2+} ions on the helix–coil transition of DNA. *Biopolymers*, **14**, 2137–2158.
106 Villain-Simonnet, A., Milas, M., and Rinaudo, M. (2000) A new bacterial polysaccharide (YAS34). II. Influence of thermal treatments on the conformation and structure. Relation with gelatin ability. *Int. J. Biol. Macromol.*, **27**, 77–87.